W9-BMB-021

SIGNALS AND LINEAR SYSTEMS

SIGNALS AND LINEAR SYSTEMS

SECOND EDITION

Robert A. Gabel

Sperry Research Center formerly of University of Colorado at Denver

Richard A. Roberts

University of Colorado at Boulder

John Wiley and Sons, Inc.

New York • Chichester • Brisbane • Toronto • Singapore

Library of Congress Cataloging in Publication Data:

Gabel, Robert A
Signals and linear systems.

Bibliography: p.
Includes index.
1. Electric networks. 2. Electric engineering—Mathematics. 3. System analysis. I. Roberts, Richard A., 1935– joint author. II. Title.
TK454.2.G22 1980 620'.7'2 80-14811
ISBN 0-471-04958-1

Printed in the United States of America

10 9 8 7 6 5

To *Elizabeth,*

Jonathan, and Alison

Judy,

Mark, John, Michele, Molly, Emily, and Mary

PREFACE

After communicating with a number of instructors who have adopted the first edition of this book and after using it ourselves for several semesters we have concluded that certain modifications would improve the presentation and have incorporated them in this new edition. As in the first edition, we stress the relationships among the various representations of a linear system: the difference or differential equation model, the block diagram or flow graph form, the impulse-response description, the state-variable formulation, and the transfer-function characterization. We emphasize throughout the book that these representations are tightly related and may be employed to great conceptual and computational advantage in the analysis and synthesis of linear systems. The presentation of the material is organized to enable the various topics to reinforce one another. New results are related to those that have been mastered previously. Additional examples have been incorporated to further illustrate the material as it is developed. Several problems are presented using comparative solution approaches, so that students may experience the various solution methods for representative applications. As previously, the material is general and chosen to lead naturally into succeeding courses in communication systems, control systems, and other areas that use these basic techniques in specific advanced applications. Prerequisites assumed of the student are a sophomore-level course covering differential equations and a course in the student's major area (such as circuits or mechanics) that deals with the derivation of mathematical models of physical systems.

The most pronounced change in the second edition is the reorganization of the time-domain material in Chapters 1 to 3. Whereas the first edition emphasized most strongly the similarities between discrete- and continuous-time system analysis, with different approaches in separate chapters, we have found the presentation to flow more smoothly if the various analysis approaches are contrasted within a body of material, with discrete- and continuous-time systems developed in separate chapters. It is our experience and that of our colleagues who have presented the material in both forms that students more readily perceive the similarities between discrete- and continuous-time system analysis than they do the relationships between, for example, analyses based on the system equation solution, the convolution sum or integral, and the state-variable description. This modified organization also allows the instructor to begin with either discrete-time or continuous-time systems. (Our suggestion, reflected in the chapter ordering, is to treat discrete-time systems first).

As an aid in motivating and interrelating the various approaches, we have introduced the concept of the system frequency response at an early point in the discussion. Evaluation of the frequency response is demonstrated with successive system models as they are developed, rather than being relegated to a later section of the transform domain discussion. This, we feel, is a major feature of the presentation offered here, and one that has greatly enlivened our classroom discussions.

We discovered that adoptions of the first edition were primarily for electrical engineering courses. Therefore and in line with the suggestions from users, we have changed the Z-transform notation of Chapter 4 to correspond with the standard used for publication in this discipline. Additional material concerning the frequency response of digital filters has been included and related to the preceding treatment.

Chapter 5 on Fourier analysis has been significantly revised and expanded to include the Fourier analysis of discrete-time signals and systems. It also includes new material on the implications of using windows in FIR filter synthesis, numerical computation of Fourier transforms, and intelligent use of the Fast Fourier

Transform. This material serves to further integrate the student's perceptions of discrete- and continuous-time system analysis.

Chapter 7 on digital filter synthesis has been reorganized to more clearly develop and contrast the standard design approaches. The material on mixed continuous- and discrete-time systems has been placed to summarize the principal developments in the text.

This book is used in a one-semester junior course in linear systems analysis at the University of Colorado. The course meets for a total of 40 lecture hours. In this time we cover most of Chapters 1 to 6 and selected portions of Chapter 7 as time permits. We present the material in the order given here, although other choices are certainly possible. For example, Chapters 2 and 3 may be interchanged if an individual would prefer to begin with continuous-time systems. Likewise, Chapters 5 and 6 may be interchanged, and Z-transforms may be introduced after Fourier and Laplace transforms, if desired. If time is a major factor, we recommend that Fourier analysis be emphasized at the expense of Z-transforms and Laplace transforms.

We thank sincerely the numerous instructors and students who have used the first edition and who have given us suggestions for improvements. In particular, we thank our colleagues, Lloyd Griffiths and Tom Mullis, who have discussed the material and its presentation with us at length and have contributed many of the problems and examples.

Robert A. Gabel
Richard A. Roberts

CONTENTS

SIGNALS AND LINEAR SYSTEMS

CHAPTER 1
LINEAR SYSTEMS

1.1 INTRODUCTION

The study of linear systems has been an essential part of formal undergraduate training for many years. Linear system analysis is useful because, even though physical systems are never completely linear, a linear model is often appropriate over certain ranges of input-output values. A large body of mathematical theory is available for engineers and scientists to use in the analysis of linear systems. In contrast, the analysis of nonlinear systems is essentially *ad hoc*. Each nonlinear system must be studied as one of a kind, since there are few general methods of analysis.

The analysis of linear systems is often facilitated by the use of a particular class of input signals. Thus it is natural to include a study of signals and their various representations in our study of linear systems. We shall find sinusoidal and impulsive signals especially useful as system inputs.

As engineers, we are interested in the synthesis as well as the analysis of systems. In fact, it is the synthesis or design of systems that is the really creative portion of engineering. Yet, as in so many creative efforts, one must learn first how to analyze systems before one can proceed with system design. The work in this book is directed primarily toward the analysis of certain classes of linear systems.

However, because design and analysis are so intimately connected, this material will also provide a basis for simple design.

We can divide the analysis of systems into three aspects:

1. The development of a suitable mathematical model for the physical problem of interest. This portion of the analysis is concerned with obtaining the "equations of motion," boundary or initial conditions, parameter values, etc. This is the process of combining judgment, experience, and experiments to develop a suitable model. This first step is the hardest to develop formally.
2. After a suitable model is obtained, one then solves the resultant equations to obtain solutions in various forms.
3. One then relates or interprets the solution to the mathematical model in terms of the physical problem. Of course, meaningful interpretations and predictions concerning the physical system can be made only if the development in **1** has been accurate enough.

The primary emphasis of this work is on the second and third aspects mentioned above. The first step is essential but is accomplished more completely and appropriately within a particular discipline. Thus, chemical engineers will learn to write equations of motion for chemical processes, electrical engineers for electrical circuits, and so on. After a model is obtained, one can consider various techniques for its analysis and provide a basis for its mathematical interpretation.

Because linear models are so often used in all disciplines of engineering and science, this material is very useful. Perhaps the best way to point out this fact is to present examples from various physical problems. The only drawback to this method is that the reader may not always possess the necessary background to perform the first step in the analysis, that is, to write the equations of motion. This problem is to be expected. As one gains familiarity with a given discipline, this first step becomes natural. We shall use linear models based on electrical engineering applications for the most part. Certain problems at the ends of the chapters present physical examples from other disciplines.

We shall present several models for analyzing linear systems. Each model is useful in its own right, but together they present a more complete view of linear systems. By considering these different techniques, we hope to unify the reader's view of the subject.

1.2 CLASSIFICATION OF LINEAR SYSTEMS

A system is a mathematical model or abstraction of a physical process that relates inputs or external forces to the output or response of the system. Input and output share a cause-effect relationship. There are several classifications or types of systems.

A *causal* or *nonanticipatory* system produces an output that at any time t_0 is a function of only those input values that have occurred for times up to, and including, t_0. In other words, the system does not respond to input values until they have been actually applied to the system. Stated in this way, it appears that all real physical systems are causal. We shall show, however, that one can use noncausal systems in many applications.

The *state* of a system is a fundamental concept. The state is a minimal set of variables chosen such that if their values are known at time t_0 and all inputs are known for times greater than t_0, one can calculate the outputs of the system for times greater than t_0. The state of a system can be thought of as the system's memory. The memory at any time t_0 summarizes the effect of all past inputs and any initial state or memory.

The input, state, and output are, in general, sets of variables that we shall represent as vector quantities. For example, an n-variable input is written as

$$\mathbf{u}(t) = \begin{bmatrix} u_1(t) \\ u_2(t) \\ \vdots \\ u_n(t) \end{bmatrix} \tag{1.1}$$

We use $\mathbf{u}$, $\mathbf{y}$, and $\mathbf{x}$ to denote input, output, and state variables, respectively. Different vectors of the same class are distinguished by superscripts, for example, $\mathbf{u}^1$, $\mathbf{u}^2$, etc.

We shall be concerned here with both continuous and discrete-time systems. Continuous time systems are systems for which the input, output, and state are all functions of a continuous real variable t. We shall also study systems in which the time variable is defined only for discrete instants of time t_k, where k is an integer. In this case the system is called a *discrete-time system.*

We shall denote a function of continuous time by $f(t)$ or, in cases where no confusion would result, merely by f. Its value at t is $f(t)$. Similarly, a discrete-time function is denoted as $f(k)$ (or f), and its value at $t = t_k$ by any of the notations $f(t_k) \equiv f(kt) \equiv f(k) \equiv f_k$. Notice that $f(k)$ is itself a continuous variable—only its argument is discrete. Discrete-time functions are commonly called *time sequences* or merely *sequences.* Thus an input sequence of n variables may be denoted as

$$\mathbf{u}(k) \equiv \mathbf{u} \equiv \begin{bmatrix} u_1(k) \\ u_2(k) \\ \vdots \\ u_n(k) \end{bmatrix} \tag{1.2}$$

Our models of physical systems are restricted to constant parameter systems. That is, we shall assume that parameters of the system do not change with time. This leads to the idea of time-invariant and shift-invariant systems. A time-invariant, continuous-time system can be characterized as follows. If an input $u(t)$ gives rise to an output $y(t)$, then a shifted version of the input $u(t \pm \tau)$ gives rise to an output $y(t \pm \tau)$. Similarly, for a shift-invariant, discrete-time system, if an input $u(k)$ produces an output $y(k)$, then $u(k \pm n)$ produces an output $y(k \pm n)$. This is another way of saying the system response does not depend on the time origin but only on the form of the input.

We shall, in the sequel, consider linear, time- (or shift) invariant systems. These systems are most often characterized by linear differential (or difference) equations with constant coefficients. These are our basic models for continuous and discrete-time systems, respectively.

1.3 LINEARITY

We have classified systems in several ways. One of the most important concepts in system theory is linearity. What precisely is a linear system? Linear systems possess the property of *superposition.* That is, if u^1 produces an output y^1 and u^2 produces an output y^2, then an input $(u^1 + u^2)$ produces the output $(y^1 + y^2)$. In symbols, if

$$u^1 \to y^1$$

and

$$u^2 \to y^2 \tag{1.3}$$

then

$$u^1 + u^2 \to y^1 + y^2$$

for some class of inputs u^j, $j = 1, 2, \ldots$. Superposition also implies that if

$$u \to y$$

then (1.4)

$$\alpha u \to \alpha y, \qquad \alpha \text{ a rational number}$$

This latter property is called *homogeneity* if it is true for all α.

A convenient notation for the arrows in (1.3) and (1.4) is to use functional notation and represent the system as a transformation T of inputs u into outputs y. A system is linear if T satisfies

$$T(\alpha u^1 + \beta u^2) = \alpha T(u^1) + \beta T(u^2) \tag{1.5}$$

where α and β are arbitrary constants. Some examples of how to verify whether or not a system is linear follow.

Example 1.1 Suppose a system has an input-output relation given by the linear equation

$$y = au + b, \qquad a, b \text{ constants} \tag{1.6}$$

Does this equation represent the input-output relation of a linear system? We can write (1.6) in the form

$$y = T(u) = au + b$$

Consider two inputs u^1 and u^2. The corresponding outputs are:

$$T(u^1) = au^1 + b$$

$$T(u^2) = au^2 + b$$

Now apply an input $(u^1 + u^2)$. The output is

$$T(u^1 + u^2) = a(u^1 + u^2) + b$$

But notice that

$$\begin{aligned} T(u^1) + T(u^2) &= au^1 + b + au^2 + b \\ &= a(u^1 + u^2) + 2b \\ &\neq T(u^1 + u^2) \end{aligned}$$

Therefore, the system is not linear! The problem is that b is added to au. This offset at the origin destroys the superposition property. ■

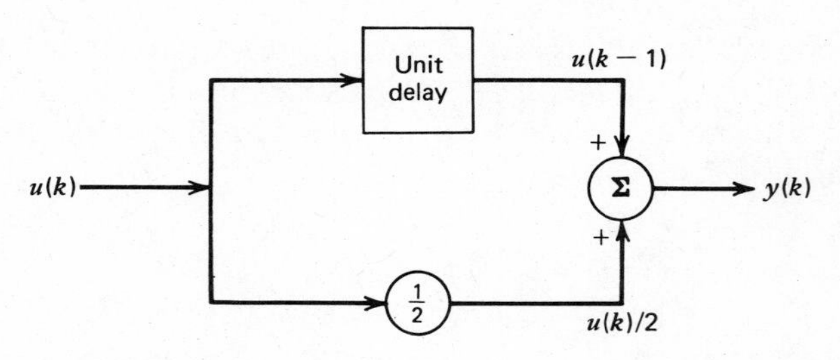

Figure 1.1

Example 1.2 Consider the discrete-time system shown in Figure 1.1. The block diagram contains a unit delay element, a multiplier of value $\frac{1}{2}$, and a summer. The delay element is a device that holds the previous value fed to it, in this case the previous value of the input. The equation for the output is

$$y(k) = \frac{1}{2}\, u(k) + u(k-1) \tag{1.7}$$

Does this system satisfy the property of superposition? An input u^1 produces an output

$$y^1(k) = \frac{1}{2}\, u^1(k) + u^1(k-1)$$

An input u^2 produces the output

$$y^2(k) = \frac{1}{2}\, u^2(k) + u^2(k-1)$$

An input made up of $\alpha u^1 + \beta u^2$ produces the output

$$\begin{aligned} y(k) &= \frac{1}{2}\,[\alpha u^1(k) + \beta u^2(k)] + [\alpha u^1(k-1) + \beta u^2(k-1)] \\ &= \frac{1}{2}\,\alpha u^1(k) + \alpha u^1(k-1) + \frac{1}{2}\,\beta u^2(k) + \beta u^2(k-1) \\ &= \alpha y^1(k) + \beta y^2(k) \end{aligned}$$

Thus this discrete-time system is linear. ■

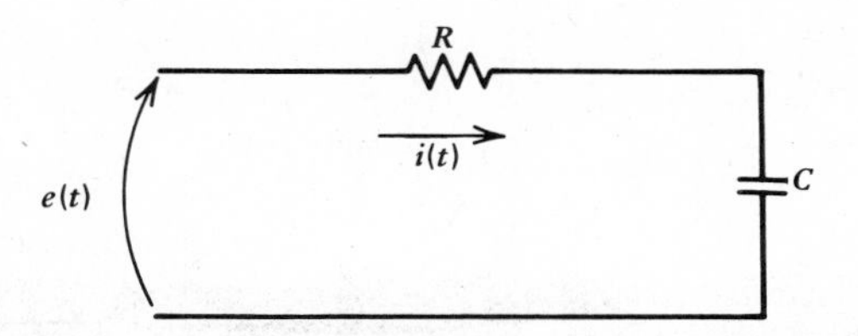

Figure 1.2

Example 1.3 The simple RC network shown in Figure 1.2 has as input the current $i(t)$ and as output the voltage $e(t)$. Does this transformation of $i(t)$ into $e(t)$ represent a linear system? We assume that the initial energy storage is zero, that is, there is no net charge on the capacitor. We use Kirchoff's voltage law to write

$$e(t) = Ri(t) + \frac{1}{C}\int_0^t i(t')\,dt' \tag{1.8}$$

Again we use superposition to verify linearity. Suppose we apply the two separate inputs $i_1(t)$ and $i_2(t)$. The corresponding outputs are

$$e_1(t) = Ri_1(t) + \frac{1}{C}\int_0^t i_1(t')\,dt'$$

$$e_2(t) = Ri_2(t) + \frac{1}{C}\int_0^t i_2(t')\,dt'$$

If we apply the input $\alpha i_1(t) + \beta i_2(t)$, the output is

$$\begin{aligned} e(t) &= R[\alpha i_1(t) + \beta i_2(t)] + \frac{1}{C}\int_0^t [\alpha i_1(t') + \beta i_2(t')]\,dt' \\ &= \alpha\left[Ri_1(t) + \frac{1}{C}\int_0^t i_1(t')\,dt'\right] + \beta\left[Ri_2(t) + \frac{1}{C}\int_0^t i_2(t')\,dt'\right] \\ &= \alpha e_1(t) + \beta e_2(t) \end{aligned}$$

This represents a linear system. Notice that in this example we have assumed the initial voltage on the capacitor to be zero. Suppose instead there is an initial voltage stored across the capacitor. Assume the sign of this initial voltage to be a voltage drop in the clockwise direction. The Kirchoff voltage equation is now

$$e(t) = Ri(t) + \frac{1}{C}\int_0^t i(t')\,dt' + v_0, \qquad t \geq 0 \tag{1.9}$$

where

$$v_0 = \frac{q}{C} = \frac{\int_{-\infty}^{0} i(t')\,dt'}{C}$$

is the initial voltage on the capacitor. Because of the constant v_0 (1.9) no longer represents a linear relation between $i(t)$ and $e(t)$. This is similar to Example 1.1. In verifying whether a system is linear one should always set initial conditions to zero. If this is not done, the resulting input-output equation will contain constant terms that arise from the nonzero initial conditions. Superposition in this case will fail. The system will appear to be nonlinear when, in fact, it is a linear system with nonzero initial conditions. ∎

Initial Energy Storage in Linear Systems

Linear systems with initial energy storage can be analyzed as linear systems by decomposing the total response into two separate responses: the response from the initial energy storage and the response resulting from the system input. The decomposition is shown schematically in Figure 1.3. In the decomposition shown in Figure 1.3, all of the systems labeled T are identical. We write the system output y as a sum of $y^{(d)}$, the output of an initially relaxed system with input u, plus $y^{(h)}$, the output of the system with zero input and with initial conditions identical to those of the original system.

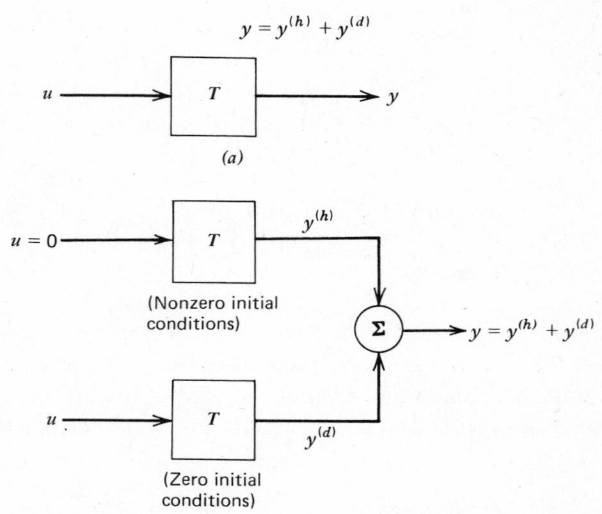

Figure 1.3

The solution $y = y^{(h)} + y^{(d)}$ of the decomposed system is identical to the solution of the original system. Both solutions satisfy the same differential or difference equation and they have the same initial conditions.

Example 1.4 The system of Example 1.1, described by the algebraic equation

$$y = au + b$$

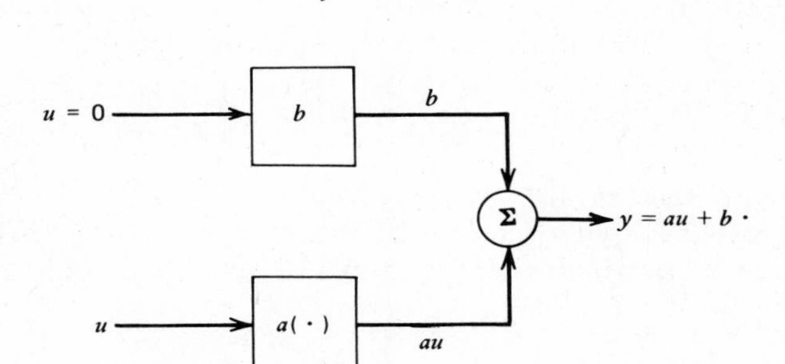

Figure 1.4

can be decomposed, as shown in Figure 1.4. Here the output resulting from the input u is $y^{(d)} = au$. The output of original system with zero input is $y^{(h)} = b$. ■

Example 1.5 Consider the *RC* network of Example 1.3 described by the differential equation

$$e(t) = Ri(t) + \frac{1}{C}\int_0^t i(t')\,dt' + v_0$$

We have defined $i(t)$ as the system's input and $e(t)$ as the system's output. In our decomposition we set $y^{(d)}(t)$ equal to the output of the system with zero initial conditions. Thus,

$$y^{(d)}(t) = Ri(t) + \frac{1}{C}\int_0^t i(t')\,dt'$$

The solution $y^{(h)}(t)$ is the output of the original system with $i(t) = 0$, but with an initial condition $e(0) = v_0$.

$$y^{(h)}(t) = v_0$$

Thus, we have

$$e(t) = y^{(d)}(t) + y^{(h)}(t)$$
$$= Ri(t) + \frac{1}{C}\int_0^t i(t')\,dt' + v_0$$

as before. By decomposing the system in this manner, we can apply superposition to the linear part, that is, the $u(t) \to y^{(d)}(t)$ relation, and then add the term $y^{(h)}(t)$ to obtain the entire output. ■

Implicit Models

In the examples presented thus far, the output y of the system was written as an explicit function of the input u. That is, we could express $y = T(u)$. There are many models that are not given in this explicit form. For example, consider the difference equation

$$y_k = u_k + \frac{1}{2} y_{k-2} \tag{1.10}$$

The output y is the sum of the present input and one-half the output two clock cycles delayed. How then do we test superposition in this case? A block diagram of (1.10) is shown in Figure 1.5.

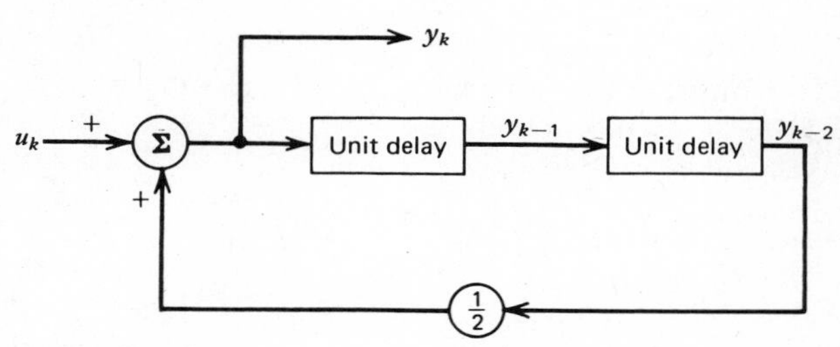

Figure 1.5

Equation 1.10 generates a transformation T. At this point we cannot develop the form of this transformation. However, if we assume that such a transformation exists, we can verify the linearity of (1.10) as before. Thus consider two inputs u^1 and u^2. The outputs are defined by

$$y_k^1 - \frac{1}{2} y_{k-2}^1 = u_k^1$$

$$y_k^2 - \frac{1}{2} y_{k-2}^2 = u_k^2$$

For an input $(\alpha u^1 + \beta u^2)$ we obtain an output y_k^3 defined as

$$y_k^3 - \frac{1}{2} y_{k-2}^3 = \alpha u_k^1 + \beta u_k^2$$

$$= \alpha\left(y_k^1 - \frac{1}{2} y_{k-2}^1\right) + \beta\left(y_k^2 - \frac{1}{2} y_{k-2}^2\right)$$

$$= (\alpha y_k^1 + \beta y_k^2) - \frac{1}{2}(\alpha y_{k-2}^1 + \beta y_{k-2}^2)$$

Thus we find that $y_k^3 = \alpha y_k^1 + \beta y_k^2$ and so the system is linear. This same process can also be used in the case of linear differential equations.

1.4 DISCRETE-TIME SYSTEMS

Discrete-time systems express a relationship between discrete-time functions or sequences. These kinds of systems are becoming increasingly important in all disciplines of engineering because of the increasing speed and low cost of digital components. The design and analysis of discrete-time circuits will continue to grow and complement the corresponding continuous-time theory. We shall introduce some terminology and notation for our study of discrete-time systems.

As stated previously, a discrete-time function is a function of a discrete argument. We shall use the notations $\{f(k)\}$ or f to denote the complete sequence. The sequence values are denoted by any of the notations $f(k)$, f_k, $f(t_k)$, or $f(Tk)$ where k is an integer. We can define a sequence f in two ways:

i. We can specify a rule for calculating f_k. For example, if

$$f_k = \begin{cases} \left(\dfrac{1}{2}\right)^k, & k \geq 0 \\ 0, & k < 0 \end{cases}$$

we generate the sequence $\left\{\begin{matrix} \cdots 0, 0, 1, \frac{1}{2}, \frac{1}{4}, \cdots, (\frac{1}{2})^k, \cdots \\ \uparrow k = 0 \end{matrix}\right\}$

ii. We can explicitly list the values of the sequence. For example,

$$f = \left\{\begin{matrix} \cdots 0, 0, 1, 5, -3, 4, 0, 0, \cdots \\ \uparrow \end{matrix}\right\}.$$

The arrow is used to denote the $k = 0$ term. We shall use the convention that if the arrow is omitted, the first value listed is the $k = 0$ term and all values are zero for $k < 0$. That is, if

$$f = \left\{1, \frac{1}{2}, \frac{1}{4}, \ldots, \left(\frac{1}{2}\right)^k, \ldots\right\}$$

then this is equivalent to

$$f_k = \begin{cases} \left(\frac{1}{2}\right)^k, & k \geq 0 \\ 0, & k < 0 \end{cases}$$

A discrete-time system is a system that transforms input sequences u into output sequences y. These systems are composed of unit delay elements, constant multipliers, and summers. They can be modeled by linear difference equations with constant coefficients of the form

$$y_k + b_1 y_{k-1} + b_2 y_{k-2} + \cdots + b_n y_{k-n} = a_0 u_k + a_1 u_{k-1} + \cdots + a_m u_{k-m} \tag{1.11}$$

or

$$y_k = \sum_{i=0}^{m} a_i u_{k-i} - \sum_{i=1}^{n} b_i y_{k-i} \tag{1.12}$$

In (1.11) and (1.12) y_k is expressed as a linear combination of present and past values of the input u_k and past values of the output. Equation 1.11 or 1.12 is one of the basic models for discrete-time systems that we shall analyze in detail.

We define the sum of two sequence $\{a_k\} + \{b_k\}$ as the sequence $\{c_k\}$, where

$$c_k = a_k + b_k$$

The product of two sequences $\{a_k\}\{b_k\}$ is the sequence $\{c_k\}$, with

$$c_k = a_k b_k$$

The product of a constant α and a sequence $\{a_k\}$ is the sequence $\{c_k\}$ with

$$c_k = \alpha a_k$$

Difference or recurrence equations are valuable models in problems in which one can express the desired output y_k (for some value of k) in terms of previous values of y_k and of the input u_k and its previous values. Some examples of physical problems that result in difference equations follow.

Example 1.6 Find the difference equation model for the discrete-time circuit shown in Figure 1.6. The unit delay elements are memory elements that hold the previous input for one clock cycle. Thus the output of the first delay unit is y_{k-1}.

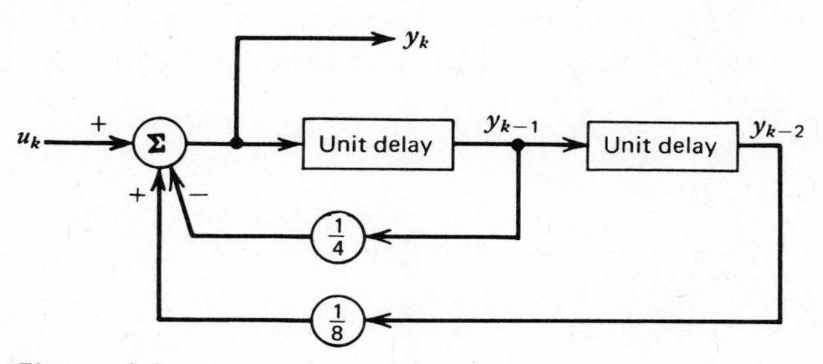

Figure 1.6

Similarly, the output of the second delay unit is y_{k-2}. Equating the output of the summer y_k to the entering arrows, we obtain

$$y_k = u_k - \frac{1}{4} y_{k-1} + \frac{1}{8} y_{k-2}$$

as the difference equation model for this circuit. Notice how quickly one can write equations for these kinds of systems. ■

Example 1.7 Once every cycle in a chemical process, u_k liters of chemical A and $100 - u_k$ liters of chemical B are added to 900 liters of a mixture in a large vat, where $0 \leq u_k \leq 100$, $k = 1, 2, 3, \ldots$. The vat contents are thoroughly mixed and 100 liters of mixture are drawn off. Let y_k be the fractional concentration of chemical A in the mixture drawn off, that is, $1000y_k$ is the amount of chemical A in the vat. To develop a recursion relation for y_k, we express the total amount of chemical A at cycle k to the amount of chemical A at cycle $k - 1$ plus any inputs. Thus,

$$1000y_k = 900y_{k-1} + u_k$$

or (1.13)

$$y_k = 0.9y_{k-1} + 0.001u_k, \qquad k = 1, 2, 3, \ldots$$

Equation 1.13 merely states that the amount of chemical A at cycle k is the amount at cycle $k - 1$ plus the amount added at cycle k. ■

Example 1.8 Geometric growth is a model used in many fields. Suppose a certain population has N_t individuals at the end of year t with $t = 0, 1, 2, \ldots$. The population is known to increase at a relative rate of 2% per year. That is, the increase in population in any year is proportional to the population at the beginning of the year. The proportionality constant is, in this case, 0.02. Thus, since the increase in population is $N_{t+1} - N_t$ we have the difference equation

$$N_{t+1} - N_t = 0.02N_t$$

or (1.14)

$$N_{t+1} - 1.02N_t = 0, \qquad t = 0, 1, 2, \ldots$$

with N_0 the initial population. Notice that in this example the input is zero. The output is growing because of the nonzero initial condition (population) N_0. ■

Example 1.9 Information theory is a discipline concerned with, among other things, how to transmit signals (information) efficiently. The medium through which the signals are transmitted is called a *channel*. Each physical channel possesses a theoretical limit (because of noise and bandwidth) on the amount of information per unit time that can be sent through the channel without error. This is called the *capacity* C of the channel. Its units are bits per second and it is defined formally as

$$C = \lim_{t \to \infty} \left[\frac{\log_2 N_t}{t} \right] \text{ bits/second} \tag{1.15}$$

In (1.15) N_t is the number of messages of duration t that can be transmitted. Thus given any channel, as the number of "messages" per unit time that can be transmitted increases, capacity increases.

Suppose we have a communication system that uses only two symbols. Call them S_1 and S_2, for example, dots and dashes. Messages are made of combinations of the symbols S_1 and S_2. Suppose that S_1 lasts t_1 seconds and that S_2 is t_2 seconds in duration. Let N_t be the number of messages of duration t. What is the capacity of this channel?

In order to calculate capacity we must calculate N_t. We can calculate N_t by developing a recursion for N_t. How many messages are there in t seconds? The total number can be divided into two classes—those that end in the symbol S_1 and those that end in the symbol S_2. How many end in S_1? S_1 lasts t_1 seconds. Thus there are $(t - t_1)$ seconds before S_1 begins. By definition there are N_{t-t_1} messages in this interval. Therefore there are exactly N_{t-t_1} messages that end in the symbol S_1. Similarly, the total number of messages of duration t that end in S_2 must be the number N_{t-t_2}. Since the total number of messages N_t ends in either S_1 or S_2, we have

$$N_t = N_{t-t_1} + N_{t-t_2} \tag{1.16}$$

To calculate channel capacity we must solve (1.16) for N_t and then use (1.15) to find C. If, for example, $t_1 = 1$ and $t_2 = 2$, then (1.16) reduces to

$$N_t - N_{t-1} - N_{t-2} = 0, \qquad t = 1, 2, 3, \ldots \tag{1.17}$$

The initial conditions are N_1 and N_2, the number of messages of duration 1 and 2 time units, respectively. In this case $N_1 = 1$, $N_2 = 2$. If $t_1 = t_2 = 1$, would you expect channel capacity to increase or decrease over the case $t_1 = 1$ and $t_2 = 2$? ■

1.5 CONTINUOUS-TIME SYSTEMS

Continuous-time systems are perhaps more familiar to undergraduate engineering students because of previous courses in electrical circuits and mechanics. The basic mathematical description of these systems is a linear differential equation with constant coefficients (assuming constant parameter systems) of the form

$$\frac{d^n y(t)}{dt^n} + b_1 \frac{d^{n-1} y(t)}{dt^{n-1}} + \cdots + b_n y(t) = a_0 \frac{d^m u(t)}{dt} + a_1 \frac{d^{m-1} u(t)}{dt^{m-1}} u(t) + \cdots + a_m u(t) \tag{1.18}$$

In this model the input $u(t)$ and the output function $y(t)$ are functions of a continuous real variable t that is usually a time variable.

Example 1.10 Electrical networks are a classical example of systems that may be modeled by a linear differential equation. The equations describing the circuit

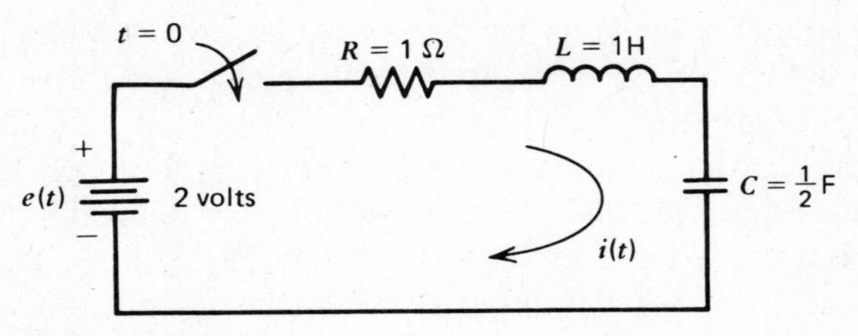

Figure 1.7

are obtained using Kirchoff's two laws. Consider the network shown in Figure 1.7. Kirchoff's voltage law taken around the single loop yields the equation

$$e(t) = Ri(t) + L\frac{di(t)}{dt} + \frac{1}{C}\int_0^t i(t')\,dt', \qquad t > 0 \tag{1.19}$$

We can convert (1.19) into an equation involving only derivatives by differentiating both sides with respect to t. We obtain

$$\frac{de(t)}{dt} = 0 = R\frac{di(t)}{dt} + L\frac{d^2i(t)}{dt} + \frac{1}{C}i(t) \tag{1.20}$$

In order to solve (1.20) we must specify initial conditions for the network. If we assume that the switch closes at time $t = 0$, then the current in the loop just after the switch closes, $i(0^+)$, is zero because current cannot change instantaneously through an inductor. Thus no current flows at $t = 0^+$ and the entire voltage appears across the inductor, that is,

$$L\frac{di(0^+)}{dt} = e(0^+)$$

or (1.21)

$$\frac{di(0^+)}{dt} = \frac{e(0^+)}{L} = 2 \text{ amp/sec}$$

Equation 1.21 and $i(0^+)$ are the initial conditions of the circuit. The problem is thus to solve:

$$\frac{d^2i(t)}{dt^2} + \frac{di(t)}{dt} + 2i(t) = 0$$

with

$$i(0^+) = 0 \qquad \text{and} \qquad \left.\frac{di(t)}{dt}\right|_{t=0^+} = 2$$

■

Example 1.11 Linear differential equations are used as models for many physical problems. Consider an idealized example of the treatment of sewage. An aeration pond contains a concentration C of a certain pollutant. Raw sewage being added to the pond contains a higher concentration of the pollutant than does the pond. After some time in the pond, bacteria digest the pollutant and the mixture is allowed to flow into a river. Figure 1.8 depicts the situation. Assume that the input rate is a steady i_1 gal/min with an input pollutant concentration of C_1 lb/gal. The output effluent is a steady i_2 gal/min. The pond contains G_0 gallons initially, with P_0 pounds of pollutant. The problem is to determine how long we can empty the aerated mixture into the river before the concentration exceeds the government standard of $0.1C_1$.

With constant rates of flow into and out of the pond, the total volume of water in the pond is

$$\text{Total volume at time } t = G_0 + (i_1 - i_2)\,t \text{ gallons}$$

Let $P(t)$ be the number of pounds of pollutant in the pond at time t. The rate of change of the amount of pollutant is $dP(t)/dt$. This rate of change is equal to the amount entering the pond, $C_1 i_1$, minus the amount leaving the pond, $i_2(P(t)/\text{total volume})$. Thus

$$\frac{dP(t)}{dt} = C_1 i_1 - i_2 \frac{P(t)}{G_0 + (i_1 - i_2)t}$$

or

$$\frac{dP(t)}{dt} + \frac{i_2}{G_0 + (i_1 - i_2)t} P(t) = C_1 i_1 \tag{1.22}$$

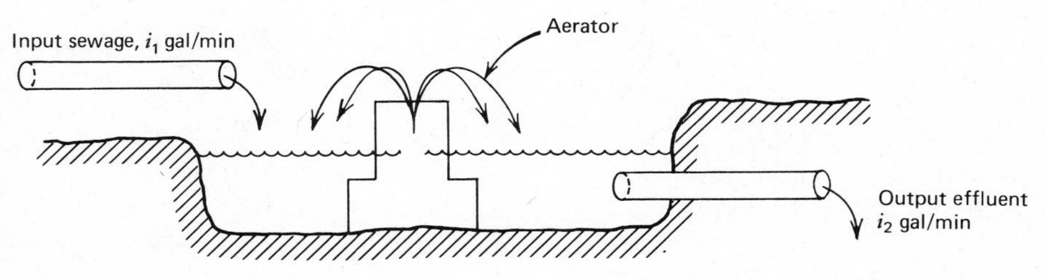

Figure 1.8

Equation 1.22 is a first order linear differential equation with time-varying coefficients. However, the equation is easily solved using standard techniques. We shall deal primarily with constant parameter systems that have constant coefficients, for example, the situation obtained with $i_1 = i_2$. ■

In the following chapters we shall present several methods for analyzing systems described by linear difference and differential equations. We shall present the theory first for discrete-time systems and then for continuous-time systems. There is a great similarity in the mathematics for these two classes of systems. It is helpful to keep in mind the solution processes for discrete-time systems when studying continuous-time systems, and vice-versa. We shall first discuss discrete-time systems because the mathematics tends to be easier.

One of the objectives of this material is to present several alternative descriptions of a linear system. We shall discuss each description in detail and some methods of expressing one description in terms of the others. As you read this material and encounter new descriptions of the same system, try to express each model in terms of parameters of the others. These kinds of transformations are useful and will aid you in understanding the attributes of the various descriptions.

PROBLEMS

√**1.1.** Are the following systems linear? Show.

(a) $T(u(t)) = \begin{cases} 0, & t < T_0 \\ \alpha u(t), & t \geq T_0 \end{cases}$, α a constant

(b) $\mathbf{u}(t) = \begin{bmatrix} u_1(t) \\ u_2(t) \end{bmatrix}$ → system → $y(t) = \min\,[u_1(t), u_2(t)]$

vector input

(c) $y(t) = \begin{cases} \alpha u(t) + u(t^2), & t \geq 0 \\ 0, & t < 0 \end{cases}$

(d) $y(k) = \sum_{n=k}^{k+2} k^2 u(n)$

(e) $y(k) = \alpha u(k) + \beta u(k-1) + \gamma[u(k-2)]^2$

1.2. A system is described by the following matrix input-output equation:

$$\begin{bmatrix} y_1(t) \\ y_2(t) \end{bmatrix} = \begin{bmatrix} 3 & 4 & 2 \\ -1 & t & 0 \end{bmatrix} \begin{bmatrix} u_1(t) \\ u_2(t) \\ u_3(t) \end{bmatrix}$$

(a) Is this system linear?
(b) Is this system time-invariant?

1.3. Give an example of a system that satisfies the property of homogeneity but not the principle of superposition.

1.4. In the following sketch a waveform $a(t)$ is switched on and off by a transistor switch. The output of this system can be represented as $y(t) = a(t)p(t)$, where $p(t)$ is a square wave as shown. Is this system linear? Is this system time-invariant?

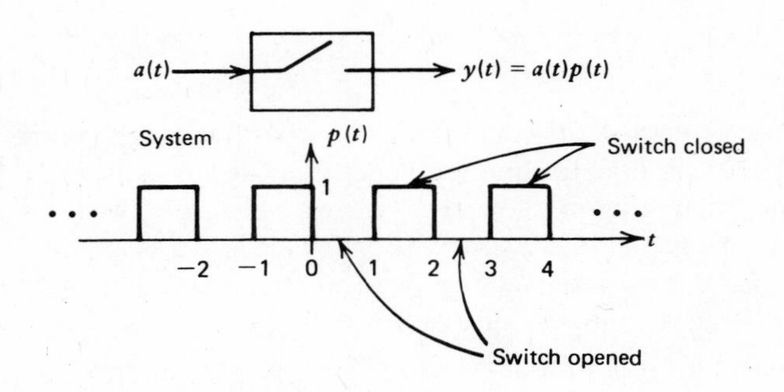

Switching characteristic

1.5. Consider a cascade of two linear systems, as shown below. System 1 is a differentiator. System 2 is an amplifier with gain proportional to time, that is, the gain is $g(t) = t$. Show that each system is linear. If systems 1 and 2 are interchanged, is the output, in general, the same? In other words, do the systems commute?

$u(t)$ → $\frac{d(\ )}{dt}$ → Amplifier $g(t) = t$ → $y(t)$

System 1 System 2

1.6. Let y_n be the number of separate regions into which n nonparallel lines divide the plane. Derive a first-order difference equation for y_n that can be used to find the number of separate regions. Assume *no three lines* intersect at a single point.

1.7. If a sum of money earns simple interest at a rate r (i.e., percentage annual rate $100r\%$), the amount on deposit at any interest date is equal to the amount on deposit 1 year before that date plus the interest earned in that year on the initial principal invested. Let s_k be the sum on deposit after year k and let s_0 be the initial deposit. Write a difference equation for s_k for $k = 0, 1, 2, \ldots$. Write down a few terms of the sequence s_k beginning with $k = 0$.

1.8. Derive a difference equation for the discrete-time system shown below in block diagram form.

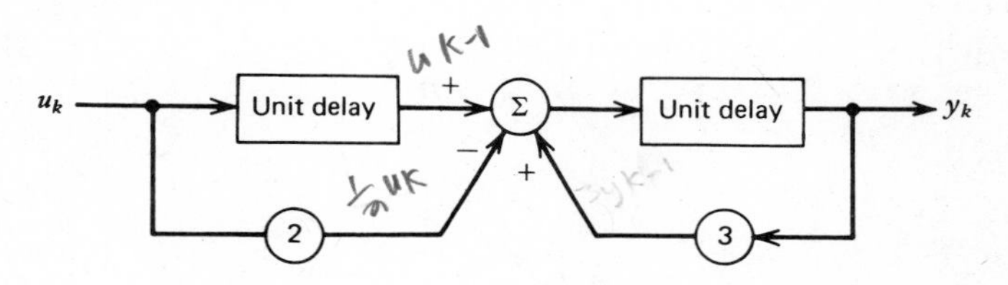

1.9. In processing a sequence of data $\{u_k\}$ in a digital computer, it is desired to "smooth" the data to eliminate sudden variations in the data. The processing consists of averaging the present sample of data u_k with the previous three values. Write the difference equation for this processing scheme. If we wished to perform a weighted average with weights $\{1, 2, 2, 1\}$, how could we write the difference equation for this processing scheme.

1.10. In Problem 9, the output is formed from present and past values of the data only. Suppose we wish to average together the present value of the data u_k, the previous value of the data, the next (or future) value of the data, and the previous value of the output (one output previous). What is the difference equation in this case?

1.11. Derive a differential equation for the following circuit relating the voltages $u(t)$ and $y(t)$.

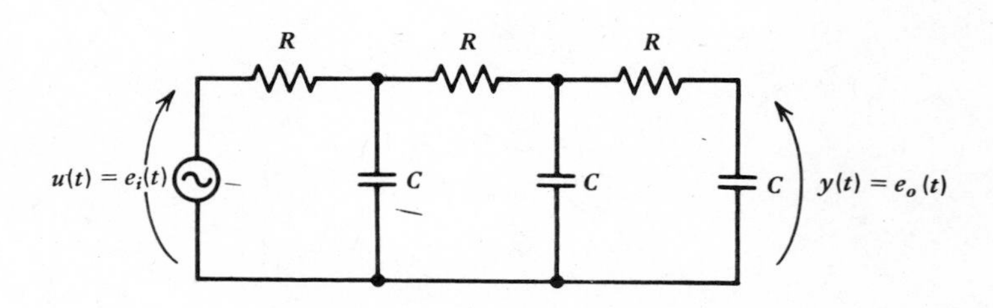

1.12. Derive a differential equation for the mechanical system relating the initial conditions (which are initial spring displacements) to the displacements $x_1(t)$ and $x_2(t)$.

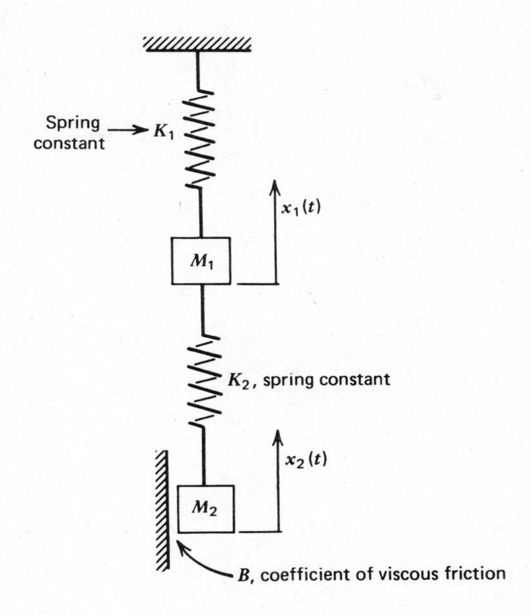

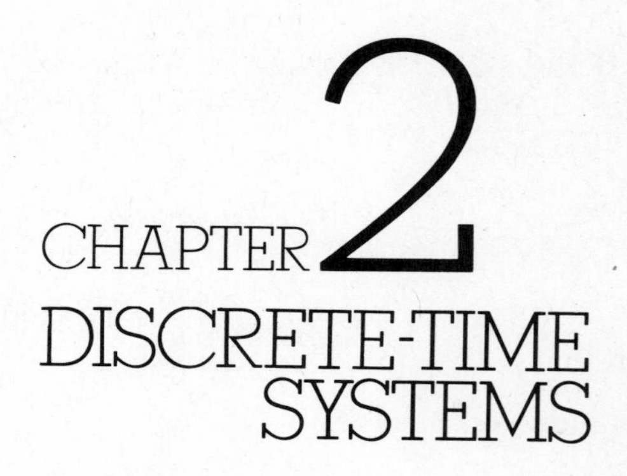

2.1 INTRODUCTION

The mathematical techniques for the analysis of linear time-invariant systems are usually classified either as *time* (or *sequence*) *domain* or as *transform domain* methods. In this chapter we shall consider three sequence domain methods for the analysis of discrete-time systems. These methods are called sequence domain techniques because the input, output, and system model are all described by using sequences of numbers. The output sequence is the response of the system to some input sequence of values. We usually interpret these sequences of numbers as indexed in time, although it is not necessary to do so.

The three time domain methods we shall study depend on the model we choose for describing a given discrete-time system. The three descriptions we shall use in this chapter are:

a. A linear difference equation

b. The impulse-response sequence

c. A state-variable or matrix description

Each of these models can be used to find the output of a discrete-time system from knowledge of the input sequence. Why then should we consider more than one model? Each model emphasizes certain aspects of the system. We can interpret the system in different ways depending on the model. Together they give us a more complete understanding of how linear discrete-time systems work. We shall begin with linear difference equation models.

2.2 LINEAR DIFFERENCE EQUATIONS

Linear discrete-time systems transform input sequences of numbers $\{u_k\}$ into output sequences $\{y_k\}$ according to some recursion formula or difference equation. For example, the equation

$$y_k = u_k + 2u_{k-1} + 3u_{k-2} \tag{2.1}$$

is a simple expression that in words can be described as follows: Form the kth member of the output sequence y_k by adding together the present input u_k, two times the previous input u_{k-1}, and three times the input delayed twice u_{k-2}. In this example, if the input sequence consisted of the numbers $\{1, 0, 1, 2, 0, 0, \ldots\}$, the output sequence would be $\{y_k\} = \{1, 2, 4, 4, 7, 6, 0, 0, \ldots\}$.

We shall often find it convenient to schematically represent a difference or recursion formula using block diagrams or signal flow graphs. For (2.1), we have the schematic shown in Figure 2.1. The schematic consists of three components: unit delays that store previous inputs or outputs, multipliers, and adders or accumulators. Given a block diagram we can easily generate the corresponding difference equation model. For example, in Figure 2.2, the recursion formula for y_k is

$$y_k = u_k + \alpha y_{k-1}, \qquad k = 0, 1, 2, \ldots \tag{2.2}$$

Notice that in (2.2), the output is formed from the present input and previous output. Assuming $y_{-1} = 0$, that is, that the value stored in the unit delay before

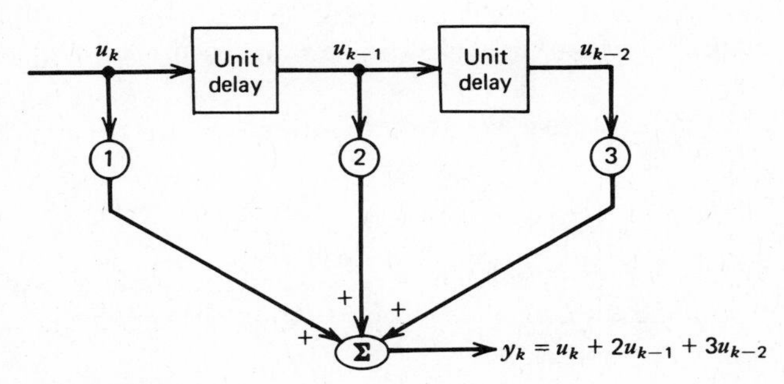

Figure 2.1

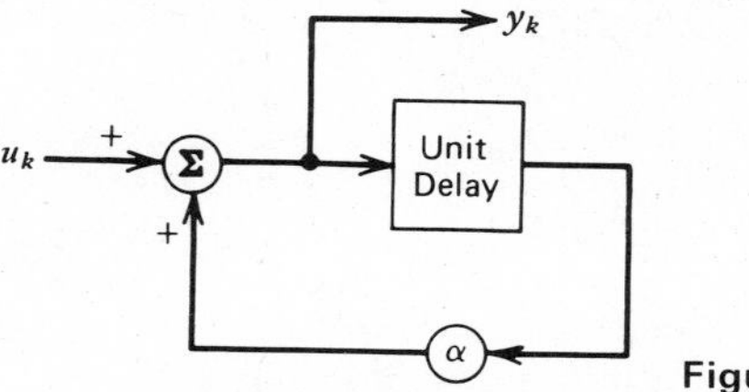

Figure 2.2

the input is applied is zero, what is y_k for any value of k? The answer to this question is given by a solution to the difference equation (2.2). We can, by brute force, solve any difference equation given the initial conditions of the unit delays that make up the system and the input sequence. In this example the initial condition or state of the system is $y_{-1} = 0$. Suppose that $u_k = 1$ for all $k \geq 0$. Then $y_0 = 1$, $y_1 = 1 + \alpha$, $y_2 = 1 + \alpha + \alpha^2$ and, in general,

$$y_k = 1 + \alpha + \alpha^2 + \cdots + \alpha^k \tag{2.3}$$

Equation 2.3 is an explicit solution to (2.2) with $y_{-1} = 0$ and $u_k = 1$. However, this solution is not in closed form. In the following we shall require all solutions to be in closed form unless otherwise stated. We can place (2.3) in closed form if we recall the formula for the partial sum of a geometric series† which is

$$y_k = 1 + \alpha + \alpha^2 + \cdots + \alpha^k = \begin{cases} \dfrac{1 - \alpha^{k+1}}{1 - \alpha}, & \alpha \neq 1 \\ k + 1, & \alpha = 1 \end{cases}, \qquad k = 0, 1, 2, \ldots$$

We shall be concerned primarily with time-invariant linear discrete-time systems. These systems can be modeled using linear difference equations with constant coefficients. In schematic form they contain only unit delays, constant multipliers, and adders. The theory of linear difference equations closely parallels the theory of linear differential equations. We can prove, for example, that solutions always exist and are unique. We shall not develop the theory here but instead use certain fundamental concepts to develop solution methods for linear difference equations.

Consider a linear difference equation of nth order with constant coefficients,

$$y_k + b_1 y_{k-1} + b_2 y_{k-2} + \cdots + b_n y_{k-n} = u_k \tag{2.4}$$

Equation 2.4, with the right-hand side zero, is called the *homogeneous* difference equation corresponding to the *nonhomogeneous* difference equation (2.4).

† See Appendix A.

The solution of linear difference equations rests primarily on the fact that if y^1 and y^2 are solutions of the homogeneous equation, then $c_1 y^1 + c_2 y^2$ is also a solution for arbitrary constants c_1 and c_2. This, coupled with the result that states if $y^{(h)}$ is a solution to the homogeneous equation and $y^{(p)}$ is a solution of the complete equation (2.4), then $y^{(h)} + y^{(p)}$ is the general solution of the complete equation. These theorems, together with existence and uniqueness of solutions, provide us with the tools needed to find closed form solutions.

To find the homogeneous solution of (2.4), we try a solution of the form $y_k = r^k$ and substitute this into the homogeneous equation. We then obtain

$$r^k + b_1 r^{k-1} + b_2 r^{k-2} + \cdots + b_n r^{k-n} = 0$$

or (2.5)

$$r^k(1 + b_1 r^{-1} + b_2 r^{-2} + \cdots + b_n r^{-n}) = 0$$

If we choose r so that (2.5) is satisfied, then $y_k = r^k$ is in fact a solution of the homogeneous equation. Equation 2.5 is satisfied if $r^k = 0$ or if $1 + b_1 r^{-1} + b_2 r^{-2} + \cdots + b_n r^{-n} = 0$. The first condition leads to a trivial solution. The second condition implies that there are n possible solutions, namely the n roots of the equation

$$1 + b_1 r^{-1} + b_2 r^{-2} + \cdots + b_n r^{-n} = 0$$

or (2.6)

$$r^n + b_1 r^{n-1} + b_2 r^{n-2} + \cdots + b_n = 0$$

If the n roots of (2.6) are distinct roots, $r_1, r_2, \ldots, r_n$, then we obtain n homogeneous solutions of the form r_i^k, $i = 1, 2, \ldots, n$. A complete homogeneous solution to (2.4) is obtained by adding the individual solutions together to obtain

$$\begin{aligned} y_k^{(h)} &= c_1 y_1(k) + c_2 y_2(k) + \cdots + c_n y_n(k) \\ &= c_1 r_1^k + c_2 r_2^k + \cdots + c_n r_n^k \end{aligned}$$

Example 2.1 Consider the discrete-time system shown in Figure 2.3. Find the difference equation model and the associated homogeneous solution.

The difference equation for the block diagram is given by

$$y_k - 5y_{k-1} + 6y_{k-2} = 0$$

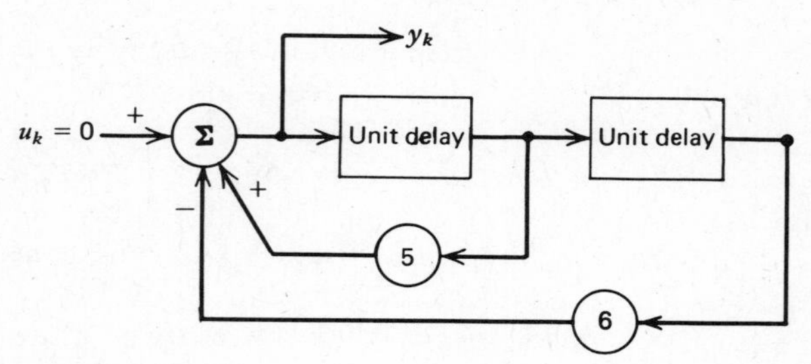

Figure 2.3

Let $y_k = r^k$ and substitute in the above equation. We then obtain

$$r^k - 5r^{k-1} + 6r^{k-2} = 0$$

or

$$r^k(1 - 5r^{-1} + 6r^{-2}) = 0$$

The auxiliary equation that defines the homogeneous solutions is therefore

$$r^2 - 5r + 6 = 0$$

This equation has roots of 2 and 3. Hence, the complete homogeneous solution is

$$y_k = c_1 2^k + c_2 3^k$$

■

Notice that we can always verify or test whether a given output sequence is, in fact, a solution to a particular difference equation. For example, in Example 2.1, if we substitute $y_k = c_1 2^k + c_2 3^k$ into $y_k - 5y_{k-1} + 6y_{k-2} = 0$, we have

$$(c_1 2^k + c_2 3^k) - 5(c_1 2^{k-1} + c_2 3^{k-1}) + 6(c_1 2^{k-2} + c_2 3^{k-2}) \stackrel{?}{=} 0$$

Collecting terms, we obtain

$$c_1 2^k(1 - 5 \cdot 2^{-1} + 6 \cdot 2^{-2}) + c_2 3^k(1 - 5 \cdot 3^{-1} + 6 \cdot 3^{-2}) \stackrel{?}{=} 0$$

Since the terms in both brackets are zero, we have $0 = 0$ and, thus, verify the solution.

Given an auxiliary equation of the form (2.6), there correspond n solution sequences y_k^i, $i = 1, 2, \ldots, n$. The form of these solution sequences depends on the multiplicity of the roots r_i^k, $i = 1, 2, \ldots, n$. The following rules summarize how one chooses the n solution sequences.

1. For each simple real root r_i, assign $y_k^i = r_i^k$.
2. For each multiple real root r of multiplicity m, assign m sequences r_i^k, $kr_i^k, \ldots, k^{m-1}r_i^k$.
3. For each pair of complex roots $a \pm jb$, assign the sequences $(a + jb)^k$ and $(a - jb)^k$. We usually write these sequences in polar form as $\rho^k \cos \phi k$ and $\rho^k \sin \phi k$ where $\rho = (a^2 + b^2)^{1/2}$ and $\phi = \tan^{-1}(b/a)$.
4. For each pair of complex roots $a \pm jb$ of multiplicity m we assign the sequences $\rho^k \cos \phi k, \rho^k \sin \phi k; k\rho^k \cos \phi k, k\rho^k \sin \phi k; \ldots, k^{m-1}\rho^k \cos \phi k, k^{m-1}\rho^k \sin \phi k$.

Notice that (3) and (4) are special cases of (1) and (2).

Example 2.2 The second-order difference equation

$$y_k - 2ay_{k-1} + y_{k-2} = 0$$

occurs rather frequently in applications. The associated auxiliary equation is

$$r^2 - 2ar + 1 = 0 \tag{2.7}$$

This equation has roots given by

$$r_1, r_2 = \frac{2a + \sqrt{4a^2 - 4}}{2} = a \pm \sqrt{a^2 - 1}$$

which implies a (homogeneous) solution of the form

$$y_k = c_1(a + \sqrt{a^2 - 1})^k + c_2(a - \sqrt{a^2 - 1})^k$$

This solution can be written more compactly.

1. Assume $|a| < 1$. The roots $a \pm \sqrt{a^2 - 1}$ are complex. In polar form we have

$$\rho = [a^2 + (1 - a^2)]^{1/2} = 1$$

$$\phi = \tan^{-1}\left(\frac{\sqrt{1 - a^2}}{a}\right)$$

Note that $\cos \phi = a$ and $\sin \phi = \sqrt{1 - a^2}$. Thus the solution for $|a| < 1$ is

$$y_k = c_1 \cos \phi k + c_2 \sin \phi k, \qquad a = \cos \phi \qquad \text{or} \qquad \phi = \cos^{-1} a$$

2. If $a = 1$, the auxiliary equation (2.7) has repeated roots at 1. Thus

$$y_k = c_1 + c_2 k$$

3. If $a = -1$, the auxiliary equation (2.7) has repeated roots at -1. Thus

$$y_k = (c_1 + c_2 k)(-1)^k$$

4. If $|a| > 1$, we leave it to the reader to show that

$$y_k = c_1 \cosh \phi k + c_2 \sinh \phi k, \qquad a = \cosh \phi; \text{ that is, } \phi = \cosh^{-1} a$$

■

Example 2.3 Consider the resistor ladder circuit shown in Figure 2.4. We wish to determine the voltages $V_1, V_2, \ldots, V_N$ at the nodes as shown. At the kth node we can use Kirchhoff's current law $i_1 = i_2 + i_3$ to obtain the difference equation

$$\frac{V_{k-1} - V_k}{R} = \frac{V_k - V_{k+1}}{R} + \frac{V_k}{aR}$$

Collecting terms we have

$$aV_{k+1} - (2a + 1)V_k + aV_{k-1} = 0, \qquad k = 1, 2, \ldots, N - 1$$

with boundary conditions

$$V_0 = E \qquad \text{and} \qquad V_N = 0$$

Suppose $a = 1$. Then we must solve

$$V_{k+1} - 3V_k + V_{k-1} = 0, \qquad k = 1, 2, \ldots, N - 1$$

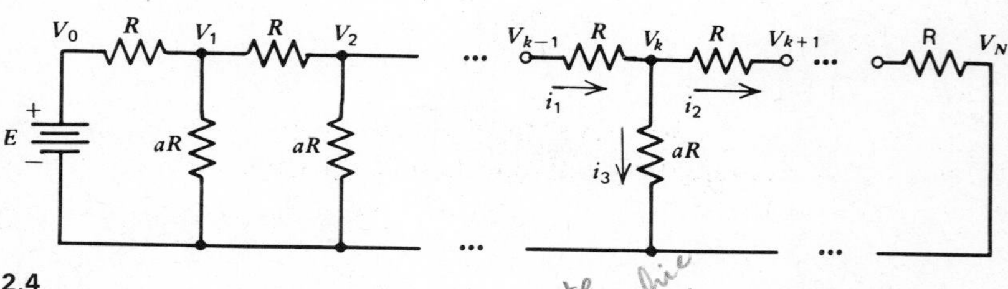

Figure 2.4

The auxiliary equation is
(characteristic)

$$r^2 - 3r + 1 = 0$$

with roots

$$r_1, r_2 \cong 2.62, 0.38$$

Thus

$$V_k = c_1(2.62)^k + c_2(0.38)^k$$

To find the constants c_1 and c_2 we use the boundary conditions

$$V_0 = E = c_1 + c_2$$
$$V_N = 0 = c_1(2.62)^N + c_2(0.38)^N$$

which imply that

$$c_1 = \frac{E(0.38)^N}{(0.38)^N - (2.62)^N}, \qquad c_2 = \frac{E(2.62)^N}{(2.62)^N - (0.38)^N}$$

from which we obtain

$$V_k = \frac{E}{(2.62)^N - (0.38)^N}[-(0.38)^N(2.62)^k + (2.62)^N(0.38)^k], \qquad k = 0, 1, \ldots, N \quad \blacksquare$$

2.3 THE GENERAL SOLUTION OF NONHOMOGENEOUS DIFFERENCE EQUATIONS

The general solution of a nonhomogeneous equation is obtained by adding to the complete homogeneous solution $y^{(h)}$ a particular solution $y^{(p)}$ of the complete difference equation. There are many methods of finding a particular solution to the complete equation. We shall use the method of undetermined coefficients. This method can be used whenever the forcing sequence is itself the solution to some linear difference equation with constant coefficients.

Consider the nth order difference equation

$$y_k + b_1 y_{k-1} + b_2 y_{k-2} + \cdots + b_n y_{k-n} = u_k \tag{2.8}$$

It is convenient to express difference equations such as (2.8) in operator notation. We shall define a shift operator S as

$$S^{\pm n}[y_k] = y_{k\pm n} \tag{2.9}$$

The symbol $S^{\pm n}$ denotes the shift operator, indicating that the sequence y_k is to be operated on (or transformed) to yield a new sequence $y_{k\pm n}$, a shifted version of the original sequence. Thus we can express (2.8) as

$$(1 + b_1 S^{-1} + b_2 S^{-2} + \cdots + b_n S^{-n})[y_k] = u_k$$

More compactly, we write

$$L[y_k] = u_k$$

where L is the linear difference operator

$$L = 1 + b_1 S^{-1} + b_2 S^{-2} + \cdots + b_n S^{-n} \tag{2.10}$$

The method of undetermined coefficients can be conveniently explained in terms of the above operator notation. Given a nonhomogeneous equation of the form (2.8), we find a linear difference operator L_A, called an annihilator operator, such that

$$L_A[u_k] = 0 \tag{2.11}$$

This operator is then applied to both sides of (2.8). The resulting homogeneous equation is then solved as we have previously discussed. The corresponding homogeneous solution with arbitrary multiplier constants is substituted into (2.11), and the undetermined coefficients arising from the L_A operator are evaluated. Some examples illustrate the procedure.

Example 2.4 Consider the discrete-time system shown in Figure 2.5. Find the general solution for an input sequence

$$u_k = \begin{cases} 3^k, & k \geq 0 \\ 0, & k < 0 \end{cases}$$

The difference equation for this system is found by equating the output of the summer to its three inputs. Thus

$$y_k = \frac{5}{6} y_{k-1} - \frac{1}{6} y_{k-2} + 3^k, \qquad k \geq 0 \tag{2.12}$$

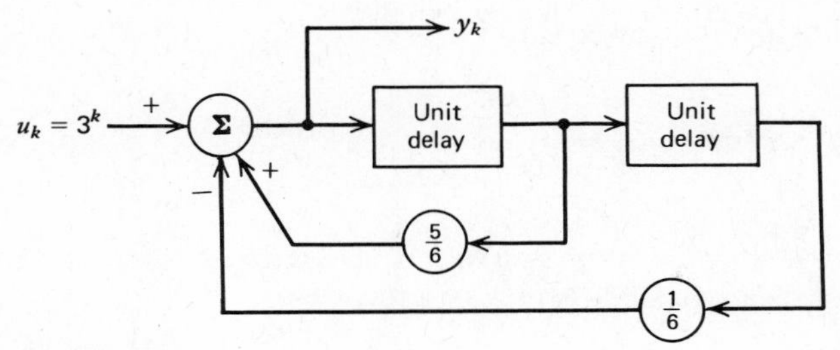

Figure 2.5

In operator notation we have

$$\left(1 - \frac{5}{6}S^{-1} + \frac{1}{6}S^{-2}\right)[y_k] = 3^k, \qquad k \geq 0$$

The auxiliary equation is

$$r^2 - \frac{5}{6}r + \frac{1}{6} = 0$$

The roots of this equation are $r_1 = \frac{1}{2}, r_2 = \frac{1}{3}$. The homogeneous solution is therefore

$$y_k = c_1\left(\frac{1}{2}\right)^k + c_2\left(\frac{1}{3}\right)^k, \qquad k \geq 0$$

To find the forced solution we seek a linear difference operator L_A so that

$$L_A[3^k] = 0$$

Another way to state this problem is to ask for the homogeneous linear difference equation for which 3^k is a solution. Since a^k is a solution of the equation $(1 - as^{-1})[y_k] = 0$, we try an operator $(1 - 3S^{-1})$. Then we have

$$(1 - 3S^{-1})[3^k] = 3^k - 3 \cdot 3^{k-1} = 0$$

Thus

$$L_A = (1 - 3S^{-1})$$

Applying $(1 - 3S^{-1})$ to both sides of the original equation yields

$$(1 - 3S^{-1})\left(1 - \frac{5}{6}S^{-1} + \frac{1}{6}S^{-2}\right)[y_k] = (1 - 3S^{-1})[3^k] = 0$$

The corresponding auxiliary equation is

$$(r-3)\left(r^2 - \frac{5}{6}r + \frac{1}{6}\right) = 0$$

with roots $r_1 = \frac{1}{2}, r_2 = \frac{1}{3}, r_3 = 3$. The root $r_3 = 3$ is due to the annihilator operator. Thus we choose as a particular solution a form

$$y_k^{(p)} = c_3 3^k$$

Now substitute $y_k^{(p)}$ into the original equation (2.12). We have

$$c_3 3^k - \frac{5}{6} c_3 3^{k-1} + \frac{1}{6} c_3 3^{k-2} = 3^k$$

Factoring out 3^k on the left side gives

$$3^k\left[c_3 - \frac{5}{6} c_3 3^{-1} + \frac{1}{6} c_3 3^{-2}\right] = 3^k$$

For $y_k^{(p)}$ to be a solution, the bracketed term must be unity; that is,

$$c_3 - \frac{5}{6} c_3 3^{-1} + \frac{1}{6} c_3 3^{-2} = 1$$

which implies

$$c_3 = \frac{27}{20}$$

The particular solution is therefore

$$y_k^{(p)} = \frac{27}{20} 3^k$$

The general solution is

$$y_k = c_1\left(\frac{1}{2}\right)^k + c_2\left(\frac{1}{3}\right)^k + \frac{27}{20} 3^k, \qquad k \geq 0$$

The constants c_1 and c_2 can now be evaluated based on the initial conditions of the system. The first two sequences of y_k, that is, $c_1(\frac{1}{2})^k + c_2(\frac{1}{3})^k$, constitute the transient response of the system. The solution $\frac{27}{20}3^k$ is the steady-state response. ■

Example 2.5 Find the steady-state solution to the difference equation

$$8y_k - 6y_{k-1} + y_{k-2} = 5 \sin\left(\frac{k\pi}{2}\right).$$

In operator notation we have

$$(8 - 6S^{-1} + S^{-2})[y_k] = 5 \sin\left(\frac{k\pi}{2}\right)$$

We seek an annihilator for the term 5 sin $(k\pi/2)$ or, in other words, what linear difference equation has 5 sin $(k\pi/2)$ as a homogeneous solution. From the previous example we saw that ca^k is annihilated by the operator $(1 - aS^{-1})$. Now 5 sin $(k\pi/2)$ can be expressed as

$$5 \sin\left(\frac{k\pi}{2}\right) = 5\left[\frac{e^{j(k\pi/2)} - e^{-j(k\pi/2)}}{2j}\right]$$

$$= \frac{5}{2j} e^{j(k\pi/2)} - \frac{5}{2j} e^{-j(k\pi/2)}$$

The annihilator for $(5/2j)[e^{j(\pi/2)}]^k$ is $[1 - e^{j(\pi/2)}S^{-1}]$, and the annihilator for $(-5/2j)[e^{-j(\pi/2)}]^k$ is $[1 - e^{-j(\pi/2)}S^{-1}]$. The annihilator for the sum of two terms is the product of the annihilators for the individual terms. Thus we have

$$L_A\left[5 \sin\left(\frac{k\pi}{2}\right)\right] = (1 - e^{j(\pi/2)}S^{-1})(1 - e^{-j(\pi/2)}S^{-1})\left[5 \sin \frac{k\pi}{2}\right] = 0$$

If we apply L_A to both sides of the original equation, we obtain the homogeneous equation

$$(1 - e^{j(\pi/2)}S^{-1})(1 - e^{-j(\pi/2)}S^{-1})(8 - 6S^{-1} + S^{-2})[y_k] = 0$$

The roots arising from the operator L_A are

$$r_3 = e^{j(\pi/2)} \qquad r_4 = e^{-j(\pi/2)}$$

Thus the steady-state solution is of the form

$$y_k^{(p)} = a \sin\left(\frac{k\pi}{2}\right) + b \cos\left(\frac{k\pi}{2}\right)$$

where a and b are constants that are determined by substituting $y_k^{(p)}$ into the original difference equation. This substitution gives

$$8\left[a \sin\left(\frac{k\pi}{2}\right) + b \cos\left(\frac{k\pi}{2}\right)\right] - 6\left[a \sin\frac{(k-1)\pi}{2} + b \cos\frac{(k-1)\pi}{2}\right] + \left[a \sin\frac{(k-2)\pi}{2} + b \cos\frac{(k-2)\pi}{2}\right] = 5 \sin\left(\frac{k\pi}{2}\right)$$

Using the trigonometric identities

$$\sin\left[\frac{(k-1)\pi}{2}\right] = \sin\frac{k\pi}{2}\cos\frac{\pi}{2} - \cos\frac{k\pi}{2}\sin\frac{\pi}{2} = -\cos\frac{k\pi}{2}$$

$$\cos\left[\frac{(k-1)\pi}{2}\right] = \cos\frac{k\pi}{2}\cos\frac{\pi}{2} + \sin\frac{k\pi}{2}\sin\frac{\pi}{2} = \sin\frac{k\pi}{2}$$

$$\sin\left[\frac{(k-2)\pi}{2}\right] = \sin\frac{k\pi}{2}\cos\pi - \cos\frac{k\pi}{2}\sin\pi = -\sin\frac{k\pi}{2}$$

$$\cos\left[\frac{(k-2)\pi}{2}\right] = \cos\frac{k\pi}{2}\cos\pi + \sin\frac{k\pi}{2}\sin\pi = -\cos\frac{k\pi}{2}$$

we can simplify our last expression to

$$8y_k^{(p)} - 6y_{k-1}^{(p)} + y_{k-2}^{(p)} = \sin\left(\frac{k\pi}{2}\right)[7a - 6b] + \cos\left(\frac{k\pi}{2}\right)[6a + 7b] = 5 \sin\left(\frac{k\pi}{2}\right)$$

This last equation implies

$$7a - 6b = 5 \qquad 6a + 7b = 0$$

Solving, we obtain $a = +\frac{7}{17}$, $b = -\frac{6}{17}$. The steady-state solution is therefore

$$y_k^{(p)} = \frac{1}{17}\left[+7 \sin\left(\frac{k\pi}{2}\right) - 6 \cos\left(\frac{k\pi}{2}\right)\right]$$

■

Table 2.1

Forcing Sequence u	Annihilator Operator L_A
a^k	$1 - aS^{-1}$
$\sin \phi k$ or $\cos \phi k$	$(1 - e^{j\phi}S^{-1})(1 - e^{-j\phi}S^{-1})$
k^n	$(1 - S^{-1})^{n+1}$
$k^n a^k$	$(1 - aS^{-1})^{n+1}$
$a^k \sin \phi k$ or $a^k \cos \phi k$	$(1 - ae^{j\phi}S^{-1})(1 - ae^{-j\phi}S^{-1})$
$k \sin \phi k$ or $k \cos \phi k$	$[(1 - e^{j\phi}S^{-1})(1 - e^{-j\phi}S^{-1})]^2$

The process of finding an annihilator operator is simplified by the use of Table 2.1 in which an annihilator operator is listed for several typical forcing sequences. In working with sums of inputs, we note that if the sequence u is annihilated by the operator $L_u(S)$ and the sequence v is annihilated by $L_v(S)$, then the sequence $c_1 u + c_2 v$ is annihilated by $L_u(S)L_v(S)$.

We can find an annihilator operator for any sequence u that is itself a solution to a linear difference equation with constant coefficients. If the roots of the augmented auxiliary equation (obtained by applying L_A to both sides of the original difference equation) repeat any of the roots in the original auxiliary equation, we must follow the rules for repeated roots. For example, if a^k is a forcing function and if "a" is a root of the auxiliary equation, then the corresponding forced solution is cka^k. The annihilator in this case is $(1 - aS^{-1})$ which, of course, creates a repeated root in the augmented auxiliary equation. The following example illustrates a case in which the L_A operator repeats roots of the auxiliary equation.

Example 2.6 Find the output of the discrete-time system shown in Figure 2.6. Assume zero initial conditions, that is, $y_k = 0$, $k < 0$. The difference equation describing this system is

$$y_k + \frac{1}{9} y_{k-2} = \left(\frac{1}{3}\right)^k \cos\left(\frac{k\pi}{2}\right), \qquad k \geq 0$$

In operator notation we have

$$\left(1 + \frac{1}{9} S^{-2}\right)[y_k] = \left(\frac{1}{3}\right)^k \cos\left(\frac{k\pi}{2}\right), \qquad k \geq 0$$

We first find the transient (homogeneous) solution. The auxiliary equation is

$$r^2 + \frac{1}{9} = 0$$

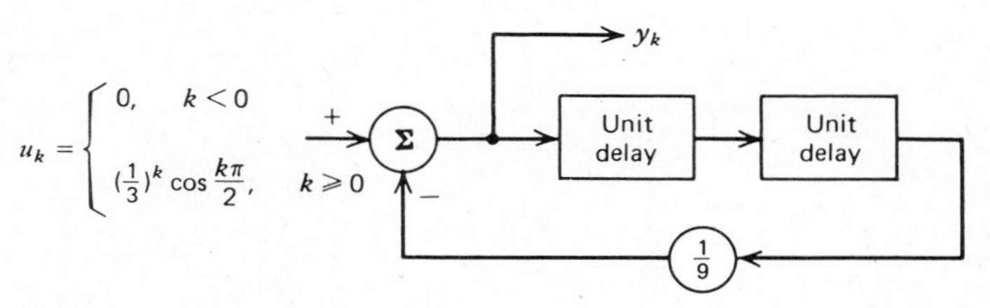

Figure 2.6

which has roots $r_1 = i/3$ and $r_2 = -i/3$. The corresponding solutions are

$$y_k^{(1)} = \left(\frac{1}{3}\right)^k \cos\left(\frac{k\pi}{2}\right), \qquad y_k^{(2)} = \left(\frac{1}{3}\right)^k \sin\left(\frac{k\pi}{2}\right)$$

[*Note*: $\rho = \sqrt{0^2 + (\frac{1}{3})^2} = \frac{1}{3}$, $\theta = \tan^{-1}(\frac{1}{3}/0) = \pi/2$.] The transient or homogeneous solution is therefore

$$y_k^{(h)} = c_1\left(\frac{1}{3}\right)^k \cos\left(\frac{k\pi}{2}\right) + c_2\left(\frac{1}{3}\right)^k \sin\left(\frac{k\pi}{2}\right)$$

c_1 and c_2 are arbitrary constants that can be evaluated using the initial conditions *after* we have determined the particular response.

To find the particular response we note from Table 2.1 that to annihilate $(\frac{1}{3})^k \cos(k\pi/2)$ we use an operator L_A given by

$$\begin{aligned} L_A &= \left(1 - \frac{1}{3}e^{j(\pi/2)}S^{-1}\right)\left(1 - \frac{1}{3}e^{-j(\pi/2)}S^{-1}\right) \\ &= 1 - \frac{S^{-1}}{3}\left(e^{j(\pi/2)} + e^{-j(\pi/2)}\right) + \frac{1}{9}S^{-2} \\ &= 1 - \frac{2}{3}\cos\left(\frac{\pi}{2}\right)S^{-1} + \frac{1}{9}S^{-2} = 1 + \frac{1}{9}S^{-2} \end{aligned}$$

Applying this operator to both sides of the original equation gives

$$\left(1 + \frac{1}{9}S^{-2}\right)\left(1 + \frac{1}{9}S^{-2}\right)[y_k] = 0$$

The roots of the auxiliary equation in this augmented equation corresponding to the L_A operator repeat the original homogeneous equation roots. Therefore, we must multiply the solutions arising from the L_A operator by k. That is, the particular solution is of the form

$$y_k^{(p)} = k\left(\frac{1}{3}\right)^k \left[c_3 \cos\left(\frac{k\pi}{2}\right) + c_4 \sin\left(\frac{k\pi}{2}\right)\right]$$

where c_3 and c_4 are to be determined. Substituting $y_k^{(p)}$ into the original difference equation and using

$$\cos\left[\frac{(k-2)\pi}{2}\right] = -\cos\left(\frac{k\pi}{2}\right), \qquad \sin\left[\frac{(k-2)\pi}{2}\right] = -\sin\left(\frac{k\pi}{2}\right)$$

we obtain

$$k\left(\frac{1}{3}\right)^k \left(c_3 \cos\left(\frac{k\pi}{2}\right) + c_4 \sin\left(\frac{k\pi}{2}\right)\right) + \frac{1}{9}(k-2)\left(\frac{1}{3}\right)^{k-2}\left[-c_3 \cos\left(\frac{k\pi}{2}\right) - c_4 \sin\left(\frac{k\pi}{2}\right)\right] = \left(\frac{1}{3}\right)^k \cos\left(\frac{k\pi}{2}\right)$$

or

$$2c_3 \cos\left(\frac{k\pi}{2}\right) + 2c_4 \sin\left(\frac{k\pi}{2}\right) = \cos\left(\frac{k\pi}{2}\right)$$

This implies that $c_3 = \frac{1}{2}$ and $c_4 = 0$. Hence, for $k \geq 0$,

$$y_k^{(p)} = \frac{k}{2}\left(\frac{1}{3}\right)^k \cos\left(\frac{k\pi}{2}\right)$$

The general solution $y_k = y_1^{(h)} + y_k^{(p)}$ is given by

$$y_k = c_1\left(\frac{1}{3}\right)^k \cos\left(\frac{k\pi}{2}\right) + c_2\left(\frac{1}{3}\right)^k \sin\left(\frac{k\pi}{2}\right) + \frac{k}{2}\left(\frac{1}{3}\right)^k \cos\left(\frac{k\pi}{2}\right), \qquad k \geq 0$$

To find the constants c_1 and c_2 we use the initial conditions $y_{-1} = y_{-2} = 0$.

$$y_{-1} = c_1\left(\frac{1}{3}\right)^{-1} \cos\left(\frac{-\pi}{2}\right) + c_2\left(\frac{1}{3}\right)^{-1} \sin\left(\frac{-\pi}{2}\right) - \left(\frac{1}{2}\right)\left(\frac{1}{3}\right)^{-1} \cos\left(\frac{-\pi}{2}\right) = 0$$

or

$$y_{-1} = -3c_2 = 0, \qquad c_2 = 0$$

And

$$y_{-2} = c_1\left(\frac{1}{3}\right)^{-2} \cos(-\pi) - \left(\frac{1}{3}\right)^{-2} \cos(-\pi) = 0$$

or

$$y_{-2} = -9c_1 + 9 = 0, \qquad c_1 = 1$$

Hence, the output sequence is

$$y_k = \begin{cases} 0, & k < 0 \\ \left(\frac{k}{2} + 1\right)\left(\frac{1}{3}\right)^k \cos\left(\frac{k\pi}{2}\right), & k \geq 0 \end{cases}$$

■

We have seen that the method of applying an annihilator operator is basically a formal procedure for guessing a trial solution. If, for example, $u_k = k^n$ and no roots of the auxilliary equation equal 1, then an appropriate annihilator is $L_A = (1 - S^{-1})^{n+1}$, yielding the trial solution $y_k = c_0 + c_1 k + c_2 k^2 + \cdots + c_n k^n$. In closing this section, we should emphasize that these results apply also to the case where the forcing function is not simply u_k but rather $L_D[u_k]$, where L_D is some driver operator. The general form for $L_D[\ ]$ is $[a_0 + a_1 S^{-1} + \cdots + a_m S^{-m}]$, yielding the forcing function $a_0 u_k + a_1 u_{k-1} + a_m u_{k-m}$. We note that the annihilator operator remains unchanged, since $L_A\{L_D[u]\} = L_D\{L_A[u]\} = L_D[0] = 0$. Thus the solution function remains unchanged as well, with only the constants differing from the earlier case. The solution may be obtained either by imposing $L[y] = L_D[u]$ to evaluate the unknown constants, or by applying the superposition property to handle the sum of forcing terms. We shall discuss this general case further in the following sections on frequency response and the state-variable formulation.

2.4 THE FREQUENCY RESPONSE OF DISCRETE-TIME SYSTEMS

Engineers are often asked to find the response of linear systems to sinusoidal inputs. The steady-state response to a spectrum of input sinusoids is called the *frequency response* of the system. The frequency response specifies the gain and

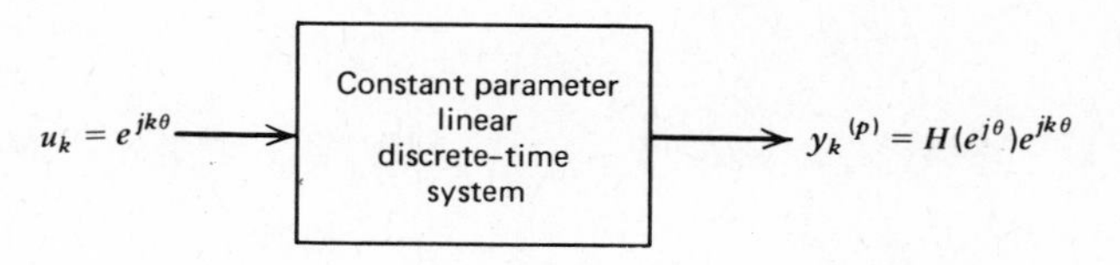

Figure 2.7

phase response of the system to input sinusoids at all frequencies. It is a fundamental characterization of linear, constant parameter systems. Many linear system requirements are specified in terms of frequency response.

The frequency response of linear, constant parameter systems can be obtained easily if we use the following fundamental property: If we force these systems with a complex exponential, say $e^{jk\theta}$, the steady-state response is always of the form $H(e^{j\theta})e^{jk\theta}$. In other words, the steady-state response is the identical exponential, modified in amplitude and phase by the system function $H(e^{j\theta})$ as shown in Figure 2.7. The quantity $H(e^{j\theta})$ as a function of θ is the frequency response of the system. In the case of discrete-time systems we need only find the steady-state solution to an input $u_k = e^{jk\theta}$ to determine the frequency response of these systems.

Example 2.7 Let us evaluate the frequency response of the first-order system shown in Figure 2.8. The difference equation for this system is

$$y_k - ay_{k-1} = u_k$$

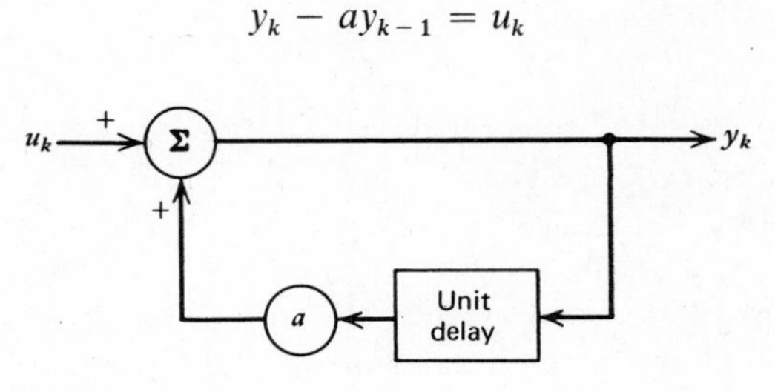

Figure 2.8

With $u_k = e^{jk\theta}$, we set $L_A = 1 - e^{j\theta}S^{-1}$ to obtain

$$(1 - e^{j\theta}S^{-1})(1 - aS^{-1})y_k = 0$$

from which

$$y_k = c_1 a^k + c_2 e^{jk\theta}$$

Applying $L[y] = u$, we find

$$(1 - aS^{-1})[c_1 a^k + c_2 e^{jk\theta}] = 0 + c_2(1 - ae^{-j\theta})e^{jk\theta} = e^{jk\theta}$$

from which

$$c_2 = \frac{1}{1 - ae^{-j\theta}}$$

Thus the steady-state solution is

$$y_k^{(p)} = \frac{1}{1 - ae^{-j\theta}} e^{jk\theta}$$

that is, the system transfer function is

$$H(e^{j\theta}) = \frac{1}{1 - ae^{-j\theta}}$$

To plot $H(e^{j\theta})$, we find the magnitude and phase terms as

$$|H(e^{j\theta})| = \left|\frac{1}{1 - a\cos\theta + ja\sin\theta}\right| = \sqrt{\frac{1}{1 - 2a\cos\theta + a^2}}$$

and

$$\arg[H(e^{j\theta})] = -\tan^{-1}\left[\frac{a\sin\theta}{1 - a\cos\theta}\right]$$

These functions are plotted in Figure 2.9. For $0 < a < 1$ as shown, the system amplifies exponentials with θ small and attenuates exponentials with θ near π.

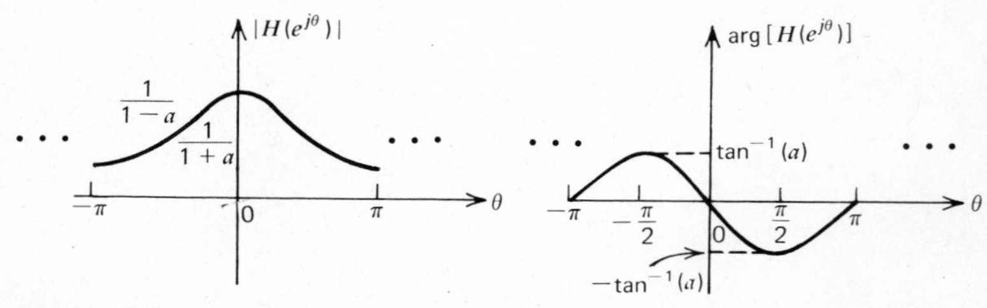

Figure 2.9

Note that $H(e^{j\theta})$ will always be periodic with period 2π in θ, since $e^{j(\theta+2\pi)} = e^{j\theta}e^{j2\pi} = 1 \cdot e^{j\theta} = e^{j\theta}$; hence $H[e^{j(\theta+2\pi)}] = H(e^{j\theta})$. ■

Example 2.8 Determine and sketch the frequency response of the discrete-time system in Figure 2.10. The difference equation for this system is

$$y_k + \frac{1}{2} y_{k-2} = u_k$$

Let $u_k = e^{jk\theta}$ and find the steady-state response. The transient response (homogeneous solution) is obtained from the auxiliary equation

$$r^2 + \frac{1}{2} = 0$$

with roots $r_1 = j/\sqrt{2}$, $r_2 = -j/\sqrt{2}$. Thus the transient response of the system is

$$y_k^{(h)} = \left(\frac{\sqrt{2}}{2}\right)^k \left[c_1 \cos\left(\frac{k\pi}{2}\right) + c_2 \sin\left(\frac{k\pi}{2}\right)\right]$$

which approaches 0 as $k \to \infty$, independently of c_1 and c_2. To find the steady-state response assume a trial solution of the form $y_k = ce^{jk\theta}$. Substituting this into the original equation yields

$$ce^{jk\theta} + \frac{1}{2} ce^{j(k-2)\theta} = e^{jk\theta}$$

or

$$e^{jk\theta}\left[c + \frac{1}{2} ce^{-2j\theta}\right] = e^{jk\theta}$$

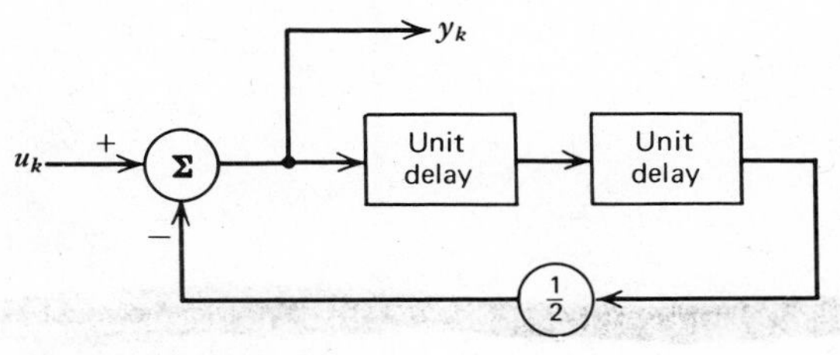

Figure 2.10

We must set the bracketed term equal to unity, yielding

$$c = \frac{1}{1 + \frac{1}{2} e^{-2j\theta}}$$

Thus the steady-state solution is

$$y_k^{(p)} = \frac{1}{1 + \frac{1}{2}e^{-2j\theta}} e^{jk\theta}$$

This solution is of the form $H(e^{j\theta})e^{jk\theta}$ where

$$H(e^{j\theta}) = \frac{1}{1 + \frac{1}{2}e^{-2j\theta}}$$

is the frequency response of the discrete-time system. It is a complex function of the normalized frequency variable θ. To evaluate the magnitude and phase response of the system, $|H(e^{j\theta})|$ and arg $[H(e^{j\theta})]$, respectively, we first rationalize $H(e^{j\theta})$:

$$H(e^{j\theta}) = \frac{1}{1 + \frac{1}{2}e^{-2j\theta}} = \frac{1}{1 + \dfrac{\cos 2\theta}{2} - j\dfrac{\sin 2\theta}{2}} = \frac{2}{2 + \cos 2\theta - j \sin 2\theta}$$

$$= \frac{2(2 + \cos 2\theta + j \sin 2\theta)}{(2 + \cos 2\theta)^2 + (\sin 2\theta)^2} = \frac{4 + 2 \cos 2\theta + j 2 \sin 2\theta}{5 + 4 \cos 2\theta}$$

The magnitude of $H(e^{j\theta})$ is therefore

$$|H(e^{j\theta})| = \frac{[(4 + 2 \cos 2\theta)^2 + (2 \sin 2\theta)^2]^{1/2}}{5 + 4 \cos 2\theta} = \frac{[20 + 16 \cos 2\theta]^{1/2}}{5 + 4 \cos 2\theta}$$

The phase response is the angle associated with $H(e^{j\theta})$:

$$\arg [H(e^{j\theta})] = \tan^{-1}\left[\frac{\sin 2\theta}{2 + \cos 2\theta}\right]$$

These functions of θ are plotted in Figure 2.11.

Notice again the periodicity of $H(e^{j\theta})$. If we evaluate $H(e^{j(\theta + 2\pi)})$ we obtain

$$H[e^{j(\theta + 2\pi}] = \frac{1}{1 + \frac{1}{2}e^{-2j(\theta + 2\pi)}} = \frac{1}{1 + \frac{1}{2}e^{-2j\theta}e^{-4j\pi}}$$

$$= \frac{1}{1 + \frac{1}{2}e^{-2j\theta}} = H(e^{j\theta})$$

Therefore $H(e^{j\theta})$ is periodic of period 2π. This is true of all discrete-time systems. (The response of this particular system has period π because θ occurs in even

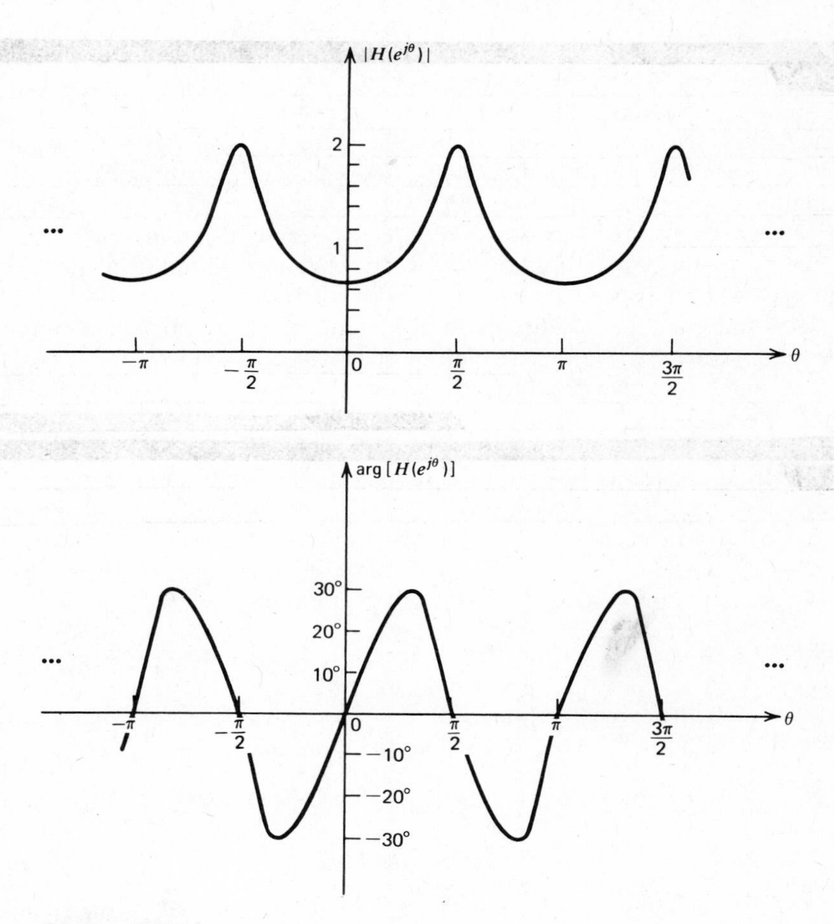

Figure 2.11

multiples.) For a difference equation with real coefficients we can show that $|H(e^{j\theta})|$ has even symmetry and arg $[H(e^{j\theta})]$ has odd symmetry about the origin. Thus a plot on the normalized frequency interval $0 \leq \theta \leq \pi$ is sufficient to define the frequency response for all θ. The points $\pm\pi$ are called the *fold-over* frequencies of the frequency response. The periodicity of the frequency response is evident in the plots of $|H(e^{j\theta})|$ and arg $[H(e^{j\theta})]$ shown in Figure 2.11. ■

Our discussion thus far has assumed implicitly a unit delay time of 1. That is, the time between samples of the input sequence, the output sequence, and the time delay of the unit delays is unity. If we assume a unit delay time (and time) between samples of T seconds, an unnormalized frequency response can be found. Let

$\theta = 2\pi f T$ where f is the frequency in Hertz and T is the sampling interval in seconds. In terms of f, the normalized interval $-\pi \leq \theta \leq \pi$ becomes $-\pi \leq 2\pi f T \leq \pi$ or $-1/2T \leq f \leq 1/2T$. The foldover frequency is now $(1/2T)$ Hertz. In general, we would like to make the fold-over frequency as large as possible in order to filter a large band of frequencies. This implies that we must choose T to be small. However, a small value of T means we must process a large number of samples per second. Thus we must balance our desire to obtain a large fold-over frequency with the cost and complexity needed to process samples quickly. Systems that operate at high speed are generally more costly.

We emphasize that all discrete-time systems have a frequency response that is periodic. Thus, for example, when we speak of a lowpass filter realized by a discrete-time system, we mean that the filter is lowpass with respect to the fundamental interval $-1/2T \leq f \leq 1/2T$. In fact, the periodicity of the response implies that we cannot build a lowpass response in the same sense as is done with analog circuits. In applications this is not a serious limitation of discrete-time circuits since one is usually interested in processing data over a finite band of frequencies.

Let us review the process by which we obtained the frequency response of a discrete-time system. Given a linear difference equation in the general form of (2.13) we set $u_k = e^{jk\theta}$ and find the steady-state response.

$$y_k + b_1 y_{k-1} + b_2 y_{k-2} + \cdots + b_n y_{k-n} = a_0 u_k + a_1 u_{k-1} + \cdots + a_m u_{k-m} \tag{2.13}$$

We know that the forced output is of the form $y_k^{(p)} = H(e^{j\theta})e^{jk\theta}$. Thus in (2.13) we substitute $H(e^{j\theta})e^{jk\theta}$ for y_k, and $e^{jk\theta}$ for u_k. This yields

$$H(e^{j\theta})e^{jk\theta} + b_1 H(e^{j\theta})e^{j(k-1)\theta} + \cdots + b_n H(e^{j\theta})e^{j(k-n)\theta} = a_0 e^{jk\theta} + \cdots + a_m e^{j(k-m)\theta} \tag{2.14}$$

In (2.14) we factor out $e^{jk\theta}H(e^{j\theta})$ on the left-hand side and $e^{jk\theta}$ on the right-hand side of the equation.

$$e^{jk\theta}H(e^{j\theta})[1 + b_1 e^{-j\theta} + \cdots + b_n e^{-jn\theta}] = e^{jk\theta}[a_0 + a_1 e^{-j\theta} + \cdots + a_m e^{-jm\theta}]$$

Cancelling $e^{jk\theta}$ on both sides and solving for $H(e^{j\theta})$ gives us

$$H(e^{j\theta}) = \frac{a_0 + a_1 e^{-j\theta} + \cdots + a_m e^{-jm\theta}}{1 + b_1 e^{-j\theta} + \cdots + b_n e^{-jn\theta}} \tag{2.15}$$

Equation 2.15 is a valuable formula. It allows us to calculate the frequency response of a discrete-time system directly from the difference equation model (2.13). We do not have to go through the process of finding steady-state solution. We conclude this section with a further example.

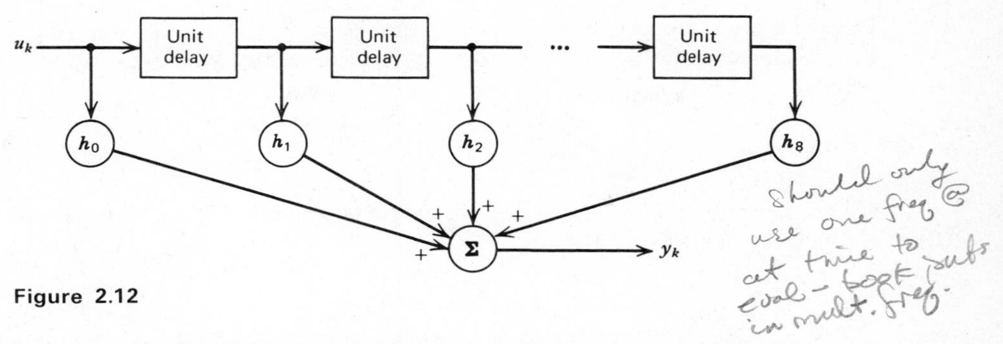

Figure 2.12

Example 2.9 Find the frequency response of the discrete-time system shown in Figure 2.12. The difference equation for this system is

$$y_k = h_0 u_k + h_1 u_{k-1} + \cdots + h_8 u_{k-8} \tag{2.16}$$

Notice that the output y_k depends only on u_k and past values of the input. This kind of circuit is called a *nonrecursive* or *finite duration impulse-response* (FIR) filter (the latter terminology is explained in Section 2.7).

Let

$$h_n = \begin{cases} \dfrac{1}{2} \operatorname{sinc}\left[(n-4)\dfrac{\pi}{2}\right], & n = 0, 1, 2, \ldots, 8 \\ 0, & \text{otherwise} \end{cases}$$

where sinc (x) is defined as $\sin x/x$ and is tabulated in Appendix B. This function is even, that is, sinc $(-x) =$ sinc (x). Calculation of $h_0, h_1, \ldots, h_8$ gives

$$h_0 = \frac{1}{2} \operatorname{sinc}\left(\frac{-4\pi}{2}\right), \qquad h_8 = \frac{1}{2} \operatorname{sinc}\left(\frac{4\pi}{2}\right)$$

$$h_1 = \frac{1}{2} \operatorname{sinc}\left(\frac{-3\pi}{2}\right), \qquad h_7 = \frac{1}{2} \operatorname{sinc}\left(\frac{-3\pi}{2}\right)$$

$$h_2 = \frac{1}{2} \operatorname{sinc}\left(\frac{-2\pi}{2}\right), \qquad h_6 = \frac{1}{2} \operatorname{sinc}\left(\frac{2\pi}{2}\right)$$

$$h_3 = \frac{1}{2} \operatorname{sinc}\left(\frac{-\pi}{2}\right), \qquad h_5 = \frac{1}{2} \operatorname{sinc}\left(\frac{\pi}{2}\right)$$

$$h_4 = \frac{1}{2} \operatorname{sinc}(0) = \frac{1}{2}$$

The evenness of the sinc function implies that $h_0 = h_8, h_1 = h_7, h_2 = h_6, h_3 = h_5$. The frequency response of this circuit can be found directly from (2.16), using the result (2.15). Thus

$$H(e^{j\theta}) = h_0 + h_1 e^{-j\theta} + \cdots + h_8 e^{-j8\theta} \tag{2.17}$$

We can rewrite (2.17) in a more convenient form by factoring $e^{-4j\theta}$ from each term on the right-hand side. Thus,

$$\begin{aligned} H(e^{j\theta}) = e^{-4j\theta}[&h_0 e^{j4\theta} + h_1 e^{3j\theta} + h_2 e^{2j\theta} + h_3 e^{j\theta} + h_4 \\ &+ h_5 e^{-j\theta} + h_6 e^{-2j\theta} + h_7 e^{-3j\theta} + h_8 e^{-4j\theta}] \end{aligned}$$

Using the symmetry in the coefficients we can combine terms like $h_0 e^{4j\theta} + h_8 e^{-4j\theta} = 2h_0 \cos(4\theta)$. Thus

$$H(e^{j\theta}) = e^{-4j\theta}[2h_0 \cos 4\theta + 2h_1 \cos 3\theta + 2h_2 \cos 2\theta + 2h_3 \cos \theta + h_4] \tag{2.18}$$

Equation 2.18 is an expression for the frequency response of the circuit. The term in brackets is purely real. Thus the amplitude response of $H(e^{j\theta})$ is

$$\begin{aligned} |H(e^{j\theta})| &= 2\left| h_0 \cos 4\theta + h_1 \cos 3\theta + h_2 \cos 2\theta + h_3 \cos \theta + \frac{h_4}{2} \right| \\ &= 2\left| \frac{1}{2} \operatorname{sinc} 2\pi \cos 4\theta + \frac{1}{2} \operatorname{sinc} \frac{3\pi}{2} \cos 3\theta + \frac{1}{2} \operatorname{sinc} \pi \cos 2\theta \right. \\ &\quad \left. + \frac{1}{2} \operatorname{sinc} \frac{\pi}{2} \cos \theta + \frac{1}{4} \right| \\ &\cong \left| 0 - 0.212206 \cos 3\theta + 0 + 0.636619 \cos \theta + \frac{1}{2} \right| \end{aligned} \tag{2.19}$$

A plot of this expression is shown in Figure 2.13. The angle or phase of $H(e^{j\theta})$ is defined by the term $e^{-4j\theta}$, that is,

$$\arg[H(e^{j\theta})] = -4\theta \tag{2.20}$$

The phase response is also shown in Figure 2.13. Notice that the phase response is piecewise linear in θ. The discontinuities in the phase response are caused by a change of sign in the amplitude response. The phase response represents a pure delay and is called a *distortionless phase response*. This type of phase response is

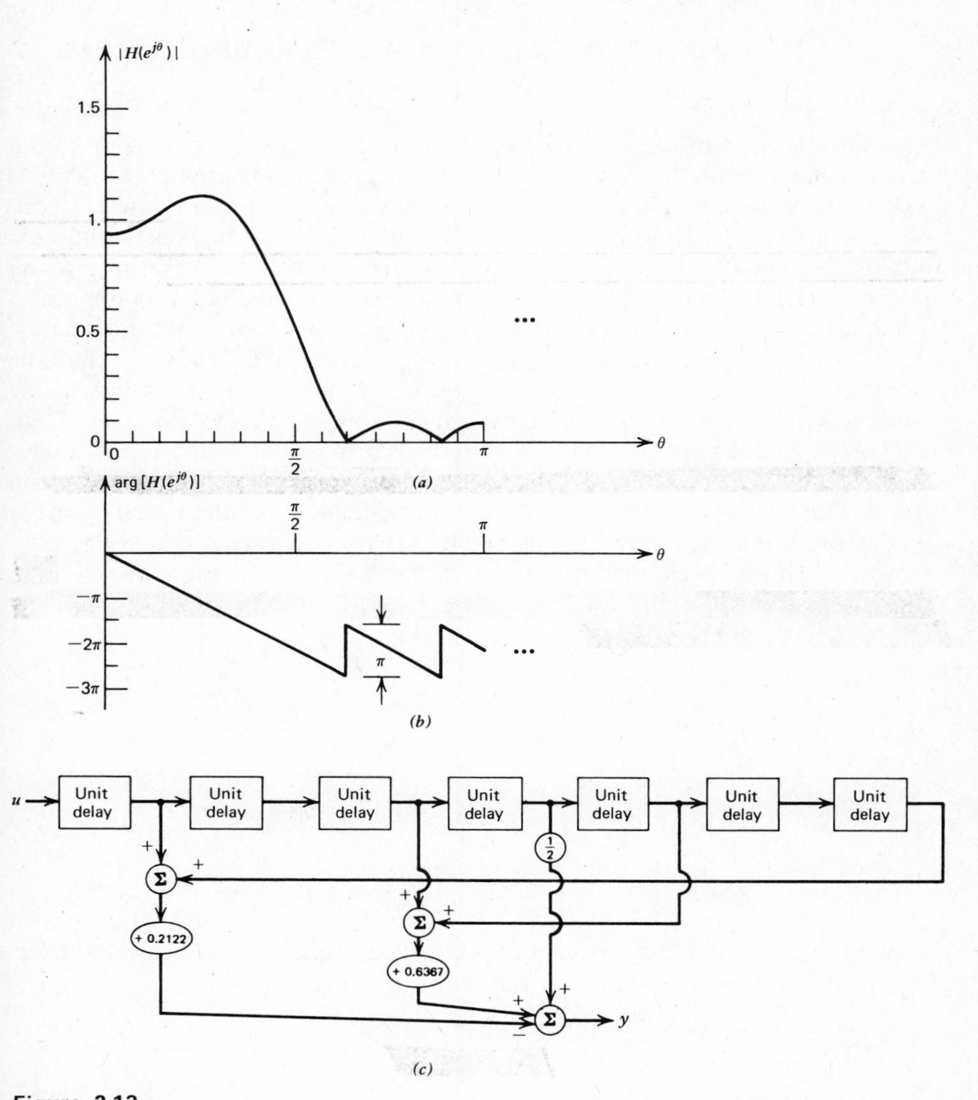

Figure 2.13

obtained whenever the coefficients of the filter are real and symmetric, that is, $h_i = h_{8-i}$, $i = 0, 1, 2, 3$.

A realization of this filter can be accomplished using only three multipliers as shown in Figure 2.13. This realization is based on combining terms as in (2.19). It is important to keep the number of multipliers as small as possible since they generally are the most costly components of a digital filter realization. ■

(ALSO SLOWEST)

2.5 CONVOLUTION AND IMPULSE RESPONSE

The difference equation model allows us to solve for the output resulting from a given input sequence. Another method of analyzing discrete-time systems is based on the so-called impulse-response sequence of the system. Using a convolutional sum operation one can find the output to any given input assuming zero initial conditions, that is, the unit delays initially contain zeros. The impulse response model thus provides us with another representation of linear systems.

Linear systems are characterized by the superposition property. If we know the response for input sequences $u_1(k)$ and $u_2(k)$ separately, then we can find the response due to $u_1(k) + u_2(k)$ by merely summing the separate outputs. If the linear system is also shift invariant, we can shift these inputs to any point along the k-axis and obtain the output by making a corresponding shift in k. In symbols, if $u(k)$ produces an output $y(k)$, then $u(k + n)$ produces the output $y(k + n)$. In this section we shall use these two properties to develop another formulation for finding the output of a linear, time-invariant discrete-time system given the input sequence. This formulation makes use of the *impulse-response sequence.* The impulse response is defined as the system output due to an impulse sequence $\{\delta_k\}$ at the system input, where

$$\delta_k = \begin{cases} 1, & k = 0 \\ 0, & k \neq 0 \end{cases} \tag{2.21}$$

Define the response to $\{\delta_k\}$ as $\{h_k\}$, the impulse response sequence:

$$\{\delta_k\} \to \{h_k\}$$

If we multiply the impulse sequence by a constant c and apply this to the system, then by linearity the output is also multiplied by c, that is,

$$c\{\delta_k\} \to c\{h_k\}$$

If we shift the position of this impulse sequence, then by the shift invariance of the system, we also shift the output sequence by the same amount, that is,

$$c\{\delta_{k\pm n}\} \to c\{h_{k\pm n}\}$$

These relationships are shown schematically in Figure 2.14.

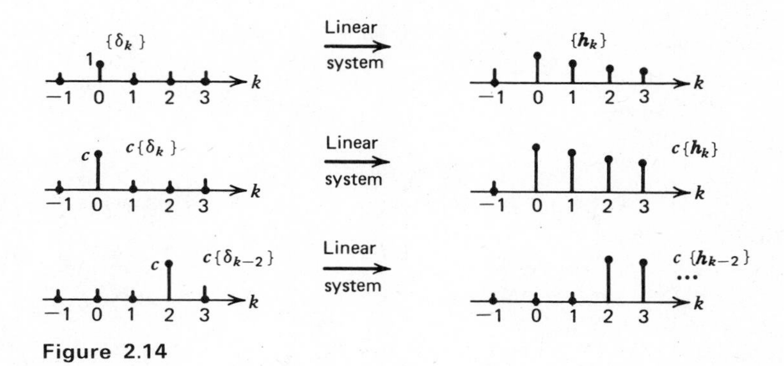

Figure 2.14

We are, of course, interested in the response resulting from arbitrary input signal sequences, say u. How can we use the impulse-response sequence to help us? Suppose we represent the input sequence as follows:

$$\{u_k\} = \cdots + u_{-2}\{\delta_{k+2}\} + u_{-1}\{\delta_{k+1}\} + u_0\{\delta_k\} + u_1\{\delta_{k-1}\} + \cdots$$

$$= \sum_{n=-\infty}^{\infty} u_n\{\delta_{k-n}\}$$

In other words, we take each point of the sequence u_j and multiply it by a shifted version of the impulse sequence $\{\delta_{k-j}\}$. Because $\{\delta_{k-j}\}$ has value unity for $k = j$ and zero otherwise, this procedure allows us to represent an arbitrary input sequence as a weighted sum of shifted impulse sequences. Thus

$$\{u_k\} = \sum_{j=-\infty}^{\infty} u_j\{\delta_{k-j}\} \tag{2.22}$$

Figure 2.15 depicts an example of the representation implied by (2.22).

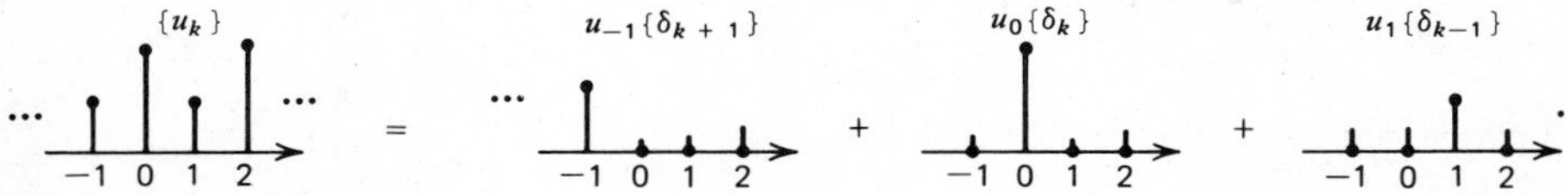

Figure 2.15

The general term $u_j\{\delta_{k-j}\}$ gives rise to a response $u_j\{h_{k-j}\}$. Applying the superposition property, we find that the total response is merely the sum of the individual responses of the form $u_j\{h_{k-j}\}$. Thus the output sequence is

$$\{y_k\} = \sum_{j=-\infty}^{\infty} u_j\{h_{k-j}\}$$

The kth term of $\{y_k\}$ is

$$y_k = \sum_{j=-\infty}^{\infty} u_j h_{k-j} \tag{2.23}$$

The sum in (2.23) is known as a *convolution sum.* The shorthand notation used for convolution is

$$\{y_k\} = \{u_k\} * \{h_k\} \qquad \text{or} \qquad y = u * h$$

If we let $m = k - j$ in (2.23), we can write (2.23) as

$$y_k = \sum_{m=-\infty}^{\infty} u_{k-m} h_m \tag{2.24}$$

which means that convolution is commutative, that is,

$$y = h * u = u * h$$

Given the impulse response h we can, using (2.23) or (2.24), obtain the output sequence for a quiescent system, that is, one with zero initial conditions. This characterization of the input-output relation for a given system is quite different from the difference equation characterization. This formulation is a useful conceptual aid in understanding how a linear system processes the input sequence to form the output sequence. The graphical interpretation given in the next section is the basis of this understanding.

2.6 THE CONVOLUTION OPERATION

Consider the convolution of two sequences $u = \left\{\cdots 1, 3, \underset{\uparrow}{1}, 3, 1, \cdots\right\}$ and $h = \left\{\underset{\uparrow}{1}, 2, 1\right\}$. Call the result y. To find y_k we use (2.23). For example, y_1 is given by

$$y_1 = \sum_{n=-\infty}^{\infty} u_n h_{1-n}$$

To find y_1 we need h_{1-n}. To find h_{1-n} we first take $\{h_n\}$ and form $\{h_{-n}\}$, which is the mirror image of $\{h_n\}$ through a vertical axis at the origin. See Figure 2.16*b*.

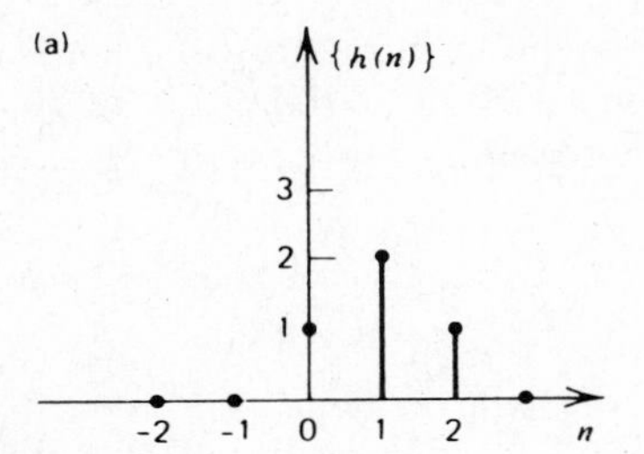
(a)
{h(n)}
3
2
1
-2
-1
0
1
2
n

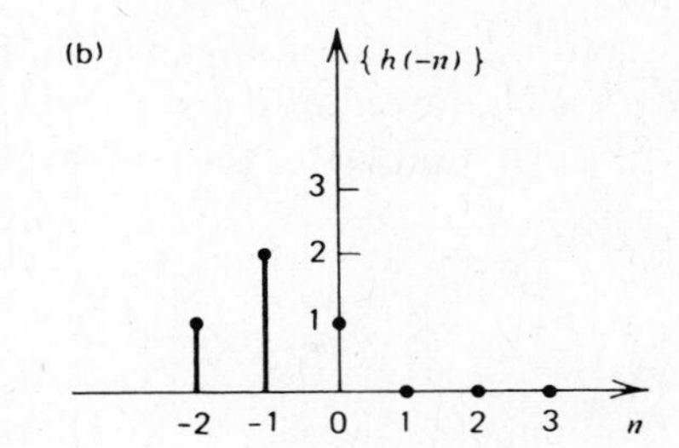
(b)
{h(-n)}
3
2
1
-2
-1
0
1
2
3
n

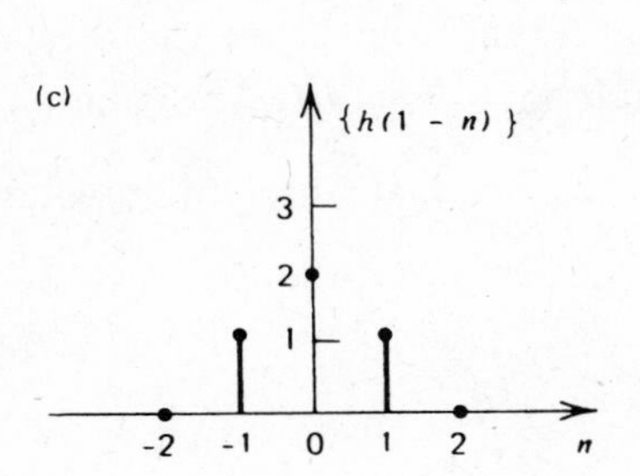
(c)
{h(1 - n)}
3
2
1
-2
-1
0
1
2
n

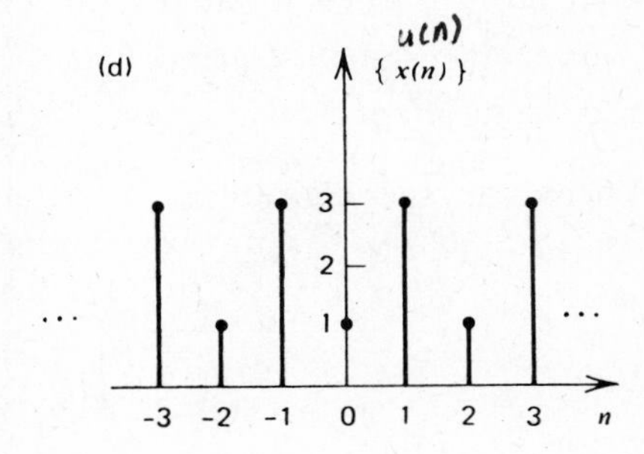
u(n)
(d)
{x(n)}
3
2
1
...
-3
-2
-1
0
1
2
3
n

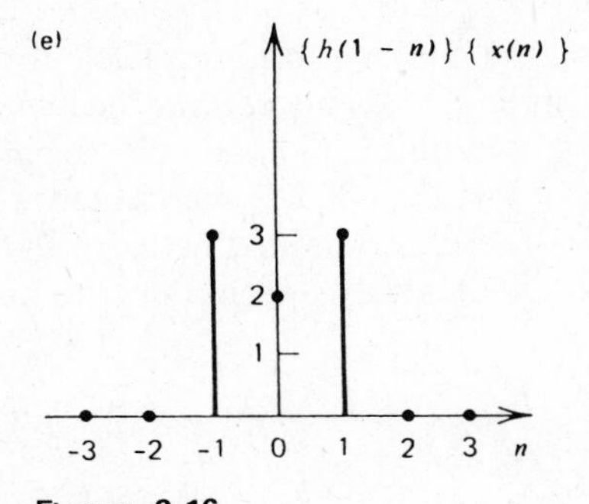
(e)
{h(1 - n)} {x(n)}
3
2
1
-3
-2
-1
0
1
2
3
n

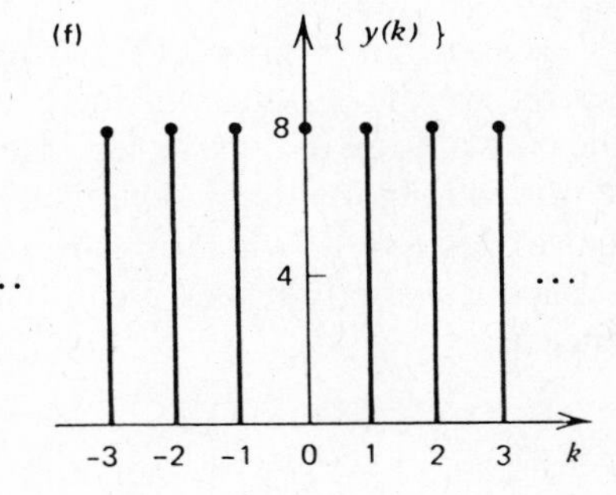
(f)
{y(k)}
8
4
...
-3
-2
-1
0
1
2
3
k

Figure 2.16

Now shift $\{h_{-n}\}$ right one unit to form $\{h_{1-n}\}$ as shown in Figure 2.16*c*. This shifted sequence is multiplied with u, depicted in Figure 2.16*d*, and the resultant sequence values, shown in Figure 2.16*e*, are then added together to obtain y_1

$$y_1 = \sum_{n=-\infty}^{\infty} u_n h_{1-n} = 3 + 2 + 3 = 8$$

To summarize, we can view the convolution sum as composed of four basic operations:

1. Take a mirror image of $\{h(n)\}$ about vertical axis through the origin to obtain $\{h(-n)\}$.
2. Shift $\{h(-n)\}$ to the right by an amount equal to the value of k where the output sequence is to be evaluated, producing $\{h(k - n)\}$.
3. Multiply this shifted sequence $\{h(k - n)\}$ and the input sequence $\{u(n)\}$.
4. Add the resultant sequence values to obtain the value of the convolution at k.

There is another algorithm that we can use to evaluate discrete convolutions. Suppose that we wish to convolve h and u, where

$$h(k) = \begin{cases} \left(\dfrac{1}{2}\right)^k, & k \leq 0 \\ 0, & k < 0 \end{cases}$$

and

$$\{u(k)\} = \{3, 2, 1\}$$

Construct a matrix with h bordering the top of the matrix and u the left side of the matrix, as shown in Figure 2.17. In this case, the matrix is infinite. The entries in the matrix are the products of the corresponding row and column headers. To find the convolution of the two sequences, we need only "fold and add" according to the dotted diagonal lines. The first term $y(0)$, is thus 3. The second term, $y(1)$, is equal to $2 + \frac{3}{2} = \frac{7}{2}$, which is the sum of the terms contained between the first and second diagonal lines. We can continue in this way and obtain the output sequence as

$$\{y_k\} = \left\{3, \frac{7}{2}, \frac{11}{4}, \frac{11}{8}, \frac{11}{16}, \ldots, \frac{11}{2^k} \cdots\right\}$$

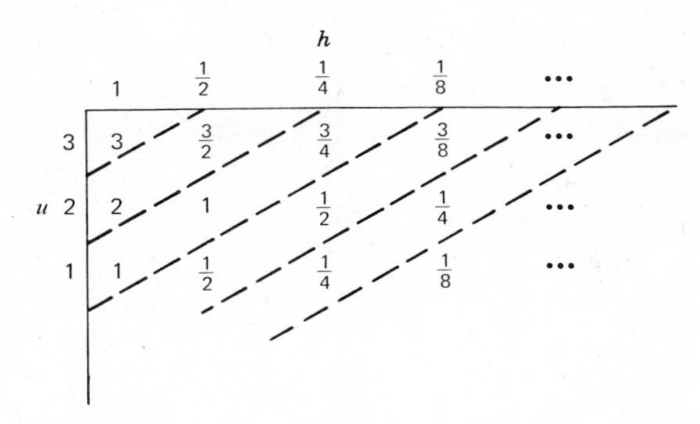

Figure 2.17

The reader can verify this result by using the formal calculations indicated in (2.23). In the case of two-sided sequences, the zeroth term in the output is contained between the diagonals containing the intersection term of the zeroth indices for the row and column sequences. This algorithm is not always a satisfactory method because the result is not easily placed in closed form. Only in simple cases can one determine the closed form solution of y_k.

Example 2.10 Convolve the impulse sequence $\{\delta_k\}$ with an arbitrary sequence $\{u_k\}$. The kth term of $\{y_k\}$ is

$$y_k = \sum_{n=-\infty}^{\infty} u_n \delta_{k-n} \tag{2.25}$$

Each term in δ_{k-n} is zero except for $k = n$. The only nonzero term in (2.25) occurs for $k = n$ and, thus, $y_k = u_k$. In other words, convolution of $\{u_k\}$ and $\{\delta_k\}$ reproduces the sequence $\{u_k\}$. ■

Example 2.11 Convolve the sequences $\{u_k\}$ and $\{h_k\}$ where

$$u_k = \begin{cases} 0, & k < 0 \\ a^k, & k \geq 0 \end{cases} \qquad h_k = \begin{cases} 0, & k < 0 \\ b^k, & k \geq 0 \end{cases}$$

Using the convolution sum, the output y_k is

$$y_k = \sum_{n=-\infty}^{\infty} u_n h_{k-n} = \sum_{n=0}^{k} u_n h_{k-n} \tag{2.26}$$

The lower limit in (2.26) is zero because $u_n = 0$ for $n < 0$. The upper limit is k because $h_{k-n} = 0$ for $n > k$, that is, whenever the index on h is negative. Thus,

$$y_k = \begin{cases} 0, & k < 0 \\ \sum_{n=0}^{k} a^n b^{k-n}, & h \geq 0 \end{cases}$$

For $k \geq 0$, we can evaluate the sum using the formula for the partial sum of a geometric sequence.

$$y_k = b^k \sum_{n=0}^{k} a^n b^{-n} = b^k \sum_{n=0}^{k} \left(\frac{a}{b}\right)^n = \begin{cases} b^k \dfrac{1-(a/b)^{k+1}}{1-(a/b)}, & a \neq b \\ b^k(1+k), & a = b \end{cases} \qquad k \geq 0$$ ■

Example 2.12 Using the convolution sum, determine the output of the digital system shown in Figure 2.18. Assume the input sequence is $\{u_k\} = \{3, -1, 3\}$.

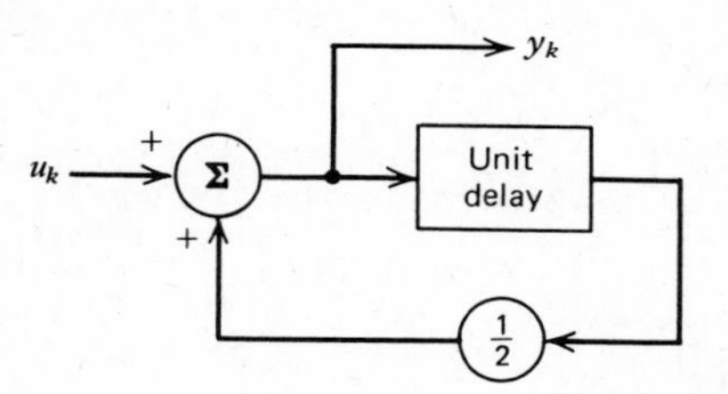

Figure 2.18

The difference equation for this system is found by equating the output of the adder y_k to the two inputs, u_k and $\frac{1}{2}y_{k-1}$. Thus,

$$y_k = \frac{1}{2} y_{k-1} + u_k$$

Assuming the system is initially quiescent, $y_{-1} = 0$, we must first find the impulse response sequence h. By definition h is the output for an input $u = \{\delta_k\}$, that is, h satisfies the difference equation

$$h_k = \frac{1}{2} h_{k-1} + \delta_k \tag{2.27}$$

Suppose we try an iterative approach. From (2.27) we have

$$h_0 = \frac{1}{2} h_{-1} + \delta_0 = 0 + 1 = 1$$

$$h_1 = \frac{1}{2} h_0 + \delta_1 = \frac{1}{2} \cdot 1 + 0 = \frac{1}{2}$$

$$h_2 = \frac{1}{2}h_1 + \delta_2 = \frac{1}{2}\cdot\frac{1}{2} + 0 = \frac{1}{4}$$

$$\cdots$$

$$h_k = \frac{1}{2}h_{k-1} + \delta_k = \begin{cases} \left(\frac{1}{2}\right)^k, & k \geq 0 \\ 0, & k < 0 \end{cases}$$

The output is therefore given by

$$y = \{3, -1, 3\} * \left\{\left(\frac{1}{2}\right)^k\right\}, \qquad k \geq 0$$

The resultant sequence values can be calculated using the "fold and add" method as depicted in Figure 2.19. The output sequence is

$$y = \left\{3, \frac{1}{2}, \frac{13}{4}, \frac{13}{8}, \frac{13}{16}, \ldots, \frac{13}{2^k}, \ldots\right\}$$

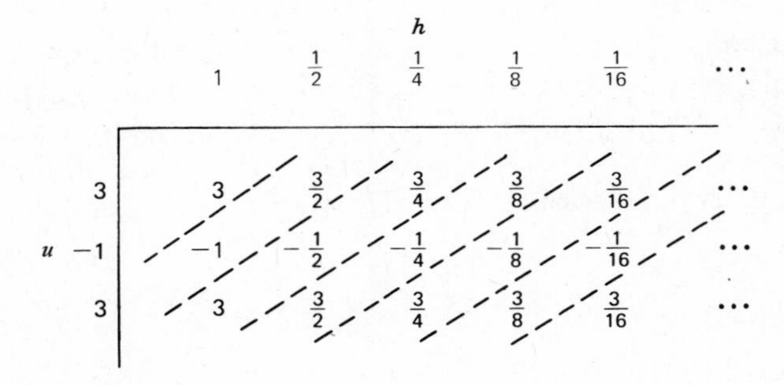

Figure 2.19

■

In this example we were able to obtain a closed form solution for the impulse-response sequence by simply recognizing the general term from our iterative expansion. In more complicated systems it would be difficult to recognize the form of the general term. In these cases we need a general method for finding the closed form expression for h_k.

2.7 FINDING THE IMPULSE-RESPONSE SEQUENCE

In order to use the convolution sum as a method of calçulating the response of a linear system, we must have a method for finding the impulse response sequence in closed form. There is such a method based on knowledge of the homogeneous solution of the difference equation model. Suppose we have a system modeled by the difference equation

$$y_k + b_1 y_{k-1} + b_2 y_{k-2} + \cdots + b_n y_{k-n} = u_k \tag{2.28}$$

By definition, the impulse-response sequence $\{h_k\}$ satisfies (2.28) for $u_k = \delta_k$. That is,

$$h_k + b_1 h_{k-1} + b_2 h_{k-2} + \cdots + b_n h_{k-n} = \delta_k \tag{2.29}$$

Now $\delta_k = 0$ for $k > 0$. Therefore, $\{h_k\}$ must satisfy the homogeneous difference equation corresponding to (2.28) for $k > 0$. This means that $\{h_k\}$ can be expressed as a sum of n linearly independent solutions $\phi_k^{(1)}, \phi_k^{(2)}, \ldots, \phi_k^{(n)}$, each of which satisfies (2.28) with $u_k = 0$. Thus,

$$h_k = c_1 \phi_k^{(1)} + c_2 \phi_k^{(2)} + \cdots + c_n \phi_k^{(n)} \tag{2.30}$$

where the constants c_i, $i = 1, 2, \ldots, n$ are evaluated using the initial conditions for h. These initial conditions are imposed on the system by the unit impulse sequence $\{\delta_k\}$. There are several equivalent sets of initial conditions obtained using (2.29) and the fact that $h_k = 0$ for $k < 0$. We need only to include the effect of the impulse on the circuit as evaluated using (2.29). Thus, for example, we can use the set of conditions:

$$h_{-1} = h_{-2} = \cdots = h_{-n+1} = 0$$

and

$$h_0 = \delta_0 - b_1 h_{-1} - b_2 h_{-2} - \cdots - b_n h_{-n} = 1$$

Or we could use the set of conditions:

$$h_{-1} = h_{-2} = \cdots = h_{-n+2} = 0$$

$$h_0 = 1$$

$$h_1 = \delta_1 - b_1 h_0 - b_2 h_{-1} - \cdots - b_n h_{-n+1} = -b_1$$

Continuing in a similar manner we can generate as many sets of initial conditions as we desire using (2.29).

Example 2.13 Find the output of the following discrete-time system for an input

$$u_k = \begin{cases} \left(\frac{1}{2}\right)^k, & k \geq 0 \\ 0, & k < 0 \end{cases}.$$

The difference equation model for this system is found by equating the output of the summer y_k, to the three inputs to the summer. Thus

$$y_k = u_k + \frac{3}{4} y_{k-1} - \frac{1}{8} y_{k-2}$$

In standard form

$$y_k - \frac{3}{4} y_{k-1} + \frac{1}{8} y_{k-2} = u_k$$

The auxiliary equation is

$$r^2 - \frac{3}{4} r + \frac{1}{8} = 0$$

with roots $r_1 = \frac{1}{2}$, $r_2 = \frac{1}{4}$. The impulse-response sequence is therefore

$$h_k = \begin{cases} c_1 \left(\frac{1}{2}\right)^k + c_2 \left(\frac{1}{4}\right)^k, & k \geq 0 \\ 0, & k < 0 \end{cases}$$

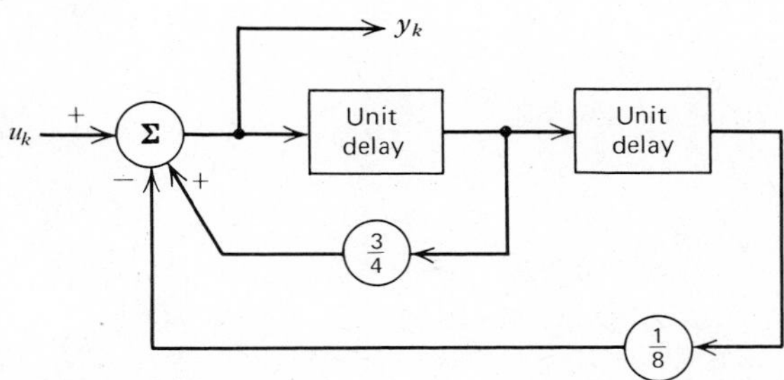

Figure 2.20

To find the initial conditions, we use the fact that

$$h_k - \frac{3}{4} h_{k-1} + \frac{1}{8} h_{k-2} = \delta_k$$

Thus, for example,

$$h_0 = \delta_0 + \frac{3}{4} h_{-1} - \frac{1}{8} h_{-2} = 1$$

$$h_1 = \delta_1 + \frac{3}{4} h_0 - \frac{1}{8} h_{-1} = \frac{3}{4}$$

Using these initial conditions, c_1 and c_2 must satisfy

$$1 = c_1 + c_2$$

$$\frac{3}{4} = \frac{1}{2} c_1 + \frac{1}{4} c_2$$

which imply $c_1 = 2$ and $c_2 = -1$. Thus, the impulse-response sequence is

$$h_k = \begin{cases} 2\left(\frac{1}{2}\right)^k + (-1)\left(\frac{1}{4}\right)^k, & k \geq 0 \\ 0, & k < 0 \end{cases}$$

The output sequence is given by

$$\begin{aligned} y_k &= \sum_{n=0}^{k} u_n h_{k-n} = \sum_{n=0}^{k} \left(\frac{1}{2}\right)^n \left[2\left(\frac{1}{2}\right)^{k-n} - \left(\frac{1}{4}\right)^{k-n}\right] \\ &= 2\left(\frac{1}{2}\right)^k \sum_{n=0}^{k} 1 - \left(\frac{1}{4}\right)^k \sum_{n=0}^{k} 2^n = 2\left(\frac{1}{2}\right)^k [1 + k] - \left(\frac{1}{4}\right)^k \left[\frac{1 - 2^{k+1}}{-1}\right] \\ &= \begin{cases} 2k\left(\frac{1}{2}\right)^k + \left(\frac{1}{4}\right)^k, & k \geq 0 \\ 0, & k < 0 \end{cases} \end{aligned}$$

Notice that we could also obtain the impulse-response sequence using other sets of initial conditions, again based on $h_k - \frac{3}{4}h_{k-1} + \frac{1}{8}h_{k-2} = \delta_k$. Equivalent sets of initial conditions are: $h_{-1} = 0, h_0 = 1$; $h_0 = 1, h_1 = \frac{3}{4}$; $h_1 = \frac{3}{4}, h_2 = \frac{7}{16}$, etc. ■

Example 2.14 Find the impulse-response sequence of the discrete-time circuit shown in Figure 2.21. The difference equation model for this system is

$$y_k = 5u_k - 2u_{k-1} + \frac{3}{2}u_{k-2} - 3u_{k-3}$$

In this system the output is formed from the present input and previous values of the input. We can find the impulse response sequence directly from the difference equation

$$h_k = 5\delta_k - 2\delta_{k-1} + \frac{3}{2}\delta_{k-2} - 3\delta_{k-3}$$

Thus, we have

$$h_0 = 5,$$

$$h_1 = -2$$

$$h_2 = \frac{3}{2}$$

$$h_3 = -3$$

$$h_k = 0, \qquad \text{otherwise}$$

These kinds of discrete-time circuits are called *Finite-duration Impulse Response* (FIR) filters for obvious reasons. In earlier terminology they were known as non-recursive filters because there is no feedback of past outputs to form the present output.

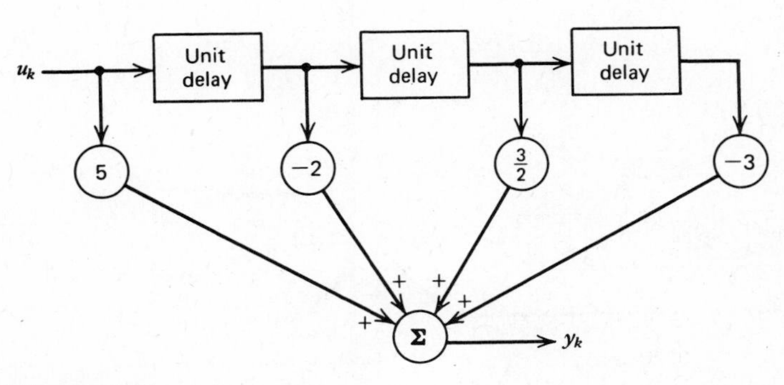

Figure 2.21

■

Example 2.15 Find the impulse-response sequence of the discrete-time system shown in Figure 2.22. The difference equation model for this system is

$$y_k = -u_k + 2u_{k-3} + \frac{1}{2} y_{k-1} - \frac{1}{18} y_{k-2}$$

or

$$y_k - \frac{1}{2} y_{k-1} + \frac{1}{18} y_{k-2} = -u_k + 2u_{k-3}$$

There is more than one method of solution. We can use superposition and find the impulse-response sequence from the input $-\delta_k$ and then find the impulse-response due to the input $2\delta_{k-3}$. The desired impulse response is then the sum of these two individual responses. The associated auxiliary equation is

$$r^2 - \frac{1}{2} r + \frac{1}{18} = 0$$

with roots $r_1 = \frac{1}{3}$, $r_2 = \frac{1}{6}$. The impulse-response sequence is thus of the form

$$h_k = c_1 \left(\frac{1}{3}\right)^k + c_2 \left(\frac{1}{6}\right)^k, \qquad k \geq 0$$

The initial conditions due to the lower branch $(-\delta_k)$ are

$$h_0 = -1, \qquad h_1 = -\frac{1}{2}$$

Thus c_1 and c_2 must satisfy

$$-1 = c_1 + c_2$$

$$-\frac{1}{2} = \frac{1}{3} c_1 + \frac{1}{6} c_2$$

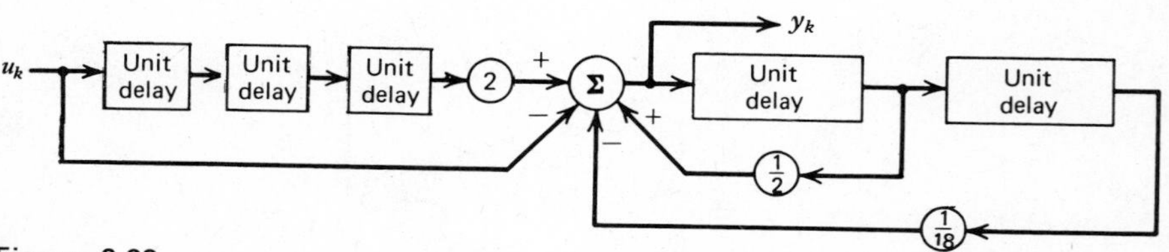

Figure 2.22

which yields $c_1 = -2$, $c_2 = 1$. Thus the impulse response due to $-\{\delta_k\}$ is $\{h_k^{(1)}\}$:

$$h_k^{(1)} = \begin{cases} -2\left(\frac{1}{3}\right)^k + \left(\frac{1}{6}\right)^k, & k \geq 0 \\ 0, & k < 0 \end{cases}$$

Call the impulse-response due to $2\{\delta_{k-3}\}$, $\{h_k^{(2)}\}$. The initial conditions for an input $-\{\delta_k\}$ were previously found to be $h_0 = -1$, $h_1 = -\frac{1}{2}$ with a corresponding impulse-response sequence $\{h_k^{(1)}\}$ as given above. The response to $2\{\delta_k\}$ is therefore $4(\frac{1}{3})^k - 2(\frac{1}{6})^k$. We obtain this result by linearity since $2\{\delta_k\}$ is -2 times $-\{\delta_k\}$. The response to $2\{\delta_{k-3}\}$ is thus

$$h_k^{(2)} = \begin{cases} 4\left(\frac{1}{3}\right)^{k-3} - 2\left(\frac{1}{6}\right)^{k-3}, & k \geq 3 \\ 0, & k < 3 \end{cases}$$

The limits result from the fact that $2\{\delta_{k-3}\}$ is zero until $k = 3$. The total response is thus

$$h_k = \begin{cases} h_k^{(1)}, & k = 0, 1, 2 \\ h_k^{(1)} + h_k^{(2)}, & k \geq 3 \\ 0, & \text{otherwise} \end{cases}$$

Hence

$$h_k = \begin{cases} -2\left(\frac{1}{3}\right)^k + \left(\frac{1}{6}\right)^k, & k = 0, 1, 2 \\ 106\left(\frac{1}{3}\right)^k - 431\left(\frac{1}{6}\right)^k, & k \geq 3 \\ 0, & \text{otherwise} \end{cases}$$

Another method of solution involves using initial conditions imposed by $-\{\delta_k\} + 2\{\delta_{k-3}\}$ in one step. One must make sure to include both impulses in the calculation of the initial conditions. Thus, to use h_0 and h_1, for example, is incorrect, since the impulse $2\{\delta_{k-3}\}$ would not be included in the calculations. One could use as initial conditions h_2 and h_3, h_3 and h_4, h_4 and h_5, etc. One cannot use any earlier sequence of initial conditions. Using h_2 and h_3 means that the values of h_0 and h_1 must be calculated as special cases. Suppose we use h_2 and h_3 as initial conditions. We have

$$h_k = \frac{1}{2} h_{k-1} - \frac{1}{18} h_{k-2} - \delta_k + 2\delta_{k-3}$$

from which

$$h_0 = \frac{1}{2} h_{-1} - \frac{1}{18} h_{-2} - \delta_0 + 2\delta_{-3} = -1$$

$$h_1 = \frac{1}{2} h_0 - \frac{1}{18} h_{-1} - \delta_1 + 2\delta_{-2} = -\frac{1}{2}$$

$$h_2 = \frac{1}{2} h_1 - \frac{1}{18} h_0 - \delta_2 + 2\delta_{-1} = -\frac{7}{36}$$

$$h_3 = \frac{1}{2} h_2 - \frac{1}{18} h_1 - \delta_3 + 2\delta_0 = \frac{139}{72}$$

Thus, c_1 and c_2 must satisfy

$$h_2 = -\frac{7}{36} = \frac{c_1}{9} + \frac{c_2}{36}$$

$$h_3 = \frac{139}{72} = \frac{c_1}{27} + \frac{c_2}{216}$$

and so

$$c_1 = 106 \qquad c_2 = -431$$

The impulse-response sequence is, therefore,

$$\begin{aligned} h_k &= -1, \qquad k = 0 \\ &= -\frac{1}{2}, \qquad k = 1 \\ &= 106\left(\frac{1}{3}\right)^k - 431\left(\frac{1}{6}\right)^k, \qquad k \geq 2 \\ &= 0, \qquad \text{otherwise} \end{aligned}$$

Example 2.16 We can also find the frequency response of a system using the impulse response and the convolution sum. Consider the system of Example 2.8 repeated below. The impulse response of this system is a linear combination of the homogeneous solutions of the difference equation

$$y_k + \frac{1}{2} y_{k-2} = u_k$$

The auxiliary equation is

$$r^2 + \frac{1}{2} = 0$$

with roots $r_1, r_2 = \pm j/\sqrt{2}$. This implies that h_k is of the form

$$h_k = \left(\frac{\sqrt{2}}{2}\right)^k \left[c_1 \cos\left(\frac{k\pi}{2}\right) + c_2 \sin\left(\frac{k\pi}{2}\right)\right], \qquad k \geq 0$$

Using the initial conditions imposed on the circuit by the impulse sequence, we have

$$h_0 = 1 = c_1 \cos 0 + c_2 \sin 0$$

$$h_1 = 0 = c_1 \cos\left(\frac{\pi}{2}\right) + c_2 \sin\left(\frac{\pi}{2}\right)$$

Thus $c_1 = 1$ and $c_2 = 0$. The impulse-response sequence is

$$h_k = \begin{cases} \left(\frac{\sqrt{2}}{2}\right)^k \cos\left(\frac{k\pi}{2}\right), & k \geq 0 \\ 0, & k < 0 \end{cases}$$

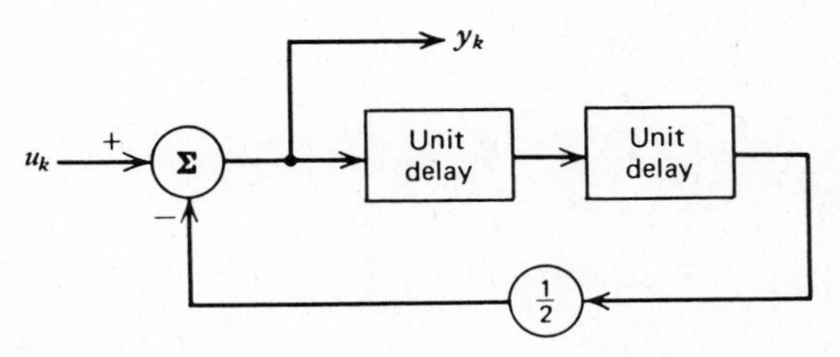

Figure 2.23

To obtain the frequency response we find the steady-state response to an input $u_k = e^{jk\omega T}$. We shall assume that u_k exists for all k, $-\infty < k < \infty$, in order to eliminate the transient response of the circuit. The output due to this input is

$$y_k = \sum_{n=-\infty}^{\infty} h_n u_{k-n} = \sum_{n=-\infty}^{\infty} \left(\frac{1}{\sqrt{2}}\right)^n \cos\left(\frac{n\pi}{2}\right) e^{j(k-n)\theta}, \qquad \theta = \omega T$$

$$= e^{jk\theta} \sum_{n=-\infty}^{\infty} e^{-jn\theta}\left[\frac{e^{jn\pi/2} + e^{-jn\pi/2}}{2}\right]\left(\frac{1}{\sqrt{2}}\right)^n$$

$$= e^{jk\theta} 2 \cdot \sum_{n=0}^{\infty} \frac{1}{2}\left\{\left[\frac{1}{\sqrt{2}} e^{-j\theta} e^{j(\pi/2)}\right]^n + \left[\frac{1}{\sqrt{2}} e^{-j\theta} e^{-j(\pi/2)}\right]^n\right\}$$

$$= e^{jk\theta}\left\{\frac{1}{1 - \dfrac{e^{-j\theta}e^{j(\pi/2)}}{\sqrt{2}}} + \frac{1}{1 - \dfrac{e^{-j\theta}e^{-j(\pi/2)}}{\sqrt{2}}}\right\}$$

$$= e^{jk\theta}\left\{\frac{1}{1 + \frac{1}{2}e^{-j2\theta}}\right\}$$

The frequency response of this circuit is thus the term in brackets,

$$H(e^{j\theta}) = \frac{1}{1 + \frac{1}{2}e^{-2j\theta}}$$

which we calculated previously from the difference equation. ■

We can generalize this discussion of frequency response by using the convolution sum. We have from convolution the output y_k as

$$y_k = \sum_{n=-\infty}^{\infty} h_n u_{k-n} \tag{2.31}$$

We also know that if $u_k = e^{jk\theta}$, then the steady-state output is of the form

$$y_k = H(e^{j\theta}) e^{jk\theta} \tag{2.32}$$

Assuming in (2.31) that the input is $e^{jk\theta}$ we can use (2.32) on the left-hand side to obtain

$$H(e^{j\theta})e^{jk\theta} = \sum_{n=-\infty}^{\infty} h_n e^{j(k-n)\theta}$$

$$= e^{jk\theta} \sum_{n=-\infty}^{\infty} h_n e^{-jn\theta} \tag{2.33}$$

Cancelling $e^{jk\theta}$ on both sides of (2.33) gives us another expression for $H(e^{j\theta})$

$$H(e^{j\theta}) = \sum_{n=-\infty}^{\infty} h_n e^{-jn\theta} \tag{2.34}$$

Equation 2.34 expresses the frequency response of a discrete-time system in terms of the impulse-response sequence. In our previous discussions we derived an expression for $H(e^{j\theta})$ in terms of the coefficients in the difference equation, Equation 2.15 repeated below.

$$H(e^{j\theta}) = \frac{a_0 + a_1 e^{-j\theta} + \cdots + a_m e^{-jm\theta}}{1 + b_1 e^{-j\theta} + \cdots + b_n e^{-jn\theta}} \tag{2.35}$$

Equations 2.34 and 2.35 are especially useful since they allow us to calculate a system's frequency response directly from either the difference equation or the impulse-response sequence. Equations 2.34 and 2.35 taken together imply that

$$\sum_{l=-\infty}^{\infty} h_l e^{-jl\theta} = \frac{a_0 + a_1 e^{-j\theta} + \cdots + a_m e^{-jm\theta}}{1 + b_1 e^{-j\theta} + \cdots + b_n e^{-jn\theta}} \tag{2.36}$$

Equation 2.36 could be used to generate an expression for the impulse-response sequence directly from the coefficients of the difference equation. The process would involve expanding the right-hand side of (2.36) in a power series in $e^{-jk\theta}$. The impulse-response sequence would then be the coefficients of $e^{-jk\theta}$. This process is not computationally efficient in most cases; however, we can illustrate the process with the following example.

Example 2.17 We shall use the same system as in Examples 2.8 and 2.16. The difference equation is

$$y_k + \frac{1}{2} y_{k-2} = u_k$$

From Example 2.16 the impulse-response sequence is

$$h_k = \begin{cases} \left(\dfrac{1}{\sqrt{2}}\right)^k \cos\left(\dfrac{k\pi}{2}\right), & k \geq 0 \\ 0, & k < 0 \end{cases} \tag{2.37}$$

From (2.35)

$$H(e^{j\theta}) = \frac{1}{1 + \frac{1}{2}e^{-2j\theta}} \tag{2.38}$$

Suppose we now generate a power series in $e^{-jk\theta}$ by dividing the denominator of (2.38) into its numerator. We have

$$\begin{array}{r|l} & 1 - \frac{1}{2}e^{-2j\theta} + \frac{1}{4}e^{-4j\theta} - + \cdots \\ \hline 1 + \frac{1}{2}e^{-2j\theta} & 1 \\ & 1 + \frac{1}{2}e^{-2j\theta} \\ \hline & -\frac{1}{2}e^{-2j\theta} \\ & -\frac{1}{2}e^{-2j\theta} - \frac{1}{4}e^{-4j\theta} \\ \hline & \qquad\quad \frac{1}{4}e^{-4j\theta} + \cdots \end{array}$$

From this process we conclude that

$$\sum_{n=-\infty}^{\infty} h_n e^{-jn\theta} = 1 - \frac{1}{2}e^{-2j\theta} + \frac{1}{4}e^{-4j\theta} - + \cdots \tag{2.39}$$

We can now equate coefficients of like powers of $e^{-j\theta}$ to obtain the first few terms of h. Equation 2.39 implies $h_0 = 1, h_1 = 0, h_2 = -\frac{1}{2}, h_3 = 0, h_4 = \frac{1}{4}$, etc. Compare this to (2.37). This process of long division to generate the power series for $e^{-j\theta}$ does not generally yield a closed form for h_k; thus this method has limited utility.

■

Thus far we have analyzed two models for linear, constant-parameter, discrete-time systems. Both of these models, though different in how one performs the actual calculations, are similar in other ways. They are both descriptions of the system in terms of the sequence or time domain. They are both input–output characterizations. They treat the system as a black box in which internal structure is ignored. In other words, they allow one to calculate outputs y given inputs u and the system model. In many cases, this kind of black-box approach to analysis is adequate. Note, however, that neither of these models is easily generalized to cover systems with multiple inputs and outputs.

In the next section we introduce a third description, again based in the sequence domain. It is a matrix description of a system, called a state-variable model. It is easily generalized to the case of multiple inputs and outputs. It not only furnishes input-output information, but also displays explicitly the internal states of a system. However, this additional information is not obtained free of cost. The state-variable model, in general, requires more computations for input-output information than either the difference equation or the impulse-response method.

2.8 STATE VARIABLES FOR DISCRETE-TIME SYSTEMS

In Chapter 1 we have defined the state of a system. Let us review the definition. "The state of a system at some index k is a set of variables $x_1(k), x_2(k), \ldots, x_n(k)$ such that knowledge of these variables and future inputs allows one to calculate all future outputs of the system." In most cases we seek a minimal set of variables to act as the state of system. The state is essentially the memory of the system.

Consider an nth order discrete-time circuit with difference equation

$$y_k + b_1 y_{k-1} + \cdots + b_n y_{k-n} = a_0 u_k + a_1 u_{k-1} + \cdots + a_m u_{k-m} \tag{2.40}$$

In (2.40) n and m are arbitrary. For purposes of discussion assume $m < n$. Two block diagrams of (2.40) are shown in Figure 2.24. (Proving the equivalence of these two block diagrams is left as a problem for the reader.) For the system described by (2.40) and represented schematically as in Figure 2.24*b*, we choose as the state at any index k, the contents of the unit delays. These unit delay elements store the value of the variable entering the unit delay for one index k or one clock cycle. Physically they are often implemented as shift registers that store a binary representation of the input to the delay element. Collectively the unit delays form the memory of the system. Their contents summarize all past behavior of the system. Let us define state variables for (2.40) as shown in Figure 2.24*b* and derive a recursion formula or description of how the states change as new inputs enter

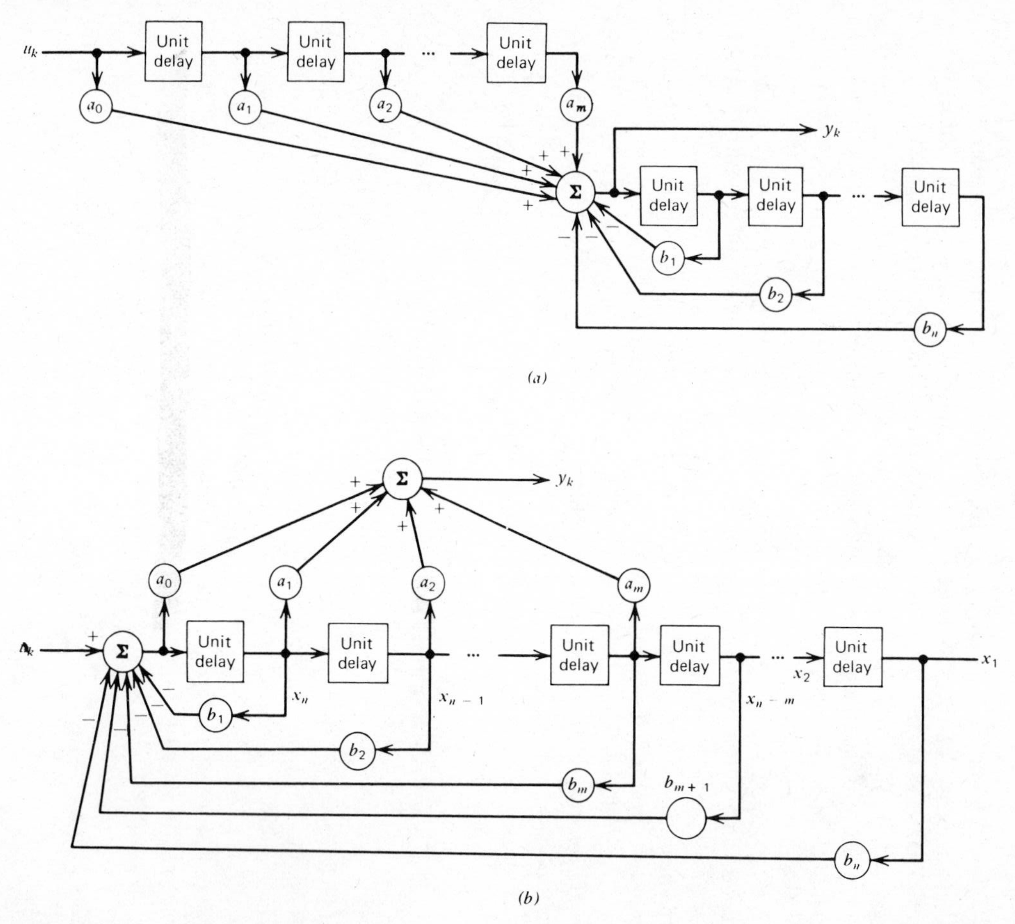

Figure 2.24

the system. In addition, we shall obtain an expression for the output y_k in terms of the state and inputs. From Figure 2.24*b* we have

$$
\begin{aligned}
x_1(k+1) &= x_2(k) \\
x_2(k+1) &= x_3(k) \\
&\vdots \\
x_{n-1}(k+1) &= x_n(k) \\
x_n(k+1) &= u(k) - b_1 x_n(k) - b_2 x_{n-1}(k) - \cdots - b_n x_1(k)
\end{aligned}
\tag{2.41}
$$

Expressing these n equations in matrix form gives us

$$
\begin{bmatrix} x_1(k+1) \\ x_2(k+1) \\ \vdots \\ x_{n-1}(k+1) \\ x_n(k+1) \end{bmatrix} = \begin{bmatrix} 0 & 1 & 0 & \cdots & 0 & 0 \\ 0 & 0 & 1 & \cdots & 0 & 0 \\ & & \cdots & & & \\ 0 & 0 & 0 & & 0 & 1 \\ -b_n & -b_{n-1} & \cdots & & -b_2 & -b_1 \end{bmatrix} \begin{bmatrix} x_1(k) \\ x_2(k) \\ \vdots \\ x_{n-1}(k) \\ x_n(k) \end{bmatrix} + \begin{bmatrix} 0 \\ 0 \\ \vdots \\ 0 \\ 1 \end{bmatrix} u(k)
\tag{2.42}
$$

The output $y(k)$ is given in terms of the input $u(k)$ and the state $x_1(k), x_2(k), \ldots, x_n(k)$ as

$$
\begin{aligned}
y(k) = {} & a_1 x_n(k) + a_2 x_{n-1}(k) + \cdots + a_m x_{n-m-1}(k) \\
& + a_0[u(k) - b_1 x_n(k) - b_2 x_{n-1}(k) - \cdots - b_n x_1(k)]
\end{aligned}
\tag{2.43}
$$

Equation 2.43 in matrix form is

$$
y(k) = [-a_0 b_n, -a_0 b_{n-1}, \ldots, a_m - a_0 b_m, \ldots, a_2 - a_0 b_2, a_1 - a_0 b_1] \begin{bmatrix} x_1(k) \\ x_2(k) \\ \vdots \\ x_n(k) \end{bmatrix} + a_0 u(k)
\tag{2.44}
$$

Defining the state matrix $\mathbf{x}(k)$ as the column vector

$$
\begin{bmatrix} x_1(k) \\ x_2(k) \\ \vdots \\ x_n(k) \end{bmatrix},
$$

we can write (2.42) and (2.43) as

$$\begin{aligned} \mathbf{x}(k+1) &= \mathbf{A}\mathbf{x}(k) + \mathbf{B}u(k) \\ y(k) &= \mathbf{C}\mathbf{x}(k) + \mathbf{D}u(k) \end{aligned} \tag{2.45}$$

where the matrices (**A**, **B**, **C**, **D**), for this example, are

$$\mathbf{A} = \begin{bmatrix} 0 & 1 & 0 & 0 & \cdots & 0 & 0 \\ 0 & 0 & 1 & 0 & \cdots & 0 & 0 \\ & & & \cdots & & & \\ 0 & 0 & 0 & \cdots & & 0 & 1 \\ -b_n & -b_{n-1} & & \cdots & & -b_2 & -b_1 \end{bmatrix}, \qquad \mathbf{B} = \begin{bmatrix} 0 \\ 0 \\ \vdots \\ 0 \\ 1 \end{bmatrix},$$

$$\mathbf{C} = [-a_0 b_n, -a_0 b_{n-1}, \ldots, a_m - a_0 b_m, \ldots, a_1 - a_0 b_1], \qquad \mathbf{D} = [a_0]$$

The matrices (**A**, **B**, **C**, **D**) constitute another description of a discrete-time system. This model is more general than our previous two models. Generality comes at the price of more computations. However, matrix formulation is ideally suited for certain applications that just cannot be modeled using difference equations or the impulse-response sequence. Any problem in which the internal states of the system must be used to optimize or analyze performance must use a model that explicitly defines the internal states. The only such analytical model is a state-variable formulation.

Example 2.18 Generate a state-variable description of a discrete-time system represented by the difference equation

$$y(k) + 2y(k-1) + y(k-2) = u(k)$$

A schematic diagram of a system described by the above is shown in Figure 2.25. There are two delay elements. Therefore, define two state variables as the contents of these two registers.

$$x_1(k) = y(k-2)$$

$$x_2(k) = y(k-1)$$

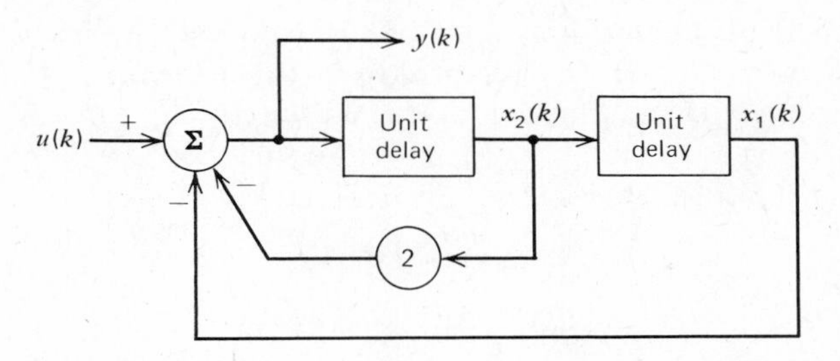

Figure 2.25

Referring to Figure 2.25, we can write a simple recursion for each state variable as

$$\begin{aligned} x_1(k+1) &= x_2(k) \\ x_2(k+1) &= y(k) = u(k) - 2x_2(k) - x_1(k) \end{aligned} \tag{2.46}$$

Equations 2.46 in matrix form are

$$\begin{bmatrix} x_1(k+1) \\ x_2(k+1) \end{bmatrix} = \begin{bmatrix} 0 & 1 \\ -1 & -2 \end{bmatrix} \begin{bmatrix} x_1(k) \\ x_2(k) \end{bmatrix} + \begin{bmatrix} 0 \\ 1 \end{bmatrix} u(k)$$

Also, the output $y(k)$ in terms of the states and the input is:

$$y(k) = [-1 \quad -2] \begin{bmatrix} x_1(k) \\ x_2(k) \end{bmatrix} + [1]u(k)$$

The state-variable matrices are, therefore,

$$\mathbf{A} = \begin{bmatrix} 0 & 1 \\ -1 & -2 \end{bmatrix}, \quad \mathbf{B} = \begin{bmatrix} 0 \\ 1 \end{bmatrix}, \quad \mathbf{C} = [-1 \quad -2], \quad \mathbf{D} = [1]$$

The matrices (**A**, **B**, **C**, **D**) for a given system are not unique. We can illustrate this by redefining the states in Example 2.18. Thus, let

$$x_1(k) = y(k-1)$$

$$x_2(k) = y(k-2)$$

With this definition of states, the reader can easily show that the state matrices are

$$\mathbf{A} = \begin{bmatrix} -2 & -1 \\ 1 & 0 \end{bmatrix}, \quad \mathbf{B} = \begin{bmatrix} 1 \\ 0 \end{bmatrix}, \quad \mathbf{C} = [-2 \quad -1], \quad \mathbf{D} = [1] \qquad \blacksquare$$

More generally, for an nth order system, choose any $n \times n$ nonsingular† matrix $\mathbf{T}$ and define a new state vector $\mathbf{q}(k) = \mathbf{T}\mathbf{x}(k)$. The new state vector $\mathbf{q}(k)$ is merely some linear transformation of the old state vector $\mathbf{x}(k)$. We have

$$\begin{aligned}\mathbf{q}(k+1) = \mathbf{T}\mathbf{x}(k+1) &= \mathbf{T}[\mathbf{A}\mathbf{x}(k) + \mathbf{B}u(k)] \\ &= \mathbf{T}\mathbf{A}\mathbf{x}(k) + \mathbf{T}\mathbf{B}u(k) \\ &= \mathbf{T}\mathbf{A}\mathbf{T}^{-1}\mathbf{q}(k) + \mathbf{T}\mathbf{B}u(k)\end{aligned}$$

and

$$\begin{aligned}y(k) &= \mathbf{C}\mathbf{x}(k) + \mathbf{D}u(k) \\ &= \mathbf{C}\mathbf{T}^{-1}\mathbf{q}(k) + \mathbf{D}u(k)\end{aligned}$$

Thus, if we let

$$\begin{aligned}\hat{\mathbf{A}} &= \mathbf{T}\mathbf{A}\mathbf{T}^{-1} \qquad \hat{\mathbf{C}} = \mathbf{C}\mathbf{T}^{-1} \\ \hat{\mathbf{B}} &= \mathbf{T}\mathbf{B} \qquad\quad\;\; \hat{\mathbf{D}} = \mathbf{D}\end{aligned} \tag{2.47}$$

then

$$\begin{aligned}\mathbf{q}(k+1) &= \hat{\mathbf{A}}\mathbf{q}(k) + \hat{\mathbf{B}}u(k) \\ \mathbf{y}(k) &= \hat{\mathbf{C}}\mathbf{q}(k) + \hat{\mathbf{D}}u(k)\end{aligned} \tag{2.48}$$

Equations 2.48 yield the same output $y(k)$ for a given input $u(k)$. The output is expressed in terms of a different state vector $\mathbf{q}$ and a different *internal* model of the system represented by the new matrices $(\hat{\mathbf{A}}, \hat{\mathbf{B}}, \hat{\mathbf{C}}, \hat{\mathbf{D}})$. The use of the nonsingular matrix $\mathbf{T}$ applied to the state vector $\mathbf{x}$ allows us to change analytically the internal connections of the system while maintaining the same input-output relation. We shall explore this property of state-variable models in more detail in Section 2.10.

This matrix formulation is easily extended to multiple input-output systems. Suppose, for example, we have a system with r inputs and s outputs. The state equations are again

$$\begin{aligned}\mathbf{x}(k+1) &= \mathbf{A}\mathbf{x}(k) + \mathbf{B}\mathbf{u}(k) \\ \mathbf{y}(k) &= \mathbf{C}\mathbf{x}(k) + \mathbf{D}\mathbf{u}(k)\end{aligned} \tag{2.49}$$

general forms

†A nonsingular matrix $\mathbf{T}$ is a matrix whose inverse $\mathbf{T}^{-1}$ exists. See Appendix C.

$$[\mathbf{x}]_{n\times 1} = \begin{bmatrix} \mathbf{A} \\ n \times n \end{bmatrix}[\mathbf{x}]_{n\times 1} + \begin{bmatrix} \mathbf{B} \\ n \times r \end{bmatrix}[\mathbf{u}]_{r\times 1}$$

$$[\mathbf{y}]_{s\times 1} = \begin{bmatrix} \mathbf{C} \\ s \times n \end{bmatrix}[\mathbf{x}]_{n\times 1} + \begin{bmatrix} \mathbf{D} \\ s \times r \end{bmatrix}[\mathbf{u}]_{r\times 1}$$

Figure 2.26

In this case $\mathbf{x}(k)$ is n-dimensional, $\mathbf{u}(k)$ is r-dimensional, with the r components the inputs $u_1(k)$, $u_2(k)$, ..., $u_r(k)$, and $\mathbf{y}(k)$ is s-dimensional with the s components the outputs $y_1(k)$, $y_2(k)$, ..., $y_s(k)$. The matrices $\mathbf{A}$, $\mathbf{B}$, $\mathbf{C}$, $\mathbf{D}$ have dimensions $n \times n$, $n \times r$, $s \times n$, $s \times r$, respectively, as shown in Figure 2.26.

Example 2.19 To illustrate the case of multiple inputs and outputs, consider the system shown in Figure 2.27 with state variables chosen as the outputs of the delay elements as indicated. Writing two state equations at the points just before the two unit delays gives

$$x_1(k+1) = a_1x_1(k) + a_{12}x_2(k) + u_1(k) + b_2u_2(k)$$

$$x_2(k+1) = a_2x_2(k) + u_2(k)$$

In matrix form we have

$$\mathbf{x}(k+1) = \begin{bmatrix} a_1 & a_{12} \\ 0 & a_2 \end{bmatrix}\mathbf{x}(k) + \begin{bmatrix} 1 & b_2 \\ 0 & 1 \end{bmatrix}\mathbf{u}(k)$$

The output equations are

$$y_1(k) = x_1(k) + u_2(k)$$

$$y_2(k) = c_1x_1(k) + x_2(k)$$

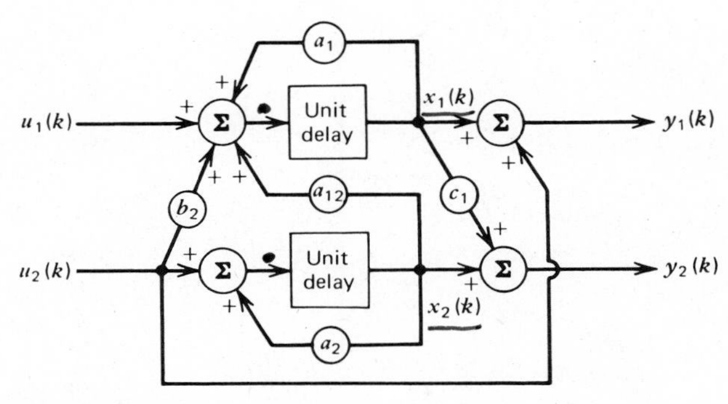

Figure 2.27

In matrix form we have

$$\mathbf{y}(k) = \begin{bmatrix} 1 & 0 \\ c_1 & 1 \end{bmatrix} \mathbf{x}(k) + \begin{bmatrix} 0 & 1 \\ 0 & 0 \end{bmatrix} \mathbf{u}(k)$$

The **A**, **B**, **C**, **D** matrices are thus

$$\mathbf{A} = \begin{bmatrix} a_1 & a_{12} \\ 0 & a_2 \end{bmatrix}, \quad \mathbf{B} = \begin{bmatrix} 1 & b_2 \\ 0 & 1 \end{bmatrix}, \quad \mathbf{C} = \begin{bmatrix} 1 & 0 \\ c_1 & 1 \end{bmatrix}, \quad \mathbf{D} = \begin{bmatrix} 0 & 1 \\ 0 & 0 \end{bmatrix} \quad \blacksquare$$

Perhaps the simplest method of writing state equations directly from a difference equation is first to sketch a block diagram model of the difference equation. Then use the block diagram, as we have demonstrated here, to find the state equations.

2.9 THE SOLUTION OF STATE-VARIABLE EQUATIONS

We have now a third model of linear discrete-time systems. This model describes the system in terms of its state, that is, the state vector **x**, and is a complete internal description of the system. We consider now the problem of finding the response $\mathbf{y}(k)$ in terms of the state and the input $\mathbf{u}(k)$. To find $\mathbf{y}(k)$ we must know the state $\mathbf{x}(k)$ where $\mathbf{x}(k)$ evolves according to

$$\mathbf{x}(k+1) = \mathbf{A}\mathbf{x}(k) + \mathbf{B}\mathbf{u}(k) \tag{2.50}$$

One method of finding $\mathbf{x}(k)$, given $\mathbf{x}(0)$ the initial state, is to solve (2.50) iteratively. Thus,

$$\mathbf{x}(1) = \mathbf{A}\mathbf{x}(0) + \mathbf{B}\mathbf{u}(0)$$

$$\mathbf{x}(2) = \mathbf{A}\mathbf{x}(1) + \mathbf{B}\mathbf{u}(1) = \mathbf{A}^2\mathbf{x}(0) + \mathbf{A}\mathbf{B}\mathbf{u}(0) + \mathbf{B}\mathbf{u}(1)$$

and, in general,

$$\mathbf{x}(k) = \mathbf{A}^k\mathbf{x}(0) + \sum_{m=0}^{k-1} \mathbf{A}^{k-1-m}\mathbf{B}\mathbf{u}(m), \qquad k > 0 \tag{2.51}$$

If the initial state is $\mathbf{x}(k_0)$ and we have $u(k)$ for $k > k_0$, then (2.51) generalizes to

$$\mathbf{x}(k) = \mathbf{A}^{k-k_0}\mathbf{x}(k_0) + \sum_{m=0}^{k-k_0-1} \mathbf{A}^{k-k_0-1-m}\mathbf{B}\mathbf{u}(k_0 + m) \tag{2.52}$$

The matrix $\mathbf{A}^k$ is the k-fold product† $\mathbf{A} \times \mathbf{A} \times \cdots \times \mathbf{A}$ and is sometimes called the *fundamental* or *transition* matrix of the system. Referring to (2.51), we recognize two kinds of terms. The first term $\mathbf{A}^k\mathbf{x}(0)$ represents an evolution of the state due only to nonzero initial conditions. The second term, which is essentially a convolution sum, corresponds to a state response due to an input and zero initial conditions. We find the output from (2.51) using

$$\begin{aligned}\mathbf{y}(k) &= \mathbf{C}\mathbf{x}(k) + \mathbf{D}\mathbf{u}(k)\\ &= \mathbf{C}\mathbf{A}^k\mathbf{x}(0) + \sum_{m=0}^{k-1} \mathbf{C}\mathbf{A}^{k-1-m}\mathbf{B}\mathbf{u}(m) + \mathbf{D}\mathbf{u}(k)\end{aligned} \tag{2.52}$$

In (2.52) there are three terms added together to form the response $\mathbf{y}(k)$. The first term $\mathbf{C}\mathbf{A}^k\mathbf{x}(0)$ is the response of the system due only to the initial state $\mathbf{x}(0)$. The third term $\mathbf{D}\mathbf{u}(k)$ represents all "straight-through paths" from input to output. The second and third terms together form the response of the system to the input $\mathbf{u}$ with zero initial conditions. Recall in the impulse response model that we obtained the output by convolving the impulse response sequence $h(k)$ with the input assuming zero initial conditions, that is, $\mathbf{x}(0) = \mathbf{0}$. If we use an input $u(k) = \delta(k)$, the output is by definition $h(k)$. Thus

$$h(k) = \mathbf{C}\mathbf{A}^k\mathbf{x}(0) + \sum_{m=0}^{k-1} \mathbf{C}\mathbf{A}^{k-1-m}\mathbf{B}\delta(m) + \mathbf{D}\delta(k) \tag{2.53}$$

The first term is zero since $\mathbf{x}(0) = \mathbf{0}$. The third term is equal to $\mathbf{D}$ for $k = 0$ and zero otherwise. The second term is $\mathbf{C}\mathbf{A}^{k-1}\mathbf{B}$ for $k > 0$. Note that the sum has no terms for $k = 0$. Thus

$$h(k) = \begin{cases} \mathbf{D}, & k = 0 \\ \sum_{m=0}^{k-1} \mathbf{C}\mathbf{A}^{k-1-m}\mathbf{B}\delta(m) = \mathbf{C}\mathbf{A}^{k-1}\mathbf{B}, & k > 0 \\ 0, & k < 0 \end{cases} \tag{2.54}$$

†$\mathbf{A}^0$ is defined as $\mathbf{I}$, the identity matrix.

In (2.53) and (2.54) we see that the essential calculation is finding $\mathbf{A}^k$ for *all values of k*. Now we could again by brute force computation find $\mathbf{A}^k$ by a k-fold matrix multiplication. This type of solution is unacceptable because it is not in closed form and offers no insight into the structure of the solution. The next section presents the details on how to compute $\mathbf{A}^k$ and other functions of a matrix in closed form for all values of k.

2.10 FUNCTIONS OF A MATRIX

In this section we shall examine two methods for computing $\mathbf{A}^k$ and other functions of a matrix $\mathbf{A}$ in closed form. We begin with a brief discussion of properties of matrices that will be useful in our development.

The *characteristic equation* of the $n \times n$ system matrix $\mathbf{A}$ is closely related to the auxiliary equation of the corresponding difference equation model for the system. The characteristic equation of $\mathbf{A}$ is defined to be

$$g(\lambda) = |\mathbf{A} - \lambda\mathbf{I}| = 0 \tag{2.55}$$

where $|\mathbf{B}|$ means the determinant of $\mathbf{B}$ and $\mathbf{I}$ is the identity matrix. For example, the characteristic equation of the matrix

$$\mathbf{A} = \begin{bmatrix} 4 & 3 \\ 1 & 2 \end{bmatrix}$$

is

$$g(\lambda) = |\mathbf{A} - \lambda\mathbf{I}| = 0$$

that is,

$$g(\lambda) = \left| \begin{bmatrix} 4 & 3 \\ 1 & 2 \end{bmatrix} - \lambda \begin{bmatrix} 1 & 0 \\ 0 & 1 \end{bmatrix} \right| = 0$$

which becomes

$$g(\lambda) = \begin{vmatrix} 4 - \lambda & 3 \\ 1 & 2 - \lambda \end{vmatrix} = 0$$

or

$$g(\lambda) = \lambda^2 - 6\lambda + 5 = (\lambda - 5)(\lambda - 1) = 0$$

Thus, $\lambda^2 - 6\lambda + 5 = 0$ is the characteristic equation of the matrix. The roots of the characteristic equation are called the *eigenvalues* of the matrix **A**. In this case, the eigenvalues are $\lambda_1 = 5$ and $\lambda_2 = 1$. The eigenvalues of the system matrix **A** and the roots of the corresponding auxiliary equation for the difference equation of the system are identical.

The first method we shall discuss for finding functions of a matrix is based on the Caley–Hamilton theorem, which states that every $n \times n$ matrix satisfies its own characteristic equation. For example, if we substitute **A** for λ in $g(\lambda) = \lambda^2 - 6\lambda + 5$, we obtain the matrix equation

$$\mathbf{g}(\mathbf{A}) = \mathbf{A}^2 - 6\mathbf{A} + 5\mathbf{A}^0 = \mathbf{0}$$

That is,

$$\begin{bmatrix} 4 & 3 \\ 1 & 2 \end{bmatrix}\begin{bmatrix} 4 & 3 \\ 1 & 2 \end{bmatrix} - 6\begin{bmatrix} 4 & 3 \\ 1 & 2 \end{bmatrix} + 5\begin{bmatrix} 1 & 0 \\ 0 & 1 \end{bmatrix} = \begin{bmatrix} 0 & 0 \\ 0 & 0 \end{bmatrix}$$

$$\begin{bmatrix} 19 & 18 \\ 6 & 7 \end{bmatrix} - \begin{bmatrix} 24 & 18 \\ 6 & 12 \end{bmatrix} + \begin{bmatrix} 5 & 0 \\ 0 & 5 \end{bmatrix} = \begin{bmatrix} 0 & 0 \\ 0 & 0 \end{bmatrix}$$

The equation is certainly satisfied in this case. The theorem states that it is true for any square matrix. In general, the characteristic equation of a square $n \times n$ matrix **A** is

$$\begin{aligned} g(\lambda) &= |\mathbf{A} - \lambda\mathbf{I}| \\ &= \lambda^n + a_{n-1}\lambda^{n-1} + \cdots + a_1\lambda + a_0 = 0 \end{aligned} \tag{2.56}$$

Now substituting **A** for λ as above, we have

$$\mathbf{g}(\mathbf{A}) = \mathbf{A}^n + a_{n-1}\mathbf{A}^{n-1} + \cdots + a_1\mathbf{A} + a_0\mathbf{I} = \mathbf{0} \tag{2.57}$$

where, again, **I** is the identity matrix, and **0** is a matrix whose elements are all zero. Equation 2.57 may be rewritten as

$$\mathbf{A}^n = -a_{n-1}\mathbf{A}^{n-1} - \cdots - a_1\mathbf{A} - a_0\mathbf{I} \tag{2.58}$$

Thus $\mathbf{A}^n$ may be expressed in terms of the matrices $\mathbf{A}^{n-1}$, $\mathbf{A}^{n-2}$, ..., **A**, and **I**. To extend this result, we multiply (2.58) by **A** to obtain

$$\mathbf{A}^{n+1} = -a_{n-1}\mathbf{A}^n - \cdots - a_1\mathbf{A}^2 - a_0\mathbf{A} \tag{2.59}$$

Substituting (2.58) for $\mathbf{A}^n$, we obtain

$$\begin{aligned}\mathbf{A}^{n+1} &= -a_{n-1}(-a_{n-1}A^{n-1} - \cdots - a_1\mathbf{A} - a_0\mathbf{I}) - \cdots - a_1\mathbf{A}^2 - a_0\mathbf{A} \\ &= (a_{n-1}^2 - a_{n-2})\mathbf{A}^{n-1} + (a_{n-1}a_{n-2} - a_{n-3})\mathbf{A}^{n-2} + \cdots + a_{n-1}a_0\mathbf{I} \qquad (2.60)\end{aligned}$$

This gives us an expression for $\mathbf{A}^{n+1}$, again in terms of $\mathbf{A}^{n-1}$, $\mathbf{A}^{n-2}$, ..., $\mathbf{A}$, and $\mathbf{I}$. Continuing this process, we see that any power of $\mathbf{A}$ can be represented as a weighted sum of matrices involving $\mathbf{A}$ to at most the $n-1$ power. Hence, functions of matrices that can be expressed as

$$\begin{aligned}\mathbf{f}(\mathbf{A}) &= \alpha_0\mathbf{I} + \alpha_1\mathbf{A} + \cdots + \alpha_k\mathbf{A}^k + \cdots \\ &= \sum_{k=0}^{\infty} \alpha_k\mathbf{A}^k \qquad (2.61)\end{aligned}$$

can be represented as

$$\begin{aligned}\mathbf{f}(\mathbf{A}) &= \beta_0\mathbf{I} + \beta_1\mathbf{A} + \cdots + \beta_{n-1}\mathbf{A}^{n-1} \\ &= \sum_{k=0}^{n-1} \beta_k\mathbf{A}^k \qquad (2.62)\end{aligned}$$

Here $\beta_0, \beta_1, \beta_2, \ldots, \beta_{n-1}$ are functions of $\alpha_0, \alpha_1, \ldots$ and n. The calculation of $\beta_0, \beta_1, \ldots, \beta_{n-1}$ can be carried out by using the iterative method used in the calculation of $\mathbf{A}^n$ and $\mathbf{A}^{n+1}$ in (2.58) and (2.59). However, although straightforward, this process can be lengthy.

To develop an easier method, let us return to the characteristic equation of the matrix $\mathbf{A}$.

$$g(\lambda) = |\mathbf{A} - \lambda\mathbf{I}| = \lambda^n + a_{n-1}\lambda^{n-1} + \cdots + a_1\lambda + a_0 = 0 \qquad (2.63)$$

Following the same steps as before, we can express the eigenvalues λ^n, λ^{n+1}, λ^{n+2}, and so on in terms of $\lambda, \lambda^2, \ldots, \lambda^{n-1}$.

$$\lambda^n = -a_{n-1}\lambda^{n-1} - a_{n-2}\lambda^{n-2} - \cdots - a_1\lambda - a_0$$

$$\lambda^{n+1} = (a_{n-1}^2 - a_{n-2})\lambda^{n-1} + (a_{n-1}a_{n-2} - a_{n-3})\lambda^{n-2} + \cdots + a_{n-1}a_0$$

As before, we can write polynomials of λ in terms of $\lambda, \lambda^2, \ldots, \lambda^{n-1}$.

$$f(\lambda) = \alpha_0 + \alpha_1\lambda + \alpha_2\lambda^2 + \cdots = \sum_{k=0}^{\infty} \alpha_k\lambda^k \qquad (2.64)$$

$$= \beta_0 + \beta_1\lambda + \cdots + \beta_{n-1}\lambda^{n-1} = \sum_{k=0}^{n-1} \beta_k\lambda^k \qquad (2.65)$$

Here we wish to find the n unknowns $\beta_0, \beta_1, \ldots, \beta_{n-1}$. We know that (2.65) holds for any λ that is a solution of the characteristic equation (2.63); that is, for any eigenvalue of the matrix A. Assume first that the eigenvalues are distinct; that is, that none is repeated. Substituting $\lambda_1, \lambda_2, \ldots, \lambda_n$ in (2.65) gives n equations in n unknowns

$$\begin{aligned} f(\lambda_1) &= \beta_0 + \beta_1\lambda_1 + \cdots + \beta_{n-1}\lambda_1^{n-1} \\ f(\lambda_2) &= \beta_0 + \beta_1\lambda_2 + \cdots + \beta_{n-1}\lambda_2^{n-1} \\ &\vdots \\ f(\lambda_n) &= \beta_0 + \beta_1\lambda_n + \cdots + \beta_{n-1}\lambda_n^{n-1} \end{aligned} \tag{2.66}$$

from which we can obtain $\beta_0, \beta_1, \ldots, \beta_{n-1}$. By comparing (2.65) and (2.62), we see that these coefficients are exactly the ones that appeared earlier in (2.62); that is,

$$\mathbf{f}(\mathbf{A}) = \sum_{m=0}^{n-1} \beta_m \mathbf{A}^m \tag{2.67}$$

Hence, our problem is solved. The coefficients needed in the matrix expression for $\mathbf{f}(\mathbf{A})$ are found as the solution to a linear system of scalar equations given by (2.66).

Example 2.20 Find $\mathbf{f}(\mathbf{A})$ where

$$\mathbf{f}(\mathbf{A}) = \mathbf{A}^k, \qquad \text{and} \quad \mathbf{A} = \begin{bmatrix} \frac{1}{2} & 0 \\ \frac{1}{4} & \frac{1}{4} \end{bmatrix}$$

The characteristic equation is

$$g(\lambda) = |\mathbf{A} - \lambda\mathbf{I}| = \begin{vmatrix} \frac{1}{2} - \lambda & 0 \\ \frac{1}{4} & \frac{1}{4} - \lambda \end{vmatrix} = 0$$

That is,

$$g(\lambda) = \left(\frac{1}{2} - \lambda\right)\left(\frac{1}{4} - \lambda\right) = 0$$

Thus the eigenvalues are

$$\lambda_1 = \frac{1}{2}, \qquad \lambda_2 = \frac{1}{4}$$

Using (2.65), we obtain

$$f(\lambda) = \lambda^k = \beta_0 + \beta_1 \lambda$$

and using (2.66), we have

$$\left(\frac{1}{2}\right)^k = \beta_0 + \beta_1\left(\frac{1}{2}\right)$$

$$\left(\frac{1}{4}\right)^k = \beta_0 + \beta_1\left(\frac{1}{4}\right)$$

Solving for the unknowns β_0 and β_1 gives

$$\beta_0 = \left(\frac{1}{4}\right)^k (2 - 2^k), \qquad \beta_1 = \left(\frac{1}{4}\right)^k (4 \cdot 2^k - 4)$$

The solution for $\mathbf{A}^k$ is found by using (2.62)

$$\mathbf{f}(\mathbf{A}) = \mathbf{A}^k = \begin{bmatrix} \frac{1}{2} & 0 \\ \frac{1}{4} & \frac{1}{4} \end{bmatrix}^k = \beta_0 \mathbf{I} + \beta_1 \mathbf{A}$$

$$= \left(\frac{1}{4}\right)^k \begin{bmatrix} 2 - 2^k & 0 \\ 0 & 2 - 2^k \end{bmatrix} + \left(\frac{1}{4}\right)^k \begin{bmatrix} 2 \cdot (2^k - 1) & 0 \\ 2^k - 1 & 2^k - 1 \end{bmatrix}$$

$$= \left(\frac{1}{4}\right)^k \begin{bmatrix} 2^k & 0 \\ 2^k - 1 & 1 \end{bmatrix}$$

■

Suppose now that the characteristic equation $g(\lambda) = 0$ has repeated roots (for example, $\lambda_1 = \lambda_2$). In this case, we would be left with fewer than n linearly independent equations in (2.66). The following theorem† extends our results to the case of repeated eigenvalues.

† See, for example, L. A. Zadeh and C. A. Desoer, *Linear System Theory*. McGraw-Hill, New York, 1963, Appendix D, and especially pp. 607–609.

Theorem: Let $\mathbf{A}$ be an $n \times n$ matrix with n_0 distinct eigenvalues, $\lambda_1, \lambda_2, \ldots, \lambda_{n0}$ (if no eigenvalue is repeated, then $n_0 = n$; otherwise $n_0 < n$). Let the eigenvalue λ_1 occur with multiplicity m_i, and define the polynomials

$$\mathbf{P}(\mathbf{A}) = \sum_{m=0}^{n-1} \beta_m \mathbf{A}^m \tag{2.68}$$

$$P(\lambda) = \sum_{m=0}^{n-1} \beta_m \lambda^m$$

$$\mathbf{f}(\mathbf{A}) = \sum_{k=0}^{\infty} \alpha_k \mathbf{A}^k \tag{2.69}$$

Then the matrix $\mathbf{f}(\mathbf{A})$ is identical with the matrix $\mathbf{P}(\mathbf{A})$ if and only if

(a)
$$f(\lambda_i) = P(\lambda_i), \qquad i = 1, 2, \ldots, n_0 \tag{2.70}$$

and

(b)
$$\left.\frac{d^q}{d\lambda^q} f(\lambda)\right|_{\lambda=\lambda_i} = \left.\frac{d^q}{d\lambda^q} P(\lambda)\right|_{\lambda=\lambda_i}, \qquad \begin{matrix} i = 1, 2, \ldots, n_0 \\ q = 1, 2, \ldots, m_i - 1 \end{matrix} \tag{2.71}$$

Note that (2.71), when rewritten in terms of the unknown coefficients $\beta_0, \beta_1, \ldots, \beta_{n-1}$, becomes

$$\left.\frac{d^q}{d\lambda^q} f(\lambda)\right|_{\lambda=\lambda_i} = \frac{d^q}{d\lambda^q} \sum_{m=0}^{n-1} \beta_m \lambda^m = \sum_{m=q}^{n-1} m(m-1)\cdots(m-q+1)\beta_m \lambda_i^{m-q} \tag{2.72}$$

Equation 2.71 yields the remaining equations needed to solve for $\beta_0, \beta_1, \ldots, \beta_{n-1}$.

Example 2.21 Let

$$\mathbf{A} = \begin{bmatrix} \frac{1}{2} & 2 \\ -\frac{1}{128} & \frac{1}{4} \end{bmatrix}$$

and suppose that we wish to compute a general form for the matrix $\mathbf{A}^k$. We find the characteristic equation

$$g(\lambda) = |\mathbf{A} - \lambda \mathbf{I}| = \lambda^2 - \frac{3\lambda}{4} + \frac{9}{64} = 0$$

with a double root $\lambda = \frac{3}{8}$. From (2.70), we have

$$f(\lambda) = \lambda^k = \left(\frac{3}{8}\right)^k = \beta_0 + \beta_1\left(\frac{3}{8}\right)$$

and from (2.71),

$$f^{(1)}(\lambda) = k\lambda^{k-1} = k\left(\frac{3}{8}\right)^{k-1} = \beta_1$$

Hence

$$\beta_1 = k\left(\frac{3}{8}\right)^{k-1}$$

$$\beta_0 = \left(\frac{3}{8}\right)^k(1 - k)$$

Thus, by (2.62), we see that

$$\mathbf{A}^k = \left(\frac{3}{8}\right)^k(1 - k)\mathbf{I} + k\left(\frac{3}{8}\right)^{k-1}\mathbf{A}$$

$$= \left(\frac{3}{8}\right)^k\begin{bmatrix} 1 + \dfrac{k}{3} & \dfrac{16k}{3} \\ -\dfrac{k}{48} & 1 - \dfrac{k}{3} \end{bmatrix}$$

As a check, we verify that $\mathbf{A}^0 = \mathbf{I}$, $\mathbf{A}^1 = \mathbf{A}$, etc. ■

Example 2.22 Find the output $y(k)$ of the discrete-time system shown in Figure 2.28 for an input

$$u(k) = \begin{cases} 0, & k < 0 \\ \left(\dfrac{1}{2}\right)^k, & k \geq 0 \end{cases}$$

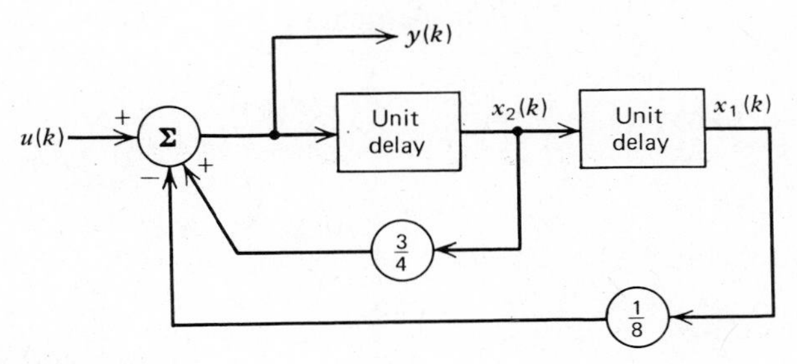

Figure 2.28

Assume zero initial conditions. This is the same system as was analyzed in Example 2.13 using the impulse response model. The initial state of the system is $\mathbf{x}(0) = \mathbf{0}$, so for $k > 0$, the state vector $\mathbf{x}(\mathrm{k})$ is given by

$$\mathbf{x}(k) = \sum_{m=0}^{k-1} \mathbf{A}^{k-1-m}\mathbf{B}u(m)$$

and the output is given by

$$y(k) = \mathbf{C}\mathbf{x}(k) + \mathbf{D}u(k)$$

where

$$\mathbf{x}(k+1) = \begin{bmatrix} 0 & 1 \\ -\frac{1}{8} & \frac{3}{4} \end{bmatrix} \mathbf{x}(k) + \begin{bmatrix} 0 \\ 1 \end{bmatrix} u(k)$$

$$y(k) = \begin{bmatrix} -\frac{1}{8} & \frac{3}{4} \end{bmatrix} \mathbf{x}(k) + [1]u(k)$$

Thus,

$$\mathbf{A} = \begin{bmatrix} 0 & 1 \\ -\frac{1}{8} & \frac{3}{4} \end{bmatrix}, \quad \mathbf{B} = \begin{bmatrix} 0 \\ 1 \end{bmatrix}, \quad \mathbf{C} = \begin{bmatrix} -\frac{1}{8} & \frac{3}{4} \end{bmatrix}, \quad \mathbf{D} = [1]$$

We need to find $\mathbf{A}^n$, $n = 1, 2, \ldots$. The characteristic equation for $\mathbf{A}$ is

$$g(\lambda) = |\mathbf{A} - \lambda\mathbf{I}| = \lambda^2 - \frac{3}{4}\lambda + \frac{1}{8} = 0$$

The eigenvalues are thus $\lambda_1 = \frac{1}{4}$, $\lambda_2 = \frac{1}{2}$. Hence

$$\mathbf{A}^n = \beta_0 \mathbf{I} + \beta_1 \mathbf{A} = \begin{bmatrix} \beta_0 & \beta_1 \\ -\frac{1}{8}\beta_1 & \beta_0 + \frac{3}{4}\beta_1 \end{bmatrix}$$

where β_0 and β_1 are solutions of

$$\beta_0 + \left(\frac{1}{4}\right)\beta_1 = \left(\frac{1}{4}\right)^n$$

$$\beta_0 + \left(\frac{1}{2}\right)\beta_1 = \left(\frac{1}{2}\right)^n$$

Solving for β_0 and β_1, we find that

$$\beta_1 = 4\left[\left(\frac{1}{2}\right)^n - \left(\frac{1}{4}\right)^n\right], \qquad \beta_0 = -\left(\frac{1}{2}\right)^n + 2\left(\frac{1}{4}\right)^n$$

And so

$$\mathbf{A}^n = \begin{bmatrix} -\left(\frac{1}{2}\right)^n + 2\left(\frac{1}{4}\right)^n & 4\left(\frac{1}{2}\right)^n - 4\left(\frac{1}{4}\right)^n \\ -\frac{1}{2}\left[\left(\frac{1}{2}\right)^n - \left(\frac{1}{4}\right)^n\right] & 2\left(\frac{1}{2}\right)^n - \left(\frac{1}{4}\right)^n \end{bmatrix}$$

or

$$\mathbf{A}^n = \left(\frac{1}{2}\right)^n \begin{bmatrix} -1 & 4 \\ -\frac{1}{2} & 2 \end{bmatrix} + \left(\frac{1}{4}\right)^n \begin{bmatrix} 2 & -4 \\ \frac{1}{2} & -1 \end{bmatrix}$$

Then

$$\mathbf{CA}^{k-1-m}\mathbf{B} = \begin{bmatrix} -\frac{1}{8} & \frac{3}{4} \end{bmatrix} \left\{ \left(\frac{1}{2}\right)^{k-1-m} \begin{bmatrix} -1 & 4 \\ -\frac{1}{2} & 2 \end{bmatrix} + \left(\frac{1}{4}\right)^{k-1-m} \begin{bmatrix} 2 & -4 \\ \frac{1}{2} & -1 \end{bmatrix} \right\} \begin{bmatrix} 0 \\ 1 \end{bmatrix}$$

$$= 2\left(\frac{1}{2}\right)^{k-m} - \left(\frac{1}{4}\right)^{k-m}$$

Thus

$$y(k) = \sum_{m=0}^{k-1} \mathbf{CA}^{k-1-m}\mathbf{B}u(m) + \mathbf{D}u(k)$$

$$= \sum_{m=0}^{k-1} \left[2\left(\frac{1}{2}\right)^{k-m} - \left(\frac{1}{4}\right)^{k-m} \right] \left(\frac{1}{2}\right)^{m} + \left(\frac{1}{2}\right)^{k}$$

$$= 2\left(\frac{1}{2}\right)^{k} \sum_{m=0}^{k-1} 1 - \left(\frac{1}{4}\right)^{k} \sum_{m=0}^{k-1} 2^{m} + \left(\frac{1}{2}\right)^{k}$$

$$= 2\left(\frac{1}{2}\right)^{k} k - \left(\frac{1}{4}\right)^{k} \frac{1-2^{k}}{1-2} + \left(\frac{1}{2}\right)^{k}$$

$$= \begin{cases} 2k\left(\frac{1}{2}\right)^{k} + \left(\frac{1}{4}\right)^{k}, & k \geq 0 \\ 0, & k < 0 \end{cases}$$

■

It is clear from this example that the calculation of the output using state variables is more complex than our two previous models. However, the state-variable method gives us information of how the internal states $x_1(k)$ and $x_2(k)$ evolve. If this information is not important in a particular case, then a simpler model may suffice.

There is a second method of finding functions of a matrix that is often faster computationally than the method presented above. It is based on the *spectral decomposition* of a matrix. This is an alternate representation of an $n \times n$ matrix in terms of n simpler matrices that we shall denote as $\mathbf{E}_1, \mathbf{E}_2, \ldots, \mathbf{E}_n$. The $\mathbf{E}_i$, $i = 1, 2, \ldots, n$, are called *constituent* matrices. It can be shown† that any $n \times n$

† B. Noble, *Applied Linear Algebra*, Prentice-Hall, Englewood Cliffs, N.J., 1969.

matrix $\mathbf{A}$ has the representation

$$\mathbf{A} = \lambda_1 \mathbf{E}_1 + \lambda_2 \mathbf{E}_2 + \cdots + \lambda_n \mathbf{E}_n$$
$$= \sum_{i=1}^{n} \lambda_i \mathbf{E}_i \tag{2.73}$$

where λ_i, $i = 1, 2, \ldots, n$ are the distinct eigenvalues of $\mathbf{A}$. (We shall discuss the repeated eigenvalue case below.) The constituent matrices $\mathbf{E}_i$, $i = 1, 2, \ldots, n$ have the following properties:

1. $$\mathbf{E}_i \mathbf{E}_j = \begin{cases} 0, & i \neq j \\ \mathbf{E}_i, & i = j \end{cases}$$

2. $$\sum_{i=1}^{n} \mathbf{E}_i = \mathbf{I} \tag{2.74}$$

3. $$\mathbf{A}\mathbf{E}_i = \mathbf{E}_i \mathbf{A} = \lambda_i \mathbf{E}_i$$

4. $\mathbf{E}_i$ have rank 1.†

Given the representation (2.73), one can also show that functions of the matrix $\mathbf{A}$ can be written as

$$f(\mathbf{A}) = \sum_{i=1}^{n} f(\lambda_i)\mathbf{E}_i \tag{2.75}$$

For example, if $f(\mathbf{A}) = \mathbf{A}^k$, then (2.75) implies that

$$\mathbf{A}^k = \lambda_1^k \mathbf{E}_1 + \lambda_2^k \mathbf{E}_2 + \cdots + \lambda_n^k \mathbf{E}_n \tag{2.76}$$

If we can find the constituent matrices $\mathbf{E}_i$, $i = 1, 2, \ldots, n$ for a matrix $\mathbf{A}$, then by using (2.75) we can find functions of a matrix $f(\mathbf{A})$.

For distinct eigenvalues we can proceed as follows. Suppose $\mathbf{A}$ is 2×2 with eigenvalues λ_1 and λ_2. From (2.75) and (2.76) we have that

$$\mathbf{A}^k = \lambda_1^k \mathbf{E}_1 + \lambda_2^k \mathbf{E}_2 \tag{2.77}$$

† The rank of a matrix is defined in Appendix C.

In (2.77) let $k = 0, 1$. We have for $k = 0$

$$\mathbf{I} = \mathbf{E}_1 + \mathbf{E}_2 \tag{2.78}$$

which is property (2) of (2.74). For $k = 1$ we have

$$\mathbf{A} = \lambda_1 \mathbf{E}_1 + \lambda_2 \mathbf{E}_2 \tag{2.79}$$

Multiply (2.78) by λ_1 and subtract the resulting matrix equation from (2.79). We obtain

$$\mathbf{A} - \lambda_1 \mathbf{I} = (\lambda_2 - \lambda_1)\mathbf{E}_2$$

which implies that

$$\mathbf{E}_2 = \frac{\mathbf{A} - \lambda_1 \mathbf{I}}{\lambda_2 - \lambda_1} \tag{2.80}$$

Similarly,

$$\mathbf{E}_1 = \frac{\mathbf{A} - \lambda_2 \mathbf{I}}{\lambda_1 - \lambda_2} \tag{2.81}$$

This method can easily be extended to higher dimensional matrices in a straightforward manner. For a 3×3 matrix, one generates 3 equations that are

$$\begin{aligned} \mathbf{I} &= \mathbf{E}_1 + \mathbf{E}_2 + \mathbf{E}_3 \\ \mathbf{A} &= \lambda_1 \mathbf{E}_1 + \lambda_2 \mathbf{E}_2 + \lambda_3 \mathbf{E}_3 \\ \mathbf{A}^2 &= \lambda_1^2 \mathbf{E}_1 + \lambda_2^2 \mathbf{E}_2 + \lambda_3^2 \mathbf{E}_3 \end{aligned} \tag{2.82}$$

Again knowing λ_1, λ_2, and λ_3 one can solve for $\mathbf{E}_1$, $\mathbf{E}_2$, and $\mathbf{E}_3$ in a straightforward manner.

Example 2.23 Find $\mathbf{A}^k$ for the $\mathbf{A}$ matrix of Example 2.22. We have

$$\mathbf{A} = \begin{bmatrix} 0 & 1 \\ -\dfrac{1}{8} & \dfrac{3}{4} \end{bmatrix}$$

with eigenvalues $\lambda_1 = \frac{1}{2}$, $\lambda_2 = \frac{1}{4}$. Using (2.76) we have

$$\mathbf{A}^k = \lambda_1^k \mathbf{E}_1 + \lambda_2^k \mathbf{E}_2$$
$$= \left(\frac{1}{2}\right)^k \mathbf{E}_1 + \left(\frac{1}{4}\right)^k \mathbf{E}_2$$

where

$$\mathbf{E}_1 = \frac{\mathbf{A} - \lambda_2 \mathbf{I}}{\lambda_1 - \lambda_2} = \frac{\begin{bmatrix} 0 & 1 \\ -\frac{1}{8} & \frac{3}{4} \end{bmatrix} - \left(\frac{1}{4}\right)\begin{bmatrix} 1 & 0 \\ 0 & 1 \end{bmatrix}}{\frac{1}{4}} = \begin{bmatrix} -1 & 4 \\ -\frac{1}{2} & 2 \end{bmatrix}$$

and

$$\mathbf{E}_2 = \frac{\mathbf{A} - \lambda_1 \mathbf{I}}{\lambda_2 - \lambda_1} = \frac{\begin{bmatrix} 0 & 1 \\ -\frac{1}{8} & \frac{3}{4} \end{bmatrix} - \left(\frac{1}{2}\right)\begin{bmatrix} 1 & 0 \\ 0 & 1 \end{bmatrix}}{-\frac{1}{4}} = \begin{bmatrix} 2 & -4 \\ \frac{1}{2} & -1 \end{bmatrix}$$

And so

$$\mathbf{A}^k = \left(\frac{1}{2}\right)^k \begin{bmatrix} -1 & 4 \\ -\frac{1}{2} & 2 \end{bmatrix} + \left(\frac{1}{4}\right)^k \begin{bmatrix} 2 & -4 \\ \frac{1}{2} & -1 \end{bmatrix}$$

as before. The calculations are generally simpler with this method. ■

The case of repeated eigenvalues is somewhat more complicated. The representation in this case for an $n \times n$ matrix $\mathbf{A}$ is

$$\mathbf{A} = \sum_{i=1}^{p} (\lambda_i \mathbf{E}_i + \mathbf{N}_i), \qquad p \leq n \tag{2.83}$$

where $\mathbf{N}_i$ is a matrix such that if r is the multiplicity of λ_i, then†

$$\mathbf{N}_i^r = 0 \tag{2.84}$$

The matrices $\mathbf{E}_i$ and $\mathbf{N}_i$ satisfy the following properties:

i. $$\mathbf{E}_i\mathbf{E}_j = \begin{cases} 0, & i \neq j \\ \mathbf{E}_i, & i = j \end{cases}$$

ii. $$\mathbf{E}_i\mathbf{N}_j = \mathbf{N}_j\mathbf{E}_i = \begin{cases} 0, & i \neq j \\ \mathbf{N}_j, & i = j \end{cases}$$

iii. $$\sum_{i=1}^{p} \mathbf{E}_i = \mathbf{I}$$

iv. $\mathbf{N}_i^r = 0$, where r is the multiplicity of λ_i.

In this case functions of a matrix are calculated using

$$\mathbf{f}(\mathbf{A}) = \sum_{i=1}^{p} \left[f(\lambda_i)\mathbf{E}_i + \sum_{k=2}^{r_i} \frac{f^{(k-1)}(\lambda_i)N_i^{(k-1)}}{(k-1)!} \right] \tag{2.85}$$

The inside sum in (2.85) must be included for multiple eigenvalues. The following example illustrates the calculation.

Example 2.24 Find $\mathbf{A}^k$ for the following 2×2 matrix

$$\mathbf{A} = \begin{bmatrix} \frac{1}{2} & 0 \\ \frac{1}{2} & \frac{1}{2} \end{bmatrix}.$$

In this case the characteristic equation is $g(\lambda) = (\frac{1}{2} - \lambda)^2 = 0$, which has two repeated eigenvalues $\lambda_1 = \lambda_2 = \frac{1}{2}$. Thus $\mathbf{A}$ has the representation

$$\mathbf{A} = \frac{1}{2}\mathbf{E}_1 + \mathbf{N}_1$$

† N_i is said to be *nilpotent* of degree r.

The matrix function $\mathbf{A}^k$ is given by

$$\mathbf{f}(\mathbf{A}) = \mathbf{A}^k = \left(\frac{1}{2}\right)^k \mathbf{E}_1 + k\left(\frac{1}{2}\right)^{k-1} \mathbf{N}_1.$$

From the equation for $\mathbf{A}^k$ we have for $k = 0, 1$

$$\mathbf{A}^0 = \mathbf{I} = \mathbf{E}_1$$

$$\mathbf{A} = \frac{1}{2}\mathbf{E}_1 + \mathbf{N}_1$$

Therefore,

$$\mathbf{E}_1 = \mathbf{I}$$

$$\mathbf{N}_1 = \mathbf{A} - \frac{1}{2}\mathbf{E}_1 = \mathbf{A} - \frac{1}{2}\mathbf{I}$$

And so

$$\begin{aligned}
\mathbf{A}^k &= \left(\frac{1}{2}\right)^k \mathbf{I} + k\left(\frac{1}{2}\right)^{k-1}\left(\mathbf{A} - \frac{1}{2}\mathbf{I}\right) \\
&= \begin{bmatrix} \left(\frac{1}{2}\right)^k & 0 \\ 0 & \left(\frac{1}{2}\right)^k \end{bmatrix} + k\left(\frac{1}{2}\right)^{k-1} \begin{bmatrix} 0 & 0 \\ \frac{1}{2} & 0 \end{bmatrix} \\
&= \begin{bmatrix} \left(\frac{1}{2}\right)^k & 0 \\ k\left(\frac{1}{2}\right)^k & \left(\frac{1}{2}\right)^k \end{bmatrix}
\end{aligned}$$

In most practical applications one does not often deal with systems with repeated eigenvalues. Usual designs separate the eigenvalues or modes of a system.

2.11 CHANGE OF INTERNAL SYSTEM STRUCTURE

Given a linear, discrete-time system (**A**, **B**, **C**, **D**) with impulse response h_k

$$h_k = \begin{cases} \mathbf{D}, & k = 0 \\ \mathbf{CA}^{k-1}\mathbf{B}, & k > 0 \end{cases} \tag{2.86}$$

we can change the state with a nonsingular linear transformation **T** and not change h. That is, we can change the state vector **x** to, say, **x**′, where

$$\mathbf{x}'(k) = \mathbf{Tx}(k) \tag{2.87}$$

With this change of coordinates the new state matrices are

$$\mathbf{A}' = \mathbf{TAT}^{-1}, \qquad \mathbf{B}' = \mathbf{TB}, \qquad \mathbf{C}' = \mathbf{CT}^{-1}, \qquad \mathbf{D}' = \mathbf{D} \tag{2.88}$$

as we derived previously (see equation 2.47). The matrices (**A**′, **B**′, **C**′, **D**′) represent a system with the same external description, that is, h is unchanged, but with a different internal structure. To demonstrate that the impulse response is unchanged we substitute into (2.86) and obtain

$$h'_k = \begin{cases} \mathbf{D}', & k = 0 \\ \mathbf{C}'(\mathbf{A}')^{k-1}\mathbf{B}', & k > 0 \end{cases} \tag{2.89}$$

In (2.89) **D**′ = **D** so that h_0 is unchanged. Also we can substitute for **C**′, **A**′, and **B**′ to obtain

$$\begin{aligned} \mathbf{C}'(\mathbf{A}')^{k-1}\mathbf{B}' &= \mathbf{CT}^{-1}(\mathbf{TAT}^{-1})^{k-1}\mathbf{TB} \\ &= \mathbf{CT}^{-1}(\mathbf{TAT}^{-1})(\mathbf{TAT}^{-1})\cdots(\mathbf{TAT}^{-1})\mathbf{TB} \\ &= \mathbf{CA}^{k-1}\mathbf{B} = h_k, \qquad k > 0 \end{aligned}$$

Example 2.25 Consider the discrete-time system with matrices

$$\mathbf{A} = \begin{bmatrix} 0 & 1 \\ -\frac{1}{8} & \frac{3}{4} \end{bmatrix}, \qquad \mathbf{B} = \begin{bmatrix} 0 \\ 1 \end{bmatrix}, \qquad \mathbf{C} = \begin{bmatrix} -\frac{1}{8} & \frac{3}{4} \end{bmatrix}, \qquad \mathbf{D} = [1]$$

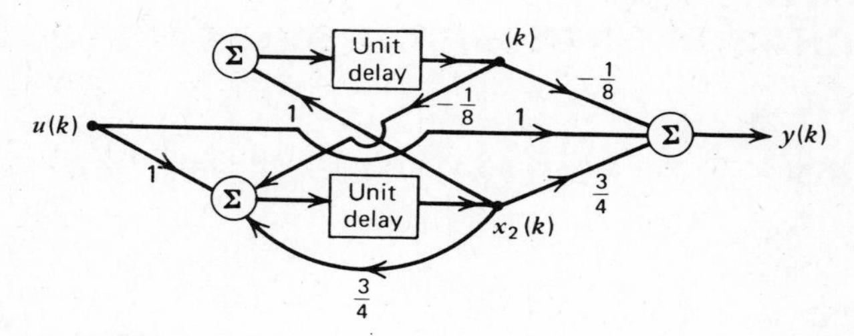

Figure 2.29

The corresponding schematic diagram is shown in Figure 2.29 where we have redrawn our usual block diagrams in a way to more easily incorporate the information contained in the model $\{\mathbf{A}, \mathbf{B}, \mathbf{C}, \mathbf{D}\}$. For example, the crossmultipliers from the unit delays correspond to a_{ij}, $i \neq j$. The loops from output to input of the unit delays correspond to a_{ii}. Similarly the inputs to the unit delays correspond to b_i and the outputs correspond to c_i.

Suppose now we change variables so that the new state is

$$\mathbf{x}' = \begin{bmatrix} 1 & 0 \\ 1 & \frac{1}{2} \end{bmatrix} \mathbf{x}$$

Then

$$\mathbf{T} = \begin{bmatrix} 1 & 0 \\ 1 & \frac{1}{2} \end{bmatrix}$$

and

$$\mathbf{T}^{-1} = \begin{bmatrix} 1 & 0 \\ -2 & 2 \end{bmatrix}.$$

The new state matrices are

$$\mathbf{A}' = \mathbf{TAT}^{-1} = \begin{bmatrix} -2 & 2 \\ -\frac{45}{16} & \frac{11}{4} \end{bmatrix}, \qquad \mathbf{B}' = \mathbf{TB} = \begin{bmatrix} 0 \\ \frac{1}{2} \end{bmatrix}, \qquad \mathbf{C}' = \mathbf{CT}^{-1} = \begin{bmatrix} -\frac{13}{8} & \frac{3}{2} \end{bmatrix}$$

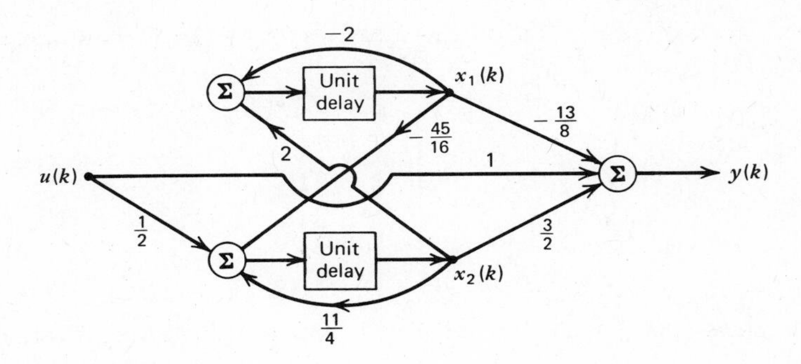

Figure 2.30

The new system structure is seen in Figure 2.30. We have changed the internal structure of this system but have retained the same input-output relation. ■

There are several reasons why one might wish to change the internal structure of a system and not change the external description of the system. For example, one might wish to minimize the number of internal multipliers because they are expensive. Or perhaps one might desire to minimize the amount of internal noise in a structure due to rounding of intermediate results. Whatever the reason for wanting to change internal connections, the state-variable method allows one to examine analytically internal structures by applying nonsingular transformations to the state vector.

One particular transformation of state coordinates is often used. It is the transformation of coordinates that makes **A** a diagonal matrix with the eigenvalues of **A** on the diagonal. One can view this change of coordinates as a "decoupling" transformation in that the natural modes of the system do not interact in forming the output. Let $\hat{\mathbf{D}}$ be the diagonal matrix obtained from **A**. Assuming that the eigenvalues of **A** are distinct, we can accomplish this transformation by using a nonsingular matrix **P** whose columns consist of the eigenvectors of the matrix **A**. The eigenvectors of **A** are those directions in the space that are not changed by the transformation **A**. In symbols, the eigenvectors $\mathbf{v}_1, \mathbf{v}_2, \ldots, \mathbf{v}_n$ of **A** are defined by

$$\mathbf{A}\mathbf{v}_i = \lambda_i \mathbf{v}_i \tag{2.90}$$

where λ_i, $i = 1, 2, \ldots, n$ are the eigenvalues of **A**. In this case, the diagonal matrix $\hat{\mathbf{D}}$ is

$$\hat{\mathbf{D}} = \mathbf{P}^{-1}\mathbf{A}\mathbf{P} \tag{2.91}$$

Example 2.26 Find the "decoupled form" of the system in Example 2.25. The **A** matrix is

$$\mathbf{A} = \begin{bmatrix} 0 & 1 \\ -\dfrac{1}{8} & \dfrac{3}{4} \end{bmatrix}$$

with eigenvalues $\lambda_1 = \frac{1}{4}$ and $\lambda_2 = \frac{1}{2}$. The corresponding eigenvectors are obtained by solving

$$(\mathbf{A} - \lambda_i \mathbf{I})\mathbf{v}_i = 0, \qquad i = 1, 2 \tag{2.92}$$

Equation 2.92 yields two equations given below.

$$\begin{bmatrix} -\dfrac{1}{4} & 1 \\ -\dfrac{1}{8} & \dfrac{1}{2} \end{bmatrix} \begin{bmatrix} v_1^{(1)} \\ v_1^{(2)} \end{bmatrix} = 0, \quad \begin{bmatrix} -\dfrac{1}{2} & 1 \\ -\dfrac{1}{8} & \dfrac{1}{4} \end{bmatrix} \begin{bmatrix} v_2^{(1)} \\ v_2^{(2)} \end{bmatrix} = 0 \tag{2.93}$$

Notice in both matrix equations in (2.93) that the rows in each matrix are the same except for a constant multiplier. If we arbitrarily set $v_1^{(1)} = v_2^{(1)} = 1$, then we find $\mathbf{v}_1$ and $\mathbf{v}_2$ within an arbitrary constant as

$$\mathbf{v}_1 = \begin{bmatrix} 1 \\ \dfrac{1}{4} \end{bmatrix}, \qquad \mathbf{v}_2 = \begin{bmatrix} 1 \\ \dfrac{1}{2} \end{bmatrix} \tag{2.94}$$

It is only the direction of $\mathbf{v}_1$ and $\mathbf{v}_2$ that we can determine from (2.92); the lengths of $\mathbf{v}_1$ and $\mathbf{v}_2$ are arbitrary. Therefore, we choose the matrix P as

$$\mathbf{P} = \begin{bmatrix} 1 & 1 \\ \dfrac{1}{4} & \dfrac{1}{2} \end{bmatrix}$$

with its inverse

$$\mathbf{P}^{-1} = 4\begin{bmatrix} \frac{1}{2} & -1 \\ -\frac{1}{4} & 1 \end{bmatrix}$$

Now use the change of state $\mathbf{x}' = \mathbf{P}^{-1}\mathbf{x}$. Note that the inverse matrix is used on the right of $\mathbf{x}$ instead of $\mathbf{P}$. It is not theoretically important whether one uses $\mathbf{P}$ or $\mathbf{P}^{-1}$ to change $\mathbf{x}$. Thus,

$$\mathbf{A}' = \hat{\mathbf{D}} = \mathbf{P}^{-1}\mathbf{A}\mathbf{P} = 4\begin{bmatrix} \frac{1}{2} & -1 \\ -\frac{1}{4} & 1 \end{bmatrix}\begin{bmatrix} 0 & 1 \\ -\frac{1}{8} & \frac{3}{4} \end{bmatrix}\begin{bmatrix} 1 & 1 \\ \frac{1}{4} & \frac{1}{2} \end{bmatrix} = \begin{bmatrix} \frac{1}{4} & 0 \\ 0 & \frac{1}{2} \end{bmatrix}$$

$$\mathbf{B}' = \mathbf{P}^{-1}\mathbf{B} = 4\begin{bmatrix} \frac{1}{2} & -1 \\ -\frac{1}{4} & 1 \end{bmatrix}\begin{bmatrix} 0 \\ 1 \end{bmatrix} = 4\begin{bmatrix} -1 \\ 1 \end{bmatrix}$$

$$\mathbf{C}' = \mathbf{C}\mathbf{P} = \begin{bmatrix} -\frac{1}{8} & \frac{3}{4} \end{bmatrix}\begin{bmatrix} 1 & 1 \\ \frac{1}{4} & \frac{1}{2} \end{bmatrix} = \begin{bmatrix} \frac{1}{16} & \frac{1}{4} \end{bmatrix}$$

The schematic of this diagonalized or decoupled circuit is shown in Figure 2.31. Notice that the two states $x_1(k)$ and $x_2(k)$ do not interact; that is, there are no cross-multipliers between the unit delay elements.

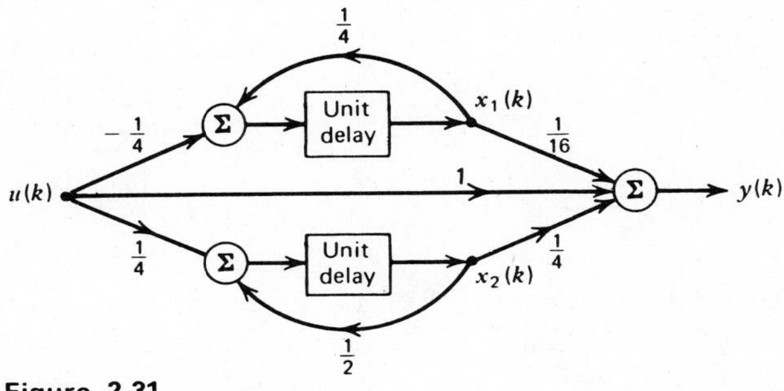

Figure 2.31

■

If one or more eigenvalues are repeated, then a form almost as simple can be obtained:

$$\mathbf{P}^{-1}\mathbf{A}\mathbf{P} = \mathbf{J} = \begin{bmatrix} \mathbf{J}_1 & 0 & \cdots & 0 \\ 0 & \mathbf{J}_2 & \cdots & 0 \\ & & \ddots & \\ 0 & 0 & \cdots & \mathbf{J}_{no} \end{bmatrix} \tag{2.95}$$

where $\mathbf{J}_1 \ldots \mathbf{J}_{no}$ are elementary square matrices of the form

$$\mathbf{J}_i = \begin{bmatrix} \lambda_i & 1 & 0 & \cdots & 0 \\ 0 & \lambda_i & 1 & \cdots & 0 \\ & & & \ddots & \\ 0 & 0 & 0 & \cdots & \lambda_i \end{bmatrix} \tag{2.96}$$

and of dimension $m_i \times m_i$, where m_i is the multiplicity of λ_i. The important point here is not how to generate these matrices, but to note how they allow us to characterize our solution.

Assume for simplicity that the eigenvalues of $\mathbf{A}$ are distinct and that we have found the matrix $\mathbf{P}$ of (2.91). Our solution for the system state $\mathbf{x}(k)$ and output $y(k)$ is expressed in terms of $\mathbf{A}^k$. Now we can write $\mathbf{A}$ as

$$\mathbf{A} = \mathbf{P}\hat{\mathbf{D}}\mathbf{P}^{-1} \tag{2.97}$$

and so

$$\begin{aligned} \mathbf{A}^k = (\mathbf{P}\hat{\mathbf{D}}\mathbf{P}^{-1})^k &= (\mathbf{P}\hat{\mathbf{D}}\mathbf{P}^{-1})(\mathbf{P}\hat{\mathbf{D}}\mathbf{P}^{-1})\cdots(\mathbf{P}\hat{\mathbf{D}}\mathbf{P}^{-1}) \\ &= \mathbf{P}\hat{\mathbf{D}}^k\mathbf{P}^{-1} \end{aligned} \tag{2.98}$$

where

$$\hat{\mathbf{D}}^k = \begin{bmatrix} \lambda_1^k & 0 & \cdots & 0 \\ 0 & \lambda_2^k & \cdots & 0 \\ & & \ddots & \\ 0 & 0 & & \lambda_n^k \end{bmatrix} \tag{2.99}$$

Thus, the elements of $\mathbf{A}^k$ and, hence, of $\mathbf{x}(k)$, are linear combinations of $\lambda_1^k, \lambda_2^k, \ldots, \lambda_n^k$. It follows that if $|\lambda_i| > 1$ for some i, then one or more of the state variables (and possibly the outputs also) will grow without bound with time. Hence, the system is unstable. Conversely, if $|\lambda_i| < 1$ for all $i = 1, \ldots, n$, and if the input is bounded, then all state variables and outputs will remain bounded. Finally, if $|\lambda_i| = 1$ for some i, then we can find a bounded input that will cause the output to increase without bound as k increases. This result gives us a very easy method to determine system stability.

The eigenvalues of the state matrix characterize the system stability. A discrete-time system will be stable if, and only if, all eigenvalues have modulus less than 1.

Example 2.27 Suppose we have a discrete-time system whose state matrix is

$$\mathbf{A} = \begin{bmatrix} 1 & a \\ 2 & \frac{1}{2} \end{bmatrix}$$

For what values of a is the system stable?

We first find the eigenvalues of this matrix as a function of the parameter a. The characteristic equation is

$$g(\lambda) = (1 - \lambda)\left(\frac{1}{2} - \lambda\right) - 2a = 0$$

That is,

$$\lambda^2 - \frac{3}{2}\lambda + \frac{1}{2} - 2a = 0$$

with the roots

$$\lambda_1 = \frac{3}{4} + \sqrt{\frac{1}{16} + 2a}$$

$$\lambda_2 = \frac{3}{4} - \sqrt{\frac{1}{16} + 2a}$$

Two cases are possible.

Case 1: Suppose $a \geq -\frac{1}{32}$. In this case, λ_1 and λ_2 are real valued, with $\lambda_1 > \lambda_2$. For the system to be stable, $|\lambda_i| < 1$, $i = 1, 2$. Thus set $\lambda_1 < 1$ to see what is implied about a.

$$\frac{3}{4} + \sqrt{\frac{1}{16} + 2a} < 1$$

That is,

$$\frac{1}{16} + 2a < \frac{1}{16}$$

which implies that $a < 0$. To insure stability, we must also have $\lambda_2 > -1$. This restriction means that

$$\frac{3}{4} - \sqrt{\frac{1}{16} + 2a} > -1$$

Thus

$$\sqrt{\frac{1}{16} + 2a} < \frac{7}{4}$$

which implies

$$a < \frac{3}{2}$$

which is assured by the preceding requirement that $a < 0$.

Case 2: Suppose now $a < -\frac{1}{32}$. In this case, λ_1 and λ_2 are complex valued. Setting $|\lambda_1| < 1$ or $|\lambda_2| < 1$ implies that

$$|\lambda_1|^2 = |\lambda_2|^2 = \left|\frac{3}{4} \pm j\sqrt{-\frac{1}{16} - 2a}\right|^2 < 1$$

or

$$\frac{9}{16} + \left(-\frac{1}{16} - 2a\right) < 1, \qquad a > -\frac{1}{4}$$

Combining these two results, we see that the system is stable for all values of a in the interval $(-\frac{1}{4}, 0)$. ■

2.12 FREQUENCY RESPONSE IN TERMS OF A, B, C, D

We have related one external description of a system, the impulse response sequence h, to the state description **A**, **B**, **C**, **D** in (2.54), repeated below.

$$h_k = \begin{cases} \mathbf{D}, & k = 0 \\ \mathbf{C}\mathbf{A}^{k-1}\mathbf{B}, & k > 0 \end{cases} \tag{2.100}$$

The frequency response $H(e^{j\theta})$ is another important external description of a system. How is it related to the state model **A**, **B**, **C**, **D**?

From (2.34) we have

$$\begin{aligned} H(e^{j\theta}) &= \sum_{n=-\infty}^{\infty} h_n e^{-jn\theta} \\ &= \sum_{n=0}^{\infty} h_n e^{-jn\theta} \end{aligned} \tag{2.101}$$

The lower limit is zero since we assume h is causal, that is, $h_n = 0$, $n < 0$. From (2.100) we can substitute for h_n in (2.101) and obtain

$$\begin{aligned} H(e^{j\theta}) &= \sum_{n=0}^{\infty} h_n e^{-jn\theta} \\ &= \mathbf{D} + \sum_{n=1}^{\infty} \mathbf{CA}^{n-1}\mathbf{B}e^{-jn\theta} \end{aligned}$$

Let $m = n - 1$. Then we have

$$\begin{aligned} H(e^{j\theta}) &= \mathbf{D} + \mathbf{C}\sum_{m=0}^{\infty} \mathbf{A}^m e^{-jm\theta}\mathbf{B}e^{-j\theta} \\ &= \mathbf{D} + \mathbf{C}\sum_{m=0}^{\infty} [\mathbf{A}e^{-j\theta}]^m \mathbf{B}e^{-j\theta} \end{aligned} \tag{2.102}$$

The expression $\sum_{m=0}^{\infty} [\mathbf{A}e^{-j\theta}]^m$ can be evaluated using the identity:

$$[\mathbf{I} - \mathbf{A}] \sum_{m=0}^{n} \mathbf{A}^m = \mathbf{I} - \mathbf{A}^{n+1} \tag{2.103}$$

Equation 2.103 is easily verified by merely expanding the left-hand side of the equation. In (2.103), if we multiply both sides by $[\mathbf{I} - \mathbf{A}]^{-1}$ and take the limit as $n \to \infty$, we obtain

$$\lim_{n\to\infty} \sum_{m=0}^{\infty} \mathbf{A}^m = \lim_{n\to\infty} \{[\mathbf{I} - \mathbf{A}^{-1}][\mathbf{I} - \mathbf{A}^{n+1}]\}$$

Assuming all the eigenvalues of A have modules less than 1, then

$$\lim_{n\to\infty} [\mathbf{I} - \mathbf{A}^{n+1}] = \mathbf{I};$$

thus we have

$$\lim_{n\to\infty} \sum_{m=0}^{n} \mathbf{A}^m = [\mathbf{I} - \mathbf{A}]^{-1}, \qquad |\lambda_i| < 1 \text{ all } i \tag{2.104}$$

Using (2.104) in (2.102) gives us

$$H(e^{j\theta}) = \mathbf{D} + \mathbf{C}[\mathbf{I} - \mathbf{A}e^{-j\theta}]^{-1}\mathbf{B}e^{-j\theta} \tag{2.105}$$

Bringing the term $e^{-j\theta}$ into the bracketed term, we have

$$H(e^{j\theta}) = \mathbf{D} + \mathbf{C}[\mathbf{I}e^{j\theta} - \mathbf{A}]^{-1}\mathbf{B} \tag{2.106}$$

Equations 2.105 or 2.106 express the external description $H(e^{j\theta})$ in terms of **A**, **B**, **C**, **D**. To illustrate the computations we shall apply (2.106) to the system of Examples 2.8 and 2.16.

Example 2.28 The system of Examples 2.8 and 2.16 is shown in Figure 2.32. The state equations are

$$\mathbf{x}(k+1) = \begin{bmatrix} 0 & 1 \\ -\frac{1}{2} & 0 \end{bmatrix} \mathbf{x}(k) + \begin{bmatrix} 0 \\ 1 \end{bmatrix} u(k)$$

$$y(k) = \begin{bmatrix} -\frac{1}{2} & 0 \end{bmatrix} \mathbf{x}(k) + [1]u(k)$$

From (2.106) the expression for $H(e^{j\theta})$ is

$$H(e^{j\theta}) = 1 + \begin{bmatrix} -\frac{1}{2} & 0 \end{bmatrix} \left\{ \begin{bmatrix} e^{j\theta} & 0 \\ 0 & e^{j\theta} \end{bmatrix} - \begin{bmatrix} 0 & 1 \\ -\frac{1}{2} & 0 \end{bmatrix} \right\}^{-1} \begin{bmatrix} 0 \\ 1 \end{bmatrix}$$

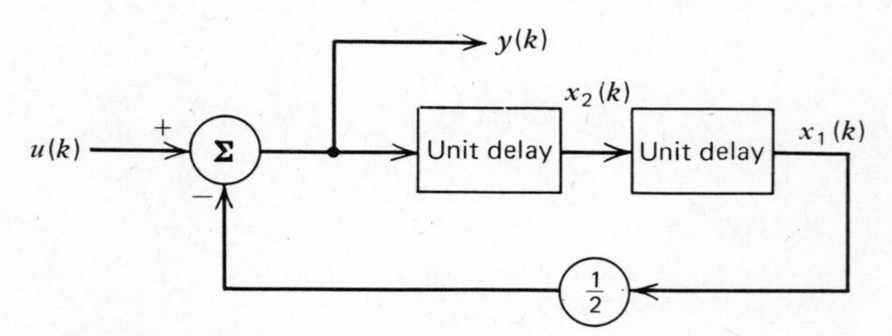

Figure 2.32

Consider the inverse matrix $[\mathbf{I}e^{j\theta} - \mathbf{A}]^{-1}$. We have

$$[\mathbf{I}e^{j\theta} - \mathbf{A}]^{-1} = \begin{bmatrix} e^{j\theta} & -1 \\ \frac{1}{2} & e^{j\theta} \end{bmatrix}^{-1} = \frac{1}{e^{j2\theta} + \frac{1}{2}} \begin{bmatrix} e^{j\theta} & 1 \\ -\frac{1}{2} & e^{j\theta} \end{bmatrix}$$

Thus, the frequency response is

$$H(e^{j\theta}) = 1 + \frac{1}{e^{j2\theta} + \frac{1}{2}} \begin{bmatrix} -\frac{1}{2} & 0 \end{bmatrix} \begin{bmatrix} e^{j\theta} & 1 \\ -\frac{1}{2} & e^{j\theta} \end{bmatrix} \begin{bmatrix} 0 \\ 1 \end{bmatrix}$$

$$= 1 + \frac{-\frac{1}{2}}{e^{j2\theta} + \frac{1}{2}} = \frac{e^{j2\theta} + \frac{1}{2} - \frac{1}{2}}{e^{j2\theta} + \frac{1}{2}} = \frac{1}{1 + \frac{1}{2}e^{-j2\theta}}$$

as we obtained previously. ■

2.13 CONCLUDING REMARKS AND FURTHER EXAMPLES

The eigenvalues of the system matrix **A** and the roots of the auxiliary equation for the corresponding difference equation model are identical. These values define the transient or homogeneous solutions to the system. These solutions added together with the correct constants form the impulse-response sequence *h*. Clearly, all three descriptions contain essentially the same information about the system. The state-variable model, as we have demonstrated, provides additional information on the evolution of internal states.

We have derived three expressions for the important frequency response characterization of a system in terms of the three models considered.

$$H(e^{j\theta}) = \frac{a_0 + a_1 e^{-j\theta} + a_2 e^{-j2\theta} + \cdots + a_m e^{-jm\theta}}{1 + b_1 e^{-j\theta} + b_2 e^{-j2\theta} + \cdots + b_n e^{-jn\theta}}$$

$$= \sum_{n=-\infty}^{\infty} h_n e^{-jn\theta}$$

$$= \mathbf{D} + \mathbf{C}[\mathbf{I}e^{j\theta} - \mathbf{A}]^{-1}\mathbf{B} \tag{2.107}$$

We shall find (2.107) a valuable asset for calculating one of the most used characterization of discrete-time circuits.

We conclude this chapter with two examples.

Example 2.29 Examples 2.13 and 2.22 are analyses of the system in Figure 2.33 using convolution and state variables, respectively. To compare our three descriptions let us determine the output of the system using the difference equation model. From the block diagram, we can write

$$y_k = u_k + \frac{3}{4} y_{k-1} - \frac{1}{8} y_{k-2}$$

The corresponding auxiliary equation is

$$r^2 - \frac{3}{4} r + \frac{1}{8} = 0$$

Note that the characteristic equation of the system matrix **A** from Example 2.22 is exactly the same equation, that is,

$$g(\lambda) = \lambda^2 - \frac{3}{4} \lambda + \frac{1}{8} = 0$$

The roots of the auxiliary equation are

$$r_1 = \frac{1}{4} \qquad r_2 = \frac{1}{2}$$

The transient solution is therefore

$$y^{(h)}(k) = \begin{cases} 0, & k < 0 \\ c_1\left(\frac{1}{4}\right)^k + c_2\left(\frac{1}{2}\right)^k, & k \geq 0 \end{cases}$$

The particular solution due to an input $(\frac{1}{2})^k$ is of the form

$$y^{(p)}(k) = c_3 k\left(\frac{1}{2}\right)^k$$

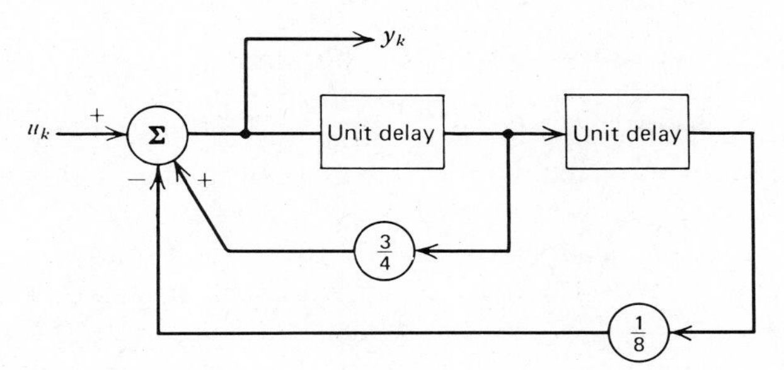

Figure 2.33

The constant c_3 is evaluated by substituting $y^{(p)}(k)$ into the original equation. We find that $c_3 = 2$. The complete solution is thus

$$y(k) = \begin{cases} 0, & k < 0 \\ 2k\left(\frac{1}{2}\right)^k + c_1\left(\frac{1}{4}\right)^k + c_2\left(\frac{1}{2}\right)^k, & k \geq 0 \end{cases}$$

To find c_1 and c_2 we use the zero initial conditions $y(-1) = 0$, $y(-2) = 0$.

$$y(-1) = 0 = -2\left(\frac{1}{2}\right)^{-1} + c_1\left(\frac{1}{4}\right)^{-1} + c_2\left(\frac{1}{2}\right)^{-1}$$

$$y(-2) = 0 = -4\left(\frac{1}{2}\right)^{-2} + c_1\left(\frac{1}{4}\right)^{-2} + c_2\left(\frac{1}{2}\right)^{-2}$$

or

$$4c_1 + 2c_2 = 4$$

$$16c_1 + 4c_2 = 16$$

which implies that $c_1 = 1$ and $c_2 = 0$. Thus,

$$y(k) = \begin{cases} 0, & k < 0 \\ 2k\left(\frac{1}{2}\right)^k + \left(\frac{1}{4}\right)^k, & k \geq 0 \end{cases}$$

Contrast this analysis with the analyses in Examples 2.13 and 2.22. Which analysis method do you prefer? ∎

Example 2.30 As an example of an application of the preceding material let us suppose we wish to design a digital circuit that will reject a very strong 60 Hz interference contaminating a 10 Hz signal of interest. Essentially we wish to build a filter that is a notch or band-stop filter that rejects the 60 Hz component while passing the 10 Hz signal with little or no alteration.

Assume that we sample the input waveform using an analog-to-digital converter at a rate of 500 samples per second. This means that the input signal level is measured every 2 milliseconds. These sampled values then constitute the input sequence $\{u(k)\}$. (We shall disregard the error incurred by representing these real values by a finite number of bits.) This input sequence is to be processed by a digital circuit to eliminate the 60 Hz interference.

This chapter covers several methods of analysis for discrete-time circuits. In this application we are asking how to *design* or *synthesize* discrete-time circuits. This is a step beyond our present background. We shall thus have to assume some basic circuit structure and see if we cannot choose the appropriate parameters (in this case the parameters are the multiplier values).

Figure 2.34 depicts a simple finite-length impulse-response filter with three unknown parameters, the multipliers a_0, a_1, and a_2. The difference equation model is $y(k) = a_0 u(k) + a_1 u(k-1) + a_2 u(k-2)$. We can find the frequency response of this structure by assuming $u(k) = e^{jk\theta}$, $\theta = \omega T$ where θ is the normalized frequency. Then,

$$\begin{aligned} y(k) &= a_0 e^{jk\theta} + a_1 e^{j(k-1)\theta} + a_2 e^{j(k-2)\theta} \\ &= e^{jk\theta}[a_0 + a_1 e^{-j\theta} + a_2 e^{-j2\theta}] \end{aligned}$$

Thus, the frequency response is

$$\begin{aligned} H(e^{jk\theta}) &= a_0 + a_1 e^{-j\theta} + a_2 e^{-j2\theta} \\ &= e^{-j\theta}[a_0 e^{j\theta} + a_1 + a_2 e^{-j\theta}] \end{aligned}$$

The magnitude or amplitude response is dependent only on the bracketed term. We want this term to be as close to zero as possible at 60 Hz. What value of θ does 60 Hz correspond to? We have that

$$\theta = \omega T = 2\pi f T$$

where $T = \frac{1}{500} = 0.002$ seconds and $f = 60$ Hz. Thus, in normalized frequency, 60 Hz (with $T = 0.002$) corresponds to

$$\theta_{60} = (2\pi \cdot 60)(0.002) = 0.75398 \text{ radians}$$

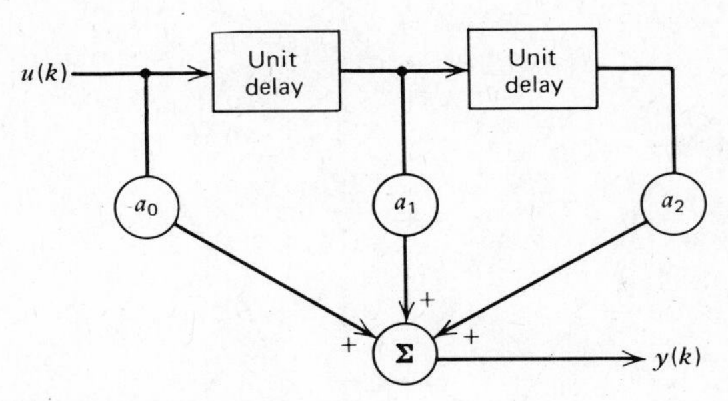

Figure 2.34

In other words, we would like

$$a_0 e^{j\theta_{60}} + a_1 + a_2 e^{-j\theta_{60}} = 0$$

We also wish to have unity gain, say, at 10 Hz. That is,

$$a_0 e^{j\theta_{10}} + a_1 + a_2 e^{-j\theta_{10}} = 1$$

where

$$\theta_{10} = (2\pi \cdot 10)(0.002) = 0.12566 \text{ radians}$$

Let's further assume $a_0 = a_2$. This means the filter has linear phase. The response is said to be phase distortionless in this case. We now have 2 equations with 2 unknowns:

$$a_1 + 2a_2 \cos \theta_{60} = 0$$

$$a_1 + 2a_2 \cos \theta_{10} = 1$$

or

$$a_1 + a_2(1.4579) = 0$$

$$a_1 + a_2(1.98439) = 1$$

Thus

$$a_1 \cong -2.77$$

$$a_2 \cong 1.90$$

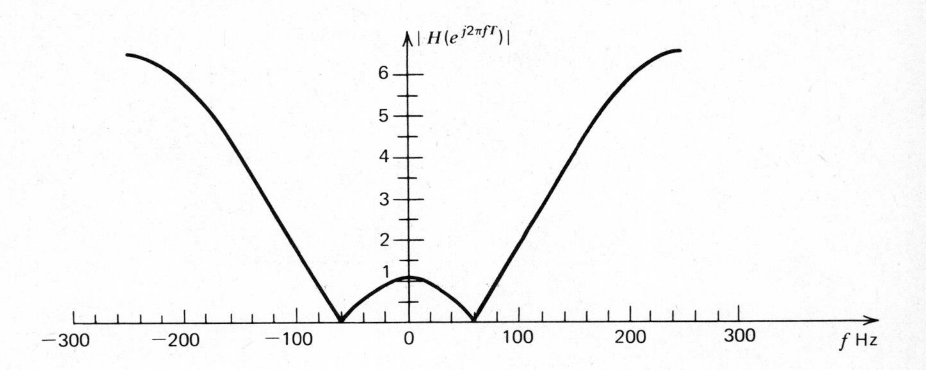

|H(e^{j2πfT})|
6
5
4
3
2
1
−300
−200
−100
0
100
200
300
f Hz

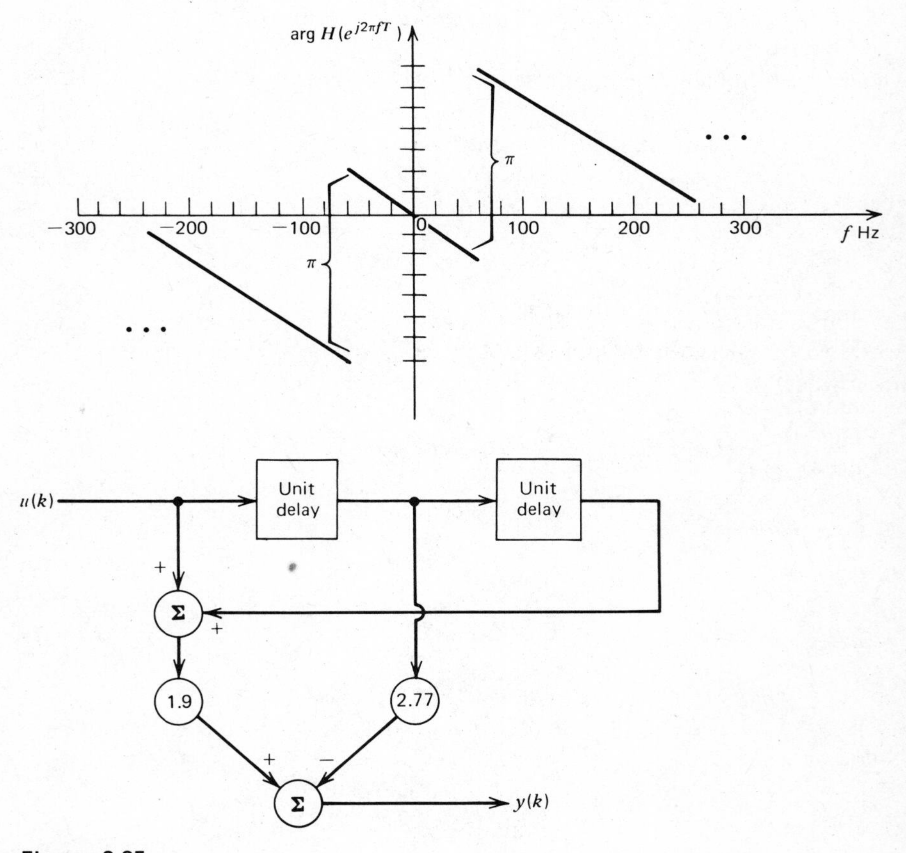

arg H(e^{j2πfT})
π
π
. . .
. . .
−300
−200
−100
0
100
200
300
f Hz
u(k)
Unit delay
Unit delay
+
Σ
+
1.9
2.77
+
−
Σ
y(k)

Figure 2.35

The difference equation model is therefore

$$y(k) = 1.9u(k) - 2.77u(k-1) + 1.9u(k-2)$$

We can check our model by finding the steady-state response to $u(k) = \cos(0.754k)$, that is, an input sinusoid at 60 Hz with $T = 0.002$ seconds. The steady-state output at 60 Hz is clearly

$$\begin{aligned} y_{ss}(k) &= 1.9\cos(0.754k) - 2.77\cos[0.754(k-1)] + 1.9\cos[0.754(k-2)] \\ &= 1.9\cos(0.754k) - 2.77\{\cos(0.754k)\cos(-0.754) \\ &\quad - \sin(0.754k)\sin(-0.754)\} \\ &\quad + 1.9\{\cos(0.754k)\cos(-1.508) - \sin(0.754k)\sin(-1.508)\} \end{aligned}$$

And so,

$$\begin{aligned} y_{ss}(k) &= \cos(0.754k)[1.9 - 2.77\cos(0.754) + 1.9\cos(1.508)] \\ &\quad + \sin(0.754k)[-2.77\sin(0.754) + 1.9\sin(1.508)] \\ &= \cos(0.754k)[1.9 - 2.019 + 0.119] \\ &\quad + \sin(0.754k)[-1.8962 + 1.8962] \\ &= 0, \qquad \text{all } k \end{aligned}$$

A similar calculation at 10 Hz will reveal that $y_{ss}(k)$ is $\cos(0.1256k)$. That is, the steady-state output is the input sinusoid with a gain of unity. Figure 2.35 depicts the frequency response of this filter. Also shown is a schematic of the circuit with the final design parameters.

Digital filters offer circuit designers certain advantages over other structures. The depth and location of the notch, for example, will not drift or change with age or temperature. These kinds of structures are inherently stable, flexible (one merely changes the coefficients to move the notch), and simple. ■

2.14 SUMMARY

We have discussed three time-domain models for the analysis of linear, shift-invariant, discrete-time systems. The first two models studied, the linear difference equation and the impulse-response sequence, are both input-output characterizations of a discrete-time system. They treat the system as a block box with no concern for the internal workings of the system. These two models cannot easily handle multiple input-output systems.

On the other hand, the third model, the state-variable or matrix model, not only yields the output for any given input sequence but also shows how the internal states of systems evolve. It is a model ideally suited to multiple inputs and outputs. Because it is a matrix formulation, dimensionality of the system does not affect the formulation. Increased dimensionality only increases the number of computations involved in solving the system equations. It is the only analytic model that explicitly defines the internal structure of a system. This increased information available from the state-variable model comes at a price: increased complexity and number of computations.

PROBLEMS

2.1. Find the appropriate annihilator for each of the following sequences.
(a) e^{bk}
(b) $B \sinh ak$
(c) $k^2 a^k + Ae^{bk}$
(d) $ka^k + A \sin bk$

2.2. Solve the following difference equations.
(a) $y_{k+2} + 7y_{k+1} + 12y_k = 0, \quad k \geq 0$
(b) $y_{k+2} + 2y_{k+1} + 2y_k = 0, \quad k \geq 0$
(c) $y_{k+2} + y_k = \sin k, \quad k \geq 0$
(d) $y_{k+2} - \frac{5}{2}y_{k+1} + y_k = 1, \quad y_0 = y_1 = 0$
(e) $u_{k+1} = u_k - y_k, \quad u_0 = 1$
$y_{k+1} = u_k + y_k, \quad y_0 = 0$

2.3. Consider the second-order difference equation $y_{k+2} - 2\tau y_{k+1} + y_k = 0$. Find solutions for the following cases.
(a) $\tau < -1$
(b) $\tau = -1$
(c) $|\tau| < 1$
(d) $\tau = +1$
(e) $\tau > 1$

2.4. The capacity of an information channel is defined as

$$C = \lim_{t \to \infty} \left[\frac{\log_2 N_t}{t} \right] \text{bits/sec}$$

where N_t is the number of possible messages of duration t (see Example 1.7).

Suppose a signaling system has two symbols S_1 and S_2: S_1 lasts t_1 seconds, and S_2 lasts t_2 seconds. Find the capacity of the channel for the cases:
(a) $S_1 = 1, S_1 = 2$
(b) $S_1 = S_2 = 1$. Explain the change in capacity between (a) and (b).

2.5. Find the steady-state outputs of the following circuits.

(a)

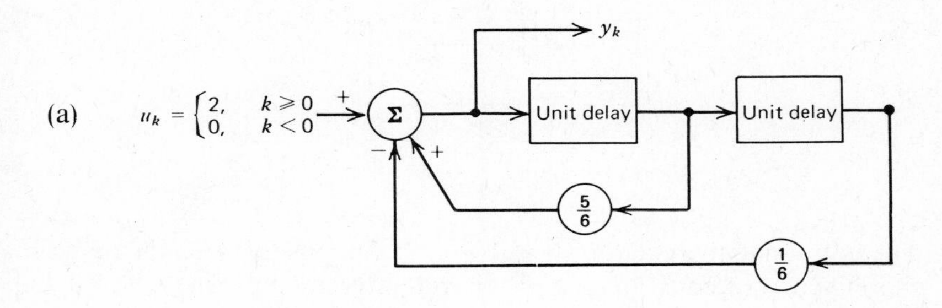

(b)

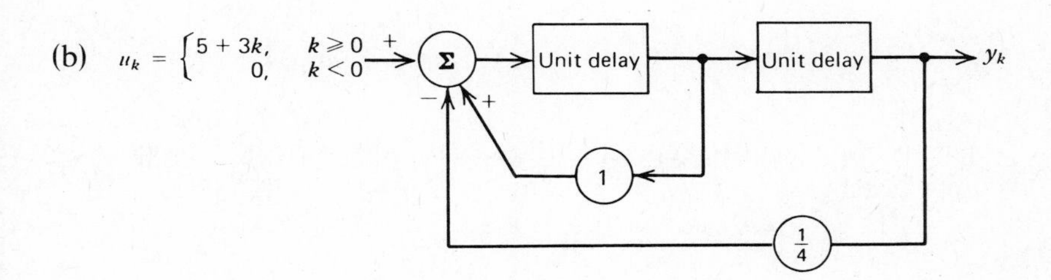

2.6. The system $y_{k+2} + a_1 y_{k+1} + a_2 y_k = 0$ is stable if all solutions satisfy $y_k \to 0$ as $k \to \infty$. This holds if and only if both roots of the auxiliary equation

$$r^2 + a_1 r + a_2 = 0$$

satisfy $|r| < 1$. Find and sketch the region in the parameter plane a_2 vs. a_1 for which the system is stable.

2.7. In an analog computer the integrator element shown below is defined by the equation

$$y(t) = y(0) + \int_0^t u(\tau)\, d\tau$$

$u(t)$ → $\int(\ )$ → $y(t)$, $y(0)$ (initial condition)

Integrator Element

The diagram shown below simulates the differential equation

$$\frac{dy^2(t)}{dt^2} + y(t) = 0, \qquad y(0) = 1, \qquad \left.\frac{dy(t)}{dt}\right|_{t=0} = 0$$

$\frac{d^2y}{dt^2}$ $\int(\,)$ $\frac{dy}{dt}$ $\int(\,)$ $y(t)$

0 (Initial condition) 1 (Initial condition)

-1

Simulation of $\frac{dy^2(t)}{dt^2} + y(t) = 0$

The solution to this equation is $y(t) = \cos t$. Suppose we wish to approximate this differential equation with a discrete-time model. One method, called Euler's method, approximates the integral with a discrete sum

$$y(0) + \int_0^t u(\tau)\, d\tau \cong y(0) + \Delta \sum_{k=0}^{n-1} u(k\Delta)$$

This is equivalent to replacing the integrators with the following discrete-time system.

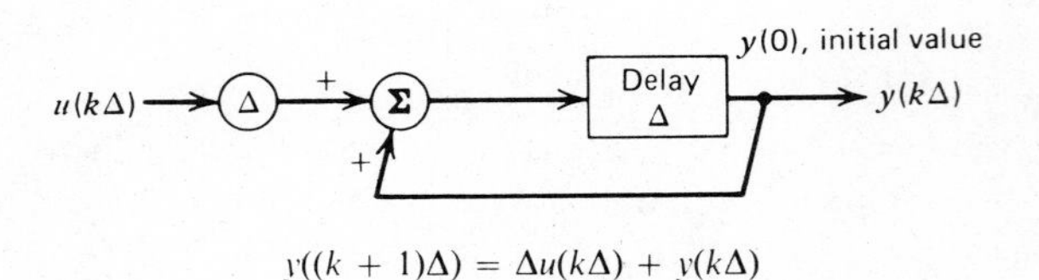

$$y((k+1)\Delta) = \Delta u(k\Delta) + y(k\Delta)$$

(a) Convert the analog diagram for the differential equation to a discrete-time system.

(b) Plot the solution $y(t)$ for $0 \le t \le 2\pi$. Plot on the same paper the following approximations.

$$y(k\Delta) \text{ for } \Delta = \frac{\pi}{2} \qquad k = 0, 1, \ldots, 4$$

$$y(k\Delta) \text{ for } \Delta = \frac{\pi}{4} \qquad k = 0, 1, \ldots, 8$$

$$y(k\Delta) \text{ for } \Delta = \frac{\pi}{8} \qquad k = 0, 1, \ldots, 16$$

2.8. In Problem 2.7, replace the discrete-time approximation to an integrator with the following.

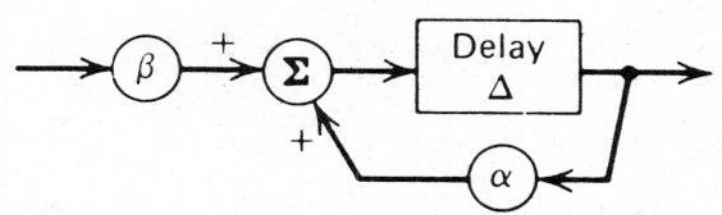

That is, consider the following discrete-time system.

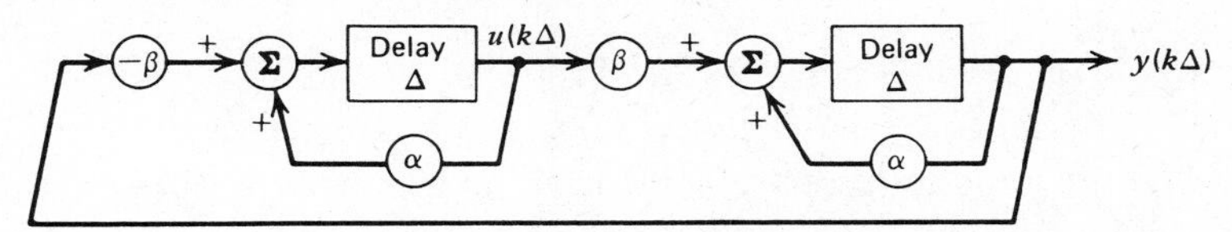

With $\alpha = 1$ and $\beta = \Delta$ the above circuit is the Euler approximation. Choose α, β, $u(0)$, $y(0)$ so that $y(k\Delta) = \cos(k\Delta)$ exactly.

2.9. Consider the following finite-length impulse-response circuit. Choose α, β, γ so that the steady-state response to $u_k = \xi_k$ is $y_k = 1$ (where $\{\xi_k\}$ is the unit step sequence) and the response to $u_k = \cos(k/10)$ is $y_k = 0$. Find and plot the magnitude and phase of the frequency-response function for the resulting system.

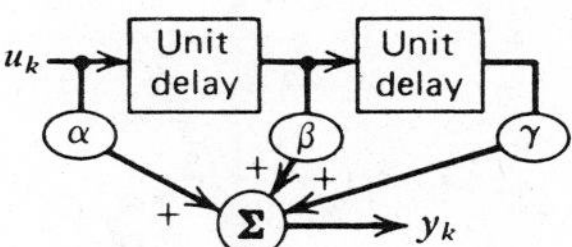

2.10. Find the difference equation relating the input $\{u_k\}$ to the output $\{y_k\}$ for the following discrete-time system. Find the frequency-response function and plot the magnitude and phase for $-\pi \le \theta \le \pi$.

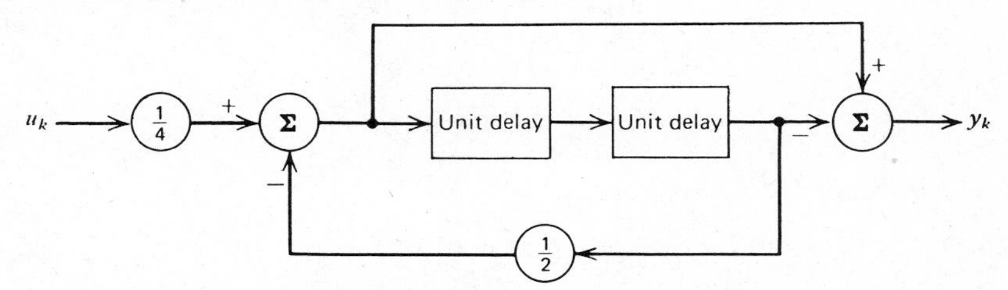

2.11. The output of a discrete-time system is y_1, where $y_1(k) = \sum_{n=0}^{\infty} u(k-n)e^{-\alpha n}$. Consider the discrete-time system shown below. Call the output y_2. Can the gains a and b be chosen so that $y_1(n) = y_2(n)$ for all n? Explain.

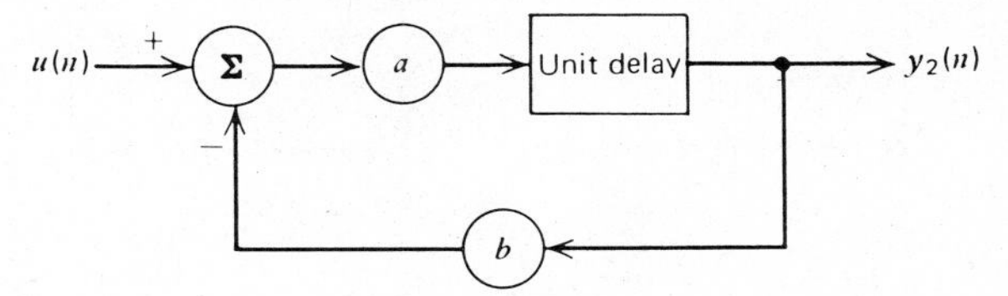

2.12. Sketch a block diagram of a system whose impulse-response sequence is:

(a) $\{1, \frac{1}{2}, \frac{1}{4}, \ldots, (\frac{1}{2})^k, \ldots\}$

(b) $\{1, 1, \frac{1}{2}, \frac{1}{2}, \frac{1}{4}, \frac{1}{4}, \ldots\}$

2.13. Find the impulse response for the discrete-time systems defined by the following difference equations. Verify your answers by substitution.

(a) $(S^2 - S + \frac{1}{4})[y_k] = u_k$

(b) $(S^2 - \frac{1}{4})[y_k] = u_k$

(c) $(1 - 3S^{-1} + 3S^{-2} - S^{-3})[y_k] = u_k$

(d) $(S^3 - 3S^2 + 3S - 1)[y_k] = u_k$

2.14. Use the impulse-response sequence to express the output of a discrete-time system described by the difference equation

$$(S^2 - 2S + 1)[y_k] = u_k$$

with $y_0 = 0$, $y_1 = 2$. Notice that the initial conditions are nonzero. One must modify our discussion of convolution to include the effects of nonzero initial conditions. Decompose the output into two outputs, one due to the impulse with zero initial conditions. Add to this output an output due entirely to the nonzero initial condition.

Answer: $y_k = 2k + \sum_{n=0}^{k-1} (k - n - 1)u_n$

2.15. Solve the following difference equation using both the direct method and the convolution method.

$$\left(1 - \frac{S^{-2}}{9}\right)[y_k] = \begin{cases} \left(\frac{1}{3}\right)^k, & k \geq 0 \\ 0, & k < 0 \end{cases}$$

with initial conditions zero.

2.16. Find the impulse-response sequences for the discrete-time systems of Problem 2.9 and 2.10.

2.17. Find the impulse-response sequence for the following discrete-time system:

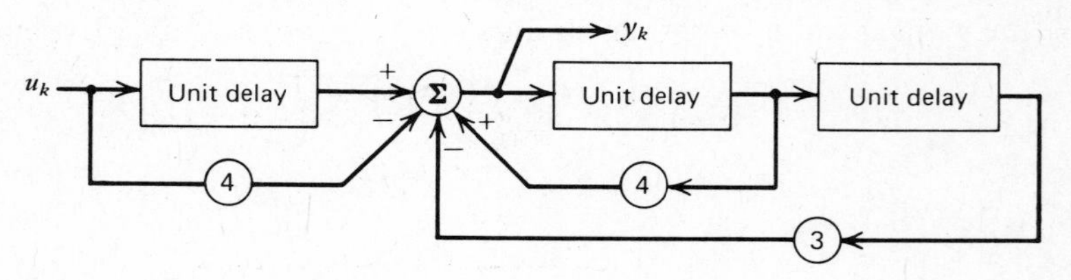

2.18. Find the response to an impulse-sequence input at points A, B, and C in the following system:

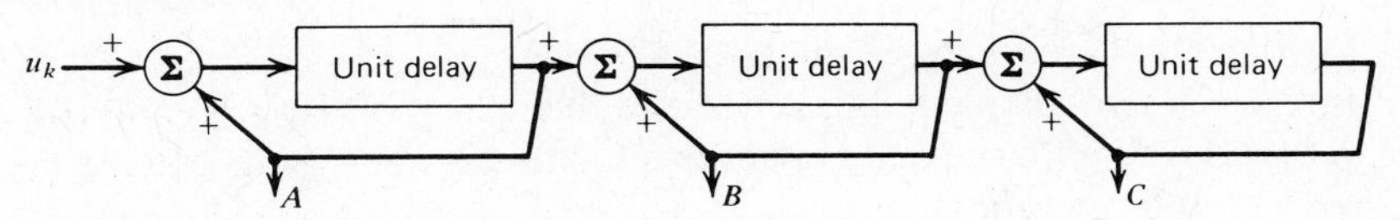

2.19. In the discrete-time system shown below, determine:

(a) The difference equation relating u and y.

(b) The impulse-response sequence.

(c) The frequency-response function $H(e^{j\theta})$. Sketch the magnitude function on the interval $[0, \pi]$.

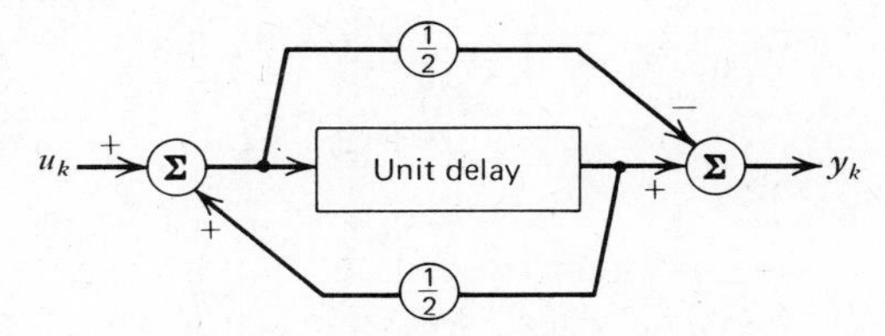

2.20. Write state-variable equations for the following discrete-time systems.

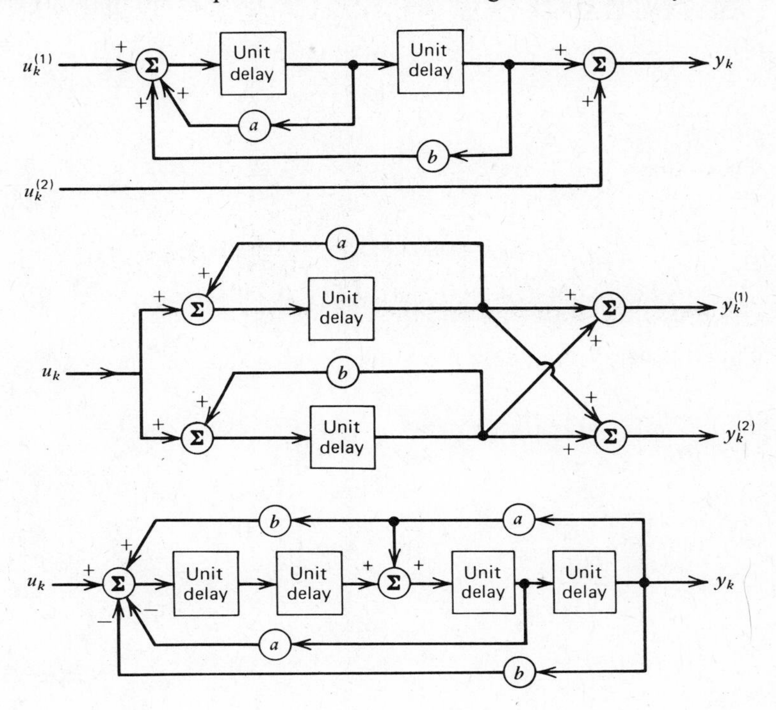

2.21. Write the state-variable equations for the following system, using a three-component state vector with the states x_1, x_2, and x_3, defined as shown.

(a) Evaluate the determinant $|\mathbf{A}|$.

(b) What are the eigenvalues of $\mathbf{A}$?

(c) Can you explain and generalize the results of parts (a) and (b)?

(d) Can you write state equations for this system using a two-component state vector? Comment.

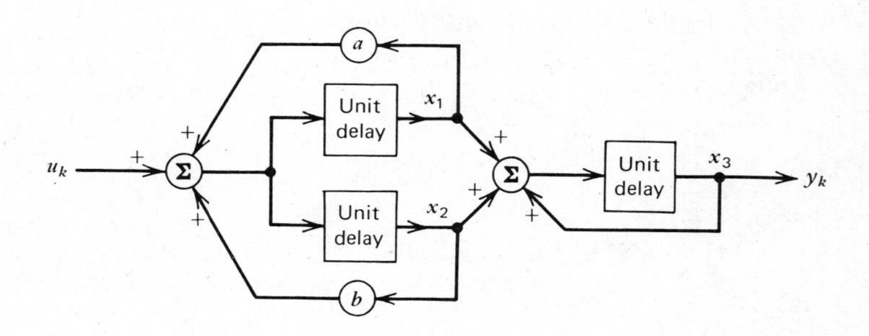

2.22. Assume that the function $f(\lambda)$ can be represented as an infinite series. Prove that $\mathbf{A}$ and $\mathbf{f}(\mathbf{A})$ commute, that is

$$\mathbf{A} \cdot \mathbf{f}(\mathbf{A}) = \mathbf{f}(\mathbf{A})\mathbf{A}$$

Prove that

$$\frac{d}{dt}[\mathbf{f}(\mathbf{A}t)] = \mathbf{A}\mathbf{g}(\mathbf{A}t), \qquad \text{where } g(t) = \frac{d}{dt} f(t)$$

2.23. Find a general expression for $\mathbf{A}^k$ for

(a) $\mathbf{A} = \begin{bmatrix} \frac{3}{4} & 0 \\ \frac{1}{2} & \frac{1}{2} \end{bmatrix}$ (b) $\mathbf{A} = \begin{bmatrix} \frac{1}{2} & \frac{1}{4} \\ \frac{1}{16} & \frac{1}{2} \end{bmatrix}$

(c) $\mathbf{A} = \begin{bmatrix} \frac{1}{2} & \frac{1}{2} \\ 1 & \frac{1}{2} \end{bmatrix}$ (d) $\mathbf{A} = \begin{bmatrix} \frac{1}{2} & 0 \\ \frac{1}{2} & \frac{1}{2} \end{bmatrix}$

(e) $\mathbf{A} = \begin{bmatrix} \frac{3}{4} & -\frac{1}{2} \\ -\frac{15}{32} & \frac{1}{2} \end{bmatrix}$

2.24. Consider the system of three difference equations:

$$x(n+1) = 3x(n) + 5y(n) + 2z(n)$$

$$y(n+1) = x(n) - y(n) + z(n), \qquad n = 0, 1, 2, \ldots$$

$$z(n+1) = 2x(n) + y(n) + 3(z)$$

Find $x(n)$, $y(n)$, $z(n)$ for all n for $x(0) = 1$, $y(0) = 0$, $z(0) = 0$.

2.25. Find the general solution for $x(n)$ and $y(n)$ for $x(0) = 1$ and $y(0) = 0$.

$$x(n+1) = x(n) + 2y(n)$$

$$n = 0, 1, 2, \ldots$$

$$y(n+1) = 3x(n) + 2y(n)$$

2.26. Find expressions for $\mathbf{A}^n$ for all n. Check your answer by explicitly calculating $\mathbf{A}$ and $\mathbf{A}^2$.

$$\mathbf{A} = \begin{bmatrix} \cos\theta & -\sin\theta \\ \sin\theta & \cos\theta \end{bmatrix}$$

$$\mathbf{A} = \begin{bmatrix} \alpha & -\beta \\ \beta & \alpha \end{bmatrix}$$

$$\mathbf{A} = \begin{bmatrix} \alpha & 0 \\ 0 & \alpha \end{bmatrix}$$

$$\mathbf{A} = \begin{bmatrix} \alpha & 1 \\ 0 & \alpha \end{bmatrix}$$

$$\mathbf{A} = \begin{bmatrix} \alpha & 1 \\ 0 & \alpha + \beta \end{bmatrix}$$

2.27. Determine the impulse response of the following discrete-time system.

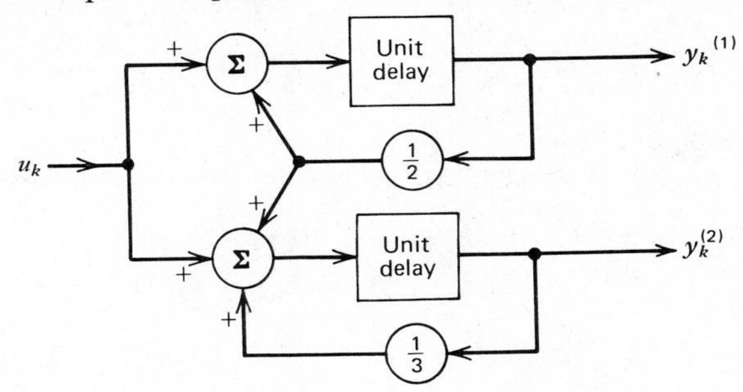

2.28. Determine and sketch a block diagram of the inverse system for the discrete-time system shown. The inverse system is defined as that system which produces u of the original system as an output for an input y.

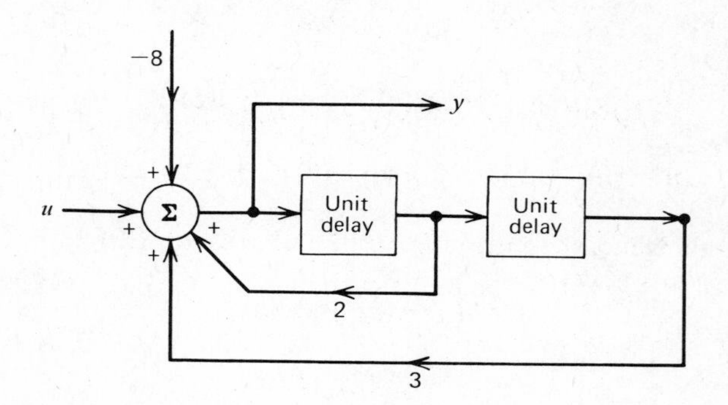

2.29. Consider a discrete-time system that is the cascade of a first- and second-order system, as shown in the sketch.

(a) Find a state-variable description for the entire system. Find a state-variable description for the second-order system.

(b) Find all values g for which the system is stable.

(c) Find the sequence at point S for an input $\{u_k\} = \{1, -a\}$.

(d) Find the output sequence for the input sequence $\{1, -a\}$.

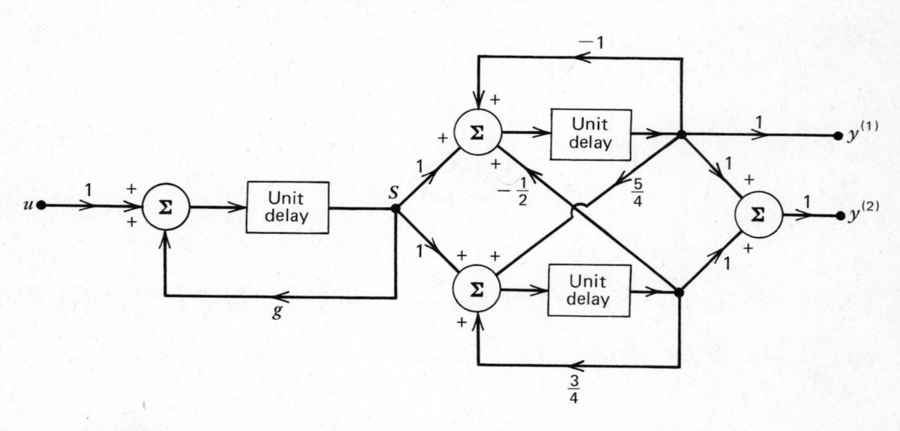

2.30. Sketch a block diagram of a discrete-time system with state-variable matrices

$$\mathbf{A} = \begin{bmatrix} 0 & 1 \\ \frac{1}{6} & \frac{5}{6} \end{bmatrix}, \qquad \mathbf{B} = \begin{bmatrix} 0 \\ 1 \end{bmatrix}, \qquad \mathbf{C} = [3, -2], \qquad \mathbf{D} = [0]$$

Suppose we change the state vector via the mapping $\mathbf{x}' = \mathbf{T}^{-1}\mathbf{x}$, where

$$\mathbf{T} = \begin{bmatrix} \frac{1}{2} & 0 \\ 0 & \frac{1}{3} \end{bmatrix}$$

Sketch a block diagram of the transformed system. Which system contains fewer multipliers?

Continuous-time linear systems can be described in several ways. In this chapter we present three time-domain descriptions. Our discussions will parallel those of discrete-time systems given in Chapter 2. The three models we shall discuss in this chapter are:

i. A linear differential equation.
ii. The impulse-response function.
iii. The state-variable formulation.

Each of these models can be used to find the output function of a continuous-time system. As is true of discrete-time systems, the state-variable model gives us additional information about the internal states of the system. We shall begin with a brief review of linear differential equations.

3.1 LINEAR DIFFERENTIAL EQUATIONS

An ordinary linear differential equation with constant coefficients is characteristic of linear, lumped, constant-parameter systems. In this section, we shall briefly review the theory of solving this class of differential equations. Suppose that we have a differential equation of order n:

$$b_n \frac{d^n y(t)}{dt^n} + b_{n-1} \frac{d^{n-1} y(t)}{dt} + \cdots + b_1 \frac{dy(t)}{dt} + y(t) = u(t) \tag{3.1}$$

It is convenient to write (3.1) in operational form as

$$(b_n D^n + b_{n-1} D^{n-1} + \cdots + b_1 D + 1)[y(t)] = u(t) \tag{3.2}$$

where

$$D \triangleq \frac{d}{dt}$$

We shall also use the notation

$$L[y] = u(t) \tag{3.3}$$

where L is the operator

$$L = b_n D^n + b_{n-1} D^{n-1} + \cdots + b_1 D + 1 \tag{3.4}$$

It is easy to verify that L is a linear operator, that is,†

$$L[c_1 y^1 + c_2 y^2] = c_1 L[y^1] + c_2 L[y^2]$$

Thus, if y^1 and y^2 are solutions to $L[y] = 0$, then $c_1 y^1 + c_2 y^2$ is also a solution of $L[y] = 0$.

The general solution of (3.1) consists of two components: (1) the homogeneous (transient, source-free, natural, complementary) solution, and (2) the component due to the source $u(t)$ (steady-state, nonhomogeneous, particular solution). The homogeneous solution $y^{(h)}$ is calculated using the homogeneous equation corresponding to (3.1). Thus $y^{(h)}$ satisfies

$$L[y^{(h)}] = 0 \tag{3.5}$$

† We shall use the notation $y^{(n)}(t)$ in this chapter to denote the nth time derivative of $y(t)$. Thus $y^{(1)}(t) = (d/dt)y(t)$, $y^{(2)}(t) = d^2 y(t)/dt^2$, etc. The context of the discussion should resolve any possible confusion of notation.

For the case of constant coefficients $y^{(h)}$ has the form

$$y^{(h)}(t) = c_1 y_1(t) + c_2 y_2(t) + \cdots + c_n y_n(t) \tag{3.6}$$

The functions $y_i(t)$, $i = 1, 2, \ldots, n$ depend on the roots of the associated auxiliary equation

$$f(r) = b_n r^n + b_{n-1} r^{n-1} + \cdots + b_1 r + 1 = 0 \tag{3.7}$$

This equation is obtained by substituting the trial solution $y(t) = e^{rt}$ into (3.5). If (3.7) has n distinct roots r_i, $i = 1, 2, \ldots, n$ the functions $y_i(t)$ are $e^{r_i t}$ and the homogeneous solution is

$$y^{(h)}(t) = c_1 e^{r_1 t} + c_2 e^{r_2 t} + \cdots + c_n e^{r_n t} \tag{3.8}$$

The functions $y_i(t)$ depend on the multiplicity of the roots in (3.7). We summarize the results below. After finding the n roots of (3.7), assign the functions $y_i(t)$ as follows:

1. For each distinct real root, the function e^{rt}.
2. For each multiple real root of multiplicity p, the functions e^{rt}, $te^{rt}, \ldots, t^{p-1}e^{rt}$.
3. For each distinct complex pair of roots $a \pm jb$, the functions $e^{at} \cos bt$, and $e^{at} \sin bt$.
4. For each complex pair of roots $a \pm jb$ of multiplicity p, the functions $e^{at} \cos bt$, $e^{at} \sin bt$, $te^{at} \cos bt$, $te^{at} \sin bt, \ldots, t^{p-1} e^{at} \cos bt$, $t^{p-1} e^{at} \sin bt$.

Notice that (3) and (4) are special cases of (1) and (2). The constants c_i, $i = 1, 2, \ldots, n$ in (3.6) are found using either initial or boundary conditions of the problem.

Example 3.1 Consider the differential equation

$$(D^3 - D^2 + D - 1)[y(t)] = 0$$

The auxiliary equation is

$$f(r) = r^3 - r^2 + r - 1 = 0$$

which has roots $r_1 = j$, $r_2 = -j$, $r_3 = 1$. The homogeneous solution function is therefore

$$\begin{aligned} y^{(h)}(t) &= c_1 e^t + c_2 e^{-jt} + c_3 e^{jt} \\ &= c_1 e^t + c_2 \cos t - jc_2 \sin t + c_3 \cos t + jc_3 \sin t \\ &= c_1 e^t + c_2' \cos t + c_3' \sin t \end{aligned}$$

We call $y^{(h)}(t)$ a *solution function* if the constants c_i, $i = 1, 2, \ldots, n$ are not specified. ■

The solution resulting from the forcing function can be found in several ways including educated guessing. We shall present the method of undetermined coefficients. This method can be used whenever successive derivatives of $u(t)$ result in a finite number of independent functions.

The process can be explained as in Section 2.3 for difference equations. We seek an operator L_A that "annihilates" $u(t)$. That is, we wish to find an L_A such that

$$L_A[u(t)] = 0 \tag{3.9}$$

Such an operator can always be found† if $u(t)$ is the solution of a homogeneous differential equation with constant coefficients. For example, if $u(t) = e^{at}$, then $L_A = D - a$ is an appropriate choice for the annihilator operator since

$$L_A[e^{at}] = (D - a)[e^{at}] = ae^{at} - ae^{at} = 0$$

If $u(t) = \alpha \cos bt + \beta \sin bt$, the annihilator is $L_A = (D - jb)(D + jb) = D^2 + b^2$. If L_{A_1} annihilates $u_1(t)$ and L_{A_2} annihilates $u_2(t)$, then $L_{A_1}L_{A_2}$ annihilates $\alpha u_1(t) + \beta u_2(t)$.

Once an annihilator operator has been found we apply it to both sides of the original differential equation. The result is a homogeneous equation. For example, suppose that L_A is the annihilator for $u(t)$ and we wish to solve

$$L[y(t)] = u(t) \tag{3.10}$$

Then

$$L_A[L[y(t)]] = L_A[u(t)] = 0 \tag{3.11}$$

† Note that L_A is not unique; generally we choose the simplest form that satisfies (3.9).

We can solve (3.11) using the method outlined previously for homogeneous equations. The solution function is of the form (cf. Equation 3.6)

$$y(t) = c_1 y_1(t) + \cdots + c_n y_n(t) + c_{p_1} y_{p_1}(t) + \cdots + c_{p_r} y_{p_r}(t) \tag{3.12}$$

The first n solutions satisfy $L[y] = 0$. Therefore, if we substitute (3.12) into (3.10), these terms will lead to zeros on the left-hand side. The remaining terms $c_{p_1} y_{p_1}(t) + \cdots + c_{p_r} y_{p_r}(t)$ arise from the forcing function or, alternately, from the operator L_A. If we now equate coefficients of like functions on both sides, we can evaluate the constants $c_{p_1}, \ldots, c_{p_r}$ and, thus, obtain the particular solution.

Example 3.2 Consider the differential equation

$$L[y(t)] = (D^2 + 1)[y(t)] = e^t$$

In this case, an annihilator for e^t is $(D - 1)$ because $(D - 1)[e^t] = 0$. Thus we operate on both sides by $(D - 1)$ and obtain the homogeneous equation

$$(D - 1)(D^2 + 1)[y(t)] = 0$$

We can solve this equation by means of the characteristic equation

$$(r - 1)(r^2 + 1) = 0$$

This equation has roots $1, j, -j$. Thus, the solution function is

$$y(t) = c_1 \cos t + c_2 \sin t + c_3 e^t$$

If we now substitute this solution function into the original equation, the first two terms are zero because they are solutions to $L[y] = 0$. We obtain one equation for c_3, the undetermined coefficient,

$$(D^2 + 1)[c_1 \cos t + c_2 \sin t + c_3 e^t] = e^t$$

Thus,

$$0 + (D^2 + 1)[c_3 e^t] = e^t$$
$$c_3 e^t + c_3 e^t = e^t$$

or

$$2c_3 e^t = e^t$$

Hence, $c_3 = \frac{1}{2}$, and the particular solution is

$$y^{(p)}(t) = \frac{1}{2}e^t$$

■

Example 3.3 Consider the differential equation

$$L[y(t)] = (D^2 + 1)[y(t)] = \sin t$$

In this problem, we choose the annihilator $(D^2 + 1)$. Thus, the homogeneous equation we wish to solve is

$$(D^2 + 1)(D^2 + 1)[y(t)] = 0$$

The corresponding auxiliary equation is

$$(r^2 + 1)(r^2 + 1) = 0$$

which has roots j, $-j$ each of multiplicity two. Hence, the solution function is

$$y(t) = c_1 \cos t + c_2 \sin t + c_3 t \cos t + c_4 t \sin t$$

Substituting this solution function into the original equation, we have

$$(D^2 + 1)[c_1 \cos t + c_2 \sin t + c_3 t \cos t + c_4 t \sin t] = \sin t$$

Performing the indicated differentiation and equating coefficients of like functions on both sides of the equation, we find that $c_4 = 0$ and $c_3 = -\frac{1}{2}$. Hence, $y^{(p)}(t) = -(t/2) \cos t$. If the initial conditions are $y(0) = 1$ and $y^{(1)}(0) = 0$, we can solve for c_1 and c_2 using our knowledge of the particular solution. We have

$$y(t) = c_1 \cos t + c_2 \sin t - \frac{1}{2} t \cos t$$

And so

$$y(0) = 1 = c_1$$

$$y^{(1)}(0) = 0 = c_2 - \frac{1}{2}$$

Thus, we know that

$$c_1 = 1$$

$$c_2 = \frac{1}{2}$$

and the complete solution is

$$y(t) = \cos t + \frac{1}{2}\sin t - \frac{t}{2}\cos t \qquad \blacksquare$$

These results can easily be generalized to treat a wider class of nonhomogeneous differential equations. As an extension of (3.1) and (3.2), suppose that we had a system, again with input $u(t)$ and output $y(t)$, described by

$$b_n \frac{d^n y(t)}{dt^n} + \cdots + b_1 \frac{dy(t)}{dt} + y(t) = a_m \frac{d^m u(t)}{dt^m} + \cdots + a_1 \frac{du(t)}{dt} + a_0 u(t)$$

or, in operational form,

$$(b_n D^n + \cdots + b_1 D + 1)[y(t)] = (a_m D^m + \cdots + a_1 D + a_0)[u(t)]$$

or

$$L[y(t)] = L_D[u(t)]$$

The solution to this modified equation may be found in two different ways. One approach is to define $\hat{u}(t) = (a_m D^m + \cdots + a_1 D + a_0)[u(t)] = L_D[u(t)]$ and then solve $L[y(t)] = \hat{u}(t)$ as outlined above. Alternatively, we could observe that if L_A annihilates $u(t)$, then, since the operators that we are discussing commute, L_A also annihilates $L_D[u(t)]$:

$$L_A\{L_D[u(t)]\} = L_D\{L_A[u(t)]\} = L_D\{0\} = 0$$

This means that the form of our solution remains exactly as given by (3.12). In this case, however, we apply $L[y(t)] = L_D[u(t)]$ to evaluate the constants $c_{p_1} \cdots c_{p_r}$.

Example 3.4 Let us modify the previous example to

$$(D^2 + 1)[y(t)] = (3D + 1)[u(t)]$$

with $u(t) = \sin t$ as before. The solution function remains:

$$y(t) = c_1 \cos t + c_2 \sin t + c_3 t \cos t + c_4 t \sin t$$

since L and L_A are unchanged. To evaluate c_3 and c_4, we set

$$(D^2 + 1)[c_3 t \cos t + c_4 t \sin t] = (3D + 1)[\sin t]$$

obtaining (after simplification)

$$2c_4 \cos t - 2c_3 \sin t = 3 \cos t + \sin t$$

Since this equality must hold for all t, we deduce that

$$2c_4 = 3$$

$$-2c_3 = 1$$

that is,

$$c_4 = \frac{3}{2}$$

$$c_3 = -\frac{1}{2}$$

As before, the initial conditions are applied to find c_1 and c_2. ■

3.2 FREQUENCY RESPONSE OF CONTINUOUS-TIME SYSTEMS

The frequency response of a continuous-time system is determined by the steady-state response to an input $e^{j\omega t}$. The output of a linear, constant-parameter system is always of the form $H(j\omega)e^{j\omega t}$. The system output is the same complex exponential modified in amplitude and phase by the system function $H(j\omega)$, which in general is complex-valued. $|H(j\omega)|$ is called the amplitude or magnitude response and $\arg [H(j\omega)]$ is the phase response.

Suppose that we have a system described by the differential equation

$$(b_n D^n + b_{n-1} D^{n-1} + \cdots + b_1 D + 1)[y(t)]$$
$$= (a_m D^m + a_{m-1} D^{m-1} + \cdots + a_0)[u(t)] \quad (3.13)$$

For an input $u(t) = e^{j\omega t}$ we claim that $H(j\omega)e^{j\omega t}$ is the steady-state solution, where $H(j\omega)$ is given by

$$H(j\omega) = \frac{a_0 + a_1 j\omega + \cdots + a_m(j\omega)^m}{1 + b_1 j\omega + \cdots + b_n(j\omega)^n} \quad (3.14)$$

In other words, we claim that $y(t) = H(j\omega)e^{j\omega t}$, with $H(j\omega)$ given by (3.14), satisfies (3.13).

$$\begin{aligned}
&(b_n D^n + \cdots + b_1 D + 1)\left[\frac{a_0 + a_1 j\omega + \cdots + a_m(j\omega)^m}{1 + b_1 j\omega + \cdots + b_n(j\omega)^n} e^{j\omega t}\right] \\
&= \frac{a_0 + a_1 j\omega + \cdots + a_m(j\omega)^m}{1 + b_1 j\omega + \cdots + b_n(j\omega)^n} \cdot (b_n(j\omega)^n + b_{n-1}(j\omega)^{n-1} + \cdots + 1)e^{j\omega t} \\
&= (a_0 + a_1 j\omega + \cdots + a_m(j\omega)^m)e^{j\omega t} \\
&= (a_m D^m + a_{m-1} D^{m-1} + \cdots + a_0)[e^{j\omega t}] \qquad (3.15)
\end{aligned}$$

Furthermore, this is the only such particular solution. Therefore, $H(j\omega)e^{j\omega t}$ is the unique steady-state solution for $u(t) = e^{j\omega t}$. Equation 3.14 is an important formula. It permits us to calculate $H(j\omega)$ directly from the differential equation model. We remind the reader that frequency-response functions are defined only for time-invariant systems.

If $u(t) = \mathrm{Re}(e^{j\omega t}) = \cos(\omega t)$, then we obtain a real output of the form

$$\begin{aligned}
y^{(p)}(t) &= \mathrm{Re}\,[H(j\omega)e^{j\omega t}] \\
&= \mathrm{Re}\,[|H(j\omega)|e^{j\arg [H(j\omega)]}e^{j\omega t}] \\
&= \mathrm{Re}\,[|H(j\omega)|e^{j(\omega t + \arg [H(j\omega)])}] \\
&= |H(j\omega)| \cos [\omega t + \arg [H(j\omega)]] \qquad (3.16)
\end{aligned}$$

Equation 3.16 is simply a statement that the input sinusoid $\cos(\omega t)$ is changed in amplitude and phase in accordance with the frequency-response function of the system.

Example 3.5 Consider the simple RC circuit shown in Figure 3.1. Assuming an input $u(t) = e_i(t)$ and an output $y(t) = e_0(t)$, what is the frequency response of the

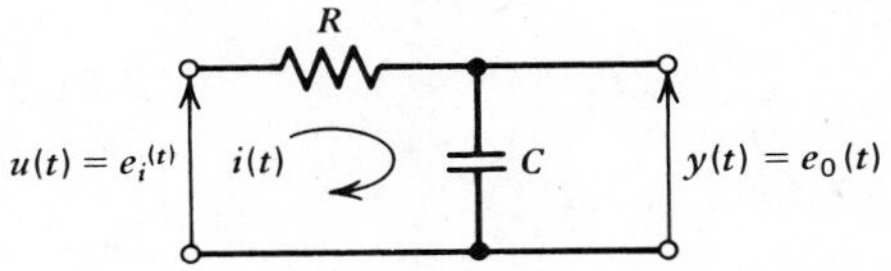

R

$u(t) = e_i(t)$ $i(t)$ C $y(t) = e_0(t)$

Figure 3.1

system? The differential equation relating $e_i(t)$ and $e_0(t)$ can be found by first writing the voltage equation around the input loop

$$
\begin{aligned}
u(t) &= Ri(t) + \frac{1}{C}\int_{-\infty}^{t} i(t')\,dt' \\
&= Ri(t) + y(t)
\end{aligned}
\tag{3.17}
$$

Solving for $i(t)$, we obtain

$$i(t) = \frac{u(t) - y(t)}{R}$$

Differentiating both sides of (3.17) gives

$$\frac{du(t)}{dt} = R\frac{di(t)}{dt} + \frac{1}{C}i(t) \tag{3.18}$$

and so

$$\frac{du(t)}{dt} = R\frac{d}{dt}\left[\frac{u(t) - y(t)}{R}\right] + \frac{1}{C}\left[\frac{u(t) - y(t)}{R}\right]$$

Collecting terms,

$$\frac{dy(t)}{dt} + \frac{1}{RC}y(t) = \frac{1}{RC}u(t)$$

The frequency-response function using (3.14) is

$$H(j\omega) = \frac{\dfrac{1}{RC}}{\dfrac{1}{RC} + j\omega} = \frac{1}{1 + j\omega RC} = \frac{1 - j\omega RC}{1 + (\omega RC)^2} \tag{3.19}$$

The amplitude response is

$$|H(j\omega)| = \left[\frac{1 + (\omega RC)^2}{[1 + (\omega RC)^2]^2}\right]^{1/2} = \left[\frac{1}{1 + (\omega RC)^2}\right]^{1/2}$$

The phase response is

$$\arg [H(j\omega)] = -\tan^{-1} (\omega RC)$$

■

3.3 CONVOLUTION—THE IMPULSE FUNCTION

We turn now to the development of convolution techniques for the analysis of continuous-time systems. As with discrete-time systems, we shall decompose the input $u(t)$ into a sum of impulse functions and then express the output $y(t)$ as a sum of the responses resulting from the individual impulses. First, however, we must define an appropriate continuous-time impulse function.

We begin by approximating an arbitrary function $u(t)$ by a series of pulses, as shown in Figure 3.2. The approximation for $u(t)$ is

$$u(t) \cong \sum_{n=-\infty}^{\infty} u(n\Delta)p_\Delta(t - n\Delta)$$

where $p_\Delta(t)$ is a pulse of unit height and width Δ as shown in Figure 3.3. This approximation of Figure 3.2 becomes better as Δ is decreased and more pulses are

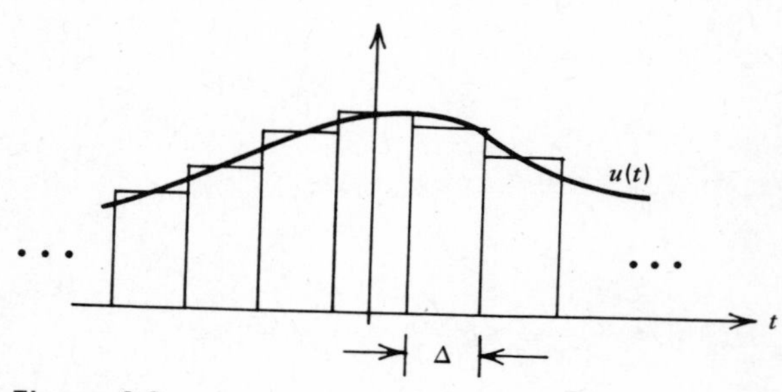

Figure 3.2

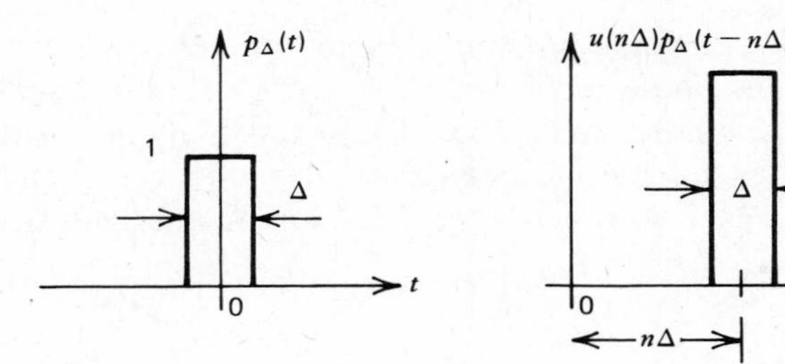

Figure 3.3

used in the representation of $u(t)$. In the limit as Δ goes to zero, we write the sum as an integral. In symbols we have

$$\begin{aligned} u(t) &= \lim_{\Delta\to 0} \sum_{n=-\infty}^{\infty} u(n\Delta)p_\Delta(t-n\Delta) \\ &= \lim_{\Delta\to 0} \sum_{n=-\infty}^{\infty} u(n\Delta)\left[\frac{1}{\Delta}p_\Delta(t-n\Delta)\right]\Delta \\ &= \int_{-\infty}^{\infty} u(\tau)\delta(t-\tau)\,d\tau \end{aligned} \tag{3.20}$$

Here $n\Delta$ is replaced by the continuous variable τ, Δ becomes $d\tau$, and we define

$$\delta(t-\tau) = \lim_{\Delta\to 0} \frac{1}{\Delta} \cdot p_\Delta(t-n\Delta)$$

or equivalently,

$$\delta(t) = \lim_{\Delta\to 0} \frac{1}{\Delta} p_\Delta(t) \tag{3.21}$$

The limiting process of (3.21) is shown schematically in Figure 3.4.

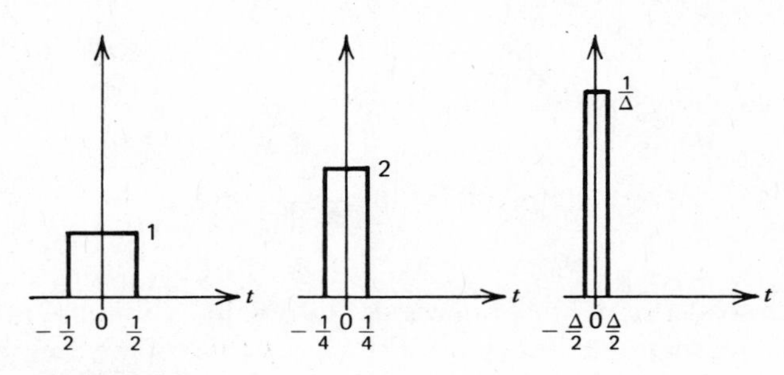

Figure 3.4

The delta function or impulse function, denoted by $\delta(t)$, is, roughly speaking, a pulse of unbounded amplitude and zero duration. It is the same kind of abstraction as a point charge or a point mass. This impulse function must be treated as a so-called generalized function, because we cannot define its value point by point as with ordinary functions. Referring to Figure 3.4, we can infer that $\delta(t)$ possesses the following properties:

1. $\delta(t) = 0$ for $t \neq 0$

2. $\delta(t)$ is undefined for $t = 0$

3. $\int_{-\infty}^{\infty} \delta(t)\, dt = 1$

One can obtain the impulse function by using a limiting process on other functions. In fact, we can base a sequence of functions $\{s_k(t)\}$ on any nonnegative function $s(t)$ for which $\int_{-\infty}^{\infty} s(t)\, dt = 1$. Defining $s_k(t) = ks(kt)$, we obtain the representation

$$\begin{aligned} \delta(t) &= \lim_{k \to \infty} s_k(t) \\ &= \lim_{k \to \infty} ks(kt) \end{aligned} \tag{3.22}$$

For example, consider the function

$$s(t) = \frac{1}{\pi(1 + t^2)}$$

This function is nonnegative for all t and has unit area. Thus the limit of the sequence of functions $s_k(t)$ given by

$$s_k(t) = ks(kt) = \frac{k}{\pi(1 + k^2 t^2)}$$

is an impulse function. That is,

$$\delta(t) = \lim_{k \to \infty} s_k(t) = \lim_{k \to \infty} \left[\frac{k}{\pi(1 + k^2 t^2)} \right]$$

The particular sequence of functions we use to obtain an impulse function is not really critical. Our main interest concerns the properties of the limiting form, that is, $\delta(t)$.

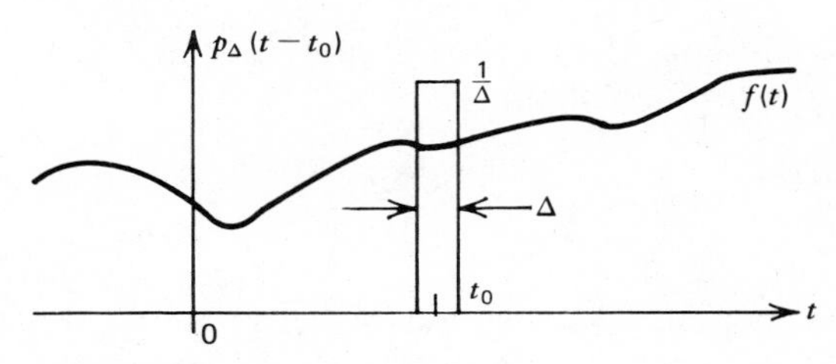

Figure 3.5

The three properties we listed previously for the impulse function can be conveniently summarized into one defining equation for $\delta(t)$. This equation is

$$\int_{-\infty}^{+\infty} f(t)\delta(t - t_0)\, dt = f(t_0) \tag{3.23}$$

provided $f(t)$ is continuous at $t = t_0$. One might justify (3.23) as follows. Let $f(t)$ be any function continuous at $t = t_0$. Consider a narrow pulse centered at $t = t_0$ as shown in Figure 3.5. For small Δ, we can write

$$\int_{-\infty}^{\infty} f(t)\left[\frac{1}{\Delta} p_\Delta(t - t_0)\right] dt = \frac{1}{\Delta}\int_{t_0 - \Delta/2}^{t_0 + \Delta/2} f(t)\, dt \cong f(t_0) \tag{3.24}$$

The approximation occurs because $f(t)$ may not have constant slope in the neighborhood of t_0. As Δ is decreased, however, the approximation becomes better, and in the limit (3.24) becomes

$$\lim_{\Delta \to 0} \int_{-\infty}^{\infty} f(t)\left[\frac{1}{\Delta} p_\Delta(t - t_0)\right] dt = f(t_0) \tag{3.25}$$

Hence, we arrive at (3.23) for our implicit definition of $\delta(t)$.

Generalized Functions

We digress briefly here to develop certain properties of the impulse function and related generalized functions. We shall use these properties in our discussion of convolution for continuous-time systems.

If we are restricted to using $\delta(t)$ only within an integral as part of the integrand, then how can we determine the properties of $\delta(t)$? One method is to use the concept of testing functions. A testing function $\theta(t)$ is a function that is continuous, has continuous derivatives of all orders, and is zero outside a finite interval. One class of testing functions is

$$\theta(t) = \begin{cases} e^{-\alpha^2/(\alpha^2 - t^2)}, & |t| < \alpha \\ 0, & \text{otherwise} \end{cases} \tag{3.26}$$

One uses testing functions as a method of "examining" the impulse function (Figure 3.5). This procedure is, roughly speaking, analogous to using the output of a measuring instrument to deduce properties about what is being measured. Consider the result of integrating the product $\theta(t) \cdot \delta(t)$. Call this result $F(\theta)$:

$$F(\theta) = \int_{-\infty}^{\infty} \theta(t)\delta(t)\, dt \tag{3.27}$$

$F(\theta)$ is a generalization of the concept of a function and is called a linear functional on the space of testing functions $\theta(t)$. Equation 3.27 assigns the number $F(\theta)$ to the function $\theta(t)$. In this case, we assign to each function $\theta(t)$ the value $F(\theta) = \theta(0)$. Let us see how this concept can be used to aid us in determining the properties of $\delta(t)$.

The Equivalence Property

One of the basic definitions we use again and again in studying impulse functions by means of testing functions is the so-called equivalence property. Suppose that $d_1(t)$ and $d_2(t)$ are expressions involving impulse functions and other functions. We define $d_1(t) = d_2(t)$ if and only if

$$\int_{-\infty}^{\infty} \theta(t)d_1(t)\, dt = \int_{-\infty}^{\infty} \theta(t)d_2(t)\, dt \tag{3.28}$$

for all testing functions $\theta(t)$ for which the integral exists. Roughly speaking, $d_1(t) = d_2(t)$ if the "examining instrument" can detect no differences between them.

Example 3.6 To illustrate the use of (3.28), let us demonstrate that $\delta(t)$ can be written as the derivative of the unit step function $\xi(t)$. The unit step function is defined as

$$\xi(t) = \begin{cases} 1, & t \geq 0 \\ 0, & t < 0 \end{cases} \tag{3.29}$$

To satisfy the equivalence property, we must show that

$$\int_{-\infty}^{\infty} \delta(t)\theta(t)\,dt = \int_{-\infty}^{\infty} \frac{d\xi(t)}{dt}\,\theta(t)\,dt \tag{3.30}$$

Integrating (3.30) on the right by parts, we have

$$\begin{aligned}\int_{-\infty}^{\infty} \frac{d\xi(t)}{dt}\,\theta(t)\,dt &= \xi(t)\theta(t)\Big|_{-\infty}^{\infty} - \int_{-\infty}^{\infty} \xi(t)\,\frac{d\theta(t)}{dt}\,dt \\ &= \xi(t)\theta(t)\Big|_{-\infty}^{\infty} - \int_{0}^{\infty} \frac{d\theta(t)}{dt}\,dt \\ &= \theta(\infty) - 0 - \theta(t)\Big|_{0}^{\infty} = \theta(0)\end{aligned}$$

Because the left-hand side is by definition also equal to $\theta(0)$, the equivalence of $\delta(t)$ and $(d\xi/dt)(t)$ is proved. ■

Example 3.7 Another useful equivalence is

$$f(t)\delta(t) = f(0)\delta(t) \tag{3.31}$$

provided $f(t)$ is continuous at $t = 0$. Equation 3.31 can be seen from

$$\begin{aligned}\int_{-\infty}^{\infty} f(t)\delta(t)\theta(t)\,dt &= \int_{-\infty}^{\infty} \delta(t)f(t)\theta(t)\,dt \\ &= f(0)\theta(0) \\ &= \int_{-\infty}^{\infty} f(0)\delta(t)\theta(t)\,dt\end{aligned}$$

Using (3.31) we find, for example, that

$$Ae^{t}\delta(t) = A\delta(t)$$

$$e^{t}\cos t\delta(t) = \delta(t)$$

$$A\sin t\delta(t) = A\sin(0)\delta(t) = 0$$

■

Higher-Order Derivatives of $\delta(t)$

Denote the nth derivative of $\delta(t)$ as $\delta^{(n)}(t)$. We claim that

$$\int_{-\infty}^{\infty} \delta^{(n)}(t)\theta(t)\,dt = (-1)^n \theta^{(n)}(t)\Big|_{t=0} = (-1)^n \theta^{(n)}(0) \tag{3.32}$$

We obtain (3.32) by integrating by parts n times. Thus

$$\begin{aligned}\int_{-\infty}^{\infty} \delta^{(n)}(t)\theta(t)\,dt &= \delta^{(n-1)}(t)\theta(t)\Big|_{-\infty}^{\infty} - \int_{-\infty}^{\infty} \delta^{(n-1)}(t)\theta^{(1)}(t)\,dt \\ &= -\delta^{(n-2)}(t)\theta^{(1)}(t)\Big|_{-\infty}^{\infty} - \int_{-\infty}^{\infty} \delta^{(n-2)}(t)\theta^{(2)}(t)\,dt \\ &\;\;\vdots \\ &= (-1)^n \theta^{(n)}(0)\end{aligned}$$

3.4 CONVOLUTION FOR CONTINUOUS-TIME SYSTEMS

We turn our attention now to a discussion of convolution in continuous-time systems. We shall again show that we can characterize the input-output relationship for linear time-invariant systems by a convolution operation involving the input and the impulse-response function of the system.

The impulse response $h(t)$ is defined as the output resulting from an input of $\delta(t)$, that is,

$$\delta(t) \to h(t)$$

If we apply an input impulse of area k, then by the linearity of the system, the output is $kh(t)$.

$$k\delta(t) \to kh(t)$$

Because the system is time-invariant,

$$k\delta(t - t_0) \to kh(t - t_0)$$

These relationships are shown schematically in Figure 3.6.

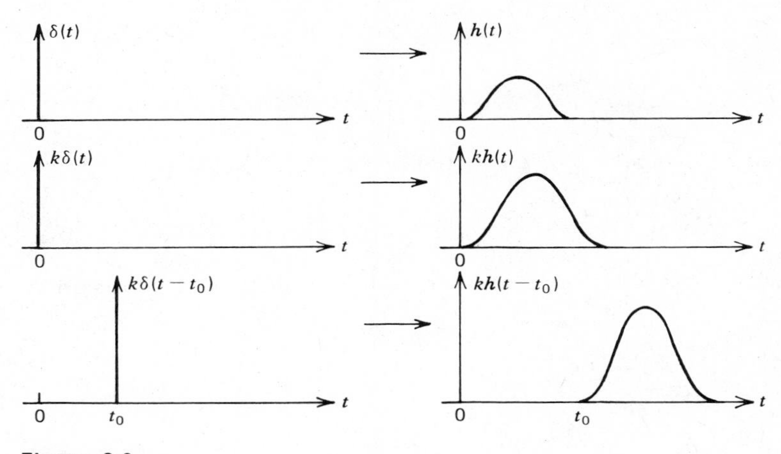

Figure 3.6

To find the response of a continuous-time system to an arbitrary input $u(t)$, we represent $u(t)$ by a train of impulse functions. Recall our representation of $u(t)$ in (3.20):

$$u(t) = \lim_{\Delta \to 0} \sum_{n=-\infty}^{\infty} u(n\Delta)\left[\frac{1}{\Delta} p_\Delta(t - n\Delta)\right]\Delta \tag{3.33}$$

Let us replace $(1/\Delta)p_\Delta(t - n\Delta)$ by its limit, $\delta(t)$. The representation for $u(t)$ is then

$$u(t) = \lim_{\Delta \to 0} \sum_{n=-\infty}^{\infty} u(n\Delta)\delta(t - n\Delta)\Delta \tag{3.34}$$

as before. Now we have the same sort of decomposition for the continuous-time signal $u(t)$ as we had for the sequence $u(k)$ in (2.22).

If we apply the impulse train representation for $u(t)$ to the system, we can evaluate the output response by calculating the response that results from each impulse separately and then adding these individual responses to obtain the complete output. This method of finding the output is possible because of the superposition property of linear systems.

The response to each impulse is quickly found as follows. The impulse at $t = 0$ gives rise to an output

$$\Delta u(0)\delta(t) \to \Delta u(0)h(t)$$

Similarly, the impulse at $t = \Delta$ gives rise to an output

$$\Delta u(\Delta)\delta(t - \Delta) \rightarrow \Delta u(\Delta)h(t - \Delta)$$

In general, the impulse at $t = n\Delta$ produces an output

$$\Delta u(n\Delta)\delta(t - n\Delta) \rightarrow \Delta u(n\Delta)h(t - n\Delta)$$

The complete response $y(t)$ is the sum of these individual responses: that is,

$$y(t) = \sum_{n=-\infty}^{\infty} \Delta u(n\Delta)h(t - n\Delta) \tag{3.35}$$

As we allow $\Delta \rightarrow 0$ and the number of impulses to grow, $n \rightarrow \infty$, so that $(n\Delta)$ becomes a continuous variable τ, the sum of (3.35) approaches an integral. The output response $y(t)$ resulting from an input $u(t)$ is thus

$$y(t) = \int_{-\infty}^{\infty} u(\tau)h(t - \tau)\, d\tau \tag{3.36}$$

Equation 3.36 is called the convolution of $u(t)$ and $h(t)$. We use the notation $y = u * h$ to denote convolution.

We can also arrive at (3.36) by a somewhat different argument. The input waveform $u(t)$ can be approximated by

$$u(t) \cong \sum_{n=-\infty}^{\infty} \Delta u(n\Delta)\delta(t - n\Delta)$$

In the limit as $\Delta \rightarrow 0$, the sum becomes an integral, so that

$$u(t) = \int_{-\infty}^{+\infty} u(\tau)\delta(t - \tau)\, d\tau \tag{3.37}$$

We interpret (3.37) as representing $u(t)$ by a continuum of impulse functions of the appropriate area. The linearity of the system allows us to calculate the system response by adding the responses that result from the continuum of input impulse functions. If $h(t)$ is the response to $\delta(t)$, then it follows that

$$y(t) = \int_{-\infty}^{+\infty} u(\tau)h(t - \tau)\, d\tau \tag{3.38}$$

The expression in (3.38) is not always easy to evaluate analytically. It is, in fact, often so involved that we use alternate solution methods based on transform techniques. Nevertheless, (3.38) is a strong conceptual aid in understanding linear systems.

As in the discrete-time case, a graphical interpretation of the convolution operation is often an aid to understanding how the system modifies input signals to obtain the output signal $y(t)$. Consider the convolution of $u(t)$ and $h(t)$

$$y = u * h$$
$$= \int_{-\infty}^{+\infty} u(\tau)h(t - \tau)\,d\tau \tag{3.39}$$

Suppose for the moment that $h(t)$ is known. Equation 3.39 then gives a function $y(t)$ for each function $u(t)$ that we substitute into this equation. To calculate a single

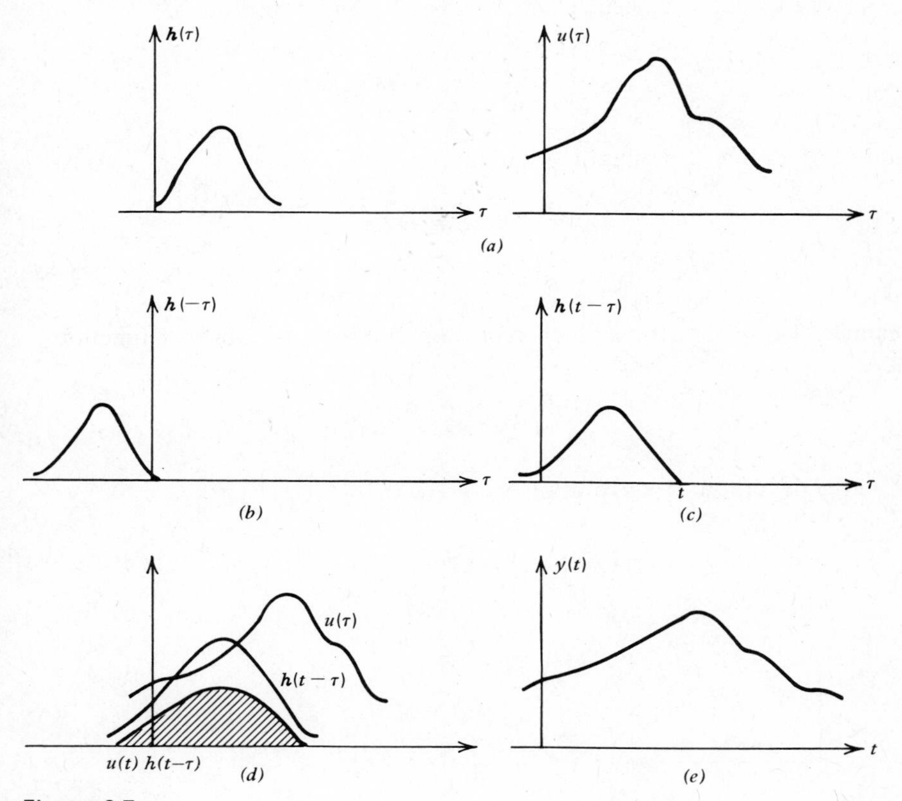

Figure 3.7

point on $y(t)$, say $y(t_1)$, we must know $u(t)$ over its complete range of t, because

$$y(t_1) = \int_{-\infty}^{+\infty} u(\tau)h(t_1 - \tau)\, d\tau$$

We can demonstrate these ideas graphically as shown in Figure 3.7. Figure 3.7*a* depicts a given $h(\tau)$ and $u(\tau)$. The convolution as written in (3.39) uses $h(t - \tau)$ in the integrand. The function $h(-\tau)$, shown in Figure 3.7*b*, is merely the mirror image of $h(\tau)$ about the line $\tau = 0$; and $h(t - \tau)$ is, for $t > 0$, the function $h(-\tau)$ shifted to the right by t. This shift is shown in Figure 3.7*c*. To calculate $y(t)$, we multiply $h(t - \tau)$ and $u(\tau)$ and integrate the product function, shown shaded in Figure 3.7*d*. The area under this shaded function is then $y(t)$. Notice that we obtain only a single value for $y(t)$ by this process. To obtain the graph of $y(t)$ for all t, we must allow the variable t in $h(t - \tau)$ to take on all values in the interval $(-\infty, \infty)$. The result is a smoothed version of $u(\tau)$ as shown in Figure 3.7*e*.

If we let $r = t - \tau$, then substituting in (3.39) we obtain

$$y(t) = \int_{-\infty}^{+\infty} u(t - r)h(r)\, dr$$

which implies that convolution is commutative: that is,

$$y = h * u = u * h$$

Example 3.8 Convolve a delta function $\delta(t)$ with an arbitrary function $f(t)$; that is, find

$$y(t) = \delta(t) * f(t)$$

From the definition of convolution, we know that

$$y(t) = \int_{-\infty}^{+\infty} f(\tau)\delta(t - \tau)\, d\tau = f(t)$$

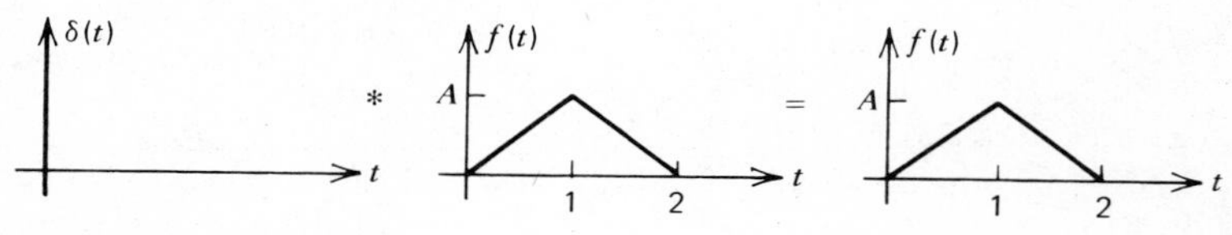

Figure 3.8

In words, convolution of a function $f(t)$ with $\delta(t)$ reproduces the function $f(t)$. This fact is shown schematically in Figure 3.8. ■

Example 3.9 Define $f(t)$ and $g(t)$ as

$$f(t) = \begin{cases} e^{-t}, & t \geq 0 \\ 0, & t < 0 \end{cases} \qquad g(t) = \begin{cases} \alpha e^{-\alpha t}, & t \geq 0 \\ 0, & t < 0 \end{cases}$$

Find $y = f * g$. From the definition we have

$$y(t) = \int_{-\infty}^{+\infty} f(\tau)g(t - \tau)\, d\tau$$

$$= \begin{cases} \displaystyle\int_0^t e^{-\tau}\alpha e^{-\alpha(t-\tau)}\, d\tau, & t \geq 0 \\ 0, & t < 0 \end{cases} = \begin{cases} \alpha e^{-\alpha t} \displaystyle\int_0^t e^{\tau(\alpha - 1)}\, d\tau, & t \geq 0 \\ 0, & t < 0 \end{cases}$$

$$= \begin{cases} \dfrac{\alpha}{\alpha - 1}(e^{-t} - e^{-\alpha t}), & \alpha \neq 1 \quad t \geq 0 \\ te^{-t}, & \alpha = 1, \\ 0, & t < 0 \end{cases}$$

In convolution problems, it is often easy to omit various cases by failing to calculate $y(t)$ for all t. A sketch of the various integrations that are implied by the convolution is often useful. We would recommend that one always sketch the two functions $f(\tau)$ and $g(t - \tau)$ to determine the correct limits and regions of integration, for example, as in Figure 3.7. ■

3.5 SOME GENERALIZATIONS OF CONVOLUTION FOR CONTINUOUS-TIME SYSTEMS

The limits of integration for convolution are, in general, $(-\infty, \infty)$. However, we often deal with causal inputs u and causal impulse functions h. If h is causal, that is, $h(t) = 0$, $t < 0$, then $h(t - \tau) = 0$ for $\tau > t$. In this case, the upper limit in (3.36) is actually t. If u is also causal, the lower limit can be replaced with 0. There are four possible cases summarized below.

$$y = u * h = \int_{-\infty}^{\infty} u(\tau)h(t - \tau)\, d\tau, \qquad \text{noncausal } u \text{ and } h$$

$$= \int_{0}^{t} u(\tau)h(t - \tau)\, d\tau, \qquad \text{causal } u \text{ and } h$$

$$= \int_{0}^{\infty} u(\tau)h(t - \tau)\, d\tau, \qquad \text{causal } u \text{ and noncausal } h$$

$$= \int_{-\infty}^{t} u(\tau)h(t - \tau)\, d\tau, \qquad \text{noncausal } u \text{ and causal } h$$

The derivation of the convolution integral for linear systems has been obtained by decomposing the input signal into a continuous sum of impulse functions. Implicit in this derivation is the assumption that the system is initially deenergized. If the system contains some initial stored energy, then an equivalent input that produces these initial stored energy sources must be included as an independent input function to the system. The total response is merely the sum of the responses that result from the input signal and the initial energy sources.

Relationships between the Step Response and Impulse Response

The step response of a linear system, denoted by $g(t)$, is the output that results from a step function input $\xi(t)$. In symbols, if $H(\cdot)$ represents the transformation performed by the linear system, then

$$g(t) = H[\xi(t)]$$

The step response $g(t)$ can be obtained by convolving $\xi(t)$ and $h(t)$. That is,

$$g = \xi * h$$

$$= \int_{-\infty}^{+\infty} \xi(\tau)h(t - \tau)\, d\tau \tag{3.40}$$

Because $\xi(t)$ is zero for $t < 0$, (3.40) can be written as

$$g(t) = \int_0^{\infty} h(t - \tau)\, d\tau$$

$$= \int_{-\infty}^{t} h(\tau)\, d\tau \tag{3.41}$$

Equation 3.41 states that the step response of a linear system is the integral of the impulse response. We can generalize (3.41) to arbitrary inputs. If $y(t)$ is the response resulting from an arbitrary input $u(t)$, then the response that results from an input $\int_{-\infty}^{t} u(t')\, dt'$ is $\int_{-\infty}^{t} y(t')\, dt'$. We can see this fact as follows. The output resulting from an input of $\int u(t')\, dt'$ is given by

$$\left(\int_{-\infty}^{t} u(t')\, dt'\right) * h(t) = [u(t) * \xi(t)] * h(t)$$

$$= u(t) * \xi(t) * h(t)$$

$$= u(t) * h(t) * \xi(t)$$

$$= y(t) * \xi(t)$$

$$= \int_{-\infty}^{t} y(t')\, dt' \tag{3.42}$$

These steps follow directly from the distributive and associative properties of convolution, namely,

$$(a * b) * c = a * (b * c)$$

and

$$(a * b) = (b * a)$$

One can use the step response $g(t)$ to characterize the input-output relationship of a linear system. Consider the convolution of an arbitrary input $u(t)$ with the impulse response $h(t)$: that is,

$$y(t) = \int_{-\infty}^{+\infty} u(\tau) h(t - \tau)\, d\tau \tag{3.43}$$

Integrate (3.43) by parts and use (3.41) to obtain

$$y(t) = -u(\tau)g(t-\tau)\Big|_{-\infty}^{\infty} + \int_{-\infty}^{\infty} \frac{du(\tau)}{d\tau} g(t-\tau)\,d\tau$$

$$= \int_{-\infty}^{\infty} \frac{du(\tau)}{d\tau} g(t-\tau)\,d\tau \tag{3.44}$$

provided that $u(t)$ and $g(t)$ are zero at $t = -\infty$. In words, the output response of a linear system is given by convolving the step response $g(t)$ with the derivative of the input, $u^{(1)}(t)$.

The function $u(t)$ may have a jump discontinuity at the origin. If it does, then $u^{(1)}(t)$ contains an impulse at the origin. This impulse can be handled by rewriting (3.44), assuming causal u and g, as

$$y(t) = \int_{0^-}^{0^+} \frac{du(\tau)}{d\tau} g(t-\tau)\,d\tau + \int_{0^+}^{t} \frac{du(\tau)}{d\tau} g(t-\tau)\,d\tau$$

$$= \int_{0^-}^{0^+} u(0)\delta(\tau)g(t-\tau)\,d\tau + \int_{0^+}^{t} \frac{du(\tau)}{d\tau} g(t-\tau)\,d\tau$$

$$= u(0)g(t) + \int_{0^+}^{t} \frac{du(\tau)}{d\tau} g(t-\tau)\,d\tau \tag{3.45}$$

where $u(0)$ is the value of the jump discontinuity at the origin. The general result is that we can express the output $y(t)$ as

$$y(t) = u^{(n+1)}(t) * h^{-(n+1)}(t) \tag{3.46}$$

where $u^{(n+1)}(t)$ and $h^{-(n+1)}(t)$ denote the $n + 1$st derivative and integral, respectively. The proof involves repeated integration by parts, as in the case of the step response. This relationship can also be easily seen in terms of a transform domain formulation (Chapters 5 and 6).

In general, an arbitrary function $u(t)$ can be expressed as a sum over n of the power functions $t^n\xi(t)$, $n = 0, 1, 2, \ldots$ (for $n = 0$ we have the step function, for $n = 1$ we have the ramp function, etc.). The output response can thus be expressed as a sum of the responses resulting from the input power functions. Since $t^n\xi(t)$ is the $(n + 1)$st integral of $\delta(t)$, the response to $t^n\xi(t)$ is the $(n + 1)$st integral of $h(t)$. Thus with

$$u(t) = \sum_{n=0}^{\infty} c_n t^n \xi(t)$$

we obtain the representation

$$y(t) = \sum_{n=0}^{\infty} c_n h^{-(n+1)}(t)$$

Example 3.10 Find the response of the system in Figure 3.9 resulting from an input pulse of the form $x(t) = a[\xi(t) - \xi(t - T)]$. We find the impulse response

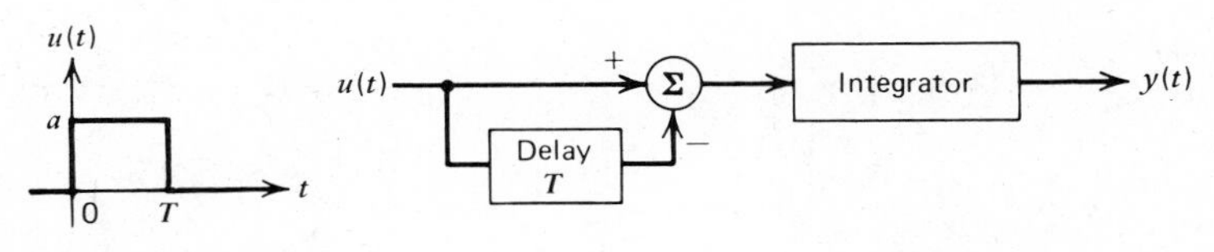

Figure 3.9

of this system by applying an input $u(t) = \delta(t)$. Because of the simple nature of this system, we can evaluate the output directly from the block diagram. Thus,

$$h(t) = \int_{-\infty}^{t} [\delta(\tau) - \delta(\tau - T)]\, d\tau$$

$$= \begin{cases} \xi(t) - \xi(t - T), & t \geq 0 \\ 0, & t < 0 \end{cases} \tag{3.47}$$

Graphically, $h(t)$ is shown in Figure 3.10. The response is therefore $y = u * h$

$$y(t) = \int_0^t a[\xi(\tau) - \xi(\tau - T)][\xi(t - \tau) - \xi(t - \tau - T)]\, d\tau$$

$$= \begin{cases} a \displaystyle\int_0^t d\tau, & 0 \leq t < T \\ a \displaystyle\int_{t-T}^{T} d\tau, & T \leq t < 2T \\ 0, & \text{otherwise} \end{cases}$$

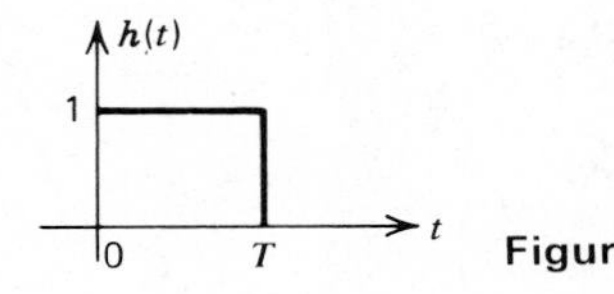

Figure 3.10

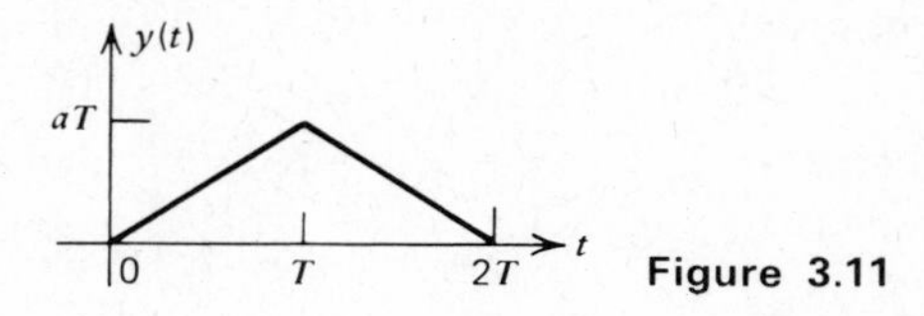

Figure 3.11

Performing the integrations, we have

$$y(t) = \begin{cases} at, & 0 \leq t < T \\ a(2T - t), & T \leq t < 2T \\ 0, & \text{otherwise} \end{cases}$$

The output response is shown in Figure 3.11. ■

This example points out the complexity involved in evaluating convolutions analytically, even for very simple functions. A sketch of the convolved functions is often a great aid in setting up the various integrals involved in calculating a convolution.

Example 3.11 Find the impulse response of the RC circuit shown in Figure 3.12. On the basis of (3.41), we can find the impulse response by first finding the step response $g(t)$ and then differentiating to obtain $h(t)$. Assuming that $RC = 1$, the differential equation relating $y(t)$ and $h(t)$ is

$$\frac{dy(t)}{dt} + y(t) = u(t)$$

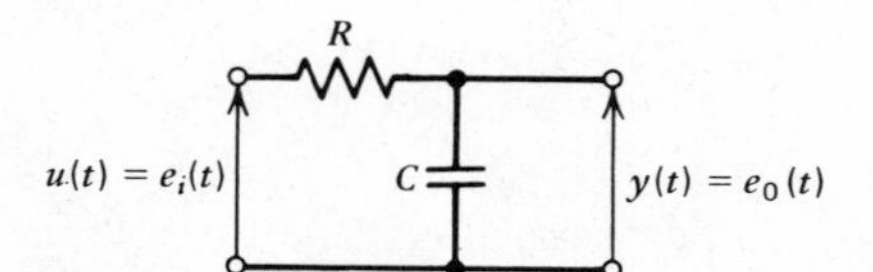

Figure 3.12

The step response is thus obtained by solving

$$\frac{dg(t)}{dt} + g(t) = \begin{cases} 1, & t \geq 0 \\ 0, & t < 0 \end{cases}$$

with

$$g(0) = 0$$

The solution is easily obtained as

$$g(t) = (1 - e^{-t})\xi(t) \tag{3.48}$$

Thus, by (3.41), we have the impulse response

$$\begin{aligned} h(t) &= \frac{d}{dt} g(t) = \frac{d}{dt} [(1 - e^{-t})\xi(t)] \\ &= e^{-t}\xi(t) + (1 - e^{-t})\delta(t) = e^{-t}\xi(t) \end{aligned}$$

■

Example 3.12 Find the output of the *RC* circuit of the previous example resulting from an input $u(t)$ given by

$$u(t) = \begin{cases} 0, & t < 0 \\ Ae^{-\beta t}, & t \geq 0 \end{cases}$$

Using the convolution integral, we have

$$\begin{aligned} y(t) &= \int_0^t Ae^{-\beta\tau}e^{-(t-\tau)}\,d\tau, \qquad t \geq 0 \\ &= Ae^{-t}\int_0^t e^{-\tau(\beta-1)}\,d\tau, \qquad t \geq 0 \\ &= \begin{cases} \left.\begin{array}{ll} \dfrac{A(e^{-\beta t} - e^{-t})}{1 - \beta}\xi(t), & \beta \neq 1 \\ Ate^{-t}\xi(t), & \beta = 1 \end{array}\right\}, & t \geq 0 \\ 0, \quad t < 0 \end{cases} \end{aligned}$$

■

Example 3.13 Find the output for the *RC* circuit of the previous two examples due to an input $u(t)$ given by

$$u(t) = \begin{cases} \sin t, \; 0 < t < \dfrac{\pi}{2} \\ 0, \qquad \text{otherwise} \end{cases}$$

The step response of this system is given by (3.48) as

$$g(t) = \begin{cases} 0, & t < 0 \\ 1 - e^{-t}, & t \geq 0 \end{cases}$$

Thus, the output $y(t)$ is

$$\begin{aligned} y &= u^{(1)} * g \\ &= \begin{cases} 0, & t < 0 \\ \displaystyle\int_0^t \left[\cos \tau - \delta\left(\tau - \frac{\pi}{2}\right)\right](1 - e^{-(t-\tau)})\, d\tau, & 0 \leq t < \dfrac{\pi}{2} \\ \displaystyle\int_0^{\pi/2} \left[\cos \tau - \delta\left(\tau - \frac{\pi}{2}\right)\right](1 - e^{-(t-\tau)})\, d\tau, & t \geq \dfrac{\pi}{2} \end{cases} \end{aligned}$$

The limits on the integrals are most easily seen by drawing some sketches of $u^{(1)}(\tau)$ and $g(t - \tau)$ for various values of t. Evaluating the above integrals, we obtain

$$y(t) = \begin{cases} 0, & t < 0 \\ \dfrac{1}{2}(\sin t - \cos t + e^{-t}), & 0 \leq t < \dfrac{\pi}{2} \\ \dfrac{e^{-t}}{2}(1 + e^{\pi/2}), & t \geq \dfrac{\pi}{2} \end{cases}$$

The input and output functions are plotted in Figure 3.13. How does this solution compare with a direct evaluation of $u * h$?

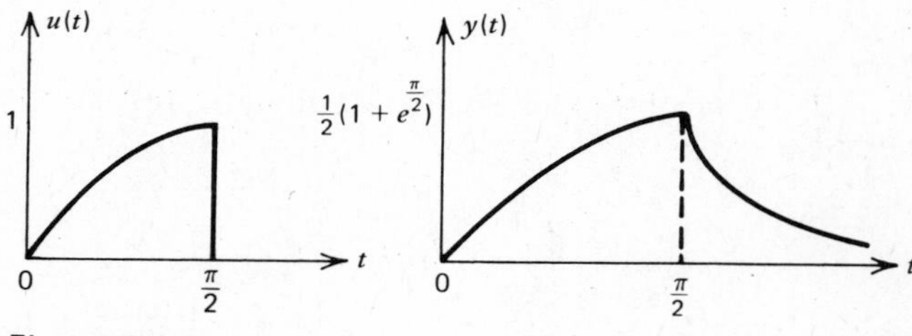

Figure 3.13

■

The step response $g(t)$ is useful from a practical point of view because a step input can easily be generated and applied to real systems. An impulse input may not be practical in many systems because a narrow pulse—our physical approximation of an impulse—contains little energy. The concept of convolution and the impulse or step response is also practical in another sense. If one must characterize a linear "black-box" system without knowledge of the internal structure of the system, then the step or impulse response is one characterization or description of the system. The frequency response function $H(j\omega)$ is an equivalent description that is also often used in these cases.

3.6 FINDING THE IMPULSE-RESPONSE FUNCTION

As with discrete-time systems, we wish to develop techniques to find the impulse-response function of a continuous-time system. In Examples 3.10 and 3.11 we have demonstrated two methods. If the system is modeled or specified by a block diagram, it may be possible to obtain the impulse response directly from the block diagram, as in Example 3.10. Usually this method will work only for very simple systems.

Example 3.11 demonstrates another technique. Suppose we have a system defined by the differential equation

$$L[y(t)] = u(t) \tag{3.49}$$

We can find the step response of this system from

$$L[g(t)] = \begin{cases} 1, & t \geq 0 \\ 0, & t < 0 \end{cases} \tag{3.50}$$

with the appropriate initial conditions. The impulse response $h(t)$ can then be obtained from

$$h(t) = \frac{d}{dt}[g(t)]$$

A third and more powerful approach is based on knowledge of the homogeneous solutions to

$$L[y(t)] = u(t)$$

To develop this method, suppose we have a second-order system of the form

$$L[y(t)] = (D^2 + a_1 D + a_0)[y(t)] = u(t) \tag{3.51}$$

Assuming that the system is initially at rest, with no energy storage, implies that the initial conditions are

$$\begin{aligned} y(0) &= 0 \\ y^{(1)}(0) &= 0 \end{aligned} \tag{3.52}$$

Now if $h(t)$ is the impulse-response function, the output is given by

$$y(t) = \int_0^t u(\tau)h(t - \tau)\, d\tau \tag{3.53}$$

Equations 3.51 and 3.53 represent two methods of finding the output response $y(t)$. Using (3.53) as a starting point, consider the conditions that (3.51) and (3.52) impose on the impulse-response function in (3.53).

We note that $y(0) = 0$ from (3.53), as required by (3.52). Differentiating (3.53) with respect to t, we have

$$\begin{aligned} y^{(1)}(t) &= h(t - \tau)u(\tau)\Big|_{\tau=t} + \int_0^t h^{(1)}(t - \tau)u(\tau)\, d\tau \\ &= h(0)u(t) + \int_0^t h^{(1)}(t - \tau)u(\tau)\, d\tau \end{aligned} \tag{3.54}$$

Equations 3.52 require that $y^{(1)}(0) = 0$. Setting $y^{(1)}(0) = 0$ in (3.54) implies that $h(0) = 0$. Differentiating again, we obtain

$$y^{(2)}(t) = h^{(1)}(0)u(t) + \int_0^t h^{(2)}(t - \tau)u(\tau)\, d\tau \tag{3.55}$$

Equations 3.53, 3.54, and 3.55 are expressions for $y(t)$, $y^{(1)}(t)$, and $y^{(2)}(t)$. Consider the result of forming the sum $y^{(2)}(t) + a_1 y^{(1)}(t) + a_0 y(t)$ using these expressions. The result is

$$h^{(1)}(0)u(t) + \int_0^t h^{(2)}(t - \tau)u(\tau)\, d\tau + a_1 \int_0^t h^{(1)}(t - \tau)u(\tau)\, d\tau + a_0 \int_0^t h(t - \tau)u(\tau)\, d\tau$$

We observe that if

a.
$$h^{(1)}(0) = 1 \tag{3.56}$$

b.
$$\int_0^t [h^{(2)}(t-\tau) + a_1 h^{(1)}(t-\tau) + a_0 h(t-\tau)]u(\tau)\,d\tau = 0 \tag{3.57}$$

then (3.53) will be a solution of (3.51)! Equation 3.57 implies that the integrand on the left-hand side is zero. If $u(t) \neq 0$, then the bracketed term is zero; that is,

$$h^{(2)}(t-\tau) + a_1 h^{(1)}(t-\tau) + a_0 h(t-\tau) = 0 \tag{3.58}$$

or

$$h^{(2)}(t) + a_1 h^{(1)}(t) + a_0 h(t) = 0 \tag{3.59}$$

Equation 3.59 is, of course, the original homogeneous differential equation (3.51). Thus the impulse response can be obtained by finding the homogeneous solutions to the original differential equation. Therefore $h(t)$ can be expressed as the sum of two linearly independent homogeneous solutions

$$h(t) = [c_1\phi_1(t) + c_2\phi_2(t)]\xi(t) \tag{3.60}$$

where

$$L[\phi_i(t)] = 0, \qquad i = 1, 2$$

The initial conditions that allow us to find c_1 and c_2 are given by (3.54) and (3.56) as

$$\begin{aligned} h(0) &= 0 \\ h^{(1)}(0) &= 1 \end{aligned} \tag{3.61}$$

Thus we can evaluate c_1 and c_2 in (3.60) from

$$h(0) = 0 = c_1\phi_1(0) + c_2\phi_2(0)$$

$$h^{(1)}(0) = 1 = c_1\phi_1^{(1)}(0) + c_2\phi_2^{(1)}(0)$$

■

Example 3.14 Consider the system represented by the differential equation

$$L[y(t)] = y^{(2)}(t) + y(t) = u(t) \tag{3.62}$$

The homogeneous solutions of (3.62) are easily found to be

$$\phi_1(t) = \sin t, \qquad \phi_2(t) = \cos t$$

Thus

$$h(t) = [c_1 \sin t + c_2 \cos t]\xi(t)$$

with

$$h(0) = 0, \qquad h^{(1)}(0) = 1$$

The constants c_1 and c_2 therefore satisfy

$$h(0) = 0 = c_1 \sin 0 + c_2 \cos 0$$

$$h^{(1)}(0) = 1 = c_1 \cos 0 - c_2 \sin 0$$

which imply that $c_1 = 1$ and $c_2 = 0$. The impulse response of the system modeled by (3.62) is thus

$$h(t) = \sin t \; \xi(t) \tag{3.63}$$

To verify this result, we substitute (3.63) into (3.62), with

$$h^{(1)}(t) = \frac{d}{dt}[\sin t \; \xi(t)] = \cos t \; \xi(t) + \sin t \; \delta(t) = \cos t \; \xi(t)$$

$$h^{(2)}(t) = \frac{d}{dt}[\cos t \; \xi(t)] = -\sin t \; \xi(t) + \cos t \; \delta(t) = -\sin t \; \xi(t) + \delta(t)$$

to obtain

$$h^{(2)}(t) + h(t) = -\sin t \; \xi(\mathrm{t}) + \delta(t) + \sin t \; \xi(t) = \delta(t)$$

Thus the output $y(t)$ for an arbitrary input $u(t)$ may be found as

$$y(t) = \int_0^t \sin(t - \tau)u(\tau)\, d\tau$$

We note that the linearly independent solutions

$$\phi_1(t) = e^{jt}, \qquad \phi_2(t) = e^{-jt}$$

could also have been chosen. The impulse-response function is

$$h(t) = (c_1 e^{jt} + c_2 e^{-jt})\xi(t)$$

with

$$h(0) = 0 = c_1 + c_2$$

$$h^{(1)}(0) = 1 = jc_1 - jc_2$$

These equations give

$$c_1 = \frac{1}{2j}, \qquad c_2 = -\frac{1}{2j}$$

from which

$$\begin{aligned} h(t) &= \left(\frac{1}{2j} e^{jt} - \frac{1}{2j} e^{-jt}\right)\xi(t) \\ &= \sin t \ \xi(t) \end{aligned}$$

as before. ■

Example 3.15 Consider a system modeled by the differential equation

$$L[y(t)] = (D^2 + 2D + 2)[y(t)] = u(t) \tag{3.64}$$

The homogeneous solutions to $L[y(t)] = 0$ are $e^{-t} \sin t$ and $e^{-t} \cos t$. Thus the impulse-response function is

$$h(t) = [c_1 e^{-t} \sin t + c_2 e^{-t} \cos t]\xi(t) \tag{3.65}$$

where the constants c_1 and c_2 are evaluated using

$$h(0) = 0 = c_2$$

$$h^{(1)}(0) = 1 = c_1$$

The impulse response of this system is

$$h(t) = e^{-t} \sin t \, \xi(t) \tag{3.66}$$

and the output response of the system for an arbitrary causal input $u(t)$ is

$$y(t) = \begin{cases} \int_0^t e^{-(t-\tau)} \sin (t - \tau) u(\tau) \, d\tau, & t \geq 0 \\ 0, & t < 0 \end{cases}$$

The reader should verify for himself that

$$(D^2 + 2D + 2)[e^{-t} \sin t \, \xi(t)] = \delta(t)$$

■

We can generalize this method to nth order systems in a straightforward manner. For the general case, we take

$$L[y(t)] = (D^n + b_{n-1}D^{n-1} + \cdots + 1)[y(t)] = u(t) \tag{3.67}$$

with initial conditions

$$y(0) = y^{(1)}(0) = \cdots = y^{(n-1)}(0) = 0$$

The output is again expressed by

$$y(t) = \int_0^t h(t - \tau)u(\tau) \, d\tau \tag{3.68}$$

Equating successive derivatives of $y(t)$ in (3.68) to zero yields

$$h(0) = h^{(1)}(0) = \cdots = h^{(n-2)}(0) = 0 \tag{3.69}$$

The nth derivative of (3.68) yields

$$y^{(n)}(t) = h^{(n-1)}(0)u(t) + \int_0^t h^{(n)}(t - \tau)u(\tau) \, d\tau \tag{3.70}$$

By exactly the same argument as we use in the second-order case discussed previously, we find that the impulse-response function for the system of (3.67) must satisfy the homogeneous equation

$$L[h(t)] = 0$$

with initial conditions $h(0) = h^{(1)}(0) = \cdots = h^{(n-2)}(0) = 0$, $h^{(n-1)}(0) = 1$. Thus

$$h(t) = c_1\phi_1(t) + c_2\phi_2(t) + \cdots + c_n\phi_n(t) \tag{3.71}$$

where these initial conditions are used to find the constants $c_1, c_2, \ldots, c_n$.

Example 3.16 Consider a system modeled by the differential equation

$$L[y(t)] = (D^2 - 1)(D^2 - 1)[y(t)] = u(t)$$

The homogeneous solutions are e^t, e^{-t}, te^t, and te^{-t}. The impulse response is therefore

$$h(t) = [c_1 e^t + c_2 e^{-t} + c_3 te^t + c_4 te^{-t}]\xi(t)$$

where the constants are evaluated by the use of

$$h(0) = 0 = c_1 + c_2$$

$$h^{(1)}(0) = 0 = c_1 - c_2 + c_3 + c_4$$

$$h^{(2)}(0) = 0 = c_1 + c_2 + 2c_3 - 2c_4$$

$$h^{(3)}(0) = 1 = c_1 - c_2 - 3c_3 + 3c_4$$

These equations yield $c_1 = \frac{1}{2}$, $c_2 = -\frac{1}{2}$, $c_3 = -\frac{1}{2}$, $c_4 = -\frac{1}{2}$. The impulse response is therefore

$$h(t) = \frac{1}{2}(e^t - e^{-t} - te^t - te^{-t})\xi(t)$$

The reader should verify that

$$(D^4 - 2D^2 + 1)\left[\frac{1}{2}(e^t - e^{-t} - te^t - te^{-t})\xi(t)\right] = \delta(t)$$ ■

To complete this section, we extend this method of finding impulse-response functions to systems that are forced by an input signal of the form $L_D[u(t)]$, rather than just $u(t)$. Consider a system of the form

$$L[y(t)] = L_D[u(t)] \tag{3.72}$$

Let the system $L[\hat{y}(t)] = u(t)$ have impulse response $\hat{h}(t)$. The response of the system modeled by $L[\hat{y}(t)] = u(t)$ is given by

$$\hat{y}(t) = \int_0^t \hat{h}(t - \tau)u(\tau)\, d\tau \tag{3.73}$$

This impulse response $\hat{h}(t)$ is found by the methods just described. However, we are forcing the system not with $u(t)$ but rather with $L_D[u(t)]$. Suppose that we apply the operator L_D to both sides of (3.74).

$$L[\hat{y}(t)] = u(t) \tag{3.74}$$

We obtain

$$L_D[L[\hat{y}(t)]] = L_D[u(t)] \tag{3.75}$$

Making use of the commutative property for linear, time-invariant differential operators, we can write (3.75) as

$$L[L_D[\hat{y}(t)]] = L_D[u(t)] \tag{3.76}$$

Let $L_D[\hat{y}(t)] = y(t)$. Then the output of the original system is merely L_D operating on $\hat{y}(t)$. Thus the overall impulse response h for (3.72) must be

$$h(t) = L_D[\hat{h}(t)] \tag{3.77}$$

We have assumed L_D to be of lower order than L. If this is not true, then terms involving $\delta(t)$ and its derivatives will be generated in (3.77) because the $(n - 1)$st and higher-order derivatives of h are not, in general, zero at $t = 0$.

Example 3.17 Consider the circuit below with a voltage source input $u(t)$ and voltage output $y(t)$ across the capacitor. Writing two loop equations we obtain the relevant differential equation relating input and output as

$$(D^2 + 2D + 2)[y(t)] = (D + 1)[u(t)] \tag{3.78}$$

The first step is to determine the impulse response $\hat{h}$ of the system

$$(D^2 + 2D + 2)[y(t)] = u(t)$$

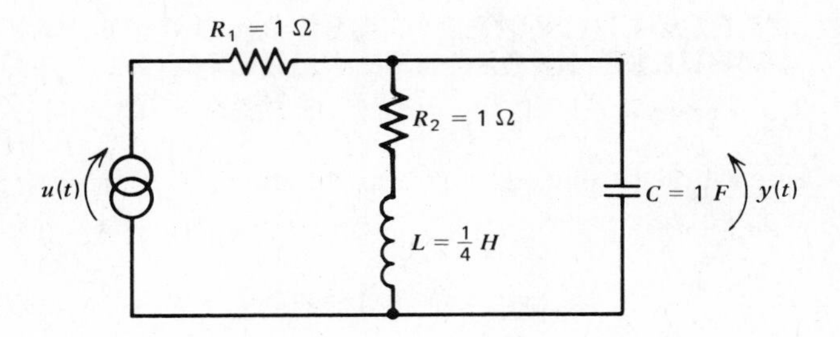

We have already solved this problem in Example 3.15. From (3.66) the impulse response function $\hat{h}$ is

$$\hat{h}(t) = e^{-t} \sin t\ \xi(t)$$

Thus, the impulse-response function h of (3.78) is

$$\begin{aligned} h(t) &= (D + 1)[\hat{h}(t)] = (D + 1)[e^{-t} \sin t\ \xi(t)] \\ &= -e^{-t} \sin t\ \xi(t) + e^{-t} \cos t\ \xi(t) + e^{-t} \sin t\ \delta(t) + e^{-t} \sin t\ \xi(t) \\ &= e^{-t} \cos t\ \xi(t) \end{aligned}$$

The output y is given by

$$y(t) = \int_0^t e^{-(t-\tau)} \cos (t - \tau) u(\tau)\, d\tau, \qquad t \geq 0$$

If $u(t)$ is a unit step input, then

$$\begin{aligned} y(t) &= \int_0^t e^{-(t-\tau)} \cos (t - \tau)\, d\tau, \qquad t \geq 0 \\ &= \int_0^t e^{-\alpha} \cos \alpha\, d\alpha \\ &= \begin{cases} \frac{1}{2}(1 + e^{-t} \sin t - e^{-t} \cos t), & t \geq 0 \\ 0, & t < 0 \end{cases} \end{aligned}$$

■

3.7 FREQUENCY RESPONSE AND THE IMPULSE-RESPONSE FUNCTION

As with discrete-time systems we can relate h and $H(e^{j\omega})$. In general, we have for any input function u,

$$y = h * u$$

In particular for $u(t) = e^{j\omega t}$, we have an output

$$\begin{aligned} y(t) &= \int_{-\infty}^{\infty} u(t - \tau)h(\tau)\, d\tau \\ &= \int_{-\infty}^{\infty} e^{j\omega(t-\tau)} h(\tau)\, d\tau \\ &= e^{j\omega t} \int_{-\infty}^{\infty} e^{-j\omega\tau} h(\tau)\, d\tau \end{aligned} \tag{3.79}$$

However, for linear, time-invariant systems an input $e^{j\omega t}$ always yields the output $H(j\omega)e^{j\omega t}$. Thus (3.79) is identically $H(j\omega)e^{j\omega t}$, that is,

$$H(j\omega) = \int_{-\infty}^{\infty} e^{-j\omega\tau} h(\tau)\, d\tau \tag{3.80}$$

Equation 3.80 allows us to calculate the frequency-response function directly in terms of the impulse-response function. For a system described by the differential equation (3.13) repeated below, we can use (3.14) to obtain (3.82).

$$\begin{aligned} &(b_n D^n + b_{n-1} D^{n-1} + \cdots + b_1 D + 1)[y(t)] \\ &\qquad = (a_m D^m + a_{m-1} D^{m-1} + \cdots + a_0)[u(t)] \end{aligned} \tag{3.81}$$

$$H(j\omega) = \frac{a_0 + a_1 j\omega + \cdots + a_m(j\omega)^m}{1 + b_1 j\omega + \cdots + b_n(j\omega)^n} = \int_{-\infty}^{\infty} h(\tau)e^{-j\omega\tau}\, d\tau \tag{3.82}$$

Example 3.18 Recall Example 3.5, which involved the calculation of a system's frequency response. In this example we shall use (3.80) and compare the results with those we found before. The system is the simple RC circuit shown in Figure 3.14. The differential equation relating u and y is

$$\frac{dy(t)}{dt} + \frac{1}{RC}y(t) = \frac{1}{RC}u(t)$$

or

$$(RCD + 1)[y(t)] = u(t)$$

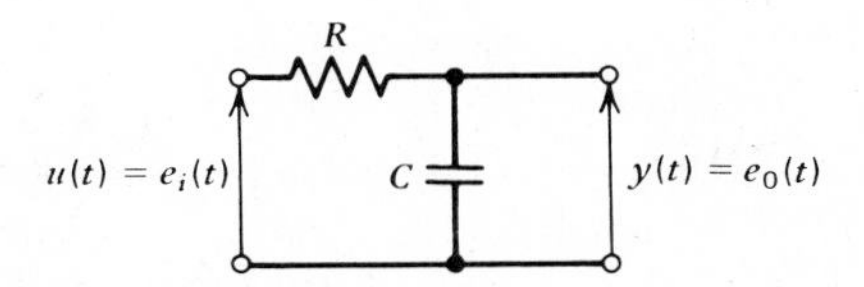

Figure 3.14

It is not difficult to find the impulse-response function h as

$$h(t) = \begin{cases} \dfrac{1}{RC}e^{-t/RC}, & t \geq 0 \\ 0, & t < 0 \end{cases} \tag{3.83}$$

The frequency response function is therefore

$$\begin{aligned} H(j\omega) &= \int_{-\infty}^{\infty} e^{-j\omega t}h(t)\,dt \\ &= \int_{0}^{\infty} \frac{1}{RC}e^{-t/RC}e^{-j\omega t}\,dt \\ &= \frac{1}{RC}\left[-\frac{e^{-t(j\omega+1/RC)}}{j\omega + \dfrac{1}{RC}}\Bigg|_{0}^{\infty}\right] = \frac{1}{1 + j\omega RC} \end{aligned}$$

as we obtained previously in Example 3.5. ■

3.8 STATE VARIABLES FOR CONTINUOUS-TIME SYSTEMS

Our discussion of state-variable descriptions for continuous-time systems is similar to the discrete-time case. We shall, in fact, find that, as before, the essential calculation involves finding a function of a matrix.

Consider a single-input, single-output system described in the equation

$$(D^n + b_{n-1}D^{n-1} + \cdots + b_1 D + 1)[y(t)] = a_0 u(t) \tag{3.84}$$

To obtain a state-variable description of this system we first sketch a block diagram of the system similar to block diagrams one might use in an analog computer simulation of the system. Integrators are the basic building blocks of the schematic. Figure 3.15 depicts a block diagram representation of (3.84). One can obtain the block diagram from (3.84) by solving for the highest order derivative of $y(t)$. This is the output of the summer. In order to find $y(t)$ we must know the initial values associated with each integrator since these values can be specified independently of the input. In other words, the integrators form the memory of the system; thus it is natural to choose the contents of integrators as the state of the system at any time t.

We define the components of the state vector $\mathbf{x}(t)$ as

$$\begin{aligned}
x_1(t) &= y(t) \\
x_2(t) &= y^{(1)}(t) = x_1^{(1)}(t) \\
x_3(t) &= y^{(2)}(t) = x_2^{(1)}(t) \\
&\vdots \\
x_n(t) &= y^{(n-1)}(t) = x_{n-1}^{(1)}(t)
\end{aligned} \tag{3.85}$$

Rewriting (3.85), we can express $\mathbf{x}^{(1)}(t)$ in terms of $\mathbf{x}(t)$ as

$$\begin{aligned}
x_1^{(1)}(t) &= x_2(t) \\
x_2^{(1)}(t) &= x_3(t) \\
&\vdots \\
x_{n-1}^{(1)}(t) &= x_n(t) \\
x_n^{(1)}(t) &= y^{(n)}(t) = a_0 u(t) - x_1(t) - \cdots - b_{n-1}x_n(t)
\end{aligned} \tag{3.86}$$

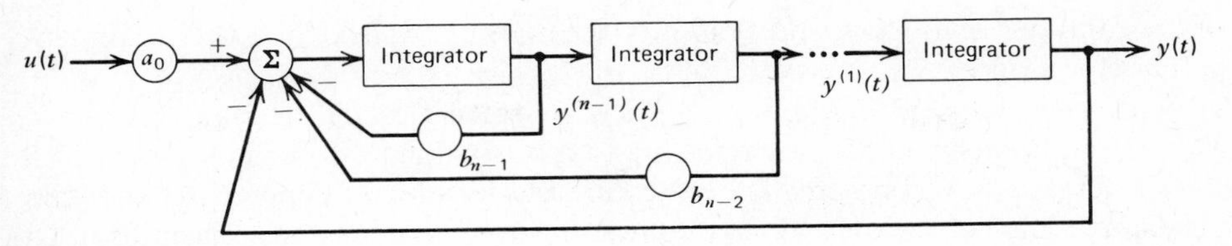

Figure 3.15

In matrix form, (3.86) can be written as

$$\mathbf{x}^{(1)}(t) = \begin{bmatrix} x_1^{(1)}(t) \\ x_2^{(1)}(t) \\ \vdots \\ x_n^{(1)}(t) \end{bmatrix} = \begin{bmatrix} 0 & 1 & 0 & \cdots & 0 \\ 0 & 0 & 1 & \cdots & 0 \\ & & \vdots & & \\ 0 & 0 & 0 & \cdots & 1 \\ -1 & -b_1 & -b_2 & & -b_{n-1} \end{bmatrix} \begin{bmatrix} x_1(t) \\ x_2(t) \\ \vdots \\ x_{n-1}(t) \\ x_n(t) \end{bmatrix} + \begin{bmatrix} 0 \\ 0 \\ \vdots \\ 0 \\ a_0 \end{bmatrix} u(t) \tag{3.87}$$

or

$$\mathbf{x}^{(1)}(t) = \mathbf{A}\mathbf{x}(t) + \mathbf{B}u(t) \tag{3.88}$$

Comparing (3.88) with (2.45), we observe one principal difference. Note that the state equation of a continuous-time system expresses the time derivative of the state vector in terms of the state vector and the input, whereas the state equation of a discrete-time system expresses the state at a given time in terms of the state at a previous time and the input.

We can, of course, express the output $y(t)$ in terms of the state $\mathbf{x}(t)$ and the input $u(t)$ as

$$y(t) = \mathbf{C}\mathbf{x}(t) + \mathbf{D}u(t) \tag{3.89}$$

where, in the formulation above,

$$\mathbf{C} = [1 \quad 0 \quad 0 \quad \cdots \quad 0], \qquad \mathbf{D} = 0$$

As before, we can handle multiple inputs and outputs with the same formulation

$$\begin{aligned} \mathbf{x}^{(1)}(t) &= \mathbf{A}\mathbf{x}(t) + \mathbf{B}\mathbf{u}(t) \\ \mathbf{y}(t) &= \mathbf{C}\mathbf{x}(t) + \mathbf{D}\mathbf{u}(t) \end{aligned} \tag{3.90}$$

where for n states, r inputs, and s outputs, the matrices **A**, **B**, **C**, and **D** have dimensions as in Figure 2.23.

As with discrete-time systems, the state variables for a continuous-time system can be easily chosen by referring to a block diagram, such as Figure 3.15 or a schematic of the system. The state variables can again be interpreted as the "memory cells" of the system. For continuous-time systems, these memory cells can be chosen to correspond to the physical energy storing components within the system. For mechanical systems, the energy-storing devices are springs and masses. For electrical systems, the energy-storing devices are capacitors and inductors. The state variables can be chosen as the appropriate physical quantity associated with the energy-storing device. For example, in springs we would choose as a state variable the contraction or expansion length. The next example demonstrates the choice in electrical systems.

Example 3.19 Consider the electrical circuit shown in Figure 3.16. We choose as state variables the voltages across the capacitors as shown in Figure 3.16.† If we now write Kirchhoff's current equations at the nodes between R_1 and R_2 and R_2 and R_3, we obtain the equations

$$
\begin{aligned}
x_1^{(1)}(t) &= \frac{1}{C_1}\left[\frac{u_1(t) - x_1(t)}{R_1} + \frac{x_2(t) - x_1(t)}{R_2}\right] \\
x_2^{(1)}(t) &= \frac{1}{C_2}\left[\frac{u_2(t) - x_2(t)}{R_3} + \frac{x_1(t) - x_2(t)}{R_2}\right]
\end{aligned}
\tag{3.91}
$$

The output $y(t)$ is clearly

$$y(t) = x_1(t) - x_2(t) \tag{3.92}$$

Equations 3.91 and 3.92 can be written in vector form as

$$
\mathbf{x}^{(1)}(t) = \begin{bmatrix} -\dfrac{1}{C_1}\left(\dfrac{1}{R_1} + \dfrac{1}{R_2}\right) & \dfrac{1}{R_2 C_1} \\ \dfrac{1}{R_2 C_2} & -\dfrac{1}{C_2}\left(\dfrac{1}{R_2} + \dfrac{1}{R_3}\right) \end{bmatrix} \mathbf{x}(t) + \begin{bmatrix} \dfrac{1}{R_1 C_1} & 0 \\ 0 & \dfrac{1}{R_3 C_2} \end{bmatrix} \mathbf{u}(t)
$$

$$y(t) = [1 \quad -1]\mathbf{x}(t)$$

† In a circuit with inductances, an appropriate choice of state variables would be the currents through these inductances.

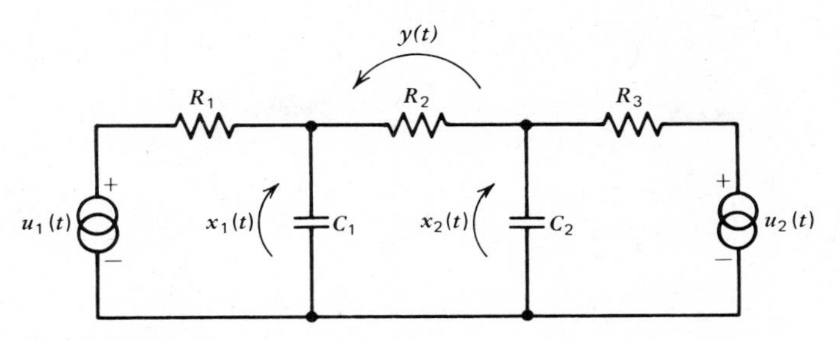

Figure 3.16

The above equations constitute the state-space formulation of the circuit in Figure 3.16. ■

3.9 SOLUTION OF THE CONTINUOUS-TIME STATE-VARIABLE EQUATION

We turn now to the solution of the state equation (3.88). We begin by considering the natural or unforced solution: that is, $\mathbf{u}(t) = \mathbf{0}$. Equation 3.88 simplifies to

$$\mathbf{x}^{(1)}(t) = \mathbf{A}\mathbf{x}(t) \tag{3.93}$$

The scalar version of the vector problem is often a good indicator of the true solution. In this case, the scalar version of (3.93) is $x^{(1)}(t) = ax(t)$, which has a solution $x(t) = e^{at}x(0)$. Suppose we try the same form for the solution of (3.93): that is,

$$\mathbf{x}(t) = \mathbf{e}^{\mathbf{A}t}\mathbf{x}(0) \tag{3.94}$$

where $\mathbf{e}^{\mathbf{A}t}$ is a function of the matrix $\mathbf{A}$ that we define as the $n \times n$ matrix

$$\mathbf{e}^{\mathbf{A}t} \triangleq \sum_{k=0}^{\infty} \frac{t^k}{k!} \mathbf{A}^k \tag{3.95}$$

The definition of $\mathbf{e}^{\mathbf{A}t}$ given in (3.95) is a special case of a function of a matrix that we have previously solved in Section 2.10. If (3.94) is a solution of (3.93), then it must, of course, satisfy (3.93). Let us check:

$$\frac{d}{dt}[\mathbf{e}^{\mathbf{A}t}\mathbf{x}(0)] \stackrel{?}{=} \mathbf{A}\mathbf{e}^{\mathbf{A}t}\mathbf{x}(0)$$

Substituting for $\mathbf{e}^{\mathbf{A}t}$ and differentiating term by term with respect to t yields

$$\left[\sum_{k=1}^{\infty} \frac{t^{k-1}}{(k-1)!} \mathbf{A}^k\right]\mathbf{x}(0) \stackrel{?}{=} \mathbf{A}\left[\sum_{k=0}^{\infty} \frac{t^k}{k!} \mathbf{A}^k\right]\mathbf{x}(0)$$

Letting $j = k - 1$ in the left-hand sum, we see that

$$\left[\sum_{j=0}^{\infty} \frac{t^j}{j!} \mathbf{A}^{j+1}\right]\mathbf{x}(0) \stackrel{?}{=} \mathbf{A}\left[\sum_{k=0}^{\infty} \frac{t^k}{k!} \mathbf{A}^k\right]\mathbf{x}(0) \tag{3.96}$$

In the left-hand side of (3.96), we can factor an $\mathbf{A}$ from $\mathbf{A}^{j+1}$ to obtain the right-hand side. Thus (3.96) is an identity and we have established that $\mathbf{x}(t) = \mathbf{e}^{\mathbf{A}t}\mathbf{x}(0)$ is the solution to (3.93). Because the solution to (3.93) is unique, $\mathbf{e}^{\mathbf{A}t}\mathbf{x}(0)$ is the unforced response of the system. To find the vector $\mathbf{x}(0)$, we evaluate (3.94) for some t_0 for which $\mathbf{x}(t)$ is known and obtain

$$\mathbf{x}(t_0) = \mathbf{e}^{\mathbf{A}t_0}\mathbf{x}(0) \tag{3.97}$$

To find $\mathbf{x}(0)$ in (3.97), we multiply both sides of (3.97) on the right by the inverse of the matrix $\mathbf{e}^{\mathbf{A}t}$; that is, $(\mathbf{e}^{\mathbf{A}t_0})^{-1}$. Thus

$$\mathbf{x}(0) = (\mathbf{e}^{\mathbf{A}t_0})^{-1}\mathbf{x}(t_0)$$

One can show (see Problem 3.30) that the inverse of $\mathbf{e}^{\mathbf{A}t}$ is

$$(\mathbf{e}^{\mathbf{A}t_0})^{-1} = \mathbf{e}^{-\mathbf{A}t_0} \tag{3.98}$$

Thus

$$\mathbf{x}(0) = \mathbf{e}^{-\mathbf{A}t_0}\mathbf{x}(t_0) \tag{3.99}$$

Substituting the results of (3.99) in (3.94), our homogeneous solution is

$$\begin{aligned}\mathbf{x}(t) &= \mathbf{e}^{\mathbf{A}t}\mathbf{e}^{-\mathbf{A}t_0}\mathbf{x}(t_0) \\ &= \mathbf{e}^{\mathbf{A}(t-t_0)}\mathbf{x}(t_0)\end{aligned} \tag{3.100}$$

where we have used

$$\mathbf{e}^{\mathbf{A}t}\mathbf{e}^{-\mathbf{A}t_0} = \mathbf{e}^{\mathbf{A}(t-t_0)} \tag{3.101}$$

Equation 3.101 is true because $\mathbf{A}$ and $-\mathbf{A}$ commute (Problem 3.30).

To find the complete solution to (3.88), we must now find a particular solution to the differential equation. To find a particular solution, assume a solution of the form

$$\mathbf{x}_p(t) = \mathbf{e}^{\mathbf{A}t}\mathbf{q}(t) \tag{3.102}$$

where $\mathbf{q}(t)$ is an unknown function to be determined. We proceed to find the unknown vector $\mathbf{q}(t)$ by substitution in (3.88).

$$\mathbf{x}_p^{(1)}(t) = \mathbf{A}\mathbf{x}_p(t) + \mathbf{B}\mathbf{u}(t)$$

$$\mathbf{A}\mathbf{e}^{\mathbf{A}t}\mathbf{q}(t) + \mathbf{e}^{\mathbf{A}t}\mathbf{q}^{(1)}(t) = \mathbf{A}\mathbf{e}^{\mathbf{A}t}\mathbf{q}(t) + \mathbf{B}\mathbf{u}(t)$$

Thus

$$\mathbf{q}^{(1)}(t) = \mathbf{e}^{-\mathbf{A}t}\mathbf{B}\mathbf{u}(t)$$

Integrating, we obtain

$$\mathbf{q}(t) = \mathbf{q}(t_0) + \int_{t_0}^{t} \mathbf{e}^{-\mathbf{A}\tau}\mathbf{B}\mathbf{u}(\tau)\, d\tau$$

Thus, the particular solution is

$$\mathbf{x}_p(t) = \mathbf{e}^{\mathbf{A}t}\mathbf{q}(t) = \mathbf{e}^{\mathbf{A}t}\mathbf{q}(t_0) + \int_{t_0}^{t} \mathbf{e}^{\mathbf{A}(t-\tau)}\mathbf{B}\mathbf{u}(\tau)\, d\tau$$

To evaluate $\mathbf{q}(t_0)$, we use the entire solution and set $\mathbf{x}(t)$ evaluated at t_0 equal to $\mathbf{x}(t_0)$. Thus, we obtain

$$\mathbf{x}(t)\Big|_{t=t_0} = \mathbf{e}^{\mathbf{A}(t-t_0)}\mathbf{x}(t_0) + \mathbf{e}^{\mathbf{A}t}\mathbf{q}(t_0) + \int_{t_0}^{t} \mathbf{e}^{\mathbf{A}(t-\tau)}\mathbf{B}\mathbf{u}(\tau)\, d\tau\Big|_{t=t_0}$$

which implies that $\mathbf{q}(t_0) = 0$. Hence, the entire solution of the state equation (3.88) is

$$\mathbf{x}(t) = \mathbf{e}^{\mathbf{A}(t-t_0)}\mathbf{x}(t_0) + \int_{t_0}^{t} \mathbf{e}^{\mathbf{A}(t-\tau)}\mathbf{B}\mathbf{u}(\tau)\, d\tau \tag{3.103}$$

The corresponding output is

$$\begin{aligned}\mathbf{y}(t) &= \mathbf{Cx}(t) + \mathbf{Du}(t) \\ &= \mathbf{C}\mathbf{e}^{\mathbf{A}(t-t_0)}\mathbf{x}(t_0) + \int_{t_0}^{t} \mathbf{C}\mathbf{e}^{\mathbf{A}(t-\tau)}\mathbf{B}\mathbf{u}(\tau)\,d\tau + \mathbf{Du}(t) \\ &= \mathbf{C}\mathbf{e}^{\mathbf{A}(t-t_0)}\mathbf{x}(t_0) + \int_{t_0}^{t} [\mathbf{C}\mathbf{e}^{\mathbf{A}(t-\tau)}\mathbf{B} + \mathbf{D}\delta(t-\tau)]\mathbf{u}(\tau)\,d\tau \end{aligned} \qquad (3.104)$$

In (3.104) the quantity $[\mathbf{C}\mathbf{e}^{\mathbf{A}t}\mathbf{B} + \mathbf{D}\delta(t)]$ is the impulse-response function $h(t)$. The matrix $\mathbf{D}$ represents straight-through paths from input to output. For multiple input-output systems $\mathbf{C}\mathbf{e}^{\mathbf{A}t}\mathbf{B}$ is an $r \times s$ matrix. The (i, j) element is the impulse response at the ith output due to an impulse applied at the jth input, with all other inputs zero. In general,

$$h(t) = \begin{cases} \mathbf{C}\mathbf{e}^{\mathbf{A}t}\mathbf{B} + \mathbf{D}\delta(t), & t \geq 0 \\ 0, & t < 0 \end{cases} \qquad (3.105)$$

To evaluate the state equation and the output in this formulation, the basic calculation is a determination of $\mathbf{e}^{\mathbf{A}t}$. This function of a matrix† can be evaluated in the manner discussed previously in Section 2.10.

Example 3.20 Evaluate $\mathbf{e}^{\mathbf{A}t}$ for the matrix

$$A = \begin{bmatrix} 3 & 0 & 0 \\ 0 & -2 & 1 \\ 0 & 4 & 1 \end{bmatrix}$$

First we find the eigenvalues of $\mathbf{A}$ from the characteristic equation

$$\begin{aligned} g(\tau) = |\mathbf{A} - \lambda\mathbf{I}| &= \begin{vmatrix} 3-\lambda & 0 & 0 \\ 0 & -2-\lambda & 1 \\ 0 & 4 & 1-\lambda \end{vmatrix} \\ &= (3-\lambda)(\lambda^2 + \lambda - 6) = 0 \end{aligned}$$

† For continuous-time systems, $\mathbf{e}^{\mathbf{A}t}$ is commonly termed the state transition matrix.

Thus, the eigenvalues of $\mathbf{A}$ are $\lambda_1 = 3$, $\lambda_2 = 2$, and $\lambda_3 = -3$. The function of a matrix we wish to find is $\mathbf{e}^{\mathbf{A}t}$. Using the results of Section 2.10 we know by the Caley-Hamilton theorem that

$$\mathbf{e}^{\mathbf{A}t} = \beta_0 \mathbf{I} + \beta_1 \mathbf{A} + \beta_2 \mathbf{A}^2 \tag{3.106}$$

We can evaluate the constants β_0, β_1, and β_2 from the equations

$$\begin{aligned} e^{3t} &= \beta_0 + 3\beta_1 + 9\beta_2 \\ e^{2t} &= \beta_0 + 2\beta_1 + 4\beta_2 \\ e^{-3t} &= \beta_0 - 3\beta_1 + 9\beta_2 \end{aligned} \tag{3.107}$$

The β's are thus found to be

$$\begin{aligned} \beta_0 &= -e^{3t} + \frac{9}{5} e^{2t} + \frac{1}{5} e^{-3t} \\ \beta_1 &= \frac{1}{6} e^{3t} - \frac{1}{6} e^{-3t} \\ \beta_2 &= \frac{1}{6} e^{3t} - \frac{1}{5} e^{2t} + \frac{1}{30} e^{-3t} \end{aligned} \tag{3.108}$$

Thus the matrix $\mathbf{e}^{\mathbf{A}t}$ is

$$\mathbf{e}^{\mathbf{A}t} = \begin{bmatrix} \beta_0 & 0 & 0 \\ 0 & \beta_0 & 0 \\ 0 & 0 & \beta_0 \end{bmatrix} + \begin{bmatrix} 3\beta_1 & 0 & 0 \\ 0 & -2\beta_1 & \beta_1 \\ 0 & 4\beta_1 & \beta_1 \end{bmatrix} + \begin{bmatrix} 9\beta_2 & 0 & 0 \\ 0 & 8\beta_2 & -\beta_2 \\ 0 & 4\beta_2 & 5\beta_2 \end{bmatrix}$$

$$= \frac{1}{5} \begin{bmatrix} 5e^{3t} & 0 & 0 \\ 0 & e^{2t} + 4e^{-3t} & e^{2t} - e^{-3t} \\ 0 & 4(e^{2t} - e^{-3t}) & 4e^{2t} + e^{-3t} \end{bmatrix} \tag{3.109}$$

If, for example, the state at $t = 0$ is

$$\mathbf{x}(0) = \begin{bmatrix} 1 \\ 1 \\ 1 \end{bmatrix},$$

then for $u(t) \equiv 0$, the state at time t is

$$\mathbf{x}(t) = \mathbf{e}^{\mathbf{A}t}\mathbf{x}(0) = \frac{1}{5}\begin{bmatrix} 5e^{3t} \\ 2e^{2t} + 3e^{-3t} \\ 8e^{2t} - 3e^{-3t} \end{bmatrix}$$

We could also calculate $\mathbf{e}^{\mathbf{A}t}$ using the spectral decomposition of $\mathbf{A}$. That is, find $\mathbf{E}_1$, $\mathbf{E}_2$, $\mathbf{E}_3$ for which

$$\mathbf{A} = \lambda_1\mathbf{E}_1 + \lambda_2\mathbf{E}_2 + \lambda_3\mathbf{E}_3$$

Then

$$\mathbf{e}^{\mathbf{A}t} = e^{\lambda_1 t}\mathbf{E}_1 + e^{\lambda_2 t}\mathbf{E}_2 + e^{\lambda_3 t}\mathbf{E}_3 \tag{3.110}$$

We can find $\mathbf{E}_1$, $\mathbf{E}_2$, $\mathbf{E}_3$ using the following

$$\mathbf{A}^k = \lambda_1^k\mathbf{E}_1 + \lambda_2^k\mathbf{E}_2 + \lambda_3^k\mathbf{E}_3 \tag{3.111}$$

And so

$$\begin{aligned} \mathbf{A}^0 &= \mathbf{I} = \mathbf{E}_1 + \mathbf{E}_2 + \mathbf{E}_3 \\ \mathbf{A} &= \lambda_1\mathbf{E}_1 + \lambda_2\mathbf{E}_2 + \lambda_3\mathbf{E}_3 \\ \mathbf{A}^2 &= \lambda_1^2\mathbf{E}_1 + \lambda_2^2\mathbf{E}_2 + \lambda_3^2\mathbf{E}_3 \end{aligned} \tag{3.112}$$

We can use the three equations in (3.112) to find $\mathbf{E}_1$, $\mathbf{E}_2$, and $\mathbf{E}_3$ and then use (3.110) to find $\mathbf{e}^{\mathbf{A}t}$. ■

In the above example, we found a general expression for the state of a system with state matrix $\mathbf{A}$ and initial state $\mathbf{x}(0)$ as given. One of the first things we should do is verify that all signals internal to system remain within those bounds within which the system is stable. We see that all states above grow without bound as e^{3t} or e^{2t}. Hence, this particular system is unstable, and we must check the suitability of our model. After a few calculations like this, we shall begin to ask whether there is not some easier way to examine system stability, given the system matrix $\mathbf{A}$. Just as with the discrete-time case, the answer is yes. We can verify that our system is either stable or unstable without solving for the state and output behavior in detail. Again, it is the set of eigenvalues of $\mathbf{A}$ that determine system behavior. The eigenvalues of $\mathbf{A}$ are identical with the roots of the associated auxiliary equation defined by the differential equation model. Since these roots define the transient response of the system, their values also define system stability.

It can be shown that the eigenvalues of **A** are the same as those of the matrix $\mathbf{WAW}^{-1}$ for any $n \times n$ matrix **W** whose inverse exists. Let us assume for convenience that the eigenvalues of the state matrix **A** are distinct, and choose for **W** the diagonalizing matrix **P** whose column are the eigenvectors of **A**. Then, as before,

$$\mathbf{P}^{-1}\mathbf{AP} \triangleq \mathbf{D} = \begin{bmatrix} \lambda_1 & 0 & \cdots & 0 \\ 0 & \lambda_2 & \cdots & 0 \\ & & \ddots & \\ 0 & 0 & & \lambda_n \end{bmatrix}$$

And so

$$\mathbf{A} = \mathbf{PDP}^{-1}$$

Substituting in the definition for $\mathbf{e}^{\mathbf{A}t}$, we find

$$\begin{aligned} \mathbf{e}^{\mathbf{A}t} &= \sum_{k=0}^{\infty} \frac{t^k}{k!} \mathbf{A}^k = \sum_{k=0}^{\infty} \frac{t^k}{k!} (\mathbf{PDP}^{-1})^k \\ &= \mathbf{P}\left[\sum_{k=0}^{\infty} \frac{t^k}{k!} \mathbf{D}^k\right]\mathbf{P}^{-1} \end{aligned}$$

However, the matrix $\sum_{k=0}^{\infty}(t^k/k!)\mathbf{D}^k$ is merely

$$\sum_{k=0}^{\infty} \frac{t^k}{k!} \begin{bmatrix} \lambda_1^k & 0 & \cdots & 0 \\ 0 & \lambda_2^k & \cdots & 0 \\ & & \ddots & \\ 0 & 0 & \cdots & \lambda_n^k \end{bmatrix} = \begin{bmatrix} e^{\lambda_1 t} & 0 & \cdots & 0 \\ 0 & e^{\lambda_2 t} & \cdots & 0 \\ & & \ddots & \\ 0 & 0 & \cdots & e^{\lambda_n t} \end{bmatrix}$$

Hence,

$$\mathbf{e}^{\mathbf{A}t} = \mathbf{P}\begin{bmatrix} e^{\lambda_1 t} & 0 & \cdots & 0 \\ 0 & e^{\lambda_2 t} & \cdots & 0 \\ & & \ddots & \\ 0 & 0 & \cdots & e^{\lambda_n t} \end{bmatrix}\mathbf{P}^{-1} \tag{3.113}$$

Therefore the elements of $\mathbf{e}^{\mathbf{A}t}$, and hence the components of $\mathbf{x}(t)$, are composed of sums of the functions $e^{\lambda_1 t}, e^{\lambda_2 t}, \ldots, e^{\lambda_n t}$. If any of these functions grows without bound, the system will be unstable. This growth will occur if the real part of any eigenvalue is greater than zero. Also, if the real part of any eigenvalue equals zero, we can find a bounded input that will cause the integral in (3.103) to become unbounded. Conversely, if the real parts of all eigenvalues are less than zero, our solution will remain unbounded for all bounded inputs. Thus our conclusion:

A continuous-time system is stable if, and only if, all eigenvalues of the state matrix have real parts less than zero.

Example 3.21 We are given a continuous-time system whose state matrix is

$$\mathbf{A} = \begin{bmatrix} \alpha & 0 & 0 \\ 0 & \beta & -1 \\ 0 & 1 & -2 \end{bmatrix}$$

For what range of values of α and β is the system stable (assume that α and β are real)?

The characteristic equation for A is

$$g(\lambda) = (\alpha - \lambda)[(\beta - \lambda)(-2 - \lambda) + 1] = 0$$

that is,

$$(\alpha - \lambda)[\lambda^2 + \lambda(2 - \beta) + 1 - 2\beta] = 0 \tag{3.114}$$

The roots of (3.114) are

$$\lambda_1 = \alpha, \qquad \lambda_2 = \frac{\beta - 2}{2} + \frac{\sqrt{\beta(\beta + 4)}}{2}, \qquad \lambda_3 = \frac{\beta - 2}{2} - \frac{\sqrt{\beta(\beta + 4)}}{2}$$

For stability, we clearly must have $\alpha < 0$. Investigating the range of β for stability involves two cases.

Case 1: λ_2 and λ_3 are complex for $-4 < \beta < 0$. In this case we must have $(\beta - 2)/2 < 0$, which means $\beta < 2$. Thus, the system is stable for all β in the range $-4 < \beta < 0$.

Case 2: λ_2 and λ_3 are real for $\beta \leq -4$ or $\beta \geq 0$. Because $\lambda_2 \geq \lambda_3$, we set

$$0 > \lambda_2 = \frac{\beta - 2}{2} + \frac{\sqrt{\beta^2 + 4\beta}}{2}$$

Thus we have

$$2 - \beta > \sqrt{\beta^2 + 4\beta}$$

from which we find

$$\beta < \frac{1}{2}$$

The conclusion is that the system is stable for all $\alpha < 0$ and $\beta < \frac{1}{2}$. ■

Example 3.22 For the circuit shown in Figure 3.17, find the step response using state-variable methods. Also calculate the impulse response using two different methods.

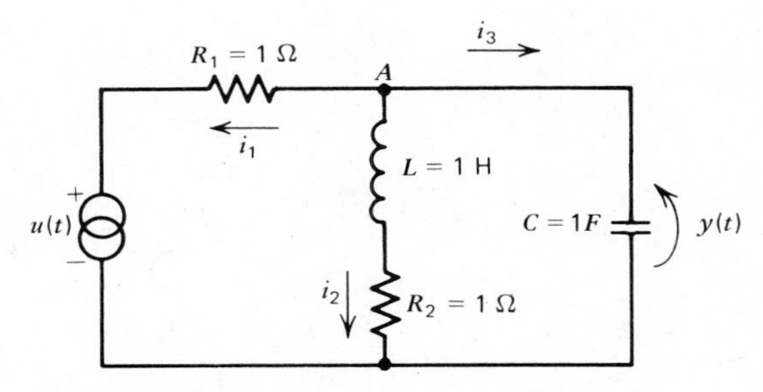

Figure 3.17

Because we wish to find the impulse response by two methods, let us first derive the differential equation relating the input $u(t)$ and the output $y(t)$. If we write Kirchhoff's current equation at node A, we have

$$i_1(t) + i_2(t) + i_3(t) = 0 \tag{3.115}$$

The various currents are given by

1. $$i_1(t) = \frac{y(t) - u(t)}{R_1}$$

2. $$i_2(t) = \frac{1}{L}\int_{-\infty}^{t} [y(\tau) - i_2(\tau)R_2]\, d\tau = \frac{1}{L}\int_{-\infty}^{t} v_L(\tau)\, d\tau$$

or
$$i_2^{(1)}(t) = \frac{1}{L}[y(t) - i_2(t)R_2]$$

where $v_L(t)$ is the voltage drop across the inductor L.

3.
$$i_3(t) = Cy^{(1)}(t)$$

If we differentiate (3.115), we have that

$$i_1^{(1)}(t) + i_2^{(1)}(t) + i_3^{(1)}(t) = 0 \tag{3.116}$$

Now use the expressions for the currents obtained above and substitute into (3.116). We have

$$\frac{y^{(1)}(t) - u^{(1)}(t)}{R_1} + \frac{1}{L}[y(t) - i_2(t)R_2] + Cy^{(2)}(t) = 0$$

Now substitute for $i_2(t) = -i_1(t) - i_3(t)$ to obtain

$$y^{(2)}(t) + 2y^{(1)}(t) + 2y(t) = u^{(1)}(t) + u(t) \tag{3.117}$$

Equation 3.117 is the differential equation that relates the input $u(t)$ to the output $y(t)$. We can use (3.117) to find the impulse response of the system using the methods of Section 3.6. However, before we do, we shall calculate the impulse response using a state-space formulation. To obtain a state-space representation, define the state variables as the current through the inductor and the voltage across the capacitor as shown in Figure 3.18.

Writing Kirchhoff's voltage equation around the loop containing $u(t)$, we obtain

$$u(t) = x_1(t) + i_1(t)R_1$$

The current $i_1(t)$ is

$$i_1(t) = x_2(t) + Cx_1^{(1)}(t)$$

Thus

$$u(t) = x_1(t) + R_1[x_2(t) + Cx_1^{(1)}(t)]$$

and so

$$x_1^{(1)}(t) = u(t) - x_1(t) - x_2(t) \tag{3.118}$$

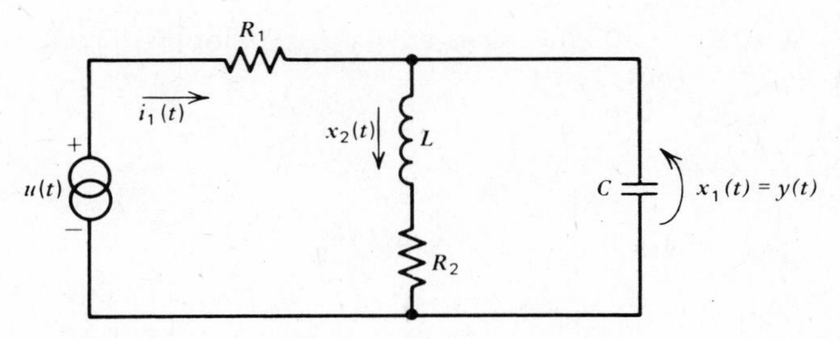

Figure 3.18

A second equation can be obtained by writing a voltage equation around the second loop.

$$x_1(t) = Lx_2^{(1)}(t) + Rx_2(t)$$

and so

$$x_2^{(1)}(t) = x_1(t) - x_2(t) \tag{3.119}$$

Combining (3.118) and (3.119), we obtain

$$\mathbf{x}^{(1)}(t) = \begin{bmatrix} -1 & -1 \\ 1 & -1 \end{bmatrix} \mathbf{x}(t) + \begin{bmatrix} 1 \\ 0 \end{bmatrix} u(t) \tag{3.120}$$

The output is clearly

$$y(t) = [1 \quad 0]\mathbf{x}(t) \tag{3.121}$$

In general, for zero initial conditions, the output $y(t)$ for an input $u(t)$ is

$$y(t) = \int_0^t \mathbf{C}e^{\mathbf{A}(t-\tau)}\mathbf{B}u(\tau)\, d\tau + \mathbf{D}u(t) \tag{3.122}$$

From (3.122) we conclude that the impulse response is (see 3.105)

$$h(t) = \begin{cases} \mathbf{C}e^{\mathbf{A}t}\mathbf{B} + \mathbf{D}\delta(t), & t \geq 0 \\ 0, & t < 0 \end{cases} \tag{3.123}$$

To calculate $h(t)$ from (3.123), we need an expression for $\mathbf{e^{At}}$. Thus, we first obtain the eigenvalues of $\mathbf{A}$ from

$$g(\lambda) = \det [\mathbf{A} - \lambda \mathbf{I}] = \lambda^2 + 2\lambda + 2 = 0$$

The roots of $g(\lambda) = 0$ are $\lambda_1 = -1 + j$ and $\lambda_2 = -1 - j$. Now

$$e^{\lambda t} = \beta_0 + \beta_1 \lambda$$

where β_0 and β_1 are found from the equations

$$e^{(-1+j)t} = \beta_0 + \beta_1(-1 + j)$$

$$e^{(-1-j)t} = \beta_0 + \beta_1(-1 - j)$$

These equations yield

$$\beta_0 = e^{-t} \sin t$$

$$\beta_1 = e^{-t}(\sin t + \cos t)$$

Thus $\mathbf{e^{At}}$ is the matrix

$$\mathbf{e^{At}} = \beta_0 \mathbf{I} + \beta_1 \mathbf{A} = e^{-t} \begin{bmatrix} \cos t & -\sin t \\ \sin t & \cos t \end{bmatrix}$$

The impulse response of the system is therefore

$$\begin{aligned} h(t) = \mathbf{Ce^{At}B} &= [1 \quad 0] e^{-t} \begin{bmatrix} \cos t & -\sin t \\ \sin t & \cos t \end{bmatrix} \begin{bmatrix} 1 \\ 0 \end{bmatrix}, \qquad t \geq 0 \\ &= e^{-t} \cos t \, \xi(t) \end{aligned} \tag{3.124}$$

Let us return to the original differential equation (3.118) that relates the input and output. We first find the homogeneous solution to (3.118). The auxiliary equation of (3.117) is

$$r^2 + 2r + 2 = 0$$

which has roots $r_1 = -1 + j$ and $r_2 = -1 - j$. Thus, the impulse-response function $\hat{h}(t)$ for the system represented by $y^{(2)}(t) + 2y^{(1)}(t) + 2y(t) = u(t)$ is given by

$$\hat{h}(t) = e^{-t}(c_1 \cos t + c_2 \sin t)\xi(t)$$

We evaluate the constants c_1 and c_2 using the initial conditions $\hat{h}(0) = 0$ and $\hat{h}^{(1)}(0) = 1$. Thus we have

$$\begin{aligned} \hat{h}(0) &= 0 = c_1 \\ \hat{h}^{(1)}(0) &= 1 = e^{-t}(-c_2 \sin 0 + c_2 \cos 0) \\ &= c_2 \end{aligned}$$

from which

$$\hat{h}(t) = e^{-t} \sin t \; \xi(t) \tag{3.125}$$

The impulse-response function corresponding to (3.117) is found by applying the differential operator $(D + 1)$ to $\hat{h}(t)$. Thus,

$$\begin{aligned} h(t) &= (D + 1)[\hat{h}(t)] \\ &= -e^{-t} \sin t \; \xi(t) + e^{-t} \cos t \; \xi(t) + e^{-t} \sin t \; \delta(t) + e^{-t} \sin t \; \xi(t) \\ &= e^{-t} \cos t \; \xi(t) \end{aligned} \tag{3.126}$$

Comparing (3.126) and (3.124), we see that they are identical.

To find the step response of this system, we can proceed in various ways. Probably the simplest method is to convolve the step input with the impulse-response function. Thus

$$\begin{aligned} y(t) &= \int_0^t h(\tau)\xi(t - \tau) \, d\tau \\ &= \int_0^t e^{-\tau} \cos \tau \, d\tau \\ &= \begin{cases} \frac{1}{2}[1 + e^{-t} \sin t - e^{-t} \cos t], & t \geq 0 \\ 0, & t < 0 \end{cases} \end{aligned} \tag{3.127}$$

We can also obtain the step response from the state-space formulation. In this case,

$$y(t) = \int_0^t \mathbf{C}e^{\mathbf{A}(t-\tau)}\mathbf{B}u(\tau)\,d\tau = \int_0^t \mathbf{C}e^{\mathbf{A}\tau}\mathbf{B}\,d\tau$$

$$= \int_0^t \mathbf{C}\begin{bmatrix} e^{-\tau}\cos\tau \\ e^{-\tau}\sin\tau \end{bmatrix} d\tau$$

$$= \mathbf{C}\begin{bmatrix} \frac{1}{2}[1 + e^{-t}\sin t - e^{-t}\cos t] \\ \frac{1}{2}[1 - e^{-t}\sin t - e^{-t}\cos t] \end{bmatrix}$$

Therefore, as before

$$y(t) = \frac{1}{2}[1 + e^{-t}\sin t - e^{-t}\cos t]\xi(t) \tag{3.128}$$

As a bonus, we find that $x_2(t)$, the current through the inductor, is

$$i_L(t) = x_2(t) = \frac{1}{2}[1 - e^{-t}\sin t - e^{-t}\cos t]\xi(t)$$

■

3.10 FREQUENCY RESPONSE IN TERMS OF A, B, C, D

The state-variable equations for continuous-time system are repeated below.

$$\mathbf{x}^{(1)}(t) = \mathbf{A}\mathbf{x}(t) + \mathbf{B}u(t) \tag{3.129}$$

$$y(t) = \mathbf{C}\mathbf{x}(t) + \mathbf{D}u(t) \tag{3.130}$$

If we assume an input of the form $u(t) = e^{j\omega t}$, the corresponding output $y(t)$ is of the form $H(j\omega)e^{j\omega t}$. If we also assume that the steady-state solution for the state $\mathbf{x}(t)$ is $\mathbf{X}(j\omega)e^{j\omega t}$, we can proceed as follows. Substitute in (3.129) $e^{j\omega t}$ for $u(t)$ and $\mathbf{X}(j\omega)e^{j\omega t}$ for $\mathbf{x}(t)$. We obtain

$$\mathbf{X}(j\omega)j\omega e^{j\omega t} = \mathbf{A}\mathbf{x}(j\omega)e^{j\omega t} + \mathbf{B}e^{j\omega t}$$

This can be written in the form

$$(\mathbf{X}(j\omega)j\omega - \mathbf{A}\mathbf{X}(j\omega))e^{j\omega t} = \mathbf{B}e^{j\omega t}$$

Cancelling $e^{j\omega t}$ on both sides gives

$$(\mathbf{I}j\omega - \mathbf{A})\mathbf{X}(j\omega) = \mathbf{B}$$

or

$$\mathbf{X}(j\omega) = (\mathbf{I}j\omega - \mathbf{A})^{-1}\mathbf{B} \tag{3.131}$$

We can now substitute in (3.130) $H(j\omega)e^{j\omega t}$ for $y(t)$, $\mathbf{X}(j\omega)e^{j\omega t}$ for $\mathbf{x}(t)$, and $e^{j\omega t}$ for $u(t)$. Then

$$H(j\omega)e^{j\omega t} = \mathbf{C}\mathbf{X}(j\omega)e^{j\omega t} + \mathbf{D}e^{j\omega t}$$

Substitute (3.131) for $\mathbf{X}(j\omega)$ and cancel the terms $e^{j\omega t}$. We then have the desired expression for $H(j\omega)$.

$$H(j\omega) = \mathbf{C}(\mathbf{I}j\omega - \mathbf{A})^{-1}\mathbf{B} + \mathbf{D} \tag{3.132}$$

Note the similarity between this formula and the corresponding expression (2.106) for discrete-time systems. The only difference is that $j\omega$ replaces $e^{j\theta}$. In continuous-time system we traverse the $j\omega$-axis to calculate frequency response. In discrete-time systems we traverse the unit circle $e^{j\theta}$ to calculate frequency response. This discussion will become more apparent when we consider frequency response in the transform domain.

3.11 SUMMARY

We have discussed three time domain models for continuous-time systems in this chapter. Our discussion is very similar to that in Chapter 2. In fact, our thought processes are identical. Only the mathematical details are changed. We have used the idea of frequency response as a means of linking the three models together. Frequency response analysis assumes a linear, constant-parameter system. If the system is time-varying, then frequency response as discussed here is no longer a valid method of characterizing the system.

The remaining chapters will take up the concepts involved in describing systems in the transform domain. We shall again consider discrete-time systems first. These transform domain characterizations are useful. They are, like the difference or differential equations and the impulse-response models, input-output characterizations of a system. They do not specify the internal structure of a system. However, they do introduce the idea of poles and zeros of a system. The concept of poles and zeros can be used to give a geometric interpretation of a system's response.

PROBLEMS

3.1. Solve the following differential equations using the given initial conditions.

(a) $(D^2 + 3D + 2)y(t) = 0$ $\qquad y(0) = 1, \left.\dfrac{dy(t)}{dt}\right|_{t=0} = 0$

(b) $(D^2 + 2D + 1)y(t) = 0$ $\qquad y(0) = 1, \left.\dfrac{dy(t)}{dt}\right|_{t=0} = 0$

(c) $(D^3 + 4D^2 + 5D + 2)y(t) = 0$ $\qquad y(0) = 1, \left.\dfrac{dy(t)}{dt}\right|_{t=0} = \left.\dfrac{d^2y(t)}{dt^2}\right|_{t=0} = 0$

(d) $(D^2 - 1)y(t) = 0$ $\qquad y(0) = 1, \left.\dfrac{dy(t)}{dt}\right|_{t=0} = 1$

(e) $(D^2 + 1)y(t) = 0$ $\qquad y(0) = 1, \left.\dfrac{dy(t)}{dt}\right|_{t=0} = 0$

(f) $(D^2 + 1)y(t) = 0$ $\qquad y(0) = 0, \left.\dfrac{dy(t)}{dt}\right|_{t=0} = 1$

(g) $(D^2 + 1)y(t) = 0$ $\qquad y(0) = \left.\dfrac{dy(t)}{dt}\right|_{t=0} = 1$

(h) $(D^4 - 1)y(t) = 0$ $\qquad y(0) = 1, \left.\dfrac{dy(t)}{dt}\right|_{t=0} = \left.\dfrac{d^2y(t)}{dt^2}\right|_{t=0} = \left.\dfrac{d^3y(t)}{dt^3}\right|_{t=0} = 0$

3.2. In the *RC* circuit shown below, the initial voltage on the capacitor is $2V$, that is, $y(0) = 2$. If the input voltage $u(t) = 0$ for $t \geq 0$ (obtained by placing a short across the input), what is the output voltage $y(t)$ for $t \geq 0$? *Answer*: $y(t) = 2e^{-t/RC}$.

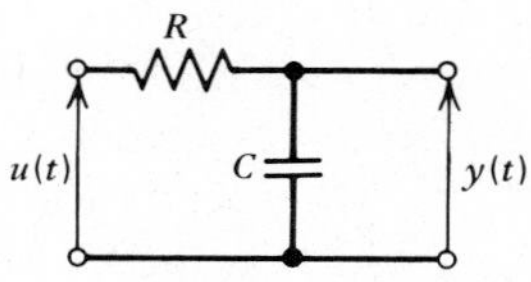

3.3. Find the appropriate annihilator for each of the following forcing functions.

(a) Ae^{at} (b) Bt^2e^{at}
(c) $A \sin \omega t$ (d) $A \sin \omega t + B \cos \omega t$
(e) $A \sin (\omega t + \phi)$ (f) $Ae^{at} + Be^{bt} + Ce^{ct}$
(g) $t^3 + B \sin t$

3.4. (a) Time-invariant linear systems forced by sinusoidal inputs form an important class of problems. One model of such systems is made by means of a linear differential equation. Show that the forced solution of the following differential equation must be of the form αe^{kt}:

$$\frac{d^2y(t)}{dt^2} + a\frac{dy(t)}{dt} + by(t) = ce^{kt}$$

where a, b, c, and k are known constants and k is not a root of the characteristic equation.

(b) In view of the result of part (a), what must the form of the forced solution be for a forcing function $c \sin \omega t$ (rather than ce^{kt})?

(c) What is the form of the forced solution if the forcing function is ce^{kt} and k satisfies the equation $k^2 + ak + b = 0$? What physical interpretation can you give to this result?

3.5. Solve the following differential equations.

(a) $(D^4 + 8D^2 + 16)y(t)] = -\sin t$

Answer. $y(t) = c_1 \cos 2t + c_2 \sin 2t + c_3 t \cos 2t + c_4 t \sin 2t - \dfrac{\sin t}{9}$

(b) $(D^3 - 2D^2 + D - 2)[y(t)] = 0, \quad y(0) = \left.\dfrac{dy(t)}{dt}\right|_{t=0} = \left.\dfrac{d^2y(t)}{dt^2}\right|_{t=0} = 1$

Answer. $y(t) = \dfrac{1}{5}(2e^{2t} + 3 \cos t + \sin t)$

(c) $(D^4 - D)[y(t)] = t^2$

Answer. $y(t) = c_1 + c_2 e^t + e^{-t/2}\left(c_3 \cos \dfrac{\sqrt{3}t}{2} + c_4 \dfrac{\sin \sqrt{3}t}{2}\right) - \dfrac{t^3}{3}$

✓ **3.6.** Solve the following differential equations.

(a) $(D^2 + 3D + 2)y(t) = 0 \qquad y(0) = \left.\dfrac{dy(t)}{dt}\right|_{t=0} = 1$

(b) $(D^2 + 3D + 2)y(t) = e^{-t} \qquad y(0) = \left.\dfrac{dy(t)}{dt}\right|_{t=0} = 0$

(c) $(D^2 + 3D + 2)y(t) = e^{-t} \qquad y(0) = \left.\dfrac{dy(t)}{dt}\right|_{t=0} = 1$

Comparing the three solutions, can you generalize the results you obtained? Compare with the solutions to 3.1 (e), (f), and (g).

3.7. Solve the following differential equations.

(a) $(D^2 + 2D + 1)y(t) = e^{-t} \qquad y(0) = 1, \left.\frac{dy(t)}{dt}\right|_{t=0} = 0$

(b) $(D^2 - 1)y(t) = e^{-t} \qquad y(0) = \left.\frac{dy(t)}{dt}\right|_{t=0} = 0$

(c) $(D^2 + 1)y(t) = \cos t \qquad y(0) = \left.\frac{dy(t)}{dt}\right|_{t=0} = 0$

3.8. Find the impulse response for the continuous-time systems defined by the following differential equations. Verify your solutions by substitution.

(a) $(D^2 + 7D + 12)[y(t)] = u(t)$
(b) $(D^2 + 6D + 9)[y(t)] = u(t)$
(c) $(D^2 + 2D + 9)[y(t)] = u(t)$
(d) $(D^3 + 6D^2 + 12D + 8)[y(t)] = u(t)$
(e) $(D^3 + 6D^2 + 12D + 8)[y(t)] = (D - 1)[u(t)]$

3.9. A system has the following response to a step input. What would be the response of this system to a ramp input, $u(t) = t\xi(t)$? (*Hint*: Assume that the system can be modeled by the use of a first-order differential equation.)

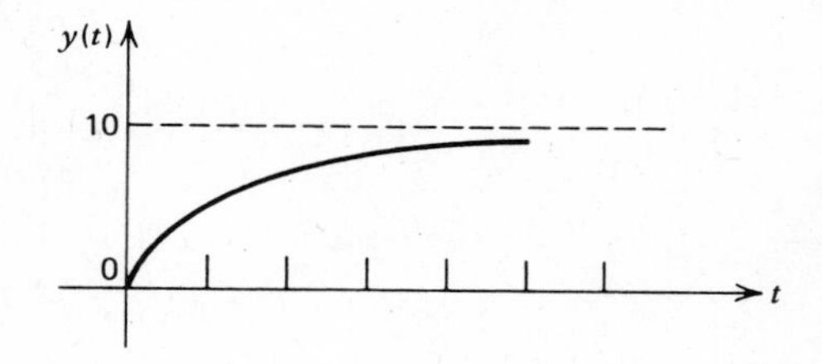

3.10. Derive a model for nanoplankton respiration that accounts for gross community photosynthesis, storage, and respiration.† The following analogies may be helpful.

The input to the system is sunlight, which can be represented by a voltage, say e_b. The production rate p of material by photosynthesis is proportional to the difference in input sunlight and the material already in the system (a "back potential" e_m). The constant of proportionality between production rate p and this difference potential is a conductance G_b. The community respiration rate p_r is assumed to be proportional to e_m. The storage rate p_c of material in the system is assumed to be proportional to the rate of change of material already in the system, and the constant of pro-

† H. T. Odum, et al., "Consequences of Small Storage Capacity in Nonoplankton Pertinent to Primary Production in Tropical Waters," *Journal of Marine Research* (June 1963).

portionality is C^{-1}, representative of the storage capacity of the system. In symbols, $p_c = (1/C)de_m/dt$. Also, the total production rate p is equal to the sum of the respiration and storage rates. (Notice that production rates are analogous to currents.)

(a) Obtain a block diagram representation of the system.
(b) Obtain an electrical schematic model for this system.
(c) Using your model, find the dependence of the respiration rate p_r upon a step input of sunlight.

3.11. Given $d^2y(t)/dt^2 + 5[dy(t)/dt] + 6y(t) = e^{-t}$, find $y(0)$ and $dy/dt(t)|_{t=0}$ such that the solution is $y(t) = ce^{-t}$. Evaluate c.

✓ **3.12.** In our treatment of convolution, we assumed that our systems were initially relaxed, with no initial energy storage. How can this method be extended to cover systems where there is some initial energy storage? Specifically, use the impulse-response function to express the output of a system described by the equation

$$(D^2 + 1)[y(t)] = u(t), \qquad t > 0$$

where $y(0) = 1$ and $y^{(1)}(0) = 0$. Notice that initial conditions are nonzero, denoting some initial energy storage in the system.

Answer. $y(t) = \cos t + \int_0^t \sin(t - \tau)u(\tau)\, d\tau$

✓**3.13.** Evaluate the following convolutions:

(a) $\xi(t) * \xi(t)$
(b) $\xi(t) * e^{at}\xi(t)$
(c) $t\xi(t) * e^{at}\xi(t)$
(d) $e^{at}\xi(t) * e^{at}\xi(t)$
(e) $e^{at}\xi(t) * e^{-at}\xi(-t)$
(f) $\sin t\xi(t) * \sin t\xi(t)$

3.14. (a) Using the convolution integral, find the output signal for the system shown. This system is often used to smooth a sequence $\{u(nT)\}$ into a continuous-time function. It is called a zero-order hold circuit.

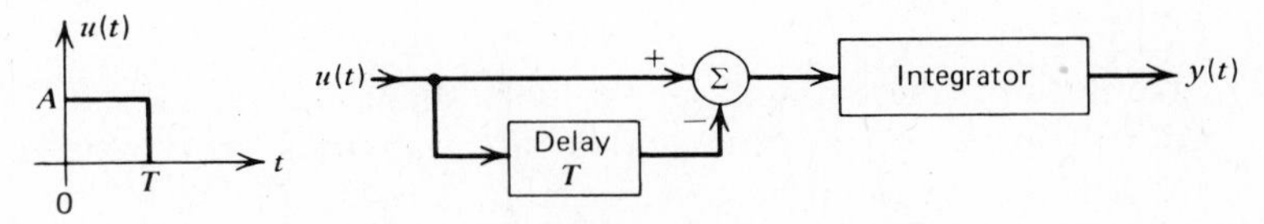

(b) What is the output if $u(t)$ is applied to two of the above systems in cascade?

3.15. Solve the following equations using both the direct method and the convolution method.

(a) $(D^2 + 7D + 12)y(t) = e^t\xi(t), \qquad y(0) = \left.\frac{dy(t)}{dt}\right|_{t=0} = 0$

(b) $(D^2 + 3D + 2)y(t) = e^{-t}\xi(t), \qquad y(0) = \left.\frac{dy(t)}{dt}\right|_{t=0} = 0$

3.16. Given a system with input $u(t)$ and output $y(t)$ such that

$$\frac{d^2y(t)}{dt^2} + 3\frac{dy(t)}{dt} + \frac{9}{4}y(t) = u(t)$$

find and sketch the corresponding impulse response $h(t)$.

3.17. Find the impulse response $h(t)$ for the system shown:

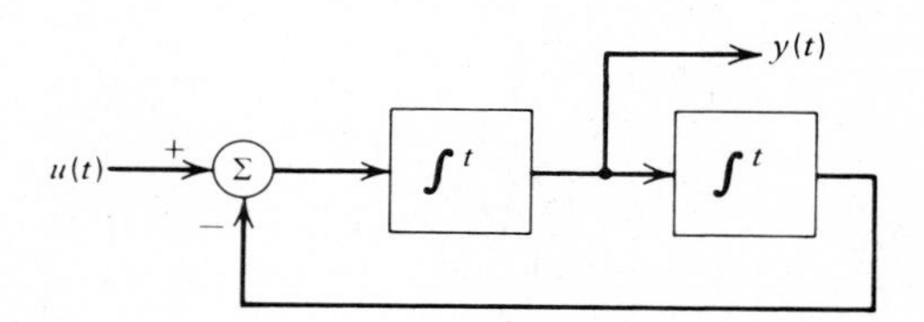

3.18. Given a system with input $u(t)$ and output $y(t)$ related by

$$\frac{d^2y(t)}{dt^2} + 3\frac{dy(t)}{dt} + 2y(t) = u(t)$$

(a) Show a block diagram of the system using integrators, summers, and coefficient multipliers.
(b) Solve for $y(t)$ if $u(t) = e^{-t}\xi(t)$ with $y(0) = dy(t)/dt|_{t=0} = 0$
(c) Find the system impulse response $h(t)$.
(d) Repeat (b) using convolution.

3.19. Repeat 3.18 (a)–(d) for a system described by the input-output relation

$$\frac{d^2y(t)}{dt^2} + 4\frac{dy(t)}{dt} + 4y(t) = \frac{d}{dt}u(t)$$

Do not use differentiators in the block diagram. If the results in (b) and (c) do not agree, find the cause and bring the answers into agreement. (*Hint*: Do the initial conditions represent a relaxed system?)

3.20. For the system shown below, find a differential equation that relates $y(t)$ and $u(t)$. Solve for $y(t)$ when $u(t) = \sin \omega t\ \xi(t)$ with a relaxed system at $t = 0$.

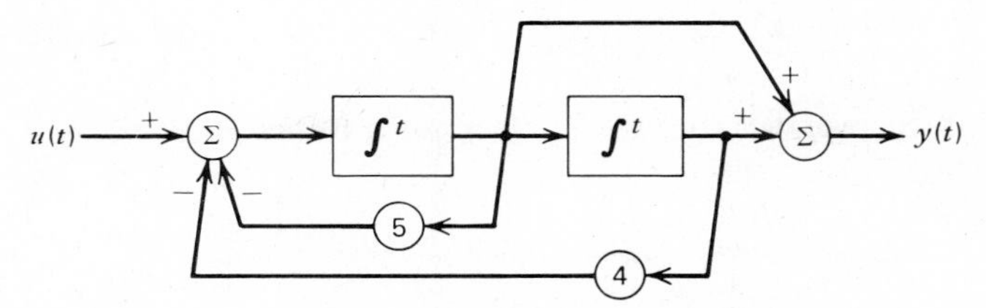

3.21. Can you find a simpler equivalent block diagram for the above system? What is the impulse response of the system? Using convolution, repeat your solution for the output $y(t)$, which results from $u(t) = \sin \omega t \xi(t)$.

3.22. Determine a system block diagram for which the input $u(t)$ and the output $y(t)$ satisfy the differential equation

$$\frac{d^2 y(t)}{dt^2} + \frac{2\,dy(t)}{dt} + 2y(t) = \frac{du(t)}{dt}$$

What is the solution to this equation when the input is $u(t) = \cos(\omega_1 t) - \sin(\omega_2 t)$?

3.23. Write state-variable equations for each of the systems shown.

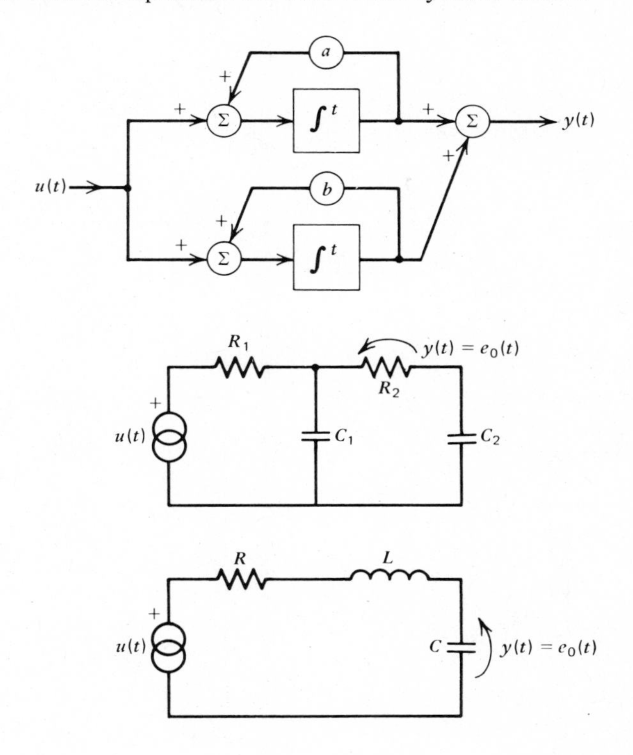

3.24. For what values of the parameters involved is each of the systems above stable?

3.25. Write a state-variable description of the following systems. For what values of K will the systems be stable?

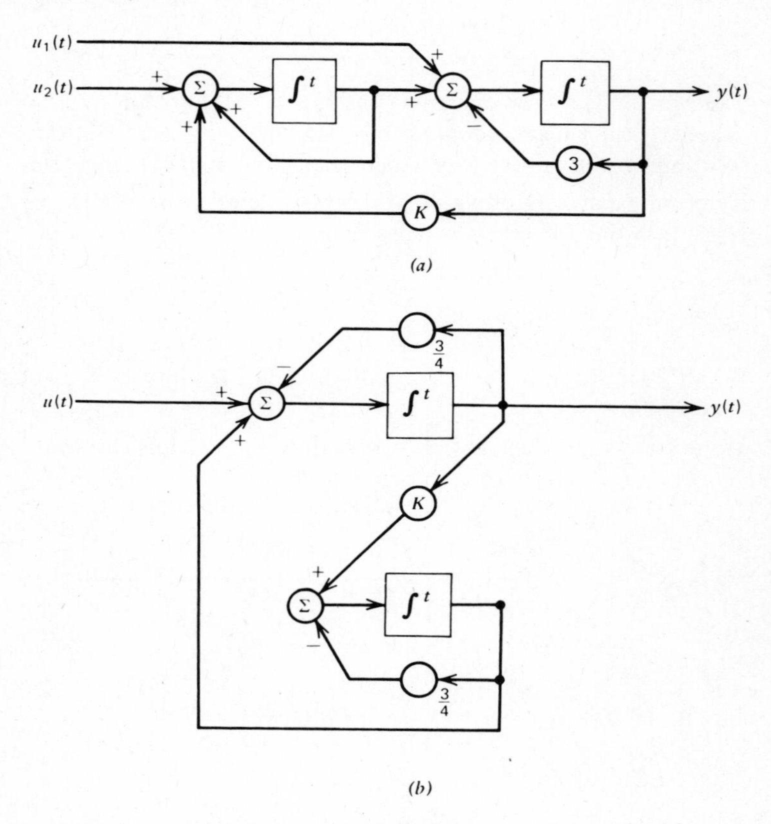

(a)

(b)

3.26. Find closed-form expressions for $\mathbf{e^{At}}$ for each of the following matrices:

(a) $\mathbf{A} = \begin{bmatrix} \frac{3}{4} & 0 \\ \frac{1}{2} & \frac{1}{2} \end{bmatrix}$ (b) $\mathbf{A} = \begin{bmatrix} \frac{1}{2} & \frac{1}{4} \\ \frac{1}{16} & \frac{1}{2} \end{bmatrix}$

(c) $\mathbf{A} = \begin{bmatrix} \frac{1}{2} & \frac{1}{4} \\ 1 & \frac{1}{2} \end{bmatrix}$ (d) $\mathbf{A} = \begin{bmatrix} \frac{1}{2} & 0 \\ 1 & \frac{1}{2} \end{bmatrix}$

(e) $\mathbf{A} = \begin{bmatrix} \frac{3}{4} & -\frac{1}{2} \\ -\frac{15}{32} & \frac{1}{2} \end{bmatrix}$ (f) $\mathbf{A} = \begin{bmatrix} 1 & -1 \\ 1 & 1 \end{bmatrix}$

(g) $\mathbf{A} = \begin{bmatrix} -1 & 1 \\ 0 & -1 \end{bmatrix}$ (h) $\mathbf{A} = \begin{bmatrix} -4 & -1 \\ 16 & 4 \end{bmatrix}$

3.27. For the block diagram systems shown below, find

(i) The matrices (**A**, **B**, **C**, **D**) of the state-variable description.
(ii) The matrix $\mathbf{e^{At}}$.
(iii) The matrix $(j\omega\mathbf{I} - \mathbf{A})^{-1}$.
(iv) The frequency-response function, with a sketch of the amplitude and phase responses.
(v) The impulse-response function, with a sketch.

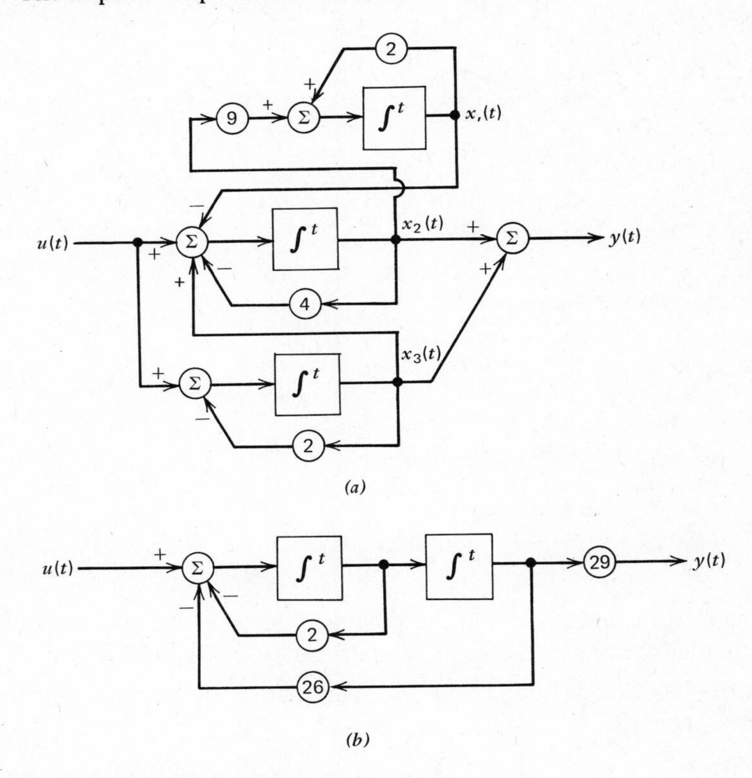

(a)

(b)

3.28. Consider the following state-variable system:

$$\begin{bmatrix} \dfrac{dx_1(t)}{dt} \\ \dfrac{dx_2(t)}{dt} \end{bmatrix} = \begin{bmatrix} 0 & 1 \\ -2 & -3 \end{bmatrix} \begin{bmatrix} x_1(t) \\ x_2(t) \end{bmatrix} + \begin{bmatrix} 0 \\ 1 \end{bmatrix} u(t)$$

$$y(t) = [c_1 \quad c_2] \begin{bmatrix} x_1(t) \\ x_2(t) \end{bmatrix} + [d]u(t)$$

(i) Find the matrix $(j\omega\mathbf{I} - \mathbf{A})^{-1}$.
(ii) Find the matrix $\mathbf{e}^{\mathbf{A}t}$.
(iii) The amplitude-response function for the system is shown below. Determine c_1, c_2, and d.

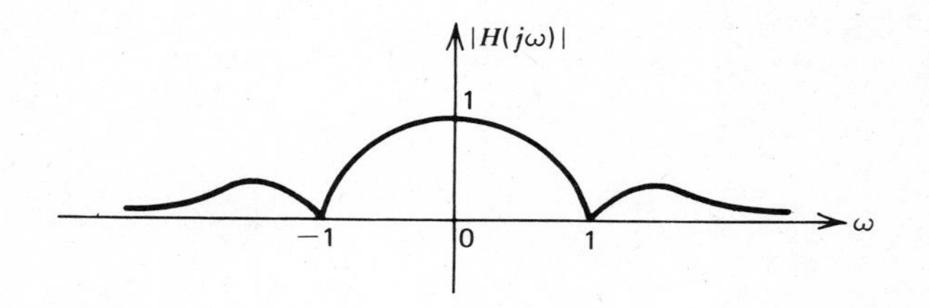

(iv) Find the impulse-response function $h(t)$.
(v) Is this system stable?

3.29. In Sections 3.8–3.9, we discussed the problem of writing and solving the state-variable equations corresponding to the differential equation

$$b_n \frac{d^n y(t)}{dt^n} + \cdots + b_1 \frac{dy(t)}{dt} + b_0 y(t) = u(t)$$

How would you approach the problem of solving the equation

$$b_n \frac{d^n y(t)}{dt^n} + \cdots + b_1 \frac{dy(t)}{dt} + b_0 y(t) = a_0 u(t) + a_1 \frac{du(t)}{dt} + \cdots + a_m \frac{d^m u(t)}{dt^m}$$

using state-variable methods? What is the *minimum* number of state variables that will be required? (Express your answer in terms of m and n.)

3.30. (a) Under what conditions on **A** and **B** (both of dimension $n \times n$) is it true that

$$\mathbf{e}^{\mathbf{A}+\mathbf{B}} = \mathbf{e}^{\mathbf{A}} \cdot \mathbf{e}^{\mathbf{B}}$$

[*Hint*: Compare $\mathbf{e}^{\mathbf{A}+\mathbf{B}}$ and $\mathbf{e}^{\mathbf{A}}\mathbf{e}^{\mathbf{B}}$ term by term using (3.95).]

(b) Using these results, prove that the inverse of $\mathbf{e}^{\mathbf{A}}$ is $\mathbf{e}^{-\mathbf{A}}$.

(c) Show that, if $\mathbf{f}(\mathbf{A})$ is given by

$$\mathbf{f}(\mathbf{A}) = \sum_{k=0}^{\infty} \alpha_k \mathbf{A}^k$$

then $\mathbf{f}(\mathbf{A})$ and **A** commute.

CHAPTER 4
THE Z-TRANSFORM

4.1 INTRODUCTION

In the preceding chapters, we have investigated various methods for formulating models of linear systems and for analyzing their behavior. In Chapter 2, we described a linear time-invariant discrete-time system by means of its difference equation, its impulse-response sequence, and its state-variable formulation. For each of these descriptions, we developed methods for evaluating the system's output given the corresponding input, and for evaluating the system's frequency response. These methods were extended in Chapter 3 to the corresponding differential equation, impulse response, and state-variable descriptions of a linear time-invariant continuous-time system. Again we focused on the computation of a system's output for a given input and on the frequency response. As with the discrete-time formulations, we found a close interrelationship among the various system descriptions. We note for both types of system that the three approaches have one common factor: they are all *time domain* descriptions of a linear system. The independent parameter is either a discrete variable k or a continuous parameter t.

We turn now to a transform-domain description of linear systems. We shall find that if we transform the time signals within our system to another form we can often express relationships more simply than we have previously. In this and the following two chapters, we develop a transform calculus for discrete-time and continuous-time systems. One important consequence of the transform-domain description of linear systems is that the convolution operation in the time domain is converted to a multiplication operation in the transform domain. The procedure is analogous to replacing the multiplication of two numbers by the addition of their logarithms. The situation is shown schematically in Figure 4.1. The familiar Laplace or Fourier transforms are used in the continuous-time domain. The Z-transform is the appropriate transformation for discrete-time systems.

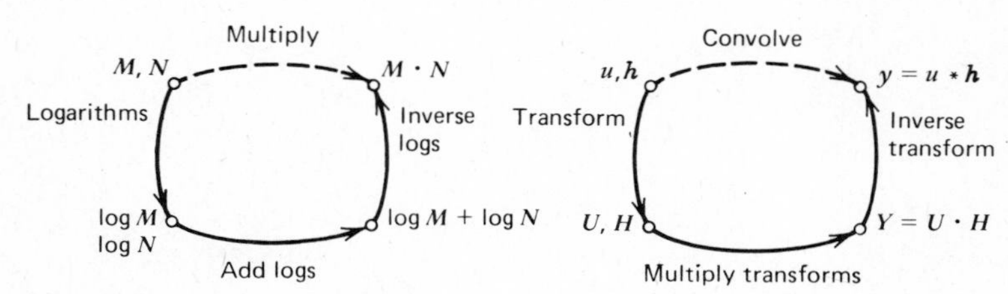

Figure 4.1

Transform methods are of most value in studying linear time-invariant systems. The transform calculus simplifies the study of these systems by:

1. Providing intuition that is not evident in the time-domain solution.
2. Including initial conditions in the solution process automatically.
3. Reducing the solution process of many problems to a simple table look-up, much as one did for logarithms before the advent of hand calculators.

However, one disadvantage of transform-domain solutions, at least pedagogically, seems to be that we often forget the physical nature of the problem in the transform domain. The solution process can become a "turn the crank" process. This tendency is unfortunate, because we then often overlook useful physical attributes of the problem.

An important aim of this study is the development of an awareness in the reader of what is involved, in a general sense, in applying the transform method. For this reason, we first examine the Z-transform, about which the reader most likely has fewer preconceived ideas. This discrete-time transform calculus has been known in probability theory as the "moment-generating function" method. Although we refer to the discrete-time domain, the index does not have to be interpreted as time. In physical applications, other interpretations may be appropriate. Many of the sequences considered here are zero for an index k less than zero. However the development is not restricted to these one-sided sequences. We have chosen to define the Z-transform in negative powers of z, conforming to the engineering literature on systems and signal processing. Publications in mathematics and geophysics, on the other hand, often use positive powers of z. Conversion between the two definitions involves a simple substitution of $1/z$ for z.

4.2 THE *Z*-TRANSFORM

Given a finite sequence x, we define a function $X(z)$ of the complex variable z by forming a polynomial

$$X(z) = x_l z^{-l} + x_{l+1} z^{-(l+1)} + \cdots + x_m z^{-m} = \sum_{k=l}^{m} x_k z^{-k} \tag{4.1}$$

The function $X(z)$ is called the *generating function* or *Z-transform* of the sequence x. (We should distinguish between the function $X(\cdot)$ and the value the function assumes for a given z. However, in agreement with standard notation, we shall follow the accepted practice of using $X(z)$ to mean either.) In (4.1), the beginning and ending indices of the sequence, l and m, can be any integers including $-\infty$ and ∞. The next examples illustrate the calculations used in obtaining the Z-transform for some simple sequences.

Example 4.1 Suppose the sequence x is the finite sequence

$$\{x_k\} = \{8, 3, \underset{\uparrow}{-2}, 0, 4, -6\}.$$

The Z-transform using (4.1) is

$$X(z) = \sum_{k=-2}^{3} x_k z^{-k} = 8z^2 + 3z - 2 + 4z^{-2} - 6z^{-3}$$ ■

Example 4.2 Consider the Z-transform of the sequence x defined by

$$x_k = \begin{cases} 0, & k < 0 \\ 2^k, & k \geq 0 \end{cases}$$

Again using the definition (4.1), we see that

$$X(z) = \sum_{k=-\infty}^{\infty} x_k z^{-k} = \sum_{k=0}^{\infty} 2^k z^{-k} = \sum_{k=0}^{\infty} (2z^{-1})^k = \frac{1}{1 - 2z^{-1}} = \frac{z}{z - 2}$$ ■

Example 4.3 Consider the Z-transform for the sequence defined by

$$x_k = \begin{cases} 2^k, & k \le 0 \\ 0, & k > 0 \end{cases}$$

In this case, the Z-transform is given by

$$X(z) = \sum_{k=-\infty}^{\infty} x_k z^{-k} = \sum_{k=-\infty}^{0} 2^k z^{-k} = \sum_{m=0}^{\infty} (2z^{-1})^{-m} \qquad \text{with} \qquad m = -k$$

$$= \sum_{m=0}^{\infty} \left(\frac{1}{2} z\right)^m = \frac{1}{1 - \frac{z}{2}} = \frac{2}{2 - z}$$

where we have used the results of Appendix A to evaluate the geometric series. ■

We can expand $X(z)$ in a power series in z (by any of several methods) to recover the sequence x. However, in cases where $\{x_k\}$ is nonzero for both positive and negative indices, it is imperative to realize that $X(z)$ itself cannot uniquely determine the sequence x. The reason is that we can expand $X(z)$ in a power series in more than one way if we have the option of considering sequences $\{x_k\}$ that can be nonzero for either positive or negative k. For example, consider the sequences x and y defined in (4.2) and (4.3).

$$x_k = \begin{cases} -a^k, & k < 0 \\ 0, & k \ge 0 \end{cases} \tag{4.2}$$

$$y_k = \begin{cases} 0, & k < 0 \\ a^k, & k \ge 0 \end{cases} \tag{4.3}$$

The Z-transform of x is

$$X(z) = Z\{\{-a^k, k < 0\}\} = \sum_{k=-\infty}^{-1} -a^k z^{-k}$$

$$= - \sum_{m=1}^{\infty} (az^{-1})^{-m}$$

$$= \frac{-1/az^{-1}}{1 - (1/az^{-1})} = \frac{-1}{az^{-1} - 1} = \frac{1}{1 - az^{-1}} \tag{4.4}$$

The Z-transform of y is

$$Y(z) = Z\{\{a^k, k \geq 0\}\} = \sum_{k=0}^{\infty} a^k z^{-k} = \frac{1}{1 - az^{-1}} \tag{4.5}$$

Comparing (4.4) and (4.5), we see that the sequences of (4.2) and (4.3) have the same Z-transforms. This result is disconcerting at first because it implies that Z-transforms and the corresponding sequences are not uniquely related. We can correct this situation by specifying the *region of convergence* for which the sums in (4.4) and (4.5) converge absolutely. We shall see that a given $X(z)$ and its region of convergence uniquely specify the sequence $\{x_k\}$ that generated the polynomial $X(z)$.

The region of convergence of $X(z)$ is the set of complex valued z for which the sum $\sum |x_k z^{-k}|$ exists: that is, the set of z for which $X(z)$ has finite value. For example, in (4.4) $X(z)$ converges absolutely for $|1/az^{-1}| < 1$ or $|z| < |a|$. Similarly, in (4.5) $Y(z)$ converges absolutely for $|az^{-1}| < 1$ or $|z| > |a|$. In Section 4.3, we show that knowledge of $X(z)$ and its region of convergence does uniquely determine the corresponding sequence $\{x_k\}$. If we are concerned only with one-sided sequences, that is, sequences nonzero for either positive or negative indices only, then the region of convergence is not needed to specify $\{x_k\}$ uniquely.

We note that certain authors define the Z-transform of a sequence $\{x_k\}$ as

$$\tilde{X}(z) = \sum_{k=-\infty}^{\infty} x_k z^k \tag{4.6}$$

We can obtain the Z-transform $\tilde{X}(z)$ of (4.6) from $X(z)$ as defined by (4.1) by substituting z for z^{-1} in (4.1) and in the expression defining the region of convergence. For example, if $X(z) = a/(a - z)$, $|z| < a$, then $\tilde{X}(z) = a/(a - z^{-1})$, $|z| > a$.

4.3 CONVERGENCE OF THE Z-TRANSFORM

Consider the Z-transform of a sequence x where

$$X(z) = \sum_{k=-\infty}^{\infty} x_k z^{-k}$$

It is important to know the region of absolute convergence for $X(z)$ in the complex

z-plane. That is, we wish to determine the complex values of z for which $\sum |x_k z^{-k}|$ has a finite value. If we represent z in polar form as $z = re^{j\theta}$, we have

$$\begin{aligned}\sum_{k=-\infty}^{\infty} |x_k z^{-k}| &= \sum_{k=-\infty}^{\infty} |x_k (re^{j\theta})^{-k}| \\ &= \sum_{k=-\infty}^{-1} |x_k r^{-k} e^{-jk\theta}| + \sum_{k=0}^{\infty} |x_k r^{-k} e^{-jk\theta}| \\ &= \sum_{m=1}^{\infty} |x_{-m}| r^m + \sum_{k=0}^{\infty} |x_k| r^{-k} \end{aligned} \tag{4.7}$$

For the infinite sum $\sum_{k=-\infty}^{\infty} |x_k z^{-k}|$ to be finite each of the two sums in (4.7) must be finite. We can insure that each of these sums is finite provided we can find three positive numbers M, R_+, and R_- such that $|x_k| \leq MR_+^k$ for $k < 0$, and $|x_k| \leq MR_-^k$ for $k \geq 0$. (We assume that R_+ and R_- are chosen so that the bounds are tight.) We can then substitute these bounds in (4.7) to obtain

$$\sum_{k=-\infty}^{\infty} |x_k z^{-k}| \leq M \left\{ \sum_{m=1}^{\infty} R_+^{-m} r^m + \sum_{k=0}^{\infty} R_-^k r^{-k} \right\} \tag{4.8}$$

The sums in (4.8) are finite if and only if $r/R_+ < 1$ in the first sum and $R_-/r < 1$ in the second. That is, the sum representing $X(z)$ converges absolutely for all z in the anulus $R_- < |z| < R_+$. Of course, if $R_+ \leq R_-$, the sum does not converge absolutely for any value of z.

The region of convergence is shown in Figure 4.2. The inner circle bounds the

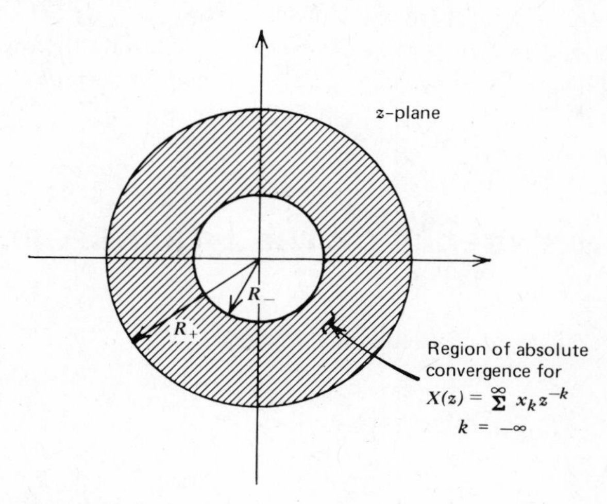

Figure 4.2

terms in negative powers of z away from the origin. The outer circle bounds the terms in positive powers of z away from large $|z|$ values.

Let us consider some examples of the region of convergence for different types of sequences.

Case 1 Finite Length Sequences: Consider the sequence

$$\{x_k\}_{k=a}^{b} = \{x_a, x_{a+1}, \ldots, x_b\}$$

In this case, $X(z) = x_a z^{-a} + x_{a+1} z^{-(a+1)} + \cdots + x_b z^{-b}$ is a polynomial in z, which converges absolutely for all z except for $z = 0$ if $b > 0$ and for $z = \infty$ if $a < 0$. For example, suppose that

$$\{x_k\} = \{1, 2, 3, 2, 1\}.$$
$$\uparrow_{k=0}$$

Then $X(z) = z^2 + 2z + 3 + 2z^{-1} + z^{-2}$, which converges absolutely for all z except $z = 0$ and $z = \infty$.

Case 2 Semi-infinite Sequences: Consider first the sequence $\{x_k\}_{k=a}^{\infty}$, with the Z-transform $X(z) = \sum_{k=a}^{\infty} x_k z^{-k}$. Here $X(z)$ converges absolutely for all z such that $R_- < |z|$, except for the point $z = \infty$ if $a < 0$. For example, suppose that $x_k = (-\frac{1}{2})^k$ for $k \geq 0$ and $x_k = 0$ for $k < 0$. Then

$$X(z) = 1 - \frac{1}{2}z^{-1} + \frac{1}{4}z^{-2} - \frac{1}{8}z^{-3} + \cdots = \frac{1}{1 + \frac{1}{2}z^{-1}} = \frac{z}{z + \frac{1}{2}} \tag{4.9}$$

Here, $R_- = \frac{1}{2}$, and $X(z)$ converges absolutely for all $\frac{1}{2} < |z|$.

Alternatively, we can consider a one sided sequence that extends back to $k = -\infty$. Suppose that we have the sequence $\{x_k\}_{k=-\infty}^{b}$ with $X(z) = \sum_{k=-\infty}^{b} x_k z^{-k}$. In this case, $X(z)$ converges absolutely for all z such that $|z| < R_+$, except for the point $z = 0$ if $0 < b$. For example, let $x_k = 2^k$ for $k < 0$ and $x_k = 0$ for $0 \leq k$. Now

$$X(z) = \cdots + \frac{1}{8}z^3 + \frac{1}{4}z^2 + \frac{1}{2}z = \frac{z/2}{1 + z/2} \tag{4.10}$$

which converges absolutely for all $|z| < 2$ since $R_+ = 2$.

Case 3 Infinite Length Sequences: Let us now examine a sequence $\{x_k\}$ that is nonzero for all k. In this case, $X(z)$ contains both negative and positive powers of z. The region of absolute convergence is the most general one, $R_- < |z| < R_+$, where again R_- is determined by the behavior of x_k for $k > 0$ [the coefficients of negative powers of z in $X(z)$] and R_+ is determined by the behavior of x_k for $k < 0$ (the coefficients of positive powers of z). For example let

$x_k = (\frac{1}{2})^{|k|} = 2^k$, $k < 0$ and $(\frac{1}{2})^k$, $k \geq 0$. Here $R_- = \frac{1}{2}$ and $R_+ = 2$. We should expect absolute convergence of $X(z)$ for $\frac{1}{2} < |z| < 2$. Indeed, we find

$$\begin{aligned} X(z) &= \sum_{k=-\infty}^{\infty} \left(\frac{1}{2}\right)^{|k|} z^{-k} = \sum_{k=-\infty}^{-1} 2^k z^{-k} + \sum_{k=0}^{\infty} \left(\frac{1}{2}\right)^k z^{-k} \\ &= \sum_{m=1}^{\infty} 2^{-m} z^m + \sum_{k=0}^{\infty} \left(\frac{1}{2z}\right)^k \\ &= \frac{z/2}{1 - z/2} + \frac{1}{1 - (1/2z)} \end{aligned} \tag{4.11}$$

where the first sum converges absolutely for $|z/2| < 1$ ($|z| < 2$) and the second for $|\frac{1}{2}z| < 1$ ($\frac{1}{2} < |z|$).

We can view these results in a more general light if we recognize that the Z-transform of a sum of sequences is the sum of the corresponding transforms, with a region of absolute convergence consisting of those values of z for which *all* of the individual transforms converge absolutely. In terms of set theory, the region of absolute convergence of a sum of transforms is the *intersection* of the individual regions of convergence.

The conclusions that we have reached in this section are portrayed in Figure 4.3.

Example 4.4 Let

$$x_k = \begin{cases} 2^k, & k < 0 \\ \left(\frac{1}{2}\right)^k, & k = 0, 2, 4, \cdots \\ \left(\frac{1}{3}\right)^k, & k = 1, 3, 5, \cdots \end{cases}$$

Then

$$\begin{aligned} X(z) &= \sum_{k=-\infty}^{\infty} x_k z^{-k} \\ &= \sum_{k=-\infty}^{-1} 2^k z^{-k} + \sum_{\substack{k=0 \\ k \text{ even}}}^{\infty} \left(\frac{1}{2}\right)^k z^{-k} + \sum_{\substack{k=0 \\ k \text{ odd}}}^{\infty} \left(\frac{1}{3}\right)^k z^{-k} \\ &= \sum_{m=1}^{\infty} 2^{-m} z^m + \sum_{n=0}^{\infty} \left(\frac{1}{2}\right)^{2n} z^{-2n} + \sum_{p=0}^{\infty} \left(\frac{1}{3}\right)^{2p+1} z^{-(2p+1)} \end{aligned}$$

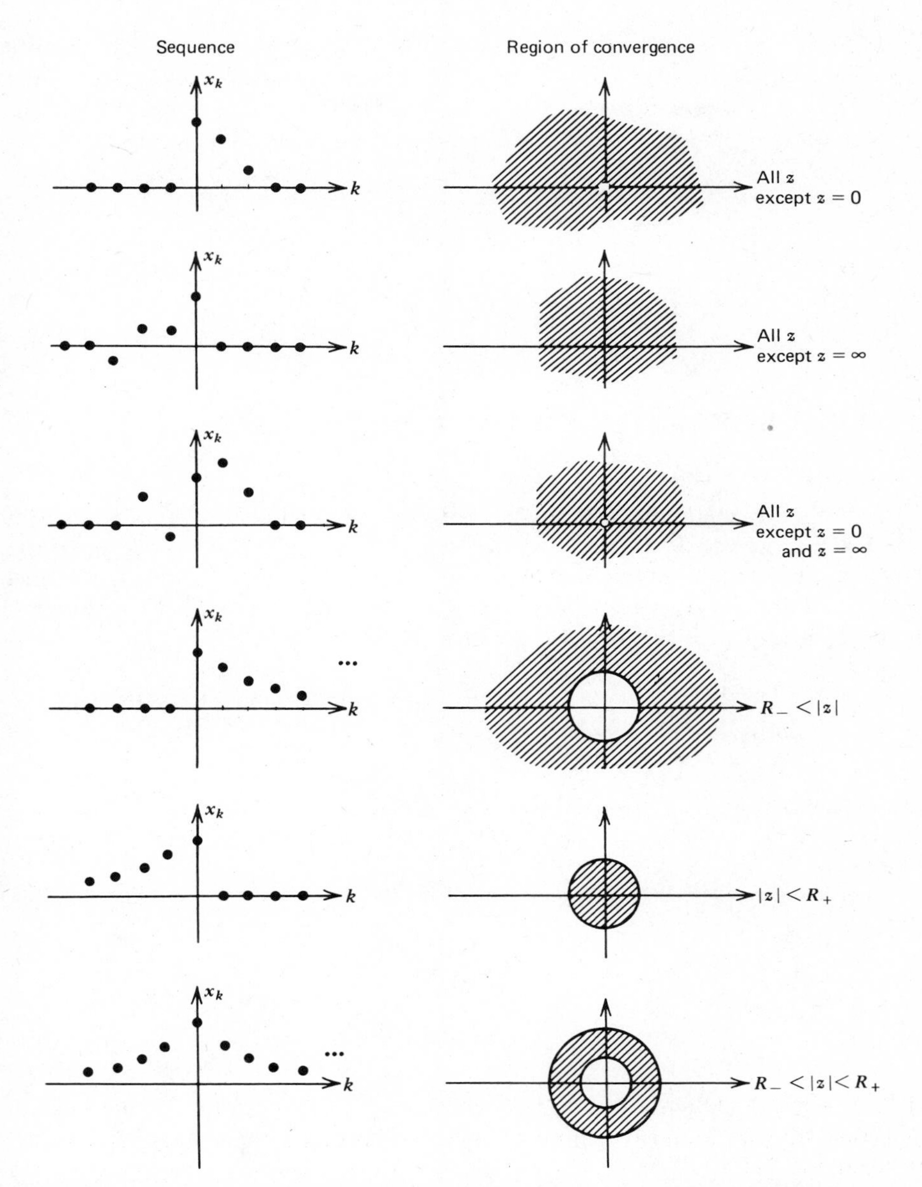

Figure 4.3 Typical sequences with the region of absolute convergence of their *Z*-Transforms

where $m = -k$, $n = k/2$, and $p = (k-1)/2$. The first sum has the closed-form representation $X_1(z) = (z/2)/[1-(z/2)]$ with a region of absolute convergence (ROC)$|z| < 2$; the second is $X_2(z) = 1/(1-\frac{1}{4}z^{-2})$ with a ROC of $\frac{1}{2} < |z|$; and the third is $X_3(z) = (\frac{1}{3})z^{-1}/(1-(\frac{1}{9})z^{-2})$ with a ROC of $\frac{1}{3} < |z|$. Thus we find

$$X(z) = \frac{z/2}{1-(z/2)} + \frac{z^2}{z^2-\frac{1}{4}} + \frac{z/3}{z^2-\frac{1}{9}}$$

with a ROC of $\frac{1}{2} < |z| < 2$. ■

4.4 PROPERTIES OF THE Z-TRANSFORM

There are many methods of finding closed-form expressions for the Z-transform of a given sequence. In this section, we develop certain useful properties of the Z-transform that can be used both in determining $X(z)$ from $\{x_k\}$ and in the inverse problem of finding $\{x_k\}$ given $X(z)$. The notation "$\{x_k\} \leftrightarrow X(z)$" is used below to denote "the sequence $\{x_k\}$ has Z-transform $X(z)$."

1. Linearity

The Z-transform is a linear operation, that is,

$$\begin{aligned} Z\{a\{x_k\} + b\{y_k\}\} &= \sum_{k=-\infty}^{\infty} (ax_k + by_k)z^{-k} \\ &= a \sum_{k=-\infty}^{\infty} x_k z^{-k} + b \sum_{k=-\infty}^{\infty} y_k z^{-k} \\ &= aX(z) + bY(z) \end{aligned} \tag{4.12}$$

Thus, if a given sequence can be written as a sum of elementary sequences, its transform may be found by summing the corresponding elementary transforms. For example, let

$$\begin{aligned} x_k &= \cos k\omega_0, \qquad k \geq 0 \\ &= \frac{1}{2} e^{jk\omega_0} + \frac{1}{2} e^{-jk\omega_0}, \qquad k \geq 0 \end{aligned}$$

Using the transform

$$a^k, \quad k \geq 0 \quad \leftrightarrow \quad \frac{1}{1 - az^{-1}}, \qquad a < |z|$$

we find

$$(e^{j\omega_0})^k, \qquad k \geq 0 \quad \leftrightarrow \quad \frac{1}{1 - e^{j\omega_0} z^{-1}}, \qquad 1 < |z|$$

and

$$(e^{-j\omega_0})^k, \qquad k \geq 0 \quad \leftrightarrow \quad \frac{1}{1 - e^{-j\omega_0} z^{-1}}, \qquad 1 < |z|$$

Adding these elemental sequences, we obtain

$$\begin{aligned} X(z) &= \frac{\frac{1}{2}}{1 - e^{j\omega_0} z^{-1}} + \frac{\frac{1}{2}}{1 - e^{-j\omega_0} z^{-1}}, \qquad 1 < |z| \\ &= \frac{1 - \cos \omega_0 \, z^{-1}}{1 - 2 \cos \omega_0 \, z^{-1} + z^{-2}} \end{aligned}$$

We note again that the ROC for the transform of a sum of sequences is the intersection of the ROCs of the elemental transforms.

2. Shifting

Let $\{x_k\} \leftrightarrow X(z)$. We wish to find the Z-transform of $\{x_{k \pm k_0}\}$, which corresponds to a delayed (by k_0 steps) version of $\{x_k\}$ if we use the minus sign, or an advanced version of $\{x_k\}$ if we use the plus sign. By definition

$$\begin{aligned} Z\{\{x_{k \pm k_0}\}\} &= \sum_{k=-\infty}^{\infty} x_{k \pm k_0} z^{-k} = \sum_{m=-\infty}^{\infty} x_m z^{-m \pm k_0} \\ &= z^{\pm k_0} \sum_{m=-\infty}^{\infty} x_m z^{-m} = z^{\pm k_0} X(z) \end{aligned} \tag{4.13}$$

A typical application of this theorem is the determination of a system's output given its difference equation description and its input u. For example, let the system be described by

$$y_k + \frac{1}{2} y_{k-1} = u_k + u_{k-1} \tag{4.14}$$

Taking the Z-transform term-by-term in this equation, we have

$$Y(z) + \frac{1}{2} z^{-1} Y(z) = U(z) + z^{-1} U(z)$$

or

$$Y(z)\left(1 + \frac{1}{2} z^{-1}\right) = U(z)(1 + z^{-1})$$

which results in

$$Y(z) = \frac{1 + z^{-1}}{1 + \frac{1}{2} z^{-1}} U(z)$$

Now from knowledge of $U(z)$, we can find $Y(z)$ and, through an inverse transform, the sequence y_k. For example, if $u_k = (\frac{1}{2})^k$, $k \geq 0$ in the case above, we have $U(z) = 1/(1 - \frac{1}{2} z^{-1})$ and

$$Y(z) = \frac{1 + z^{-1}}{1 + \frac{1}{2} z^{-1}} \cdot \frac{1}{1 - \frac{1}{2} z^{-1}}$$

which may be written as

$$Y(z) = \frac{-\frac{1}{2}}{1 + \frac{1}{2} z^{-1}} + \frac{\frac{3}{2}}{1 - \frac{1}{2} z^{-1}}$$

The details of obtaining this partial fraction expansion will be developed in the next section. For now, we can verify that this decomposition is indeed correct. Taking the inverse transform term-by-term (i.e., again applying the linearity property of Z-transforms), we obtain

$$\begin{aligned} y_k &= -\frac{1}{2}\left(-\frac{1}{2}\right)^k + \frac{3}{2}\left(\frac{1}{2}\right)^k, \qquad k \geq 0 \\ &= \left(-\frac{1}{2}\right)^{k+1} + 3\left(\frac{1}{2}\right)^{k+1}, \qquad k \geq 0 \end{aligned} \tag{4.15}$$

Note that the ROC $\frac{1}{2} < |z|$ was assumed implicitly in order that y_k be one-sided to the right, as we determine from physical reasoning.

A special case of this application is $u_k = \delta_k$, yielding $y_k = h_k$. Note that $H(z)$, the Z-transform of h_k, is found directly from the coefficients that define the difference equation. For the system just treated, we have

$$H(z) = \frac{1 + z^{-1}}{1 + \frac{1}{2}z^{-1}}$$

$$= \frac{1}{1 + \frac{1}{2}z^{-1}} + \frac{z^{-1}}{1 + \frac{1}{2}z^{-1}}$$

Assuming again a causal system, we have the ROC $\frac{1}{2} < |z|$; using the linearity and shift properties we obtain

$$h_k = \left\{\left(-\frac{1}{2}\right)^k\right\}_{k=0}^{\infty} + \left\{\left(-\frac{1}{2}\right)^{k-1}\right\}_{k-1=0}^{\infty}$$

$$= \delta_k + \left\{\left(-\frac{1}{2}\right)^k\right\}_{k=1}^{\infty} - 2\left\{\left(-\frac{1}{2}\right)^k\right\}_{k=1}^{\infty}$$

$$= \delta_k - \left\{\left(-\frac{1}{2}\right)^k\right\}_{k=1}^{\infty}$$

$$= \left\{1, \frac{1}{2}, -\frac{1}{4}, \frac{1}{8}, \cdots\right\}$$

Note that we have taken initial conditions to be zero in finding y_k and h_k above via application of the shift property. This choice of an initially relaxed system is necessary if the input-output relation of our systems is to obey superposition. Recall that h is the response of an initially relaxed system to an impulse input.

3. Shifting (Unilateral Transform)

We can modify the definition of our Z-transform so that initial conditions may be handled directly in a difference equation such as (4.14). Suppose that the lower limit on the summation of (4.1) were changed to $k = 0$. This defines a new transform, which we shall call the one-sided, or *unilateral* Z-transform, $X_u(z) = Z_u\{x_k\}$ defined as

$$X_u(z) = Z_u\{x_k\} = \sum_{k=0}^{\infty} x_k z^{-k}$$

$$= x_0 + x_1 z^{-1} + x_2 z^{-2} + \cdots \tag{4.16}$$

Let us apply the unilateral transform to a shifted sequence. Taking k_0 to be non-negative, we first treat the delayed sequence x_{k-k_0}:

$$\begin{aligned}
Z_u\{x_{k-k_0}\} &= \sum_{k=0}^{\infty} x_{k-k_0} z^{-k} \\
&= \sum_{m=-k_0}^{\infty} x_m z^{-(m+k_0)} \\
&= z^{-k_0} \sum_{m=-k_0}^{\infty} x_m z^{-m} \\
&= z^{-k_0}\{x_{-k_0} z^{k_0} + x_{-k_0+1} z^{k_0-1} + \cdots + x_{-1}z + x_0 + x_1 z^{-1} + \cdots\} \\
&= x_{-k_0} + x_{-k_0+1} z^{-1} + \cdots + x_{-1} z^{-k_0+1} + z^{-k_0}\{x_0 + x_1 z^{-1} + \cdots\} \\
&= x_{-k_0} + x_{-k_0+1} z^{-1} + \cdots + x_{-1} z^{-k_0+1} + z^{-k_0} X_u(z)
\end{aligned} \tag{4.17}$$

Thus, for example, if $y_{-1} = 2$ is an initial condition in (4.14) with $u_k = (\frac{1}{2})^k$, $k \geq 0$ as before, we obtain

$$Z_u\left\{y_k + \frac{1}{2} y_{k-1}\right\} = Z_u\{u_k + u_{k-1}\}$$

substituting $y_{-1} = 2$ and $u_{-1} = 0$, and collecting terms, we obtain

$$Y_u(z) + \frac{1}{2} y_{-1} + \frac{1}{2} z^{-1} Y_u(z) = U_u(z) + u_{-1} + z^{-1} U_u(z)$$

$$\begin{aligned}
Y_u(z)\left(1 + \frac{1}{2} z^{-1}\right) + \frac{1}{2} \cdot 2 &= U_u(z)(1 + z^{-1}) \\
&= \frac{1 + z^{-1}}{1 - \frac{1}{2} z^{-1}}
\end{aligned}$$

which results in

$$Y_u(z) = \frac{1 + z^{-1}}{(1 - \frac{1}{2} z^{-1})(1 + \frac{1}{2} z^{-1})} - \frac{1}{1 + \frac{1}{2} z^{-1}}$$

The first term on the right is the response due to u_k alone. The second term is the response due to the initial condition acting alone. Thus our output is the sum of two sequences: the response of an initially relaxed system to u_k plus the homogeneous solution for the given initial conditions

$$y_k = y_k^1 + y_k^2$$

where

$$y_k^1 = \left(-\frac{1}{2}\right)^{k+1} + 3\left(\frac{1}{2}\right)^{k+1}, \qquad k \geq 0$$

as we found in (4.15), and

$$y_k^2 = Z_u^{-1}\left\{\frac{-1}{1 + \frac{1}{2}z^{-1}}\right\}$$
$$= -\left(-\frac{1}{2}\right)^k, \qquad k \geq 0$$

Summing, we obtain the entire solution

$$y_k = \left(-\frac{1}{2}\right)^{k+1} + 3\left(\frac{1}{2}\right)^{k+1} + 2\left(-\frac{1}{2}\right)^{k+1}, \qquad k \geq 0$$
$$= 3\left[\left(-\frac{1}{2}\right)^{k+1} + \left(\frac{1}{2}\right)^{k+1}\right], \qquad k \geq 0$$
$$= \left\{0, \frac{3}{2}, 0, \frac{3}{8}, \ldots\right\}$$

Now let us examine the unilateral Z-transform of an advanced sequence. Again taking $k_0 \geq 0$, we have

$$Z_u\{x_{k+k_0}\} = \sum_{k=0}^{\infty} x_{k+k_0} z^{-k}$$
$$= \sum_{m=k_0}^{\infty} x_m z^{-(m-k_0)}$$
$$= z^{k_0} \sum_{m=k_0}^{\infty} x_m z^{-m}$$
$$= z^{k_0}[x_{k_0} z^{-k_0} + x_{k_0+1} z^{-(k_0+1)} + \cdots]$$
$$= z^{k_0}[X_u(z) - (x_0 + x_1 z^{-1} + \cdots + x_{k_0-1} z^{-(k_0-1)})]$$
$$= z^{k_0} X_u(z) - x_0 z^{k_0} - x_1 z^{k_0-1} - \cdots - x_{k_0-1} z \tag{4.18}$$

Returning to the previous example, we can rewrite the given equation in the form

$$y_{k+1} + \frac{1}{2} y_k = u_{k+1} + u_k$$

where $u_k = (\frac{1}{2})^k$, $k \geq 0$. Using the given initial condition $y_{-1} = 2$ we obtain $u_0 = 1$ and $y_0 = 0$. Applying the unilateral transform to this equation, we obtain

$$zY_u(z) + \frac{1}{2} Y_u(z) = zU_u(z) - z + U_u(z)$$

which yields

$$\begin{aligned} Y_u(z) &= \frac{(z+1)U_u(z)}{z+\frac{1}{2}} - \frac{z}{z+\frac{1}{2}} \\ &= \frac{1+z^{-1}}{1+\frac{1}{2}z^{-1}} U_u(z) - \frac{1}{1+\frac{1}{2}z^{-1}} \\ &= \frac{1+z^{-1}}{(1+\frac{1}{2}z^{-1})(1-\frac{1}{2}z^{-1})} - \frac{1}{1+\frac{1}{2}z^{-1}} \end{aligned}$$

as before.

4. Convolution

One of the most important properties of Z-transforms in applications involves the convolution property. Let $\{u_k\}_{k=0}^{\infty} \leftrightarrow U(z)$ and $\{h_k\}_{k=0}^{\infty} \leftrightarrow H(z)$. Then the convolution of u and h is $\{y_k\}_{k=0}^{\infty}$, where

$$\{y_k\} = \{u_k\} * \{h_k\}$$

with

$$y_k = \sum_{m=0}^{k} u_m h_{k-m} \tag{4.19}$$

The Z-transform of y_k is $Y(z)$, which is equal to

$$Y(z) = U(z)H(z) \tag{4.20}$$

A simple way to show (4.20) is by induction. Consider the product $U(z)H(z)$.

$$\begin{aligned} U(z)H(z) &= [u_0 + u_1 z^{-1} + u_2 z^{-2} + \cdots][h_0 + h_1 z^{-1} + h_2 z^{-2} + \cdots] \\ &= u_0 h_0 + (u_0 h_1 + u_1 h_0) z^{-1} + (u_0 h_2 + u_1 h_1 + u_2 h_0) z^{-2} + \cdots \\ &= y_0 + y_1 z^{-1} + y_2 z^{-2} + \cdots = Y(z) \end{aligned} \tag{4.21}$$

By equating like coefficients of z, we obtain

$$\begin{aligned} y_0 &= u_0 h_0 \\ y_1 &= u_0 h_1 + u_1 h_0 \\ y_2 &= u_0 h_2 + u_1 h_1 + u_2 h_0 \\ &\vdots \\ y_k &= u_0 h_k + u_1 h_{k-1} + \cdots + u_{k-1} h_1 + u_k h_0 \\ &= \sum_{m=0}^{k} u_m h_{k-m} \end{aligned} \tag{4.22}$$

In the case of general sequences that are nonzero for both positive and negative k, we have

$$y_k = \sum_{m=-\infty}^{\infty} u_m h_{k-m} \tag{4.23}$$

Taking the Z-transform of (4.23), we obtain

$$Y(z) = \sum_{k=-\infty}^{\infty} \sum_{m=-\infty}^{\infty} u_m h_{k-m} z^{-k} \tag{4.24}$$

Any power series that converges absolutely also converges uniformly within its radius of convergence; this allows us to interchange the order of summation in (4.24) to give

$$\begin{aligned} Y(z) &= \sum_{m=-\infty}^{\infty} \left[u_m z^{-m} \sum_{k=-\infty}^{\infty} h_{k-m} z^{-(k-m)} \right] \\ &= \sum_{m=-\infty}^{\infty} [u_m z^{-m} H(z)] \\ &= H(z) \sum_{m=-\infty}^{\infty} u_m z^{-m} \\ &= H(z)U(z) \end{aligned} \tag{4.25}$$

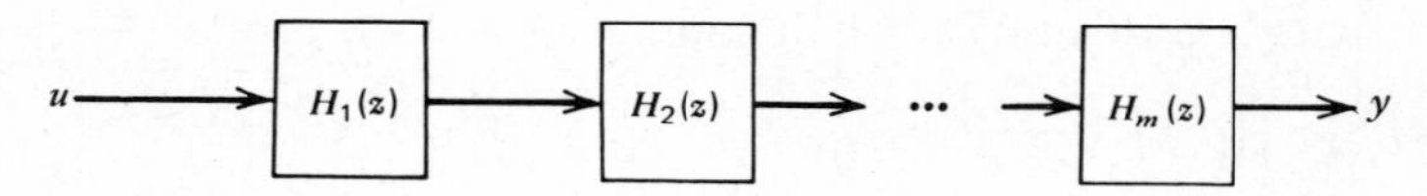

Figure 4.4

The transform $Y(z)$ converges in the common region of convergence of $U(z)$ and $H(z)$, provided that such a region exists. The transform $H(z)$ is called the *system transfer function.*

If a system can be broken down into a cascade of noninteracting systems, as shown in Figure 4.4, then the Z-transform of the output takes on a particularly simple form. If we have m systems in cascade with transfer functions $H_1(z)$, $H_2(z), \ldots, H_m(z)$, then the output in the transform domain is

$$Y(z) = U(z)H_1(z)H_2(z) \cdots H_m(z) \tag{4.26}$$

The overall system transfer function is

$$H(z) = \frac{Y(z)}{U(z)} = H_1(z)H_2(z) \cdots H_m(z) \tag{4.27}$$

For example, suppose that we wish to find the step response of the system shown in Figure 4.5. The given system is a cascade of one subsystem whose impulse response is $h_1(k) = (\frac{1}{2})^k$, $k \geq 0$ with a second whose impulse response is $h_2(k) = (-\frac{1}{2})^k$, $k \geq 0$. Setting $u_k = 1$, $k \geq 0$, we find the corresponding output as

$$y(k) = u(k) * h_1(k) * h_2(k) \tag{4.28}$$

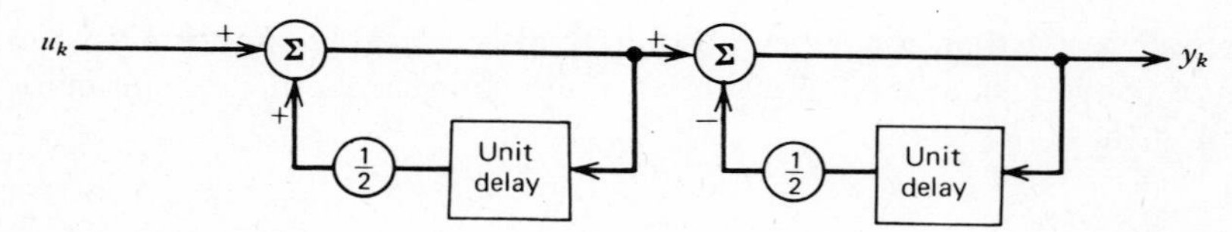

Figure 4.5

As an alternative to evaluating the indicated convolutions explicitly we may take Z-transforms in (4.28) to obtain

$$\begin{aligned} Y(z) &= Z\{u(k)\} \cdot Z\{h_1(k)\} \cdot Z\{h_2(k)\} \\ &= U(z) \cdot H_1(z) \cdot H_2(z) \end{aligned}$$

$$Y_z = \frac{1}{1-z^{-1}} \cdot \frac{1}{1-\frac{1}{2}z^{-1}} \cdot \frac{1}{1+\frac{1}{2}z^{-1}}$$

$$= \frac{\frac{4}{3}}{1-z^{-1}} - \frac{\frac{1}{2}}{1-\frac{1}{2}z^{-1}} + \frac{\frac{1}{6}}{1+\frac{1}{2}z^{-1}} \tag{4.29}$$

(In a later section, we shall formalize the procedure used to obtain this partial fraction expansion. For now the reader may verify that the decomposition is correct.) Taking inverse transforms term-by-term, with $y(k)$ chosen to be one-sided to the right by physical reasoning, we obtain

$$y_k = \frac{4}{3} - \frac{1}{2}\left(\frac{1}{2}\right)^k + \frac{1}{6}\left(-\frac{1}{2}\right)^k, \qquad k \geq 0 \tag{4.30}$$

The reader may wish to compare this solution procedure to a direct evaluation of the convolutions in (4.28) above.

5. Multiplication by k

like $\mathcal{L}\{\frac{d\,(x(t))}{dt}\}$

(k is not constant, but a variable)

Let $\{x_k\} \leftrightarrow X(z)$ and suppose that we wished to find the transform of $\{kx_k\}$. From the definition of the Z-transform we have

$$Z\{kx_k\} = \sum_{k=-\infty}^{\infty} kx_k z^{-k}$$

$$= z \sum_{k=-\infty}^{\infty} kx_k z^{-k-1}$$

$$= z \sum_{k=-\infty}^{\infty} x_k(kz^{-k-1})$$

$$= z \sum_{k=-\infty}^{\infty} x^k \left[-\frac{d}{dz} z^{-k} \right]$$

$$= -z \frac{d}{dz} \sum_{k=-\infty}^{\infty} x_k z^{-k}$$

$$= -z \frac{d}{dz} X(z) \tag{4.31}$$

The interchange of summation and differentiation in (4.31) is valid because one can always differentiate or integrate a power series term-by-term to obtain the derivative or integral of the series within its region of convergence. The derivative and integral of a power series are other power series with the same convergence region.

We can generalize the above result to multiplication by any positive power of k. Thus, by a similar argument, we obtain the transform pair†

$$\{k^n x_k\} \leftrightarrow \left[-z \frac{d}{dz}\right]^n X(z) \tag{4.32}$$

The above transform pair is valid for both one-sided and two-sided transforms.

This property can be used to extend our table of Z-transforms. Recalling that

$$a^k, \qquad k \geq 0 \quad \leftrightarrow \quad \frac{1}{1 - az^{-1}}$$

we find that

$$\begin{aligned} ka^k, \;\; k \geq 0 \;\; \leftrightarrow \;\; & -z \frac{d}{dz}(1 - az^{-1})^{-1} \\ &= (-z)(-1)(1 - az^{-1})^{-2}(-a)(-1)z^{-2} \\ &= \frac{az^{-1}}{(1 - az^{-1})^2} \end{aligned}$$

Applying the linearity and shift properties to this result, we find that

$$ka^{k-1}, \qquad k \geq 0 \quad \leftrightarrow \quad \frac{z^{-1}}{(1 - az^{-1})^2}$$

$$(k + 1)a^k, \qquad k \geq -1 \quad \leftrightarrow \quad \frac{1}{(1 - az^{-1})^2} \tag{4.33}$$

Note that this sequence is still causal, since $k + 1 = 0$ at $k = -1$. Thus we could just as well use the lower limit $k \geq 0$.

Continuing, we find that

$$\begin{aligned} k(k + 1)a^k, \qquad k \geq 0 \;\; \leftrightarrow \;\; & -z \frac{d}{dz}(1 - az^{-1})^{-2} \\ &= (-z)(-2)(1 - az^{-1})^{-3}(-a)(-1)z^{-2} \\ &= \frac{2az^{-1}}{(1 - az^{-1})^3} \end{aligned}$$

† The notation $[-z\,(d/dz)]^n$ means $-z\,(d/dz)(\cdots(-z\,(d/dz)(-z(d/dz)(\quad))))$.

from which

$$\frac{1}{2}k(k+1)a^{k-1}, \qquad k \geq 0 \quad \leftrightarrow \quad \frac{z^{-1}}{(1-az^{-1})^3}$$

and

$$\frac{1}{2}(k+1)(k+2)a^k, \qquad k \geq 0 \quad \leftrightarrow \quad \frac{1}{(1-az^{-1})^3} \tag{4.34}$$

by the same reasoning as above.

For a geometric sequence that is one-sided to the left, we start with

$$-a^k, \qquad k \leq -1 \quad \leftrightarrow \quad \frac{1}{1-az^{-1}} \tag{4.35}$$

to obtain

$$-ka^k, k \leq -1 \quad \leftrightarrow \quad \frac{az^{-1}}{(1-az^{-1})^2}$$

that is

$$-(k+1)a^k, \qquad k \leq -1 \quad \leftrightarrow \quad \frac{1}{(1-az^{-1})^2} \tag{4.36}$$

Continuing, we obtain

$$-k(k+1)a^k, \qquad k \leq -1 \quad \leftrightarrow \quad \frac{2az^{-1}}{(1-az^{-1})^3}$$

that is,

$$-\tfrac{1}{2}(k+1)(k+2)a^k, \qquad k \leq -1 \quad \leftrightarrow \quad \frac{1}{(1-az^{-1})^3} \tag{4.37}$$

Note that $k \leq -1$ was chosen for uniformity in the expressions. We could have written $k \leq -2$ in (4.36) and $k \leq -2$ or $k \leq -3$ in (4.37) with equal validity.

6. Division by $k + a$

We shall find it useful also to have a transform pair for $x_k/(k + a)$, where a is any real number. Let $X(z)$ be the Z-transform of $\{x_k\}$. Then by definition,

$$
\begin{aligned}
Z\left\{\frac{x_k}{k + a}\right\} &= \sum_{k=-\infty}^{\infty} \frac{x_k}{k + a} z^{-k} \\
&= \sum_{k=-\infty}^{\infty} x_k z^a\left(-\int_0^z \hat{z}^{-k-a-1}\, d\hat{z}\right) \\
&= -z^a \int_0^z \hat{z}^{-a-1} \sum_{k=-\infty}^{\infty} x_k \hat{z}^{-k}\, d\hat{z} \\
&= -z^a \int_0^z \hat{z}^{-a-1} X(\hat{z})\, d\hat{z} \qquad (4.38)
\end{aligned}
$$

We note that $\hat{z}$ in this expression may be treated as if it were a real variable.

For example, consider the sequence $\{x_k\} = \{1/k\}, k \geq 1 = \{1, \frac{1}{2}, \frac{1}{3}, \ldots\}$. Beginning with $v_k = 1, k \geq 1$, with the transform

$$V(z) = \sum_{k=1}^{\infty} 1 \cdot z^{-1} = \frac{z^{-1}}{(1 - z^{-1})}, \qquad 1 < |z|$$

we have

$$X(z) = -z^0 \int_0^z \hat{z}^{-1} V(\hat{z})\, d\hat{z}$$

which yields

$$
\begin{aligned}
X(z) &= -\int_0^z \frac{\hat{z}^{-2}}{1 - \hat{z}^{-1}}\, d\hat{z} \\
&= -\ln(1 - z^{-1}), \qquad 1 < |z|
\end{aligned}
$$

7. Scale Change

Let $\{x_k\} \leftrightarrow X(z)$, where $X(z)$ has the region of absolute convergence $R_- < |z| < R_+$. If a is any (possibly complex-valued) number, then the transform of $a^k x_k$

is found as

$$Z\{a^k x_k\} = \sum_{k=-\infty}^{\infty} a^k x_k z^{-k} = \sum_{k=-\infty}^{\infty} x_k \left(\frac{z}{a}\right)^{-k}$$

$$= X\left(\frac{z}{a}\right) \tag{4.39}$$

with the modified ROC

$$R_- < \left|\frac{z}{a}\right| < R_+$$

that is,

$$|a|R_- < |z| < |a|R_+$$

For example, recall that $\cos k\omega_0$, $k \geq 0$ has the transform

$$\cos k\omega_0, \quad k \geq 0 \quad \leftrightarrow \quad \frac{1 - z^{-1}\cos\omega_0}{1 - 2z^{-1}\cos\omega_0 + z^{-2}}, \qquad 1 < |z|$$

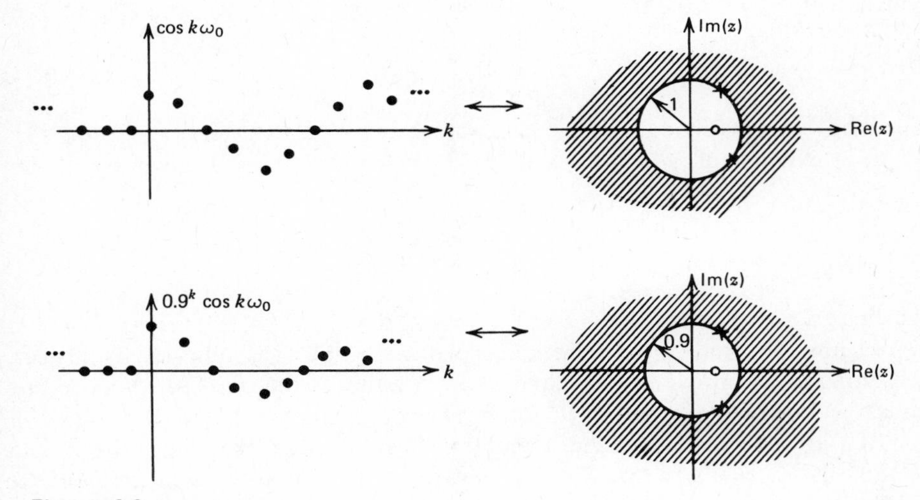

Figure 4.6

The transform of the damped cosine $0.9^k \cos k\omega_0$, $k \geq 0$ is therefore

$$0.9^k \cos k\omega_0, \quad k \geq 0 \leftrightarrow \frac{1 - 0.9z^{-1}\cos\omega_0}{1 - 1.8z^{-1}\cos\omega_0 + 0.81z^{-2}}, \qquad 0.9 < |z|$$

as illustrated in Figure 4.6.

8. Initial Value

Given a sequence $\{x_k\}$ that is zero for $k < k_0$, we can evaluate the point x_{k_0} directly from the transform $X(z)$. From

$$X(z) = \sum_{k=-\infty}^{\infty} x_k z^{-k}$$

$$= \sum_{k=k_0}^{\infty} x_k z^{-k} = x_{k_0} z^{-k_0} + x_{k_0+1} z^{-(k_0+1)} + \cdots \tag{4.41}$$

we have

$$z^{k_0} X(z) = x_{k_0} + x_{k_0+1} z^{-1} + x_{k_0+2} z^{-2} + \cdots$$

which, with $z = \infty$ becomes

$$z^{k_0} X(z)|_{z=\infty} = x_{k_0} \tag{4.42}$$

since the terms involving z^{-1}, z^{-2}, etc., vanish as $z \to \infty$.

For example, if $X(z) = (1 - \alpha z^{-1})^{-1}$, $|\alpha| < |z|$, then x_0 is given by

$$x_0 = X(z)\Big|_{z=\infty} = \frac{1}{1 - \alpha z^{-1}}\Big|_{z=\infty} = 1$$

which agrees with the known form $x_k = \{a^k\}_0^\infty = \{1, \alpha, \alpha^2, \ldots\}$. *It must first be known*, however, that $\{x_k\}$ is in fact one-sided to the right—as indicated by physical reasoning (for example, if the sequence in question is the impulse response of a causal system) or by the ROC of $X(z)$. For example, the sequence $\{x_k\} = \{\cdots - 1/\alpha^2, \ -1/\alpha, \ 0\}$ has the same transform as was examined above, but in this case $x_0 = 0 \neq X(z)|_{z=\infty}$. This initial value theorem is applicable only to an $X(z)$ that converges absolutely for arbitrarily large values of z.

As a second example, consider $X(z) = 2az^{-1}/(1 - az^{-1})^3$, $|a| < |z|$. Here we have

$$x_0 = X(z)|_{z=\infty} = 0$$

and

$$x_1 = zX(z)|_{z=\infty} = 2a$$

which agree with the known form

$$\begin{aligned}\{x_k\} &= \{k(k+1)a^k\} \qquad k \geq 0 \\ &= \{0, 2a, 6a^2, \ldots\}\end{aligned}$$

Note that we cannot determine x_2, the *second* nonzero sequence member, this way, since

$$\lim_{z \to \infty} z^2 X(z) = \lim_{z \to \infty} \frac{2az}{(1 - az^{-1})^3}$$

does not exist.

9. Final Value

Another useful result involves determining the asymptotic value of x_k directly from $X(z)$. Again we assume that x_k is one-sided to the right; taking the transform of $\{x_k - x_{k-1}\}$, we have

$$Z\{\{x_k - x_{k-1}\}\} = X(z) - z^{-1}X(z) = \sum_{k=-\infty}^{\infty} (x_k - x_{k-1})z^{-k}$$

that is,

$$(1 - z^{-1})X(z) = \lim_{N \to \infty} \sum_{k=-\infty}^{N} (x_k - x_{k-1})z^{-k} \tag{4.43}$$

Taking the limit on both sides of (4.43) as $z \to 1$, we have

$$\begin{aligned}\lim_{z\to 1}(1 - z^{-1})X(z) &= \lim_{z\to 1}\lim_{N\to\infty}\sum_{k=-\infty}^{N}(x_k - x_{k-1})z^{-k}\\ &= \lim_{N\to\infty}\lim_{z\to 1}\sum_{k=-\infty}^{N}(x_k - x_{k-1})z^{-k}\\ &= \lim_{N\to\infty}\sum_{k=-\infty}^{N}(x_k - x_{k-1})\\ &= \lim_{N\to\infty} x_N \qquad (4.44)\end{aligned}$$

where the interchange of limits is allowable within the ROC of $X(z)$. Thus the asymptotic value of x_k as $k \to \infty$ is given by the limit of $(1 - z^{-1})X(z)$ as $z \to 1$.

For example, if we take $X(z) = (1 - a)z^{-1}/[(1 - z^{-1})(1 - az^{-1})]$, $1 < |z|$ then

$$\begin{aligned}\lim_{N\to\infty} x_N &= \lim_{z\to 1}(1 - z^{-1})X(z)\\ &= \lim_{z\to 1}\frac{(1 - a)z^{-1}}{1 - az^{-1}}\\ &= 1\end{aligned}$$

which is consistent with the inverse transform $x_k = 1 - a^k$, $k \geq 0$.

It can occur that x_k may have no limit as $k \to \infty$, even though

$$\lim_{z\to 1}(1 - z^{-1})X(z)$$

is well behaved. For example, if $|a| > 1$ above with the ROC of $X(z)$ equal to $|a| < |z|$, then $x_k \to -\infty$ as $k \to \infty$. We must always check that the point $z = 1$ lies within the ROC of $X(z)$, so that the limit as $z \to 1$ can be taken for z within the ROC.

The initial and final value theorems can be generalized to sequences that are two-sided or one-sided to the left. These extensions are left as an exercise for the reader.

10. Partial Sum

Suppose that g_k is the summation of a second sequence over indices $-\infty$ up to k, that is,

$$g_k = \sum_{m=-\infty}^{k} x_m \qquad (4.45)$$

We can find the Z-transform of g_k by observing that

$$g_k - g_{k-1} = \sum_{m=-\infty}^{k} x_m - \sum_{m=-\infty}^{k-1} x_m = x_k$$

from which

$$G(z) - z^{-1}G(z) = X(z)$$

or

$$G(z) = \frac{X(z)}{1 - z^{-1}} \tag{4.46}$$

We can use this result to independently obtain the transform of $\{x_k\} = \{0, 1, 2, 3, \ldots\}$ by observing that

$$k + 1 = \sum_{m=0}^{k} 1$$

Since

$$\{1\}, \qquad k \geq 0 \quad \leftrightarrow \quad \frac{1}{1 - z^{-1}}$$

it follows that

$$\{k + 1\}, \qquad k \geq 0 \quad \leftrightarrow \quad \frac{1}{(1 - z^{-1})^2}$$

and applying a shift to the right,

$$\{k\}, \qquad k \geq 1 \quad \leftrightarrow \quad \frac{z^{-1}}{(1 - z^{-1})^2}$$

As another application, we note that

$$\lim_{k \to \infty} g_k = \sum_{m=-\infty}^{\infty} x_m$$

Table 4.1 Properties of the Z-Transform

Let $\{f_k\} \leftrightarrow F(z) \triangleq \sum_{k=-\infty}^{\infty} f_k z^{-k}$

and $\{g_k\} \leftrightarrow G(z)$

1. $\alpha\{f_k\} + \beta\{g_k\} \leftrightarrow \alpha F(z) + \beta G(z)$

2. $f_{k \pm ko} \leftrightarrow z^{\pm ko} F(z)$

3. $Z_u\{f_{k-ko}\} = z^{-ko} F_u(z) + z^{-ko+1} f_{-1} + \cdots + z^{-1} f_{-ko+1} + f_{-ko}$

   $Z_u\{f_{k+ko}\} = z^{ko} F_u(z) - z^{ko} f_0 - \cdots - z f_{ko-1}$

   where $F_u(z) = \sum_{k=0}^{\infty} f_k z^{-k}$

4. $\{f_k\} * \{g_k\} \leftrightarrow F(z) \cdot G(z)$

5. $\{kf_k\} \leftrightarrow -z \dfrac{d}{dz} F(z)$

   $\{k^n f_k\} \leftrightarrow \left(-z \dfrac{d}{dz}\right)^n F(z)$

6. $\left\{\dfrac{f_k}{(k+a)}\right\} \leftrightarrow -z^a \int z^{-a-1} F(z)\, dz$

7. $\{a^k f_k\} \leftrightarrow F\left(\dfrac{z}{a}\right)$

8. $x_0 = \lim_{z \to \infty} X(z)$ if x_k is one-sided to the right

9. $\lim_{N \to \infty} x_N = \lim_{z \to 1} (1 - z^{-1}) X(z)$ if the point $z = 1$ lies on the boundary of, or within the ROC of $X(z)$, and x_k is one-sided to the right.

10. $\sum_{m=-\infty}^{k} f_m \leftrightarrow \dfrac{F(z)}{1 - z^{-1}}$

    $\sum_{m=-\infty}^{\infty} f_m = F(1)$

Note. Properties 1 and 4 to 9 hold for both the bilateral and unilateral transforms.

Table 4.2 Z-Transforms of Selected Sequences

	f_k	$F(z)$
1.	δ_k	$1, \quad \text{all } z$
2.	$1, \quad k \geq 0$	$(1 - z^{-1})^{-1}, \quad 1 < \lvert z \rvert$
3.	$k, \quad k \geq 0$	$z^{-1}(1 - z^{-1})^{-2}, \quad 1 < \lvert z \rvert$
4.	$k^n, \quad k \geq 0$	$\left(-z \dfrac{d}{dz}\right)^n (1 - z^{-1})^{-1}, \quad 1 < \lvert z \rvert$
5.	$\dbinom{k}{n}, \quad n \leq k$	$z^{-n}(1 - z^{-1})^{n+1}, \quad 0 < \lvert z \rvert$
6.	$\dbinom{n}{k}, \quad 0 \leq k \leq n$	$(1 + z^{-1})^n, \quad 0 < \lvert z \rvert$
7.	$\alpha^k, \quad k \geq 0$	$(1 - \alpha z^{-1})^{-1}, \quad \lvert \alpha \rvert < \lvert z \rvert$
8.	$k^n \alpha^k, \quad k \geq 0$	$\left(-z \dfrac{d}{dz}\right)^n (1 - \alpha z^{-1})^{-1}, \quad \lvert \alpha \rvert < \lvert z \rvert$
9.	$\alpha^k, \quad k < 0$	$-(1 - \alpha z^{-1})^{-1}, \quad \lvert z \rvert < \lvert \alpha \rvert$
10.	$k^n \alpha^k, \quad k < 0$	$-\left(-z \dfrac{d}{dz}\right)^n (1 - \alpha z^{-1})^{-1}, \quad \lvert z \rvert < \lvert \alpha \rvert$
11.	$\alpha^{\lvert k \rvert}, \quad \text{all } k$	$(1 - \alpha^2)[(1 - \alpha z)(1 - \alpha z^{-1})]^{-1}, \quad \lvert \alpha \rvert < \lvert z \rvert < \left\lvert \dfrac{1}{\alpha} \right\rvert$
12.	$\dfrac{1}{k}, \quad k > 0$	$-\ln(1 - z^{-1}), \quad 1 < \lvert z \rvert$
13.	$\cos \alpha k, \quad k \geq 0$	$(1 - z^{-1} \cos \alpha)(1 - 2z^{-1} \cos \alpha + z^{-2})^{-1}, 1 < \lvert z \rvert$
14.	$\sin \alpha k, \quad k \geq 0$	$z^{-1} \sin \alpha (1 - 2z^{-1} \cos \alpha + z^{-2})^{-1}, \quad 1 < \lvert z \rvert$
15.	$a \cos \alpha k + b \sin \alpha k, \quad k \geq 0$	$[a + z^{-1}(b \sin \alpha - a \cos \alpha)] (1 - 2z^{-1} \cos a + z^{-2})^{-1}, \quad 1 < \lvert z \rvert$
16.	$c \cos ak + \left(\dfrac{d + c \cos \alpha}{\sin \alpha}\right) \sin \alpha k, \quad k \geq 0$	$(c + dz^{-1})(1 - 2z^{-1} \cos \alpha + z^{-2})^{-1}, \quad 1 < \lvert z \rvert$
17.	$\cosh \alpha k, \quad k \geq 0$	$(1 - z^{-1} \cosh \alpha)(1 - 2z^{-1} \cosh \alpha + z^{-2})^{-1}, \quad \max\left\{\lvert \alpha \rvert, \left\lvert \dfrac{1}{\alpha} \right\rvert\right\} < \lvert z \rvert$
18.	$\sinh \alpha k, \quad k \geq 0$	$(z^{-1} \sinh \alpha)(1 - 2z^{-1} \cosh \alpha + z^{-2})^{-1}, \quad \max\left\{\lvert \alpha \rvert, \left\lvert \dfrac{1}{\alpha} \right\rvert\right\} < \lvert z \rvert$

applying the final value property we obtain

$$\begin{aligned}\sum_{m=-\infty}^{\infty} x_m &= \lim_{k\to\infty} g_k \\ &= \lim_{z\to 1}(1 - z^{-1})G(z) \\ &= \lim\,(1 - z^{-1})\frac{X(z)}{1 - z^{-1}} \\ &= \lim_{z\to 1} X(z) \\ &= X(1)\end{aligned}$$

This can also be seen from the definition of $X(z)$:

$$\sum_{k=-\infty}^{\infty} x_k = \lim_{z\to 1} \sum_{k=-\infty}^{\infty} x_k z^{-k} = X(1)$$

Thus we have, for example,

$$\begin{aligned}\sum_{k=0}^{\infty} a^k \cos k\omega_0 &= \left.\frac{1 - az^{-1}\cos\omega_0}{1 - 2az^{-1}\cos\omega_0 + a^2 z^{-2}}\right|_{z=1} \\ &= \frac{1 - a\cos\omega_0}{1 - 2a\cos\omega_0 + a^2}\end{aligned}$$

The Z-transform properties discussed in this section are summarized in Table 4.1. Table 4.2 presents the transforms of some common sequences.

4.5 INVERSION OF THE *Z*-TRANSFORM

We turn now to the important problem of recovering the sequence $\{x_k\}$ from its Z-transform $X(z)$. As we have discussed previously, because we have not restricted the domain of the index set to be either positive or negative integers, we must have knowledge of the region of convergence of $X(z)$ to determine $\{x_k\}$ uniquely. We consider three methods of inversion: inversion by series expansion, by partial fraction expansion, and by use of an inversion integral. We assume that $X(z)$ is a rational function, as given in (4.47):

$$X(z) = \frac{a_0 z^m + a_1 z^{m-1} + \cdots + a_m}{b_0 z^n + b_1 z^{n-1} + \cdots + b_n} \tag{4.47}$$

Inversion of $X(z)$ by Direct Division

We can generate a series in z by dividing the numerator of (4.47) by the denominator. For example, if $X(z)$ converges for $|z| < R_+$, we obtain the series

$$X(z) = \frac{a_m}{b_n} + \left(\frac{a_{m-1}}{b_n} + \frac{a_m b_{n-1}}{b_n^2}\right)z + \cdots \tag{4.48}$$

which converges for the same range of z. Equation 4.48 is obtained by beginning the division with the lowest order of z: that is, with b_n. We then identify the coefficient of z^k as x_{-k}. If, however, $X(z)$ were known to converge for $R_- < |z|$, then the series in (4.48) would diverge for some z in the region of convergence of $X(z)$ and, hence, cannot be the correct representation. In this case, we should begin our division with the highest power of z to obtain

$$X(z) = \frac{a_0}{b_0} z^{m-n} + \left(\frac{a_1}{b_0} + \frac{a_0 b_1}{b_0^2}\right)z^{m-n-1} + \cdots \tag{4.49}$$

From (4.49) we can identify the nonzero terms of the sequence.

Example 4.5. Let $X(z) = a/(z - a)$ for $|z| < a$.

Here we expand in positive powers of z because $X(z)$ converges for $|z|$ less than some radius of convergence. Divide $z - a$ into a to obtain

$$\begin{array}{r|l} & -1 - (z/a) - (z^2/a^2) - \cdots \\ \hline -a + z & a \\ & \underline{a - z} \\ & \quad z \\ & \quad \underline{z - (z^2/a)} \\ & \qquad (z^2/a) \cdots \end{array}$$

and so

$$X(z) = -1 - \frac{z}{a} - \frac{z^2}{a^2} - \cdots$$

from which

$$\{x_k\} = \left\{\cdots -\frac{1}{a^3}, -\frac{1}{a^2}, -\frac{1}{a}, \underset{\uparrow\, k=0}{-1}\right\}$$

The series $X(z)$ converges absolutely for $|z| < a$. ■

Example 4.6 Let

$$X(z) = \frac{a}{z-a}, \qquad |z| > |a|$$

In this case, we expand $X(z)$ in negative powers of z. The division is

$$\begin{array}{r l}
 & (a/z) + (a^2/z^2) + (a^3/z^3) + \cdots \\
\cline{2-2}
z - a \,\big) & a \\
 & \underline{a - a^2/z} \\
 & \qquad\; a^2/z \\
 & \qquad\; \underline{a^2/z^2 - a^2/z^2} \\
 & \qquad\qquad\qquad a^3/z^2 \;\cdots
\end{array}$$

Thus

$$X(z) = \frac{a}{z} + \frac{a^2}{z^2} + \frac{a^3}{z^3} + \cdots$$

from which

$$x_k = \{0, a, a^2, \ldots\}$$
$$\uparrow_{k=0}$$

■

Example 4.7 Consider inversion of the Z-transform

$$X(z) = \frac{5z}{6z^2 - z - 1}, \qquad \frac{1}{3} < |z| < \frac{1}{2}$$

Because $X(z)$ converges in an annulus, the series expansion of $X(z)$ consists of both negative and positive powers of z. Thus, the sequence $\{x_k\}$ is nonzero for both positive and negative k. If we expand $X(z)$ in a power series in positive powers of z, we obtain

$$X(z) = -5z + 5z^2 - 35z^3 + 65z^4 - \cdots$$

This series converges absolutely for $|z| < \frac{1}{3}$. Because these values are not in the stated region of convergence, this expansion cannot be used to represent $X(z)$. If we expand $X(z)$ in negative powers of z, we obtain

$$X(z) = \frac{5}{6} z^{-1} + \frac{5}{36} z^{-2} + \frac{35}{216} z^{-3} + \cdots$$

This series converges absolutely for $|z| > \frac{1}{2}$. Again, this series cannot be used to represent $X(z)$ because the series does not converge in the stated region of convergence.

To obtain the correct expansion, we write $X(z)$ as a sum of two functions $X_1(z)$ and $X_2(z)$ so that $X_1(z)$ converges for $|z| > \frac{1}{3}$ and $X_2(z)$ converges for $|z| < \frac{1}{2}$:

$$X(z) = \frac{5z}{6z^2 - z - 1} = \frac{1}{3z + 1} + \frac{1}{2z - 1} \triangleq X_1(z) + X_2(z)$$

Expanding $X_1(z)$ in negative powers of z, we obtain

$$X_1(z) = \frac{1}{3} z^{-1} - \frac{1}{9} z^{-2} + \frac{1}{27} z^{-3} + \cdots - \frac{(-1)^k}{(3z)^k} + \cdots, \qquad |z| > \frac{1}{3}$$

Expanding $X_2(z)$ in positive powers of z, we obtain

$$X_2(z) = -1 - 2z - 4z^2 - \cdots - (2z)^k - \cdots, \qquad |z| < \frac{1}{2}$$

Thus the sum $X_1(z) + X_2(z) = X(z)$ converges for $\frac{1}{3} < |z| < \frac{1}{2}$, and the sequence $\{x_k\}$ is

$$x_k = \begin{cases} -\left(-\dfrac{1}{3}\right)^k, & k > 0 \\ -\left(\dfrac{1}{2}\right)^k, & k \leq 0 \end{cases}$$

■

It is usually difficult to determine the correct decomposition of $X(z)$ by inspection. The search for a more general procedure for finding $\{x_k\}$ from $X(z)$ leads us to the method of partial fraction expansions.

Inversion of $X(z)$ by Partial Fractions

Let us assume that the degree of the numerator in (4.47) is no greater than the degree of the denominator: that is, $m \leq n$. If this is not the case, then we can write $X(z)$ as the sum of a polynomial $Q(z)$ of degree $m - n$ plus a ratio of polynomials with a degree of the numerator one less than the degree of the denominator. That is

$$X(z) = q_0 z^{m-n} + q_1 z^{m-n-1} + \cdots + q_{m-n} + \frac{\hat{a}_0 z^{n-1} + \hat{a}_1 z^{n-2} + \cdots + \hat{a}_{n-1}}{b_0 z^n + b_1 z^{n-1} + \cdots + b_n}$$

Example 4.8 Let

$$X(z) = \frac{z^5 + 3.5z^4 - 4.5z^3 - 29.5z^2 - 44.5z - 33.5}{z^3 + 1.5z^2 - 8.5z - 15}$$

By long division, beginning with the highest powers of z, we obtain

$$X(z) = z^2 + 2z + 1 + \frac{z^2 - 6z - 18.5}{z^3 + 1.5z^2 - 8.5z - 15}$$ ■

The polynomial term, if any, corresponds to the left-sided sequence

$$x_1(k) = \{q_0, q_1, q_2, \ldots, \underset{\uparrow k=0}{q_{m-n}}\}$$

We now concentrate on the term involving the ratio of polynomials, that is, the form given by (4.47) with $m \leq n$. Our approach is motivated by the desire to write this ratio as a sum of simpler forms whose inverses can be recognized by inspection. We begin by identifying the roots of the denominator polynomial, that is, the values of z for which

$$b_0 z^n + b_1 z^{n-1} + b_2 z^{n-2} + \cdots + b_n = 0 \tag{4.50}$$

These values, called the *poles* of $X(z)$, will determine the form of the sequence x_k. Denote the poles as $p_1, p_2, \ldots, p_n$ and write (4.47) in the form

$$X(z) = \frac{a_0 z^m + a_1 z^{m-1} + a_2 z^{m-2} + \cdots + a_m}{(z - p_1)(z - p_2)\cdots(z - p_n)} \tag{4.51}$$

Suppose first that the poles of $X(z)$ are distinct; that is, that none is repeated. We begin with an expansion of $X(z)$ into the sum of terms:

$$X(z) = c_0 + \frac{c_1 z}{z - p_1} + \frac{c_2 z}{z - p_2} + \cdots + \frac{c_n z}{z - p_n} \tag{4.52}$$

where

$$c_0 = X(z)|_{z=0} = X(0) = \frac{a_m}{b_n} \tag{4.53}$$

and

$$c_i = \frac{z - p_i}{z} X(z)|_{z=p_i} \qquad \text{for} \quad i = 1, 2, \ldots, n \tag{4.54}$$

Example 4.9 Let

$$H(z) = \frac{1 + z^{-1}}{1 - \frac{5}{6}z^{-1} + \frac{1}{6}z^{-2}}$$

We multiply numerator and denominator by z^2 to obtain the form

$$H(z) = \frac{z^2 + z}{z^2 - \frac{5}{6}z + \frac{1}{6}}$$

$$= \frac{z^2 + z}{(z - \frac{1}{2})(z - \frac{1}{3})} = c_0 + \frac{c_1 z}{z - \frac{1}{2}} + \frac{c_2 z}{z - \frac{1}{3}}$$

where

$$c_0 = H(0) = 0$$

$$c_1 = \left.\frac{z^2 + z}{z(z - \frac{1}{3})}\right|_{z=1/2} = 9$$

$$c_2 = \left.\frac{z^2 + z}{z(z - \frac{1}{2})}\right|_{z=1/3} = -8$$

Thus we obtain

$$H(z) = \frac{1 + z^{-1}}{1 - \frac{5}{6}z^{-1} + \frac{1}{6}z^{-2}} = \frac{9z}{z - \frac{1}{2}} - \frac{8z}{z - \frac{1}{3}}$$

■

Example 4.10 Let

$$X(z) = \frac{z^2 - 6z - 18.5}{z^3 + 1.57z^2 - 8.5z - 15}$$

The roots of $z^3 + 1.5z^2 - 8.5z - 15 = 0$ are $p_1 = -2$, $p_2 = -2.5$, and $p_3 = 3$. Thus, we write

$$X(z) = \frac{z^2 - 6z - 18.5}{(z + 2)(z + 2.5)(z - 3)}$$

$$= c_0 + \frac{c_1 z}{z + 2} + \frac{c_2 z}{z + 2.5} + \frac{c_3 z}{z - 3}$$

where

$$c_0 = X(0) = -\frac{18.5}{-15} = 1.233$$

$$c_1 = \left.\frac{z^2 - 6z - 18.5}{z(z + 2.5)(z - 3)}\right|_{z=-2} = -0.5$$

$$c_2 = \left.\frac{z^2 - 6z - 18.5}{(z + 2)z(z - 3)}\right|_{z=-2\cdot 5} = -0.4$$

$$c_3 = \left.\frac{z^2 - 6z - 18.5}{(z + 2)(z + 2.5)z}\right|_{z=3} = -0.333$$

thus

$$X(z) = 1.233 - \frac{0.5z}{z + 2} - \frac{0.4z}{z + 2.5} - \frac{0.333z}{z - 3}$$ ■

In the case where $X(z)$ has repeated poles, we must modify the form of the expansion somewhat. Suppose that $X(z)$ has a pole of multiplicity r at $(z - p_i)$. Then the partial fraction expansion of $X(z)$ must include r terms of the form

$$c_{i_1} \frac{z}{z - p_i} + c_{i_2}\left(\frac{z}{z - p_i}\right)^2 + \cdots + c_{i_r}\left(\frac{z}{z - p_i}\right)^r \tag{4.55}$$

The coefficients c_{ij}, $j = 1, \ldots, r$, may be evaluated in several ways, such as:

1. Place the right-hand side of the partial fraction expansion over a common denominator, then evaluate the unknown coefficients by equating coefficients of like powers of z.
2. Evaluate c_{i_r} from

$$c_{i_r} = \left(\frac{z - p_i}{z}\right)^r X(z)\Big|_{z = p_i} \tag{4.56}$$

   Then equate $H(z)$ to its expanded form for a number of values of z (for example $z = 0$, $z = 1$, etc.) to obtain the remaining coefficients.
3. Evaluate c_{i_r} as above, then obtain the remaining coefficients successively as

$$c_{i_r} = \left(\frac{z - p_i}{z}\right)^r X(z)\Big|_{z = p_i} \tag{4.57}$$

$$c_{i_{r-1}} = \left(\frac{z - p_i}{z}\right)^{r-1}\left[X(z) - \frac{c_{i_r} z}{z - p_i}\right]\Bigg|_{z = p_i} \tag{4.58}$$

$$c_{i_{r-2}} = \left(\frac{z - p_i}{z}\right)^{r-2}\left[X(z) - \frac{c_{i_r} z^2}{(z - p_i)^2} - \frac{c_{i_{r-1}} z}{z - p_i}\right]\Bigg|_{z = p_i} \tag{4.59}$$

$$c_{i_1} = \left(\frac{z - p_i}{z}\right)\left[X(z) - \frac{c_{i_r} z^{r-1}}{(z - p_i)^{r-1}} - \frac{c_{i_{r-1}} z^{r-2}}{(z - p_i)^{r-2}} - \cdots - c_{i_2}\frac{z}{z - p_i}\right]\Bigg|_{z = p_i} \tag{4.60}$$

Example 4.11 Take $Y(z) = 1/[(1 - z^{-1})(1 - \frac{1}{2}z^{-1})^3]$, which is the transform of the step response of the cascade of three systems, each having transfer function $1/(1 - \frac{1}{2}z^{-1})$. We write

$$\begin{aligned} Y(z) &= \frac{z^4}{(z - 1)(z - \frac{1}{2})^3} \\ &= c_0 + \frac{c_1 z}{z - 1} + \frac{c_2 z}{z - \frac{1}{2}} + \frac{c_3 z^2}{(z - \frac{1}{2})^2} + \frac{c_4 z^3}{(z - \frac{1}{2})^3} \end{aligned}$$

where

$$c_0 = Y(0) = 0$$

$$c_1 = \frac{z-1}{z} Y(z)\Big|_{z=1} = \frac{z^3}{(z-\frac{1}{2})^3}\Bigg|_{z=1} = 8$$

and

$$c_4 = \left(\frac{z-\frac{1}{2}}{z}\right)^3 Y(z)\Bigg|_{z=1/2} = \frac{z}{z-1}\Bigg|_{z=1/2} = -1$$

We now must find c_2 and c_3. Using approach (1) above, we would write

$$Y(z) = \frac{8z}{z-1} + \frac{c_2 z}{z-\frac{1}{2}} + \frac{c_3 z^2}{(z-\frac{1}{2})^2} - \frac{z^3}{(z-\frac{1}{2})^3}$$

$$\equiv \frac{[8(z^4 - \frac{3}{2}z^3 + \frac{3}{4}z^2 - \frac{1}{8}z) + c_2(z^4 - 2z^3 + \frac{5}{4}z^2 - \frac{1}{4}z) + c_3(z^4 - \frac{3}{2}z^3 + \frac{1}{2}z^2) - (z^4 - z^3)}{(z-1)(z-\frac{1}{2})^3}$$

$$\equiv \frac{[z^4(8 + c_2 + c_3 - 1) + z^3(-12 - 2c_2 - \frac{3}{2}c_3 + 1) + z^2(6 + \frac{5}{4}c_2 + \frac{1}{2}c_3) + z(-1 - \frac{1}{4}c_2)]}{(z-1)(z-\frac{1}{2})^3}$$

$$\equiv \frac{z^4}{(z-1)(z-\frac{1}{2})^3}$$

where $\equiv$ means "is identically equal to," or "equals for all values of z." Equating numerator coefficients yields the equations

$$c_2 + c_3 = -6$$

$$2c_2 + \frac{3}{2}c_3 = -11$$

$$\frac{5}{4}c_2 + \frac{1}{2}c_3 = -6$$

$$\frac{1}{4}c_2 = -1$$

From the last and first equations, we see immediately that

$$c_2 = -4$$

and

$$c_3 = -2$$

which are consistent with the remaining two equations. Thus we have

$$Y(z) = \frac{8z}{z-1} - \frac{4z}{z-\frac{1}{2}} - \frac{2z^2}{(z-\frac{1}{2})^2} - \frac{z^3}{(z-\frac{1}{2})^3}$$

If we had chosen approach (2), we might choose c_2 and c_3 by forcing equality at $z = 2$ and $z = -1$ (recall that $z = 0$ has already been used to find c_0). Thus

$$Y(2) = \frac{16}{1(\frac{27}{8})} = \frac{16}{1} + \frac{2c_2}{\frac{3}{2}} + \frac{4c_3}{\frac{9}{4}} - \frac{8}{\frac{27}{8}}$$

that is,

$$\frac{128}{27} = 16 + \frac{4}{3}c_2 - \frac{16}{9}c_3 - \frac{64}{27}$$

or

$$9c_2 + 12c_3 = -60$$

and

$$Y(-1) = \frac{1}{-2(-\frac{27}{8})} = \frac{8}{2} + \frac{c_2}{\frac{3}{2}} + \frac{c_3}{\frac{9}{4}} - \frac{1}{\frac{27}{8}}$$

that is,

$$-\frac{4}{27} = 4 + \frac{2}{3}c_2 + \frac{4}{9}c_3 - \frac{8}{27}.$$

or

$$3c_2 + 2c_3 = -16$$

From these two relations, we obtain $c_2 = -4$ and $c_3 = -2$ as before.

To illustrate method (3), we first find c_0, c_1, and c_4 as before. Now c_3 is obtained from

$$c_3 = \left(\frac{z - \frac{1}{2}}{z}\right)^2 \left[Y(z) + \frac{z^3}{(z - \frac{1}{2})^3} + \text{other terms}\right]$$

Evaluating at $z = \frac{1}{2}$ to eliminate the "other terms," we obtain

$$\begin{aligned}
c_3 &= \left(\frac{z - \frac{1}{2}}{z}\right)^2 \left[Y(z) + \frac{z^3}{(z - \frac{1}{2})^3}\right]\Bigg|_{z=1/2} + \left(\frac{z - \frac{1}{2}}{z}\right)^2 [\text{other terms}]\big|_{z=1/2} \\
&= \left(\frac{z - \frac{1}{2}}{z}\right)^2 \left[\frac{z^4 + z^4 - z^3}{(z-1)(z - \frac{1}{2})^3}\right]\Bigg|_{z=1/2} + 0 \\
&= \left(\frac{z - \frac{1}{2}}{z}\right)^2 \left[\frac{2(z - \frac{1}{2})z^3}{(z-1)(z - \frac{1}{2})^3}\right]\Bigg|_{z=1/2} \\
&= \frac{2z}{z-1}\bigg|_{z=1/2} = -2
\end{aligned}$$

Continuing, we now set

$$c_2 = \frac{z - \frac{1}{2}}{z}\left[Y(z) + \frac{z^3}{(z - \frac{1}{2})^3} + \frac{2z^2}{(z - \frac{1}{2})^2} + \text{other terms}\right]$$

and again evaluate at $z = \frac{1}{2}$ to eliminate from consideration as many terms as possible. This yields

$$\begin{aligned}
c_2 &= \frac{z - \frac{1}{2}}{z}\left[\frac{z^4 + z^3(z-1) + 2z^2(z^2 - \frac{3}{2}z + \frac{1}{2})}{(z-1)(z - \frac{1}{2})^3}\right]\Bigg|_{z=1/2} + 0 \\
&= \frac{z - \frac{1}{2}}{z}\left[\frac{4(z - \frac{1}{2})^2 z^2}{(z-1)(z - \frac{1}{2})^3}\right]\Bigg|_{z=1/2} \\
&= \frac{4z}{z-1}\bigg|_{z=1/2} = -4
\end{aligned}$$

and again we obtain

$$Y(z) = \frac{8z}{z-1} - \frac{4z}{z - \frac{1}{2}} - \frac{2z^2}{(z - \frac{1}{2})^2} - \frac{z^3}{(z - \frac{1}{2})^3} \qquad \blacksquare$$

Having now obtained a partial fraction expansion for a given rational Z-transform, we take the inverse transform term-by-term to obtain the corresponding sequence. Here we must know the ROC of our transform, either explicitly or implicitly through the behavior of the corresponding sequence (e.g., if the sequence is one-sided to the right, then ROC extends from the furthest pole to infinity; if the sequence is absolutely summable, the ROC includes the unit circle, etc.) If $X(z)$ converges absolutely for some value of z for which $|z| > |p_i|$, then terms like $c[z/(z - p_i)]^m$ give rise to power series expansions in powers of z^{-1}, corresponding to sequences that are one-sided to the right. If, on the other hand, $X(z)$ converges absolutely for some value of z for which $|z| < |p_i|$, these terms give rise to power series expansions in powers of z, corresponding to sequences that are one-sided to the left. Table 4.3 summarizes the sequences corresponding to the individual terms in our partial fraction expansions.

Table 4.3 Inverse Transforms of the Partial Fraction Terms of $X(z)$

Partial Fraction Term	Inverse Transform Term If $X(z)$ Converges Absolutely for Some $\|z\| > \|a\|$
$\frac{z}{z-a}$	$a^k, \quad k \geq 0$
$\frac{z^2}{(z-a)^2}$	$(k+1)a^k, \quad k \geq 0$
$\frac{z^3}{(z-a)^3}$	$\frac{1}{2}(k+1)(k+2)a^k, \quad k \geq 0$
$\vdots$	$\vdots$
$\frac{z^n}{(z-a)^n}$	$\frac{1}{(n-1)!}(k+1)(k+2)\cdots(k+n-1)a^k, \quad k \geq 0$

Partial Fraction Term	Inverse Transform Term If $X(z)$ Converges Absolutely for Some $\|z\| < \|a\|$
$\frac{z}{z-a}$	$-a^k, \quad k \leq -1$
$\frac{z^2}{(z-a)^2}$	$-(k+1)a^k, \quad k \leq -1$
$\frac{z^3}{(z-a)^3}$	$-\frac{1}{2}(k+1)(k+2)a^k, \quad k \leq -1$
$\vdots$	$\vdots$
$\frac{z^n}{(z-a)^n}$	$-\frac{1}{(n-1)!}(k+1)(k+2)\cdots(k+n-1)a^k, \quad k \leq -1$

Example 4.12 In Example 4.9 we took the Z-transform $H(z) = (1 + z^{-1})/(1 - \frac{5}{6}z^{-1} + \frac{1}{6}z^{-2})$ with the partial fraction expansion

$$H(z) = \frac{9z}{z - \frac{1}{2}} - \frac{8z}{z - \frac{1}{3}}$$

Since we know that h_k is one-sided to the right, corresponding to the ROC $|z| > \frac{1}{2}$, we have

$$h_k = 9\left(\frac{1}{2}\right)^k - 8\left(\frac{1}{3}\right)^k, \qquad k \geq 0$$

■

Example 4.13 In Example 4.10 we took

$$X(z) = \frac{z^2 - 6z - 18.5}{z^3 + 1.57z^2 - 8.5z - 15}$$

with its partial fraction expansion

$$X(z) = 1.233 - \frac{0.5z}{z + 2} - \frac{0.4z}{z + 2.5} - \frac{0.333z}{z - 3}$$

Assuming here the ROC is $2.5 < |z| < 3$, we obtain

$$x_k = \begin{cases} 1.233\delta_k - 0.5(-2)^k - 0.4(-2.5)^k, & k \geq 0 \\ 0.333(3)^k, & k \leq -1 \end{cases}$$

If, on the other hand, we knew that the inverse transform of $X(z)$ was absolutely summable, we must choose a ROC of $|z| < 2$, which leads to the sequence

$$x_k = \begin{cases} 0, & k > 0 \\ 1.233, & k = 0 \\ 0.5(-2)^k + 0.4(-2.5)^k + 0.333(3)^k, & k \leq -1 \end{cases}$$

which may also be written in the form

$$x_k = 0.5(-2)^k + 0.4(-2.5)^k + 0.333(3)^k, \qquad k \leq 0$$

■

Example 4.14 In Example 4.11 we found the partial fraction expansion

$$Y(z) = \frac{1}{(1 - z^{-1})(1 - \frac{1}{2}z^{-1})^3}$$

$$= \frac{8z}{z - 1} - \frac{4z}{z - \frac{1}{2}} - \frac{2z^2}{(z - \frac{1}{2})^2} - \frac{z^3}{(z - \frac{1}{2})^3}$$

Taking the ROC $1 < |z|$, we obtain the inverse transform

$$y_k = 8 - 4\left(\frac{1}{2}\right)^k - 2(k + 1)\left(\frac{1}{2}\right)^k - \frac{1}{2}(k + 1)(k + 2)\left(\frac{1}{2}\right)^k, \qquad k \geq 0$$

$$= 8 - \left(4 + 2k + 2 + \frac{k^2}{2} + \frac{3k}{2} + 1\right)\left(\frac{1}{2}\right)^k, \qquad k \geq 0$$

$$= 8 - \left(\frac{k^2}{2} + \frac{7k}{2} + 7\right)\left(\frac{1}{2}\right)^k, \qquad k \geq 0$$

$$= 0, \quad k < 0$$

■

Alternative Partial Fraction Expansion Forms

The partial fraction expansion form shown in (4.52) and discussed above is not our only choice. We could also use the expansion (again, with $m \leq n$)

$$X(z) = \frac{a_0 z^m + a_1 z^{m-1} + a_2 z^{m-2} + \cdots + a_m}{(z - p_1)(z - p_2)\cdots(z - p_n)}$$

$$= c_0 + \frac{c_1}{z - p_1} + \frac{c_2}{z - p_2} + \cdots + \frac{c_n}{z - p_n} \tag{4.61}$$

for simple poles, with

$$c_0 = \lim_{z \to \infty} X(z) = \begin{cases} a_0 & \text{if} \quad m = n \\ 0 & \text{if} \quad m < n \end{cases} \tag{4.62}$$

and

$$c_i = (z - p_i)X(z)|_{z = p_i} \qquad \text{for} \quad i = 1, 2, \ldots, n \tag{4.63}$$

With repeated roots, we obtain the terms

$$\frac{c_{i_1}}{z - p_i} + \frac{c_{i_2}}{(z - p_i)^2} + \cdots + \frac{c_{ir}}{(z - p_i)^r}$$

where now the differentiation formula may conveniently be applied to yield

$$\begin{aligned} c_{i_r} &= (z - p_i)^r X(z)|_{z=p_i} \\ c_{i_{r-1}} &= \frac{d}{dz} (z - p_i)^r X(z)|_{z=p_i} \\ &\vdots \\ c_{i_{r-k}} &= \frac{1}{k!} \frac{d^k}{dz^k} (z - p_i)^r X(z)|_{z=p_i} \\ &\vdots \\ c_{i_1} &= \frac{1}{(r-1)!} \frac{d^{r-1}}{dz^{r-1}} (z - p_i)^r X(z)|_{z=p_i} \end{aligned} \tag{4.64}$$

In taking the inverse transform with this expansion, we must use the forms

a. if the ROC includes some $|z| > |a|$, then

$$\begin{aligned} \frac{1}{z-a} &\to a^{k-1}, \quad k \geq 1 \\ \frac{1}{(z-a)^2} &\to (k-1)a^{k-2}, \quad k \geq 1 \\ \frac{1}{(z-a)^3} &\to \frac{1}{2!}(k-1)(k-2)a^{k-3}, \quad k \geq 1 \\ &\text{etc.} \end{aligned} \tag{4.65}$$

b. If the ROC includes some $|z| < |a|$, then

$$\begin{aligned} \frac{1}{z-a} &\to -a^{k-1}, \quad k \leq 0 \\ \frac{1}{(z-a)^2} &\to -(k-1)a^{k-2}, \quad k \leq 0 \\ \frac{1}{(z-a)^3} &\to \frac{1}{2!}(k-1)(k-2)a^{k-3}, \quad k \leq 0 \\ &\text{etc.} \end{aligned} \tag{4.66}$$

Still other partial fraction expansions are in use† but will not be explored further here.

4.6 EVALUATING A SYSTEM'S FREQUENCY RESPONSE

As an application of Z-transforms, let us return to the discussion of frequency responses in Chapter 2. Recall that if a linear shift-variant discrete-time system is foced with the exponential input sequence $u_k = e^{jk\theta}$, then the resulting output is of the form $y_k = H(e^{j\theta})e^{jk\theta}$. The complex-valued function $H(e^{j\theta})$ is termed the frequency response of the system. Recall also that $H(e^{j\theta})$ can be obtained in two alternative forms. From the difference equation we obtain

$$H(e^{j\theta}) = \frac{a_0 + a_1 e^{-j\theta} + \cdots + a_m e^{-jm\theta}}{1 + b_1 e^{-j\theta} + \cdots + b_n e^{-jn\theta}} \tag{4.67}$$

where a_i, $i = 0, \ldots, m$ represent coefficients of the feedforward terms in the difference equation, and b_i, $i = 1, \ldots, n$ represent the feedback terms. Alternatively, we can write

$$H(e^{j\theta}) = \sum_{n=-\infty}^{\infty} h_n e^{-jn\theta} \tag{4.68}$$

where $\{h_n\}$ is the system's impulse response sequence.

We can easily generalize these results and obtain a relationship between a system's transfer function $H(z)$ and its frequency response $H(e^{j\theta})$. Let the system input be chosen as $u_k = z^k$ where z is a (in general) complex valued constant. The output is then given by

$$\begin{aligned} \{y_k\} &= \{z^k\} * \{h_k\} \\ &= \sum_{m=-\infty}^{\infty} h_m z^{k-m} \\ &= z^k \sum_{m=-\infty}^{\infty} h_m z^{-m} \triangleq z^k H(z) \end{aligned} \tag{4.69}$$

† For example, see H. Freeman, *Discrete-Time Systems*, Wiley, New York, 1965, pp. 48–50.

That is, y_k is simply the complex number $H(z)$ multiplied by the input sequence $u_k = z^k$. Using this fact, we can form an alternative expression for $H(z)$. Substituting $u_k = z^k$ and $y_k = H(z)z^k$ into the difference equation

$$y_k + b_1 y_{k-1} + \cdots + b_n y_{k-n} = a_0 u_k + a_1 u_{k-1} + \cdots + a_m u_{k-m}$$

we obtain

$$H(z)z^k + b_1 H(z)z^{k-1} + \cdots + b_n H(z)z^{k-n} = a_0 z^k + a_1 z^{k-1} + \cdots + a_m z^{k-m}$$

which leads to

$$z^k H(z)[1 + b_1 z^{-1} + \cdots + b_n z^{-n}] = z^k[a_0 + a_1 z^{-1} + \cdots + a_m z^{-m}]$$

from here we obtain the transfer function

$$H(z) = \frac{a_0 + a_1 z^{-1} + \cdots + a_m z^{-m}}{1 + b_1 z^{-1} + \cdots + b_n z^{-n}} \tag{4.70}$$

in terms of the difference equation coefficients.

We also see a relationship between the frequency response as expressed in (4.67) and (4.68) and the transfer function in (4.69) and (4.70). Comparing these relations, we see that

$$H(e^{j\theta}) = H(z)|_{z = e^{j\theta}} \tag{4.71}$$

In other words, the frequency response $H(e^{j\theta})$ of a discrete-time system is found by evaluating the corresponding transfer function on the unit circle at angle θ, as shown in Figure 4.7

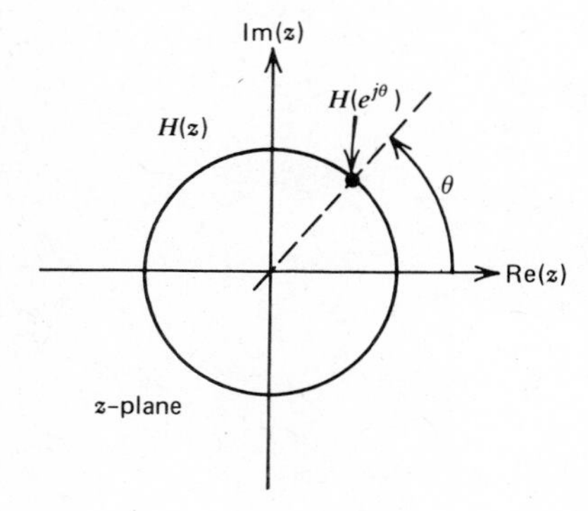

Figure 4.7 Evaluation of $H(e^{j\theta})$

Example 4.15 The block diagram of a lowpass filter is shown in Figure 4.8*a*. From the corresponding difference equation

$$y_k - 0.9630y_{k-1} + 0.4485y_{k-2} = 0.1394(u_k + 1.1349u_{k-1} + u_{k-2})$$

(or directly from the block diagram) we obtain the transfer function

$$H(z) = 0.1394\,\frac{1 + 1.1349z^{-1} + z^{-2}}{1 - 0.9630z^{-1} + 0.4485z^{-2}}$$

The frequency response is found by evaluating $H(z)$ at $z = e^{j\theta}$

$$H(e^{j\theta}) = 0.1394\,\frac{1 + 1.1349e^{-j\theta} + e^{-j2\theta}}{1 - 0.9630e^{-j\theta} + 0.4485e^{-j2\theta}}$$

The amplitude response $|H(e^{j\theta})|$ and phase response $\arg[H(e^{j\theta})]$ are shown in Figure 4.8*b* and 4.8*c*.

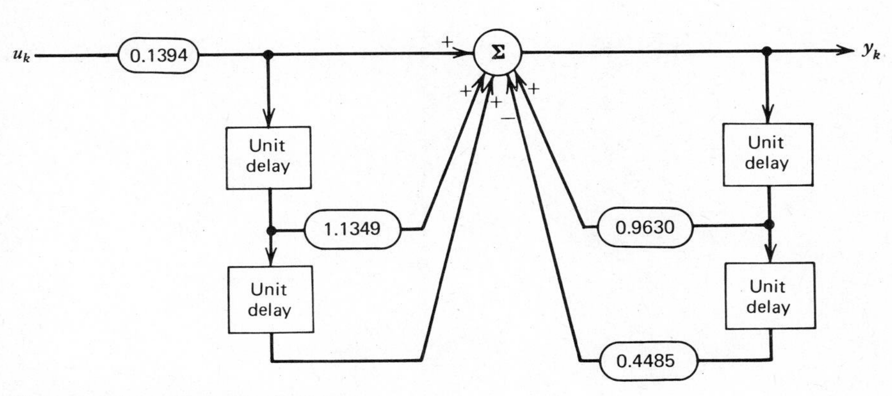

Figure 4.8*a* **Filter block diagram**

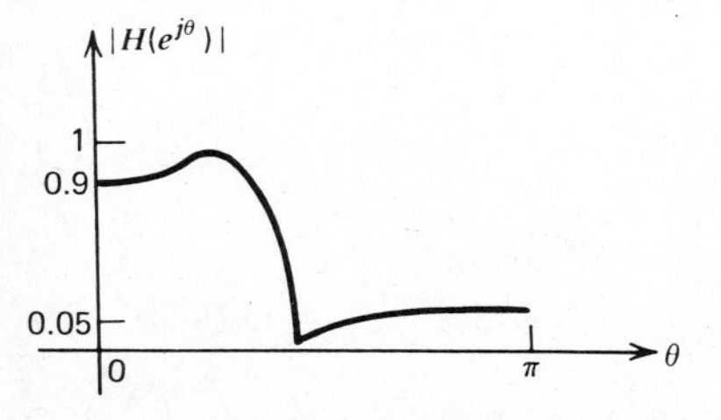

Figure 4.8*b* **Filter amplitude response**

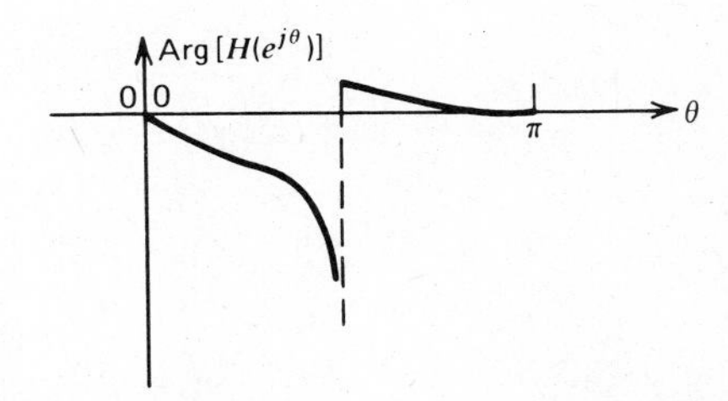

Figure 4.8*c* ■

Graphical Determination of the Frequency Response

We can give a graphical interpretation to the evaluation of $H(e^{j\theta})$ in (4.71). Let us write $H(z)$ in factored form as

$$H(z) = A\frac{(z-\zeta_1)(z-\zeta_2)\cdots(z-\zeta_m)}{(z-p_1)(z-p_2)\cdots(z-p_n)} \tag{4.72}$$

where ζ_i, $i = 1, \ldots, m$ are the *zeros* of $H(z)$ and $p_i = 1, \ldots, n$ are, as before, the corresponding poles. A term of the form $z - a$ can be viewed as a vector in the z-plane, with its head at z and tail at a. This vector will be of length $|z - a|$ and will make an angle $\phi = \arg[z - a]$ with the horizontal, as shown below.

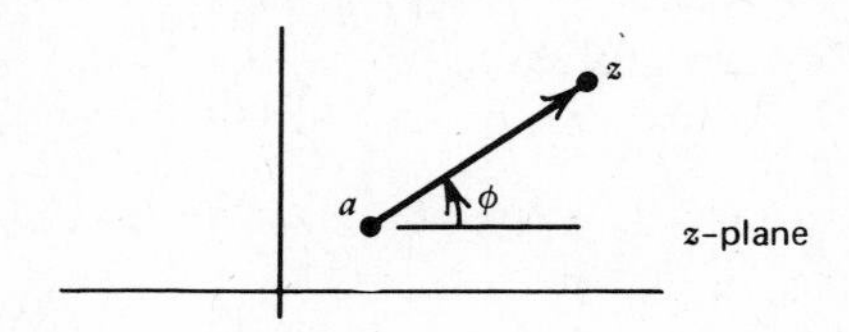

If we now let $z = e^{j\theta}$ and write the terms $(z - \zeta_i)$ and $(z - p_i)$ in polar form, we have

$$= A\frac{(e^{j\theta}-\zeta_1)(e^{j\theta}-\zeta_2)\cdots(e^{j\theta}-\zeta_m)}{(e^{j\theta}-p_1)(e^{j\theta}-p_2)\cdots(e^{j\theta}-p_n)}$$

$$= A\frac{\prod_{i=1}^{m}(e^{j\theta}-\zeta_i)}{\prod_{i=1}^{n}(e^{j\theta}-p_i)}$$

$$= A\frac{\prod_{i=1}^{m}\{|e^{j\theta}-\zeta_i|e^{j\arg(e^{j\theta}-\zeta_i)}\}}{\prod_{i=1}^{n}\{|e^{j\theta}-p_i|e^{j\arg(e^{j\theta}-p_i)}\}}$$

$$= A\frac{\prod_{i=1}^{m}|e^{j\theta}-\zeta_i|}{\prod_{i=1}^{n}|e^{j\theta}-p_i|}e^{j[\sum_{i=1}^{m}\arg(e^{j\theta}-\zeta_i)-\sum_{i=1}^{n}\arg(e^{j\theta}-p_i)]} \tag{4.73}$$

$$\triangleq |H(e^{j\theta})|e^{j\arg[H(e^{j\theta})]} \tag{4.74}$$

Thus we see that the amplitude transfer function is found from

$$|H(e^{j\theta})| = A\frac{\prod_{i=1}^{m}|e^{j\theta}-\zeta_i|}{\prod_{i=1}^{n}|e^{j\theta}-p_i|} \tag{4.75}$$

which is the scale factor A multiplied by the product of the lengths of the vectors from the zero locations ζ_i, $i = 1, \ldots, m$ to the point $z = e^{j\theta}$, divided by the product of the lengths of the vectors from the pole locations p_i, $i = 1, \ldots, n$ to the point $z = e^{j\theta}$. The phase transfer function, on the other hand, is given by

$$\arg [H(e^{j\theta})] = \sum_{i=1}^{m} \arg (e^{j\theta} - \zeta_i) - \sum_{i=1}^{n} \arg (e^{j\theta} - p_i) \tag{4.76}$$

which is the sum of the angles of the vectors from ζ_i, $i = 1, \ldots, m$ to $e^{j\theta}$ minus the sum of the angles from p_i, $i = 1, \ldots, n$ to $e^{j\theta}$. This graphical calculation is illustrated in the next example.

Example 4.16 Let us take the filter discussed in the previous example and demonstrate the ideas discussed above. Here we have

$$\begin{aligned} H(z) &= 0.1394 \frac{1 + 1.1349z^{-1} + z^{-2}}{1 - 0.9630z^{-1} + 0.4485z^{-2}} \\ &= 0.1394 \frac{[z - (-0.567 + j0.823)][z - (-0.567 - j0.823)]}{[z - (0.481 + j0.465)][z - (0.481 - j0.465)]} \end{aligned}$$

The pole-zero plot of $H(z)$ is shown in Figure 4.9. In terms of the vector lengths

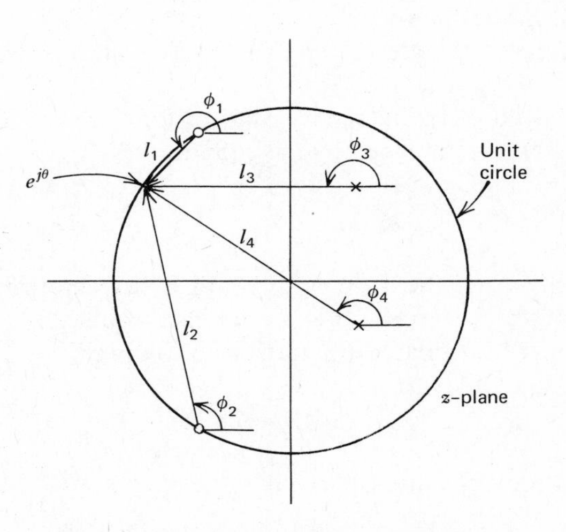

Figure 4.9 Pole-zero plot of filter

$l_i, i = 1, \ldots, 4$ and phase angles ϕ_i, $i = 1, \ldots, 4$ shown in the figure, we find the amplitude and phase responses as

$$|H(e^{j\theta})| = 0.1394 \frac{l_1 \cdot l_2}{l_3 \cdot l_4}$$

and

$$\arg [H(e^{j\theta})] = \phi_1 + \phi_2 - \phi_3 - \phi_4$$

When this evaluation is performed for θ ranging from 0 to π, we obtain the plots of Figures 4.8*b* and 4.8*c*, which were determined analytically in the preceding example. Here, however, we obtain additional insight. Observe, for example, how $|H(e^{j\theta})|$ gets smaller as we approach a zero of $H(z)$. Note, too how $\arg [H(e^{j\theta})]$ has a discontinuity of π radians as we pass through a zero and the vector $(e^{j\theta} - \zeta_1)$ "flips" and reverses its direction. ■

4.7 FURTHER APPLICATIONS OF THE *Z*-TRANSFORM

A problem that we treated in Chapter 2 involved determining the state vector of a system. Recall that in the state-variable formulation the essential calculation concerned the evolution of the state vector. We can use Z-transforms to perform this calculation.

Consider the state-variable representation for a linear stationary system. The state of the system is governed by the equation

$$\mathbf{x}_{k+1} = \mathbf{A}\mathbf{x}_k + \mathbf{B}\mathbf{u}_k \tag{4.77}$$

where $\mathbf{u}_k$ is the input, $\mathbf{x}_k$ is the state, and $\mathbf{A}$ and $\mathbf{B}$ characterize the system as we discussed in Chapter 2. Since (4.77) is a difference equation with constant coefficients, we can apply the unilateral Z-transform to obtain

$$z[\mathbf{X}_u(z) - \mathbf{x}_0] = \mathbf{A}\mathbf{X}_u(z) + \mathbf{B}\mathbf{U}_u(z) \tag{4.78}$$

Multiplying through by z^{-1} and collecting terms yields

$$[\mathbf{I} - z^{-1}\mathbf{A}]\mathbf{X}_u(z) = \mathbf{x}_0 + z^{-1}\mathbf{B}\mathbf{U}_u(z)$$

from which we obtain the solution for $\mathbf{X}_u(z)$,

$$\mathbf{X}_u(z) = [\mathbf{I} - z^{-1}\mathbf{A}]^{-1}\mathbf{x}_0 + z^{-1}[\mathbf{I} - z^{-1}\mathbf{A}]^{-1}\mathbf{B}\mathbf{U}_u(z) \tag{4.79}$$

By taking the inverse transform of both sides of (4.79), we can obtain a solution for the state vector x_k in terms of its initial value x_0 and the input sequence $\{u_k\}$. Suppose for example that $\mathbf{U}(z) = 0$ in (4.79) so that we have an unforced system. The solution reduces to

$$\mathbf{X}_u(z) = [\mathbf{I} - z^{-1}\mathbf{A}]^{-1}\mathbf{x}_0 \tag{4.80}$$

However, we know the solution for this case to be

$$\mathbf{x}_k = \mathbf{A}^k\mathbf{x}_0 \tag{4.81}$$

Thus it must be the case that

$$Z^{-1}\{[\mathbf{I} - z^{-1}\mathbf{A}]^{-1}\} = \mathbf{A}^k \tag{4.82}$$

This result furnishes us with another means for evaluating the matrix $\mathbf{A}^k$, as illustrated in the next example.

Example 4.17 Suppose that a manufacturing line is producing either calculators or CB radio sets. If calculators are being produced at a given time, we will say that the line is in state 1; if radio sets are being produced, we call this state 2. A change from one product to another is made only at the end of a production week. From observing the production cycles, we deduce that if the line is in state 1 during a given week, it will be in state 1 the next week with probability 0.5 and in state 2 with probability 0.5. If on the other hand, the line is in state 2, it will remain in state 2 with probability 0.6 and return to state 1 with probability 0.4. A graphical representation of this system is shown in Figure 4.10. The numbers on the branches in Figure 4.10 are the state transition probabilities p_{ij}. These state transition probabilities are the probability that the system now in state i will be in state j after one transition. Thus, for example, $p_{21} = \frac{2}{5}$ is the probability of

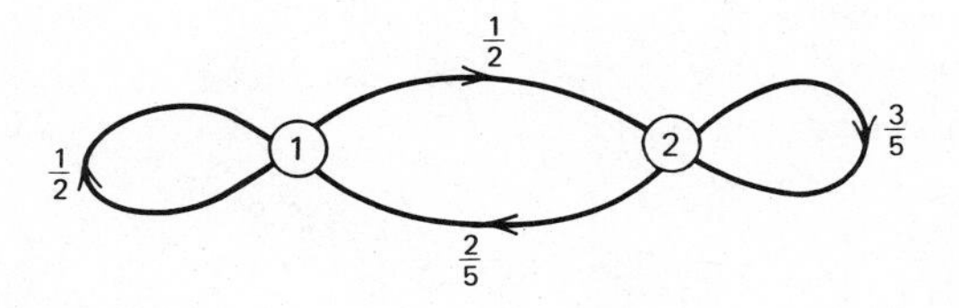

Figure 4.10

going to state 1 in one transition, given that the system is presently in state 2. It is convenient to summarize the state transition probabilities in the matrix **P** as

$$\mathbf{P} = [p_{ij}] = \begin{bmatrix} \frac{1}{2} & \frac{1}{2} \\ \frac{2}{5} & \frac{3}{5} \end{bmatrix}$$

The matrix **P** is a complete description of the Markov process. This matrix plays the same role as the matrix **A** in our discussion of linear systems. For example, if we define $\mathbf{x}_n$ as

$$\mathbf{x}_n = \begin{bmatrix} x_n^1 \\ x_n^2 \end{bmatrix}$$

where x_n^i = probability that the system is in state i at stage n, $i = 1, 2$, then the state probabilities at time $n + 1$ are

$$\mathbf{x}_{n+1} = \mathbf{P}^T \mathbf{x}_n \tag{4.83}$$

Thus, if $\mathbf{x}_n = \begin{bmatrix} 0 \\ 1 \end{bmatrix}$, which implies with probability one we manufacture CB sets,

$$\mathbf{x}_{n+1} = \begin{bmatrix} \frac{1}{2} & \frac{2}{5} \\ \frac{1}{2} & \frac{3}{5} \end{bmatrix} \begin{bmatrix} 0 \\ 1 \end{bmatrix} = \begin{bmatrix} \frac{2}{5} \\ \frac{3}{5} \end{bmatrix}$$

One might ask what is the manufacturing policy after a large number of weeks, given some initial probabilities $\mathbf{x}_0$? The answer to this question involves the calculation that we discussed above.

If we apply the transition matrix to the initial state $\mathbf{x}_0$ several times, we can, of course, calculate the final state probabilities. We also make use of (4.83). Thus applying $\mathbf{P}^T$ to $\mathbf{x}_0$ over and over, we obtain

$$\mathbf{x}_n = (\mathbf{P}^T)^n \mathbf{x}_0$$

Now consider the Z-transform of (4.83). We see that

$$z^{-1}[\mathbf{X}_u(z) - \mathbf{x}_0] = \mathbf{P}^T\mathbf{X}_u(z)$$

If we solve for $\mathbf{X}_u(z)$, we obtain

$$\mathbf{X}_u(z) = [\mathbf{I} - z\mathbf{P}^T]^{-1}\mathbf{x}_0 \tag{4.84}$$

Hence, the inverse Z-transform of $[\mathbf{I} - z\mathbf{P}^T]^{-1}$ must be identically $(\mathbf{P}^T)^n$. To carry out the calculations, we decompose (4.84) into partial fraction form. In this case the coefficients are matrices rather than scalars. Thus we write

$$[\mathbf{I} - z^{-1}\mathbf{P}^T]^{-1} = \begin{bmatrix} 1 - \frac{1}{2}z^{-1} & -\frac{2}{5}z^{-1} \\ -\frac{1}{2}z^{-1} & 1 - \frac{3}{5}z^{-1} \end{bmatrix}^{-1}$$

$$= \frac{1}{1 - \frac{11}{10}z^{-1} + \frac{1}{10}z^{-2}} \begin{bmatrix} 1 - \frac{3}{5}z^{-1} & \frac{2}{5}z^{-1} \\ \frac{1}{2}z^{-1} & 1 - \frac{1}{2}z^{-1} \end{bmatrix}$$

$$= \frac{\mathbf{C}_1}{1 - z^{-1}} + \frac{\mathbf{C}_2}{1 - \frac{1}{10}z^{-1}}$$

where

$$\mathbf{C}_1 = \frac{1}{1 - \frac{1}{10}z^{-1}} \begin{bmatrix} 1 - \frac{3}{5}z^{-1} & \frac{2}{5}z^{-1} \\ \frac{1}{2}z^{-1} & 1 - \frac{1}{2}z^{-1} \end{bmatrix}\Bigg|_{z=1}$$

$$= \frac{10}{9}\begin{bmatrix} \frac{2}{5} & \frac{2}{5} \\ \frac{1}{2} & \frac{1}{2} \end{bmatrix} = \begin{bmatrix} \frac{4}{9} & \frac{4}{9} \\ \frac{5}{9} & \frac{5}{9} \end{bmatrix}$$

and

$$\mathbf{C}_2 = \frac{1}{1 - z^{-1}} \left. \begin{bmatrix} 1 - \frac{3}{5} z^{-1} & \frac{2}{5} z^{-1} \\ \frac{1}{2} z^{-1} & 1 - \frac{1}{2} z^{-1} \end{bmatrix} \right|_{z=1/10}$$

$$= \frac{1}{-9} \begin{bmatrix} -5 & 4 \\ 5 & -4 \end{bmatrix} = \begin{bmatrix} \frac{5}{9} & -\frac{4}{9} \\ -\frac{5}{9} & \frac{4}{9} \end{bmatrix}$$

Thus

$$(\mathbf{P}^T)^n = Z^{-1} \left\{ \frac{\mathbf{C}_1}{1 - z^{-1}} + \frac{\mathbf{C}_2}{1 - \frac{1}{10} z^{-1}} \right\}$$

$$= \mathbf{C}_1 + \mathbf{C}_2 \left(\frac{1}{10} \right)^n, \qquad n \geq 0$$

$$= \begin{bmatrix} \frac{4}{9} & \frac{4}{9} \\ \frac{5}{9} & \frac{5}{9} \end{bmatrix} + \left(\frac{1}{10} \right)^n \begin{bmatrix} \frac{5}{9} & -\frac{4}{9} \\ -\frac{5}{9} & \frac{4}{9} \end{bmatrix}$$

Thus the state probability vector $\mathbf{x}_n$ can be found for any n by multiplying $(\mathbf{P}^T)^n$ by the initial state probability vector $\mathbf{x}_0$. The (i, j)th element of $(\mathbf{P}^T)^n$ represents the probability that the system will be in state i at time n, given that the initial state is j at time $n = 0$. Notice that $(\mathbf{P}^T)^n$ consists of a steady-state matrix independent of n and a transient matrix that dies away as n increases. The columns of the steady-state matrix will be identical and represent the limiting-state probabilities, in our case, the probabilities of manufacturing one product or the other. Thus, in steady state, the manufacturing line will be producing calculators with probability $\frac{4}{9}$ and CB radios with probability $\frac{5}{9}$. The effect of the initial product choice dies away with increasing n. ■

As a final application, we suggest a technique that is useful in two-dimensional problems. Here we compute transforms in terms of two independent variables.

This kind of calculation can be used in the numerical solution of partial differential equations, which always have two or more independent variables.

Example 4.18 Consider the following difference equation on the integer variables n and m:

$$c(n+1, m+1) = c(n, m+1) + c(n, m) \tag{4.85}$$

with boundary data

$$c(n, 0) = 1, \qquad n \geq 0$$
$$c(0, m) = 0, \qquad m > 0$$

Let

$$\Gamma_u(n, z) = Z_u\{\{c(n, m)\}|_{n \text{ fixed}}\}$$

Taking the Z-transform on the discrete variable m in (4.85) yields

$$z[\Gamma_u(n+1, z) - c(n+1, 0)] = z[\Gamma_u(n, z) - c(n, 0)] + \Gamma_u(n, z)$$

which, with $c(n+1, 0) = c(n, 0) = 1$, $n \geq 0$, yields

$$z\Gamma_u(n+1, z) = z\Gamma_u(n, z) + \Gamma_u(n, z) \tag{4.86}$$

Now taking the Z-transform on n in (4.86), we obtain

$$z\{y[\Omega_u(y, z) - \Gamma_u(0, z)]\} = (z+1)\Omega_u(y, z)$$

But the boundary data yield

$$\begin{aligned}\Gamma_u(0, z) &= Z_u\{\{c(0, m)\}\} \\ &= Z_u\{\{\delta_m\}\} = 1\end{aligned}$$

Thus we obtain

$$zy[\Omega_u(y, z) - 1] = (z+1)\Omega_u(y, z)$$

that is,

$$[zy - (z + 1)]\Omega_u(y, z) = zy$$

or

$$\Omega_u(y, z) = \frac{zy}{(yz - z - 1)} = \frac{1}{1 - y^{-1}(1 + z^{-1})}$$

Now taking the inverse transform with respect to y, we obtain

$$Z^{-1}\{\Omega_u(y, z)\}|_{z \text{ fixed}} = \Gamma_u(n, z) = (1 + z^{-1})^n, \qquad n \geq 0$$

that is,

$$\Gamma_u(n, z) = 1 + nz^{-1} + \frac{n(n-1)}{2!}z^{-2} + \frac{n(n-1)(n-2)}{3!}z^{-3} + \cdots$$

$$= \sum_{m=0}^{\infty} \binom{n}{m} z^{-m} \qquad \text{where} \qquad \binom{n}{m} = \frac{n!}{m!(n-m)!}$$

But by definition

$$\Gamma_u(n, z) = \sum_{m=0}^{\infty} c(n, m) z^{-m}$$

and thus, equating coefficients of the two polynomials in z^{-1}, we obtain

$$c(n, m) = \binom{n}{m} \qquad \begin{array}{l} n \geq 0 \\ 0 \leq m \leq n \end{array}$$

as shown in Figure 4.11. Note that in this two-dimensional problem the boundary consists of two lines rather than points; thus the boundary data must be specified

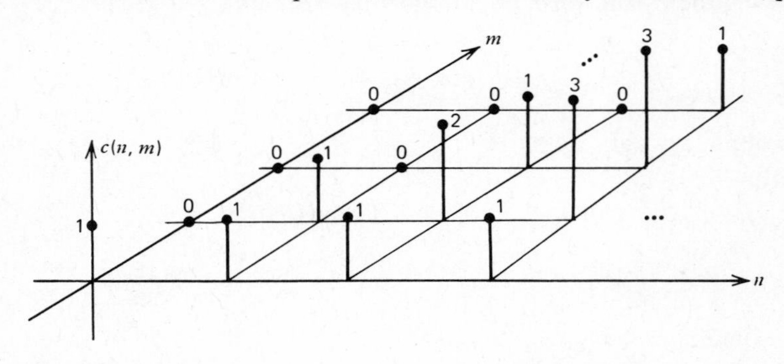

Figure 4.11

along these two lines. As a check, we can verify our solution by substituting in the given difference equation to obtain

$$\binom{n+1}{m+1} \stackrel{?}{=} \binom{n}{m+1} + \binom{n}{m}$$

$$\frac{(n+1)!}{(m+1)!(n-m)!} \stackrel{?}{=} \frac{n!}{(m+1)!(n-m-1)!} + \frac{n!}{m!(n-m)!}$$

$$\frac{(n+1)n!}{(m+1)m!(n-m)!} \stackrel{?}{=} \frac{n!(n-m)}{(m+1)m!(n-m)!} + \frac{n!}{m!(n-m)!}$$

$$\frac{n+1}{m+1} \stackrel{?}{=} \frac{n-m}{m+1} + 1 = \frac{n-m+m+1}{m+1}$$

$$\stackrel{\checkmark}{=} \frac{n+1}{m+1}$$

■

4.8 SUMMARY

This chapter has presented yet another formulation for the solution of linear, stationary, discrete-time problems. This formulation is basically different from those presented in the first three chapters in that the formulation is given in a transform domain rather than in the time domain. The properties of the Z-transform that we derived are remarkably similar to properties that occur in other transform domains (like the Fourier and Laplace domains). One of the advantages of the transform method is that the convolution process in the time domain is transformed into a multiplication process. Thus, the output of a stationary linear system can be calculated by taking the inverse transform of the product of the transform of the input and the system impulse response. This method of calculating the output is often simpler than a direct convolution of the input and the impulse-response function. By transforming the impulse-response sequence to obtain the system transfer function, we obtain another characterization of the input-output relation that is especially useful in analyzing cascaded systems.

PROBLEMS

4.1. For each of the sequences below, evaluate the corresponding Z-transform directly from the defining summation. Sketch the sequence and the ROC of the transform.

(a) $f_k = (\frac{1}{2})^k, \quad k \geq 0$
(b) $g_k = (\frac{1}{2})^k, \quad k < 0$
(c) $h_k = 2^k, \quad k < 0$
(d) $p_k = (\frac{1}{2})^{|k|}, \quad \text{all } k$
(e) $q_k = (\frac{1}{2})^k + 2(\frac{1}{3})^k, \quad k \geq 0$
(f) $r_k = 2^k + 3^k, \quad k \geq 0$

(g) $$s_k = \begin{cases} 2^k, & k \geq 0 \\ (\frac{1}{3})^k, & k < 0 \end{cases}$$

(h) $t_k = (\frac{1}{2})^k, \quad \text{all } k$
(i) $u_k = \cos k\pi/8, \quad k \geq 0$
(j) $v_k = 3\cos(k\pi/8 + \pi/8), \quad k \geq 0$

(k) $$w_k = \begin{cases} 2^k, & k = 0, 2, 4, 6, \ldots \\ 0, & \text{otherwise} \end{cases}$$

4.2. Given the following forced difference equations, find the transform $Y(z)$ of the sequence $\{y_k\}$. Assume an initially relaxed system.

(a) $$y_k - 2y_{k-1} + y_{k+2} = \begin{cases} 1, & k \geq 0 \\ 0, & k < 0 \end{cases}$$

(b) $$y_{k+2} - 2y_{k+1} + y_k = \begin{cases} 1, & k \geq 0 \\ 0, & k < 0 \end{cases}$$

(c) $$y_k - 4y_{k-2} = \begin{cases} (\frac{1}{2})^k, & k \geq 0 \\ 0, & k < 0 \end{cases}$$

(d) $$y_{k+1} - y_{k-2} = \begin{cases} 2, & k \geq 0 \\ 0, & k < 0 \end{cases}$$

(e) $$y_{k+1} + 3y_k = \begin{cases} k, & k \geq 0 \\ 0, & k < 0 \end{cases}$$

(f) $$y_{k+1} - 5y_k = \begin{cases} \sin k, & k \geq 0 \\ 0, & k < 0 \end{cases}$$

4.3. Each of the difference equations below represents an unforced system with nonzero initial conditions. In each case, determine the input sequence at the input to an *initially relaxed* system such that the output will be the same as for the first case.

(a) $y_k - ay_{k-1} = 0, \quad y_0 = 1$
(b) $y_k - 2y_{k-1} + y_{k-2} = 0, \quad y_{-1} = 0, \quad y_0 = 1$
(c) $y_k - 2y_{k-1} + y_{k-2} = 0, \quad y_{-1} = 1, \quad y_0 = 0$
(d) $y_k - 2y_{k-1} + y_{k-2} = 0, \quad y_{-1} = y_0 = 1$

4.4. Using the unilateral transform, find the general solution to the second-order difference equation

$$f_{k+1} - 2f_k + f_{k-1} = \phi_k$$

for the inputs
(a) $\phi_k = a^k, \quad a \neq 1, \quad k \geq 0$
(b) $\phi_k = 1, \quad k \geq 0$
(c) $\phi_k = k, \quad k \geq 0$

4.5. Consider the generation of Fibonacci numbers, which occur in such unsuspected places as the number of ancestors in succeeding generations of the male bee, the input impedance of a resistor ladder network, and the spacing of buds on the branch of a tree.† A generalized Fibonacci sequence is a sequence of real numbers $\{f_k\}$ satisfying the difference equation

$$f_{k+2} = f_k + f_{k+1}, \quad k \geq 0$$

The classical Fibonacci sequence $\{f_k\}$ satisfies this difference equation with $f_0 = 0$ and $f_1 = 1$.
(a) Find a general expression for f_k.
(b) Show that the ratio f_k/f_{k+1} approaches the limit $2/(1 + \sqrt{5})$ as $k \to \infty$. This ratio is known as the "golden mean" and is said to be the ratio of the sides of that rectangle of most pleasing proportions.

4.6. Let f_k be the Fibonacci sequence. Verify the following:
(a) $f_k^2 + f_{k+1}^2 = f_{2k+1}$
(b) $f_{k+2}^2 - f_{k+1}^2 = f_k f_{k+3}$
(c) Verify the following parlor trick. Turn your back and tell someone to write two positive integers in a column. Increase the column to ten numbers by adding each two successive numbers to obtain the next one. You turn around and write the sum by inspection. (The sum is 11 times the 7th number.)

4.7. A conducting rod of cross-sectional area A, thermal conductivity k, and length $L = (n + 1)h$ connects n mass points of equal mass M and specific

† "Fibonacci Numbers: Their History through 1900," Maxey Brooke, *Fibonacci Quarterly* 2, 149–152.

heat c. The masses are located at a constant separation h, and the rod extends a distance h beyond each of the extreme masses (see below):

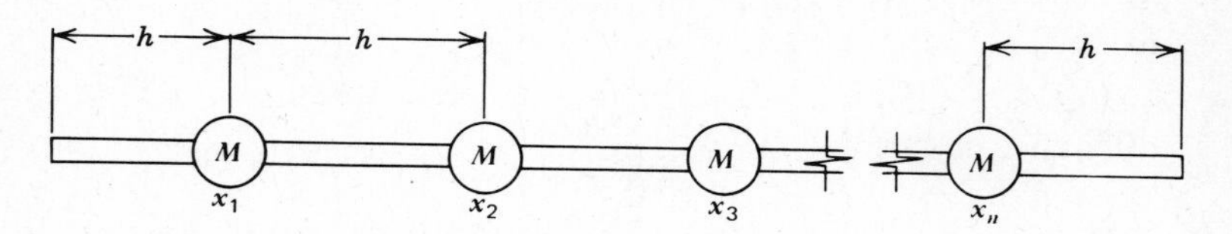

(a) If q_k is the rate of flow of heat into the kth mass and T_k is the temperature of the kth mass, show that q_k satisfies

$$q_k = cM \frac{dT_k(t)}{dt}$$

$$q_k = \frac{kA}{h} \{(T_{k+1} - T_k) - (T_k - T_{k-1})\}$$

Thus deduce that T_k satisfies

$$\frac{dT_k(t)}{dt} = \frac{kA}{hcM} [T_{k+1}(t) - 2T_k(t) + T_{k-1}(t)]$$

with the appropriate boundary conditions.

(b) Find the general solution to the temperature equation.

4.8. Sum the series

(a) $\sum_{n=0}^{\infty} e^{-x(2n+1)}$ (b) $\sum_{k=0}^{\infty} \alpha^k \sinh(kx)$

4.9. Consider the series

$$\frac{1}{1 - x^m} = 1 + x^m + x^{2m} + \cdots$$

Examine the coefficient of x^k in the series for $[1/(1 - x^m)][1/(1 - x^n)]$. Show that the coefficient of x^k is the number of ways one can solve the equation

$$k = l_1 m + l_2 n$$

for l_1 and l_2 nonnegative integers. Thus, for example, the number of ways a dollar can be changed is the 100th coefficient of the series

$$\frac{1}{1 - x} \cdot \frac{1}{1 - x^5} \cdot \frac{1}{1 - x^{10}} \cdot \frac{1}{1 - x^{25}} \cdot \frac{1}{1 - x^{50}}$$

Let x_n be the number of solutions to the equation

$$n = l_1 + 2l_2$$

(a) What is the Z-transform of $\{x_n\}$?
(b) What difference equation does x_n satisfy?
(c) What is the solution to this equation?

4.10. Given the following difference equations, evaluate the unilateral transform $Y_u(z)$ of the sequence $\{y_k\}$, using the given initial conditions.

(a) $y_k - 2y_{k-1} + y_{k-2} = 1, \quad k \geq 0, \quad y_{-1} = y_{-2} = 1$
(b) $y_k - \frac{1}{4}y_{k-2} = (\frac{1}{2})^k, \quad k \geq 0, \quad y_{-1} = 0, \quad y_{-2} = 1$
(c) $y_{k+2} - \frac{1}{4}y_k = (\frac{1}{2})^k, \quad k \geq 0, \quad y_0 = y_1 = 1$
(d) $y_{k+2} + y_{k+1} - y_k = 1, \quad k \geq 0, \quad y_0 = 1, \quad y_1 = 2$
(e) $y_{k+1} + y_k - y_{k-1} = 1, \quad k \geq 0, \quad y_0 = 1, \quad y_{-1} = -1$

4.11. Sketch a block diagram representation for each of the systems above.

4.12. Write a computer program to generate each of the $\{y_k\}$ sequences above, incorporating the given initial conditions.

4.13. Evaluate the Z-transform of each of the sequences below by using the transform Table 4.2 and the properties of the Z transform.

(a) $\{a_k\} = \{(\frac{1}{2})^k\}_0^\infty * \{(\frac{1}{2})^k\}_0^\infty$
(b) $\{b_k\} = \underbrace{\{(\frac{1}{2})^k\}_0^\infty * \{(\frac{1}{2})^k\}_0^\infty * \cdots * \{(\frac{1}{2})^k\}_0^\infty}_{M \text{ terms}}$
(c) $\{c_k\} = \{(\frac{1}{2})^{-k}\}_{-\infty}^0$
(d) $\{d_k\} = \{(\frac{1}{2})^k\}_0^\infty * \{(\frac{1}{2})^{-k}\}_{-\infty}^0$
(e) $e_k = \sum_{m=0}^k (\frac{1}{2})^m \quad k \geq 0$
(Check by evaluating the sum explicitly and transforming the result.)
(f) $f_k = \cos[(k-2)\pi/8], \quad k \geq 2$
(g) $\{g_k\} = \{k \cos k\theta\}_0^\infty$
(h) $\{h_k\} = \{ka^k \cos k\theta\}_0^\infty$

4.14. Evaluate the inverse of the following Z-transforms:

(a) $$A(z) = \frac{z^2}{(z - \frac{1}{2})(z - \frac{1}{3})}, \qquad |z| < \frac{1}{3}$$

(b) $$B(z) = \frac{z^2}{(z - \frac{1}{2})(z - \frac{1}{3})}, \qquad \frac{1}{3} < |z| < \frac{1}{2}$$

(c) $$C(z) = \frac{z^2}{(z - \frac{1}{2})(z - \frac{1}{3})}, \qquad \frac{1}{2} < |z|$$

(d) $$D(z) = \frac{1}{(z - \frac{1}{2})(z - \frac{1}{3})}, \qquad \frac{1}{2} < |z|$$

(e) $E(z) = \dfrac{-z}{(z - \frac{1}{2})(z - 2)}$, where $e_k = 0, \quad k < 0$

(f) $F(z) = \dfrac{-z}{(z - \frac{1}{2})(z - 2)}$, where $\{f_k\}$ has finite energy

(g) $G(z) = \dfrac{z^3}{(z - \frac{1}{4})^2(z - 1)}$, $|z| < \dfrac{1}{4}$

(h) $H(z) = \dfrac{z^3}{(z - \frac{1}{4})^2(z - 1)}$, $\dfrac{1}{4} < |z| < 1$

(i) $P(z) = \dfrac{z^3}{(z - \frac{1}{4})^2(z - 1)}$, $1 < |z|$

(j) $Q(z) = \dfrac{1}{(z - \frac{1}{2})^2(z - 2)}$, where $\{q_k\}$ has finite energy

4.15. The data sequence $x(n)$ is passed through a system with impulse response $g(n)$, producing the output $v(n)$. Now $v(n)$ is time-reversed, and $v(-n)$ is passed through the identical filter, producing the output $w(n)$. Finally, $w(n)$ is time-reversed to generate the sequense $y(n) = w(-n)$.

(a) Find the impulse response $h(n)$ and corresponding transfer function $H(z)$ that relate the output $y(n)$ to the input $x(n)$:

$$y(n) = x(n) * h(n); \qquad Y(z) = X(z) \cdot H(z)$$

(b) If $g(n)$ represents a stable casual filter,

(i) does $h(n)$ represent a stable filter?

(ii) does $h(n)$ represent a casual filter?

Defend your answers!

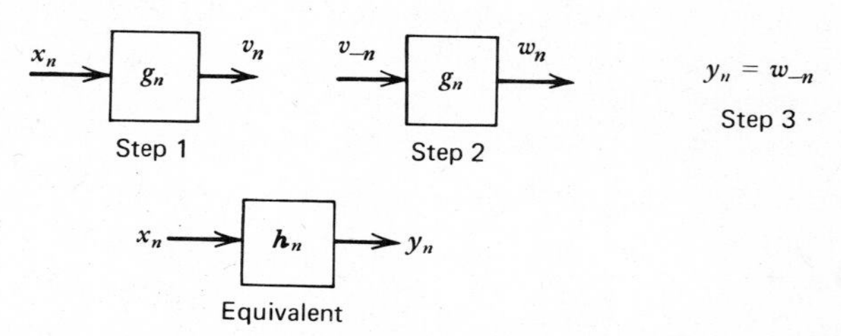

4.16. Let the system G of the previous problem be as sketched below. Evaluate $G(z)$, $H(z)$, $\{g_k\}$, and $\{h_k\}$ for the system.

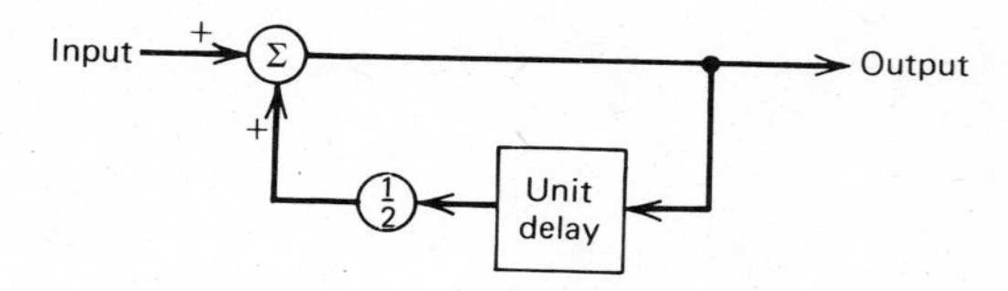

4.17. The autocorrelation r_{xx} of a sequence x may be defined as

$$r_{xx}(k) = \sum_{m=-\infty}^{\infty} x_m x_{m-k}$$

(a) Look carefully at the subscripts above.
(b) Express r_{xx} as a convolution of two sequences.
(c) Using the result of (b), what is the transform $R_{xx}(z)$?
(d) Evaluate $R_{xx}(z)$ and $r_{xx}(k)$ for the case $x_k = (\frac{1}{2})^k$, $k \geq 0$.

4.18. Find the impulse response of the system below by using Z-transforms. Repeat using another method, and compare.

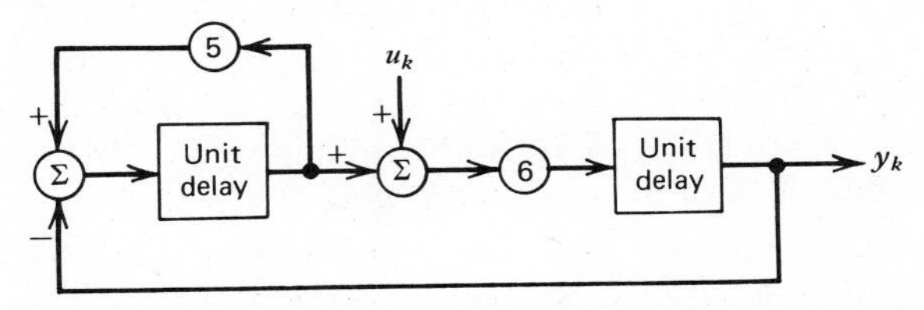

4.19. Two identical systems, H_1 and H_2, with impulse response $\{h_k\} = (\frac{1}{2})^k$, $k \geq 0$ are cascaded. If the input to the first system is a step, $u_k = 1$, $k \geq 0$:
(a) What is the input to the second?
(b) What is the output from the second? (You may wish to check your answers using convolution.)

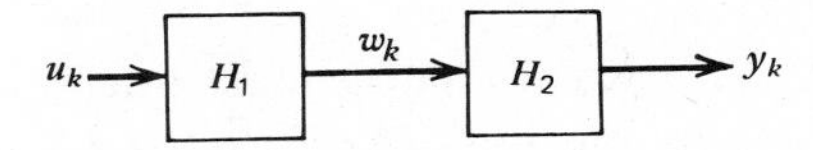

4.20. (a) Find the output of the following discrete-time system for an input sequence $\{u_n\} = \{2^{-n}\}$, $n \geq 0$.

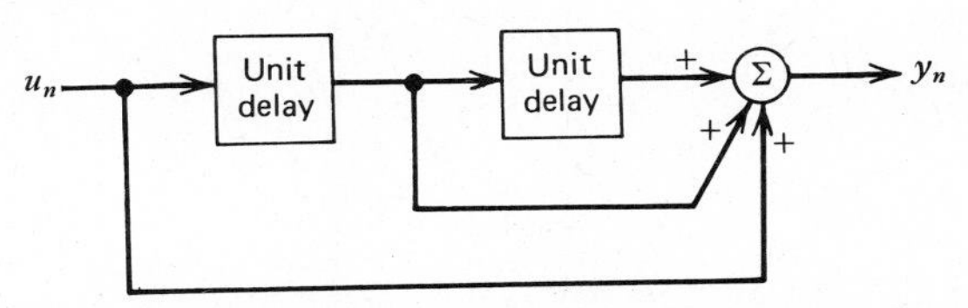

(b) What is the output sequence $\{y_n\}$ if two such systems are cascaded? Assume the same input sequence, $\{u_n\} = \{2^{-n}\}$, $n \geq 0$.

4.21. Consider the following problem taken from economics.† Let $r(n)$ be the amount of money demanded for a commodity and $p(n)$ the price of the commodity in period $(n - 1, n)$. Assume a market equation of the form

† K. E. Boulding, *Economic Analysis*, Harper, New York, 1955.

$p(n) = kr(n)$, where k is a constant. Suppose further that "people extrapolate the trend of prices so that rising prices lead to the expectation of further rise, and thus to an increase in demand, while falling prices lead to an expectation of further fall, and thus to a decrease in demand." To model this last statement, assume $r(n)$ is equal to some "rest" level $r(0)$ plus a factor proportional to the increase of the price $p(n)$ over $p(n-1)$: that is, $r(n) = r(0) + \alpha[p(n) - p(n-1)]$. Derive a difference equation for the price at $n+1$. Show that if $k\alpha > 1$, then $\{p(n)\}$ diverges to $+\infty$ for $p(0) > kr(0)$ or diverges to $-\infty$ for $p(0) < kr(0)$. For $0 < k\alpha < 1$, find the behavior of $\{p(n)\}$. What conditions are needed to obtain a limiting (equilibrium) price of $kr(0)$?

4.22. John Good-Fisher is an avid fisherman, and he has found through many days of fishing that his fishing is either good or bad. If fishing is good, he has found that 60% of the time the next day is also good and 40% of the time the next day is bad. If fishing is bad, his data indicate that the next day is also bad 30% of the time and is good 70% of the time. John is beginning a 20-day fishing trip, and his first day's fishing is bad. What is the probability John will have good fishing on his second day? On his third day? On his fifth day? On his last day?

4.23. A coin is thrown n times. The probability that it will turn up heads on the first throw is p'. On any subsequent throw, the probability the coin shows the same face as on the previous toss is p. Derive a difference equation that can be solved to obtain the probability the coin shows heads at the nth toss. Solve this difference equation using Z-transforms.

4.24. N integers are selected at random (each integer is equally likely to be chosen). These N integers are multiplied together. Show that the probability of obtaining a two in the *units* digit is $\frac{1}{4}\{(\frac{4}{5})^N - (\frac{2}{5})^N\}$.

4.25. A random variable x has a geometric probability distribution if it can assume any positive integer value with probabilities given by

$$Pr\{x = k\} = p^{k-1}q, \qquad k = 1, 2, \ldots$$

where $p = 1 - q$ and p and q are positive constants. If $P(z)$ is the Z-transform of $Pr\{x = k\}$, then show that $P^{(1)}(1)$ is $-E\{x\}$ and, thus, evaluate $E\{x\}$. Assume $Pr\{x = 0\} = 0$.

4.26. The variance σ^2 of a random variable x is given by $\sigma^2 = E\{x^2\} - E^2\{x\}$ where

$$E\{x^2\} = \sum_{k=0}^{\infty} k^2 Pr\{x = k\}$$

If $P(z)$ is the Z-transform of the sequence $\{Pr\{x = k\}\}$, then show that

$$E\{x^2\} = P^{(2)}(1) + P^{(1)}(1)$$

and thus obtain the formula

$$\sigma^2 = P^{(2)}(1) + P^{(1)}(1) - [P^{(1)}(1)]^2$$

4.27. Assume gerbils reproduce at a rate at which one pair each month is born to each pair of adults, provided the adults are two or more months old. Assume one pair is present initially and none of the gerbils die. How many pairs of gerbils are there at the end of the first year? How many pairs of gerbils are there at the end of the nth month?

4.28. Show that the transfer function of the feedback system below is given by

$$H(z) = \frac{Y(z)}{U(z)} = \frac{G(z)}{1 - F(z)G(z)}$$

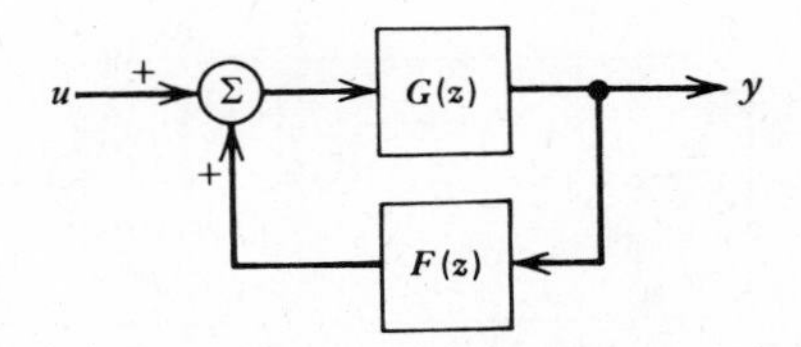

4.29. Use the result above to obtain $H(z)$ for each of the systems shown below. $H(z)$ should be simplified to a ratio of polynomials, $H(z) = P(z)/Q(z)$. In each case, identify $G(z)$ and $F(z)$ corresponding to the block diagram in Problem 4.28.

(a)

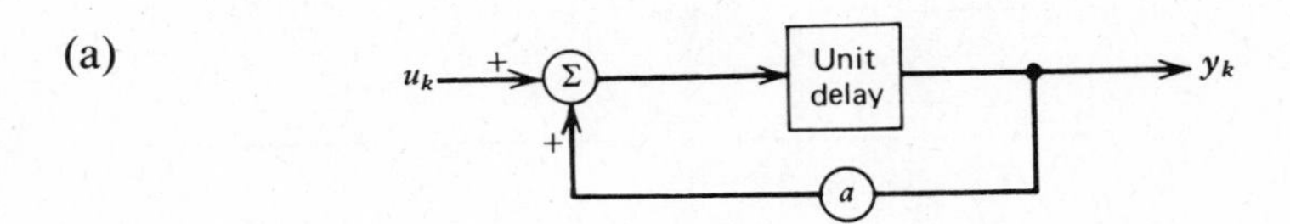

(b)

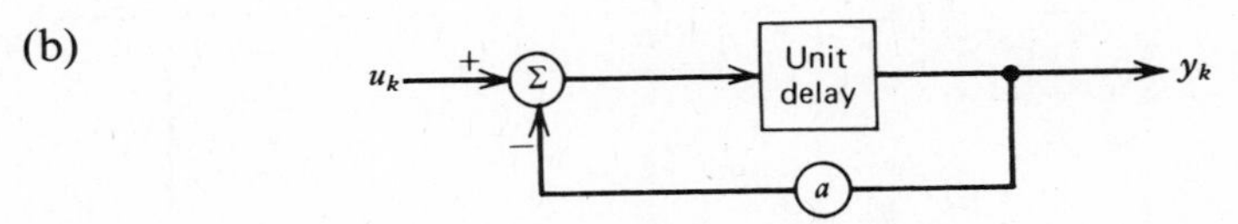

(c)

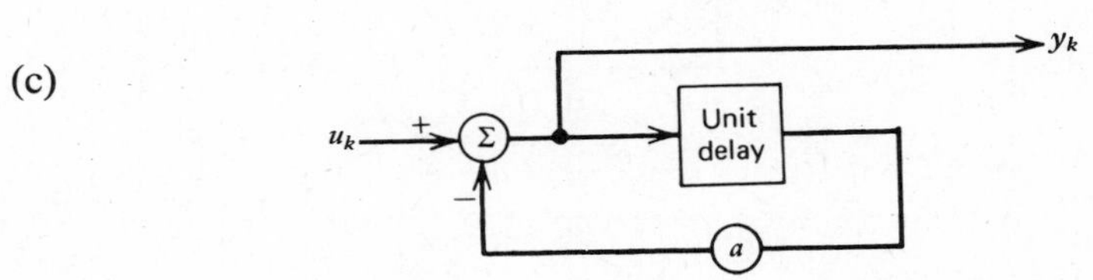

(continued)

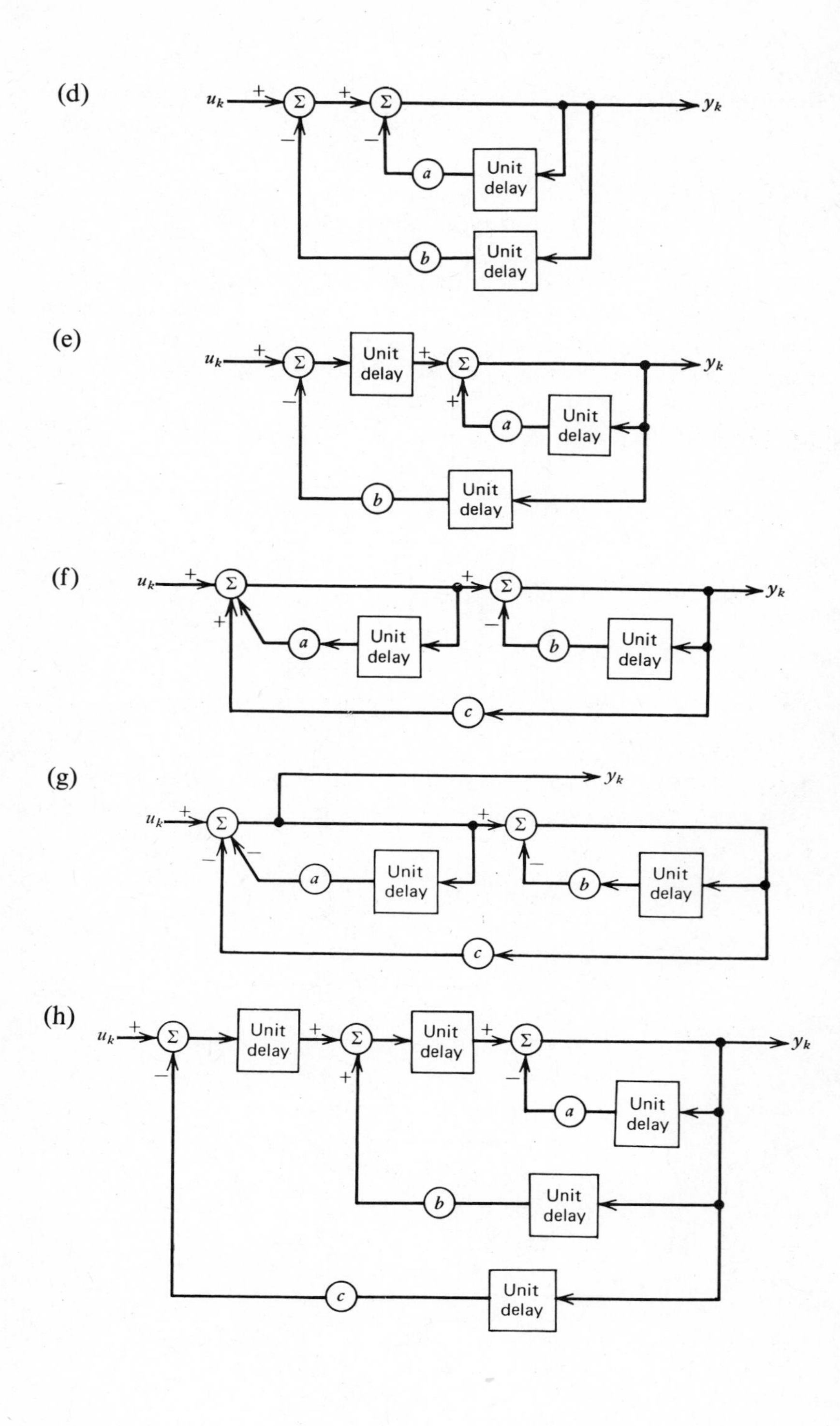
(d)
u_k
+
−
Σ
+
−
Σ
y_k
a
Unit delay
b
Unit delay
(e)
u_k
+
−
Σ
Unit delay
+
+
Σ
y_k
a
Unit delay
b
Unit delay
(f)
u_k
+
+
Σ
+
−
Σ
y_k
a
Unit delay
b
Unit delay
c
(g)
y_k
u_k
+
−
−
Σ
+
−
Σ
a
Unit delay
b
Unit delay
c
(h)
u_k
+
−
Σ
Unit delay
+
+
Σ
Unit delay
+
−
Σ
y_k
a
Unit delay
b
Unit delay
c
Unit delay

4.30. Write a difference equation relating the output sequence $\{y_k\}$ to the input sequence $\{u_k\}$ for each of the block diagrams sketched in Problem 4.29. [*Hint*: A possible approach would be to begin with the transfer function $H(z)$.]

4.31. Show the block diagram of a system whose impulse response is $h(n) = 2\cos(\pi n/4)\xi(n)$.

4.32. Find the transfer function $H(z) = Y(z)/U(z)$ and corresponding impulse response $h(n)$ for the system shown.

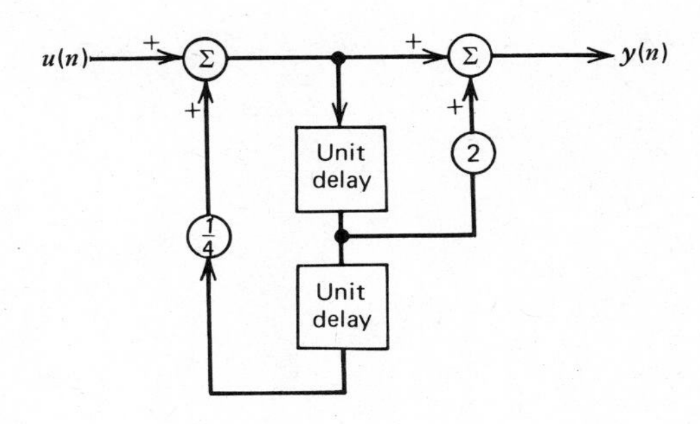

4.33. For what ranges of the parameter K will the systems below be stable?

(a)

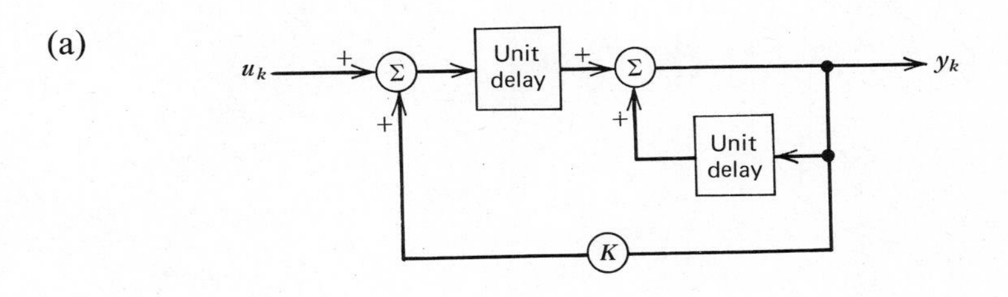

(b)

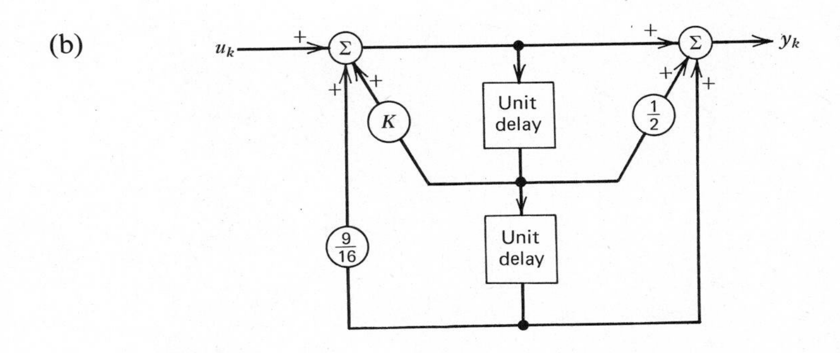

(c)

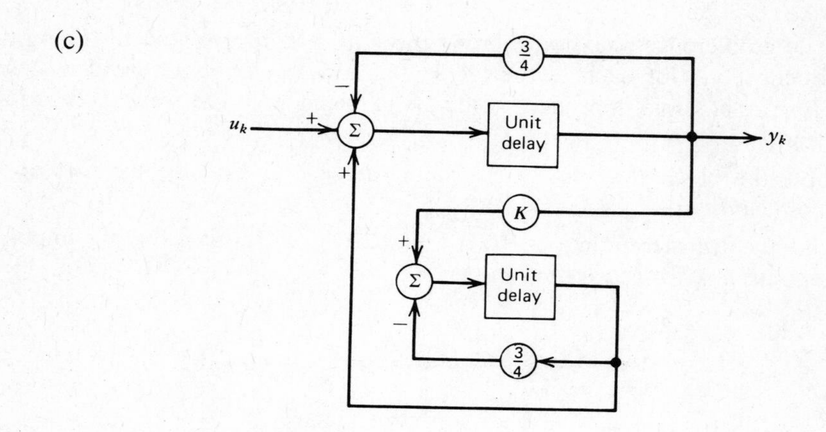

CHAPTER 5
FOURIER ANALYSIS

We have discussed several methods for finding the response of a linear system given the excitation or forcing function. In the convolution integral (or sum) we decomposed the forcing function into a sum of impulses and then found the response to each of the impulses separately. By superposition, the total response was merely the sum of the individual responses and resulted in a convolution integral (or sum). This is conceptually a very useful and powerful method of analysis. However, it is also computationally involved. In addition, it depends on knowledge of the impulse response of the system.

We can, as discussed in Chapter 4, convert the convolution operation in the time domain to a multiplication of transforms in the transform domain. This computational simplification is significant. Transform methods also provide an intuitive understanding of signals and systems. Just as we decomposed an excitation into impulses, we can in principle use other basis functions to decompose the input and response of a system. The choice of a basis set of functions depends primarily on how easy it is to decompose the input and reconstruct the response in terms of the basis functions. Using impulses the decomposition is simple. The impulses are merely amplitude scaled and time-shifted. The reconstruction of the output is more difficult. Here we must use a convolution integral or sum.

In this chapter we consider the Fourier transform methods for both discrete-time and continuous-time signals and systems. Fourier methods are based on using real or complex sinusoids as basis functions. As we have already discussed, for linear constant-parameter systems the response to a sinusoid of frequency ω is another sinusoid of exactly the same frequency modified in amplitude and phase by the linear system. This basic property is the primary reason Fourier analysis is so often used. Fourier analysis can be performed using real sinusoids of the form $\cos(\omega t + \theta)$ or with complex sinusoids of the form $e^{j\omega t}$. We shall use complex sinusoids for the most part because they are easy to work with mathematically. In contrast, the Laplace transform uses e^{st}, $s = \sigma + j\omega$, as basis functions. This set of basis functions is more general than the complex sinusoids and allows one to consider a larger class of time functions. However, for real systems Fourier methods are adequate in most cases and, in fact, preferred because the physical interpretation of the decomposition and reconstruction of the response is more intuitive.

Table 5.1

	Continuous-Time	Discrete-Time (Periodic-Frequency (Function)
Discrete frequency (periodic-time function)	*Fourier Series* $F(n) = \frac{1}{T}\int_{-T/2}^{T/2} f(t)e^{-jn\omega_0 t}\,dt$ $f(t) = \sum_{n=-\infty}^{\infty} F(n)e^{jn\omega_0 t}$ $\omega_0 = \frac{2\pi}{T}$ Section 5.1–5.4	*Discrete Fourier Transform* $F(u) = \sum_{k=0}^{N-1} f(k)W^{-kn}$ $f(k) = \frac{1}{N}\sum_{n=0}^{N-1} F(u)W^{kn}$ $W = \exp\left(j\frac{2\pi}{N}\right)$ Section 5.15–5.16
Continuous frequency	*Fourier Transform* $F(j\omega) = \int_{-\infty}^{\infty} f(t)e^{-j\omega t}\,dt$ $f(t) = \frac{1}{2\pi}\int_{-\infty}^{\infty} F(j\omega)e^{j\omega t}\,d\omega$ Section 5.8–5.14	*Discrete-Time Fourier Transform* $F(e^{j\theta}) = \sum_{k=-\infty}^{\infty} f(k)e^{-jk\theta}$ $f(k) = \frac{1}{2\pi}\int_{-\pi}^{\pi} F(e^{j\theta})e^{j\theta k}\,d\theta$ $\theta = \omega T =$ normalized frequency Section 5.5–5.7

We shall call the Fourier transform or series representation of a time function or sequence the *spectrum* of the function or sequence. The spectrum of a function can be measured with a spectrum analyzer. Many physical processes are more easily understood using the concept of the spectrum of a function or a sequence. This chapter discusses four models of Fourier analysis as shown in Table 5.1. We shall begin with Fourier series, then discuss the discrete-time Fourier transform, followed by a discussion of the Fourier transform, and finally the discrete-Fourier transform.

5.1 GENERALIZED FOURIER SERIES: ORTHOGONAL FUNCTIONS

The key to the analysis of many linear systems is a correct representation of the input signal. The Fourier series representation of a function can often be used to great advantage. Fourier series representations are based on using orthogonal functions as a basis of expansion. Suppose that we have a function $f(t)$ that we wish to represent on a finite interval $[t_1, t_2]$ by a set of n functions $\phi_1(t), \phi_2(t), \ldots, \phi_n(t)$. We assume that these n functions are orthogonal on $[t_1, t_2]$; that is

$$\int_{t_1}^{t_2} \phi_i(t)\phi_j(t)\,dt = \begin{cases} 0, & i \neq j \\ k_i, & i = j \end{cases} \tag{5.1}$$

We shall use the notation (ϕ_i, ϕ_j) to denote the integral on the left-hand side of (5.1). This idea of orthogonality is the same as the one applied to vectors. Our representation in terms of the functions $\phi_i(t)$, $i = 1, 2, \ldots, n$ is in principle the same as representing a vector $\mathbf{f}$, say, in terms of an orthogonal set of vectors that span the space containing $\mathbf{f}$ as sketched in Figure 5.1. $\mathbf{f}$ expressed in terms of the

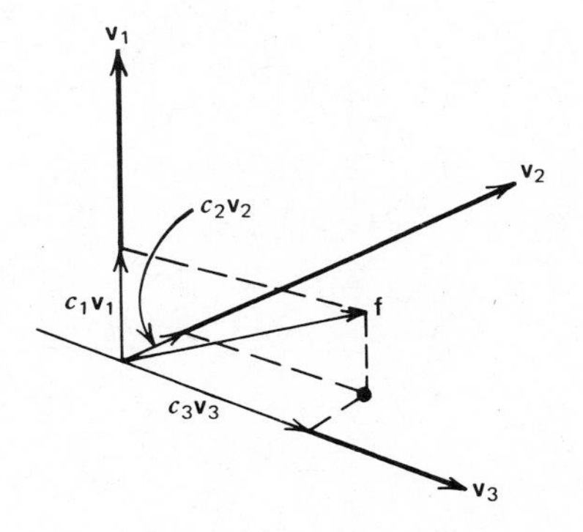

Figure 5.1

orthogonal basis, $\mathbf{v}_1, \mathbf{v}_2, \mathbf{v}_3$ is given by

$$\mathbf{f} = c_1\mathbf{v}_1 + c_2\mathbf{v}_2 + c_3\mathbf{v}_3$$

Now we ask, how should we represent $f(t)$ on $[t_1, t_2]$ in terms of the set of functions $\{\phi_i(t)\}_{i=1}^{n}$? Let us assume a representation or approximation of $f(t)$ by a linear combination of the functions $\phi_i(t)$, $i = 1, 2, \ldots, n$. That is, the representation for $f(t)$ on $[t_1, t_2]$ is of the form

$$f(t) \sim c_1\phi_1(t) + c_2\phi_2(t) + \cdots + c_n\phi_n(t) \tag{5.2}$$

We do not use an equal sign in (5.2) because, in general, the representation $\sum_{i=1}^{n} c_i\phi_i(t)$ is in error. We want the representation or approximation to be "close" to $f(t)$ in some sense. One criterion that is often used is to require that the approximation be chosen so as to minimize the mean square error (MSE) between the true value of $f(t)$ and the approximation $\sum_{i=1}^{n} c_i\phi_i(t)$. In symbols, the c_i, $i = 1, 2, \ldots, n$ are chosen to minimize

$$\text{MSE} = \frac{1}{t_2 - t_1}\int_{t_1}^{t_2}\left[f(t) - \sum_{i=1}^{n} c_i\phi_i(t)\right]^2 dt \tag{5.3}$$

The integrand of (5.3) is, of course, the error squared. The integral and constant $1/(t_2 - t_1)$ average this square error over the interval $[t_1, t_2]$. We can rewrite (5.3) as

$$\text{MSE} = \frac{1}{t_2 - t_1}\int_{t_1}^{t_2}\left[f(t) - c_1\phi_1(t) - c_2\phi_2(t) - \cdots - c_n\phi_n(t)\right]^2 dt \tag{5.4}$$

In (5.4), we now square the integrand to obtain the following expression for the MSE.

$$\begin{aligned}\text{MSE} &= \frac{1}{t_2 - t_1}\int_{t_1}^{t_2}\Big[f^2(t) + c_1^2\phi_1^2(t) + c_2^2\phi_2^2(t) + \cdots + c_n^2\phi_n^2(t) - 2c_1 f(t)\phi_1(t) \\ &\quad - 2c_2 f(t)\phi_2(t) - \cdots - 2c_n f(t)\phi_n(t)\Big] dt \\ &= \frac{1}{t_2 - t_1}\left\{\int_{t_1}^{t_2} f^2(t)\,dt + c_1^2k_1 + c_2^2k_2 + \cdots + c_n^2k_n - 2c_1\gamma_1\right. \\ &\quad \left. - 2c_2\gamma_2 - \cdots - 2c_n\gamma_n\right\}\end{aligned} \tag{5.5}$$

where we have defined γ_i, $i = 1, 2, \ldots, n$ to be

$$\gamma_i = \int_{t_1}^{t_2} f(t)\phi_i(t)\, dt = (f, \phi_i) \tag{5.6}$$

In the expression given in (5.5), we complete the square in each of the terms $(c_i^2 k_i - 2c_i\gamma_i)$ by adding and subtracting γ_i^2/k_i. That is, we write

$$c_i^2 k_i - 2c_i\gamma_i = \left(c_i\sqrt{k_i} - \frac{\gamma_i}{\sqrt{k_i}}\right)^2 - \frac{\gamma_i^2}{k_i} \tag{5.7}$$

The expression for the MSE in (5.5) can thus be written as

$$\text{MSE} = \frac{1}{t_2 - t_1}\left\{\int_{t_1}^{t_2} f^2(t)\, dt + \sum_{i=1}^{n}\left(c_i\sqrt{k_i} - \frac{\gamma_i}{\sqrt{k_i}}\right)^2 - \sum_{i=1}^{n}\frac{\gamma_i^2}{k_i}\right\} \tag{5.8}$$

It is clear from (5.3) that the MSE is always greater than or equal to zero; that is, MSE ≥ 0. From (5.8), it follows that the MSE has its least value when

$$c_i\sqrt{k_i} = \frac{\gamma_i}{\sqrt{k_i}}, \qquad i = 1, 2, \ldots, n \tag{5.9}$$

That is, the c_i's should be chosen as

$$c_i = \frac{\gamma_i}{k_i} = \frac{\int_{t_1}^{t_2} f(t)\phi_i(t)\, dt}{\int_{t_1}^{t_2} \phi_i^2(t)\, dt} = \frac{(f, \phi_i)}{(\phi_i, \phi_i)} \tag{5.10}$$

To summarize: given n mutually orthogonal functions $\phi_1(t), \phi_2(t), \ldots, \phi_n(t)$ on an interval $[t_1, t_2]$, the best approximation of an arbitrary function $f(t)$ on $[t_1, t_2]$ of the form $\sum_{i=1}^{n} c_i\phi_i(t)$ is given by choosing the c_i's according to (5.10). The criterion used in choosing this approximation is to minimize the mean square error between $f(t)$ and $\sum_{i=1}^{n} c_i\phi_i(t)$.

In (5.10), the coefficient c_i can be interpreted, loosely speaking, as the projection of $f(t)$ in the "direction" of the function $\phi_i(t)$. The analogous vector interpretation is that of a vector, say $\mathbf{f}$, projected onto some basis set of orthogonal vectors $\mathbf{v}_i$, $i = 1, 2, \ldots, n$.

The Mean Square Error

The coefficients c_i chosen according to (5.10) guarantee a minimum mean square error between $f(t)$ and its approximation $\sum_{i=1}^{n} c_i \phi_i(t)$. How small is the error? To answer this question, we need only to consider (5.8) with the optimal values of the c_i, $i = 1, 2, \ldots, n$. The minimum MSE (MMSE) is

$$\text{MMSE} = \frac{1}{t_2 - t_1}\left\{\int_{t_1}^{t_2} f^2(t)\,dt - \sum_{i=1}^{n} \frac{\gamma_i^2}{k_i}\right\}$$

From (5.9), $\gamma_i^2/k_i = c_i^2 k_i$. Therefore,

$$\text{MMSE} = \frac{1}{t_2 - t_1}\left\{\int_{t_1}^{t_2} f^2(t)\,dt - [c_1^2 k_1 + c_2^2 k_2 + \cdots + c_n^2 k_n]\right\} \tag{5.11}$$

Equation 5.11 suggests that as we increase n, the number of orthogonal functions, the minimum value of the mean square error decreases. This idea certainly seems reasonable, because as we increase n we "fill out more directions" in the space containing $f(t)$. Equation 5.3 implies that the MSE is always non-negative. Thus, as n increases without limit, the sum $\sum_{i=1}^{n} c_i^2 k_i$ may converge to the integral $\int_{t_1}^{t_2} f^2(t)\,dt$, in which case the MSE is zero and

$$\int_{t_1}^{t_2} f^2(t)\,dt = \sum_{i=1}^{\infty} c_i^2 k_i \tag{5.12}$$

If (5.12) holds for a particular $f(t)$, then the sum $\sum_{i=1}^{\infty} c_i \phi_i(t)$ is said to *converge in the mean* to $f(t)$. This equality is known as Parseval's relation; if it holds for all $f(t)$ of a certain class, then the set $\{\phi_i(t)\}$ is said to be *complete* for that class of functions. Stated differently, a complete set of functions $\{\phi_i(t)\}$ is one for which there exists no function outside the set that is orthogonal to all members of the set. Notice that complete is defined here with respect to convergence in the mean, which does not guarantee ordinary convergence at any point. That is,

$$\lim_{n\to\infty} \int_{t_1}^{t_2}\left[f(t) - \sum_{i=1}^{n} c_i \phi_i(t)\right]^2 dt = 0$$

is quite different from ordinary convergence,

$$\lim_{n \to \infty} \sum_{i=1}^{n} c_i \phi_i(t) = f(t) \tag{5.13}$$

This is a technical point. In most engineering applications, the functions we encounter that converge in the mean also converge pointwise.

Other Types of Orthogonality

Our discussion thus far has been restricted to the case where the basis functions $\{\phi_i(t)\}$ are real valued. If the basis functions are instead complex valued functions of the real variable t, orthogonality is defined as

$$(\phi_i, \phi_j^*) = \int_{t_1}^{t_2} \phi_i(t)\phi_j^*(t)\,dt = \begin{cases} 0, & i \neq j \\ k_i, & i = j \end{cases} \tag{5.14}$$

where $\phi_j^*(t)$ is the complex conjugate of $\phi_j(t)$. The generalized Fourier coefficients in this case are

$$c_i = \frac{(f, \phi_i^*)}{(\phi_i, \phi_i^*)} = \frac{\int_{t_1}^{t_2} f(t)\phi_i^*(t)\,dt}{\int_{t_1}^{t_2} \phi_i(t)\phi_i^*(t)\,dt} \tag{5.15}$$

One can also define orthogonality with respect to a weight function $w(t)$. The set of functions $\{\phi_i(t)\}$ is said to be orthogonal with respect to $w(t)$ on $[t_1, t_2]$ if

$$\int_{t_1}^{t_2} \phi_i(t)\phi_j(t)w(t)\,dt = \begin{cases} 0, & i \neq j \\ k_i, & i = j \end{cases} \tag{5.16}$$

This type of orthogonality can be reduced to the ordinary kind with weight function 1 by defining the orthogonal set to be the set of functions $\{\phi_i(t)\sqrt{w(t)}\}$.

A set of orthogonal functions $\{\phi_i(t)\}$ is said to be *orthonormal* if the constant k_i is 1 for all i. An orthonormal set can be obtained from an orthogonal set of functions by normalization of each function $\phi_i(t)$ in the set so that its norm (or length squared) is unity. This is accomplished by dividing each function $\phi_i(t)$ by its length $\sqrt{k_i}$.

Example 5.1 Consider a representation of the periodic waveform shown in Figure 5.2 by the set of mutually orthogonal functions $\{\sin n\omega_0 t\}$, $n = 1, 2, \ldots$. This set of functions is orthogonal on $(t_0, t_0 + 2\pi/\omega_0)$ for any t_0. This fact can be seen as follows:

$$(\sin n\omega_0 t, \sin m\omega_0 t) = \int_{t_0}^{t_0 + 2\pi/\omega_0} \sin n\omega_0 t \sin m\omega_0 t \, dt$$

$$= \int_{t_0}^{t_0 + 2\pi/\omega_0} \left[\frac{1}{2}\cos (n - m)\omega_0 t - \frac{1}{2}\cos (n + m)\omega_0 t\right] dt$$

$$= \begin{cases} 0, & n \neq m \\ \pi, & n = m \end{cases} \tag{5.17}$$

The set of functions $\{\sin nt\}n = 1, 2, \ldots,$ is thus orthogonal on $[0, 2\pi]$. Because $f(t)$ is periodic of period 2π, this set is one possible basis set to use in representing $f(t)$. In order to see how the approximation error varies with the number of basis functions, consider the approximation of $f(t)$ based on one term, then two terms, three terms, etc. The approximation $\hat{f}(t)$ is of the form

$$\hat{f}(t) = c_1 \sin t + c_2 \sin 2t + \cdots + c_n \sin nt \tag{5.18}$$

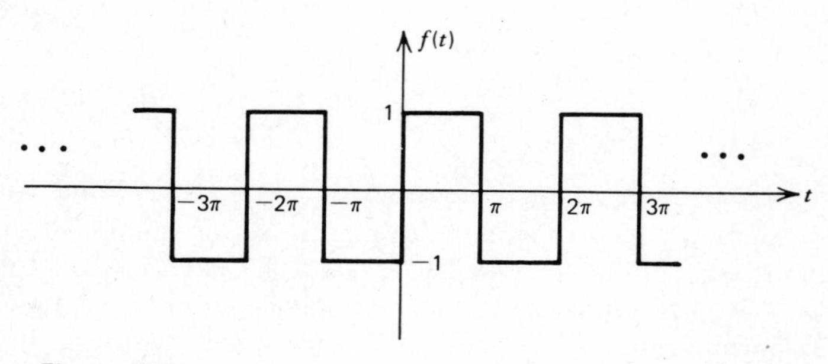

Figure 5.2

The values of the coefficients are

$$c_n = \frac{\int_0^{2\pi} f(t) \sin nt \, dt}{\int_0^{2\pi} \sin^2 nt \, dt} = \frac{1}{\pi}\left\{\int_0^{\pi} \sin nt \, dt - \int_{\pi}^{2\pi} \sin nt \, dt\right\}$$

$$= \begin{cases} \dfrac{4}{\pi n}, & n \text{ odd} \\ 0, & n \text{ even} \end{cases} \tag{5.19}$$

The mean square error is obtained from (5.11) as

$$\text{MSE} = \frac{1}{2\pi}\left\{\int_0^{2\pi} f^2(t) \, dt - c_1^2 k_1 - c_2^2 k_2 - \cdots\right\} \tag{5.20}$$

where $k_1 = k_2 = \cdots = \pi$ in this case, Thus, the mean square error is

$$\text{MSE} = \frac{1}{2\pi}\{2\pi - \pi(c_1^2 + c_2^2 + \cdots)\}$$

$$= 1 - \frac{1}{2}(c_1^2 + c_2^2 + \cdots) \tag{5.21}$$

where the c_i's are given by (5.19). Thus, if $n = 1$ (one term), the mean square error is

$$\text{MSE} = 1 - \frac{1}{2}\left(\frac{4}{\pi}\right)^2 \cong 0.19$$

For two terms, the mean square error is

$$\text{MSE} = 1 - \frac{1}{2}\left(\frac{4}{\pi}\right)^2 - \frac{1}{2}\left(\frac{4}{3\pi}\right)^2 \cong 0.10$$

For three terms, the mean square error is

$$\text{MSE} = 1 - \frac{1}{2}\left(\frac{4}{\pi}\right)^2 - \left(\frac{1}{2}\right)\left(\frac{4}{3\pi}\right)^2 - \frac{1}{2}\left(\frac{4}{5\pi}\right)^2 \cong 0.0675$$

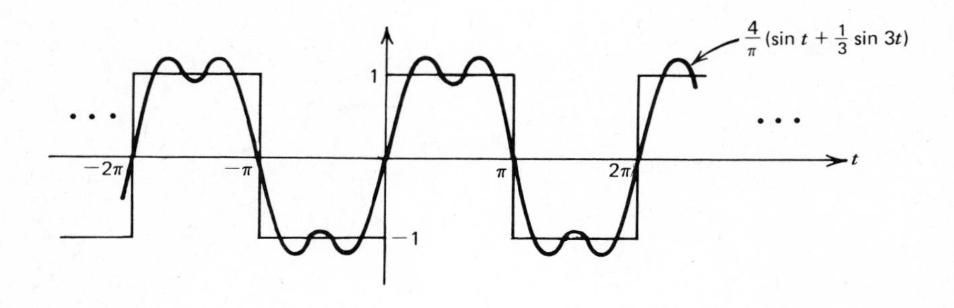

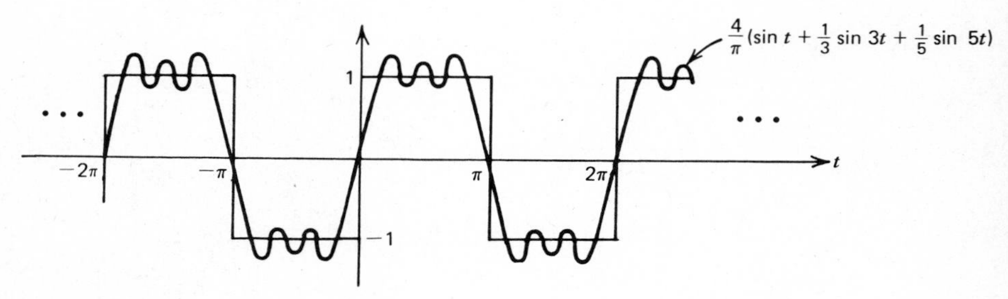

Figure 5.3

Graphically, the approximations appear as shown in Figure 5.3. Although the functions $\{\sin n\omega_0 t\}$ are orthogonal on $(t_0, t_0 + 2\pi/\omega_0)$, this basis set is not, in general, complete. Because of the odd symmetry of $f(t)$ about the vertical axis through the origin, this particular function can be accurately represented in terms of the set $\{\sin n\omega_0 t\}$. More complex functions would require the use of $\{\cos n\omega_0 t\}$ as well as $\{\sin n\omega_0 t\}$. ■

5.2 EXAMPLES OF ORTHOGONAL FUNCTIONS

There are many sets of functions that can be used to expand or represent a function on an interval $[t_1, t_2]$. We shall be concerned primarily with the set of complex exponentials $\{e^{jn\omega_0 t}\}$, $n = 0, \pm 1, \pm 2, \ldots$. Before considering this set in detail we shall briefly discuss some alternate basis functions. The choice of a basis set depends on the application, the function to be represented, and the ease of reconstruction.

Walsh Functions

The class of orthogonal functions known as Walsh functions consists of piecewise constant functions. These functions are defined by

$$\phi_0(t) = 1, \quad 0 \le t \le 1$$

$$\phi_1(t) = \begin{cases} 1, & 0 \le t < \frac{1}{2} \\ -1, & \frac{1}{2} < t \le 1 \end{cases}$$

$$\phi_2^{(1)}(t) = \begin{cases} 1, & 0 \le t < \frac{1}{4}, \quad \frac{3}{4} < t \le 1 \\ -1, & \frac{1}{4} < t < \frac{3}{4} \end{cases}$$

$$\phi_2^{(2)}(t) = \begin{cases} 1, & 0 \le t < \frac{1}{4}, \quad \frac{1}{2} < t < \frac{3}{4} \\ -1, & \frac{1}{4} < t < \frac{1}{2}, \quad \frac{3}{4} < t \le 1 \end{cases}$$

$$\phi_{m+1}^{(2k-1)}(t) = \begin{cases} \phi_m^{(k)}(2t), & 0 \le t < \frac{1}{2} \\ (-1)^{k+1}\phi_m^{(k)}(2t-1), & \frac{1}{2} < t \le 1 \end{cases}$$

$$m = 1, 2, 3, \ldots$$
$$k = 1, 2, \ldots, 2^{m-1}$$

$$\phi_{m+1}^{(2k)}(t) = \begin{cases} \phi_m^{(k)}(2t), & 0 \le t < \frac{1}{2} \\ (-1)^{k}\phi_m^{(k)}(2t-1), & \frac{1}{2} < t \le 1 \end{cases}$$

These functions are easily generated by digital logic circuitry. Multiplication using these functions is simple since all that is needed is a polarity-reversing switch. Graphs for the first six Walsh functions are shown in Figure 5.4.

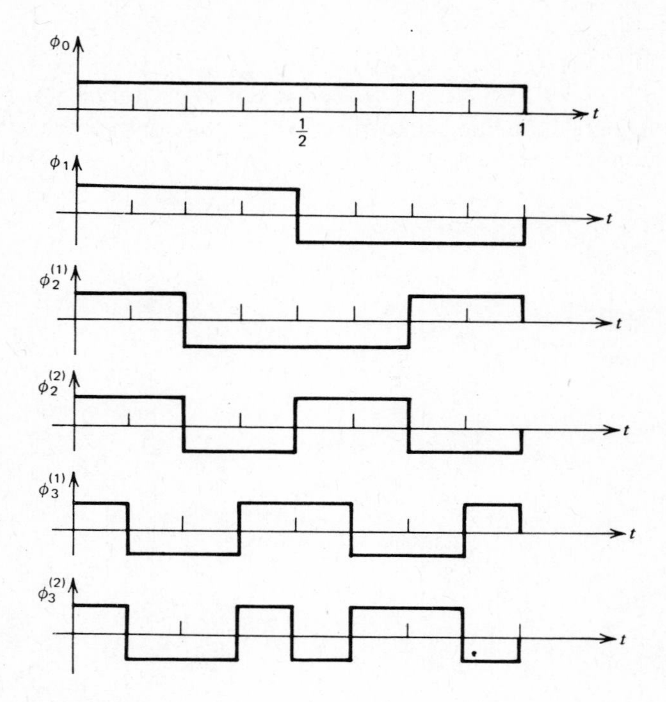

Figure 5.4

Legendre Polynomials

The Legendre polynomials form an orthogonal set on the interval $[-1, 1]$. The corresponding normalized basis functions are

$$\phi_0(t) = \frac{1}{\sqrt{2}}, \qquad \phi_1(t) = t\sqrt{\frac{3}{2}}, \qquad \phi_2(t) = \sqrt{\frac{5}{2}}\left(\frac{3}{2}t^2 - \frac{1}{2}\right)$$

$$\phi_3(t) = \sqrt{\frac{7}{2}}\left(\frac{5}{2}t^3 - \frac{3}{2}t\right), \ldots, \phi_n(t) = \left(\frac{2n+1}{2}\right)^{1/2} P_n(t) \tag{5.23}$$

The $P_n(t)$ are the Legendre polynomials. These polynomials can be generated by the formula

$$P_n(t) = \frac{1}{2^n n!} \frac{d^n}{dt^n} (t^2 - 1)^n \tag{5.24}$$

or by using the difference equation

$$nP_n(t) = (2n - 1)tP_{n-1}(t) - (n - 1)P_{n-2}(t) \tag{5.25}$$

Laguerre Functions

For the interval $[0, \infty]$ the Laguerre functions

$$\phi_n(t) = \frac{e^{-t^2/2}}{n!} L_n(t), \qquad n = 0, 1, 2, \ldots \tag{5.26}$$

form an orthonormal set. In (5.26), $L_n(t)$ are the Laguerre polynomials given by

$$L_n(t) = e^t \frac{d^n}{dt^n} (t^n e^{-t}) \tag{5.27}$$

The Laguerre functions can also be generated by the difference equation

$$L_n(t) = (2n - 1 - t)L_{n-1}(t) - (n - 1)^2 L_{n-2}(t) \tag{5.28}$$

Laguerre functions are interesting in that they can also be generated as the impulse responses of relatively simple networks.†

Example 5.2 Consider again a representation of the function $f(t)$ of Figure 5.2. In this example, we represent $f(t)$ over one period by a set of Legendre functions. Because the Legendre functions defined in (5.23) are orthonormal on $[-1, 1]$, we need to redefine $f(t)$ as shown in Figure 5.5 by letting $t' = t/\pi$. The generalized Fourier coefficients are given by

$$c_n = \int_{-1}^{1} f(t)\phi_n(t)\, dt, \qquad n = 0, 1, 2, \ldots \tag{5.29}$$

† Y. W. Lee. *Statistical Theory of Communication*, Wiley, New York, 1960.

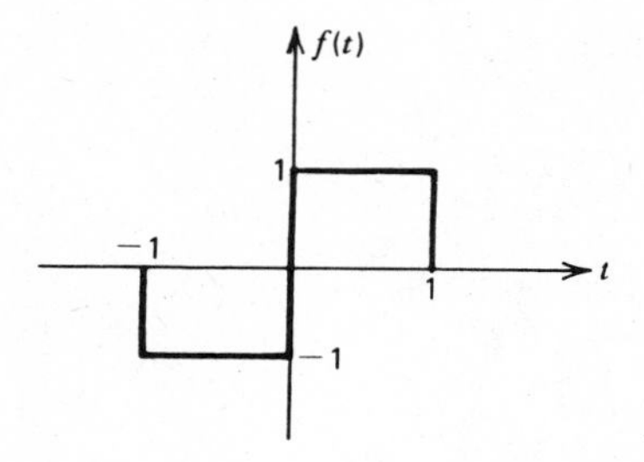

Figure 5.5

where the functions $\phi_n(t)$ are given by (5.23). Thus we obtain the generalized Fourier coefficients as

$$
\begin{aligned}
c_0 &= \int_{-1}^{1} \frac{f(t)}{\sqrt{2}}\, dt = 0 \\
c_1 &= \int_{-1}^{1} \sqrt{\frac{3}{2}}\, t f(t)\, dt = +\sqrt{\frac{3}{2}} \\
c_2 &= \int_{-1}^{1} \sqrt{\frac{5}{2}} \left(\frac{3}{2} t^2 - \frac{1}{2}\right) f(t)\, dt = 0 \\
c_3 &= \int_{-1}^{1} \sqrt{\frac{7}{2}} \left(\frac{5}{2} t^3 - \frac{3}{2} t\right) f(t)\, dt = \sqrt{\frac{7}{16}}
\end{aligned}
\tag{5.30}
$$

In general, the even-numbered coefficients c_n, $n = 0, 2, 4, \ldots$ are zero. Thus we have representation for $f(t)$ in terms of Legendre polynomials as

$$f(t) = \sum_{i=0}^{\infty} c_i \phi_i(t) = +\frac{3}{2} t + \frac{7}{4\sqrt{2}} \left(\frac{5}{2} t^3 - \frac{3}{2} t\right) + \cdots + c_n \phi_n(t) + \cdots \tag{5.31}$$

The reader should calculate the MSE for a finite expansion and compare this error to that obtained in Example 5.1. ■

5.3 THE EXPONENTIAL FOURIER SERIES

In applications, the most used orthogonal functions are the complex sinusoids, yielding the exponential Fourier Series. The reasons for this are partly historical in that these functions were the first to be used to represent an arbitrary function.

Joseph Fourier presented a paper on heat conduction to the Paris Academy of Science in 1807 in which he stated that any bounded function $f(t)$ defined on $(-a, a)$ can be expressed as

$$f(t) = \sum_{n=-\infty}^{+\infty} F_n \exp\left(\frac{jn\pi t}{a}\right)$$

where the coefficients F_n are calculated as

$$F_n = \frac{1}{2a} \int_{-a}^{a} f(t) \exp\left(-\frac{jn\pi t}{a}\right) dt$$

However, historical reasons are not the primary grounds for using the exponential function $\{e^{jn\omega_0 t}\}$ as a basis set. The exponential functions are relatively easy to manipulate mathematically. Moreover, the physical interpretation of the Fourier representation is meaningful and often very useful.

Consider the set of exponential functions $\{e^{jn\omega_0 t}\}$, $n = 0, \pm 1 \pm 2, \ldots$ These functions are orthogonal on the interval $[t_0, t_0 + 2\pi/\omega_0]$ for any value of t_0. We can easily demonstrate the orthogonality by calculating the integral I below.

$$I = (e^{jn\omega_0 t}, e^{-jm\omega_0 t}) = \int_{t_0}^{t_0 + 2\pi/\omega_0} e^{jn\omega_0 t} e^{-jm\omega_0 t}\, dt \tag{5.32}$$

If $n = m$, the integral is

$$I = \int_{t_0}^{t_0 + 2\pi/\omega_0} dt = \frac{2\pi}{\omega_0}$$

If $n \neq m$, then the integral is

$$I = \frac{1}{j(n-m)\omega_0} e^{j(n-m)\omega_0 t} \Bigg|_{t_0}^{t_0 + 2\pi/\omega_0}$$

$$= \frac{1}{j(n-m)\omega_0} e^{j(n-m)\omega_0 t_0} [e^{j2\pi(n-m)} - 1] = 0$$

since $e^{j2\pi k}$ is always unity for any integer k. Thus we have

$$(\phi_n, \phi_m^*) = \int_{t_0}^{t_0 + 2\pi/\omega_0} e^{jn\omega_0 t}\, e^{-jm\omega_0 t}\, dt = \begin{cases} 0, & n \neq m \\ \dfrac{2\pi}{\omega_0}, & n = m \end{cases} \tag{5.33}$$

The set of functions $\{e^{jn\omega_0 t}\}$, $n = 0, \pm 1, \pm 2, \ldots$ forms a complete orthogonal set on $(t_0, t_0 + T)$ where $T = 2\pi/\omega_0$. We can represent a function $f(t)$ on this interval $(t_0, t_0 + T)$ as

$$f(t) = \sum_{n=-\infty}^{\infty} F_n \cdot e^{jn\omega_0 t} \tag{5.34}$$

where the coefficients are given by

$$F_n = \frac{(f, \phi_n^*)}{(\phi_n, \phi_n^*)} = \frac{\int_{t_0}^{t_0+T} f(t)e^{-jn\omega_0 t}\, dt}{\int_{t_0}^{t_0+T} e^{jn\omega_0 t} e^{-jn\omega_0 t}\, dt} = \frac{1}{T}\int_{t_0}^{t_0+T} f(t)e^{-jn\omega_0 t}\, dt \tag{5.35}$$

We can obtain the coefficients F_m directly from (5.34) by multiplying both sides of (5.34) by $e^{-jm\omega_0 t}$ and integrating with respect to t over the interval $(t_0, t_0 + T)$. The orthogonality condition applied to the sum on the right-hand side eliminates all terms in the sum except the mth term. This expression is then solved for F_m.

To summarize: Any function $f(t)$ may be represented on the interval $(t_0, t_0 + T)$ as an infinite sum of exponentials.

$$f(t) = \sum_{n=-\infty}^{\infty} F_n e^{jn\omega_0 t}, \qquad t_0 < t < t_0 + T \tag{5.36}$$

The coefficients F_n, $n = 0, \pm 1, \pm 2, \ldots$ are found in terms of the basis set $\{e^{jn\omega_0 t}\}$, $n = 0, \pm 1, \pm 2, \ldots$, as

$$F_n = \frac{1}{T}\int_{t_0}^{t_0+T} f(t)e^{-jn\omega_0 t}\, dt \tag{5.37}$$

Example 5.3 Consider an expansion of the function $f(t) = e^{-t}$ in the interval $(-1, 1)$ by an exponential Fourier Series. Because the period $T = 2$, the Fourier coefficients are

$$F_n = \frac{1}{2}\int_{-1}^{1} f(t)e^{-jn\omega_0 t}\, dt, \qquad \omega_0 = \frac{2\pi}{T} = \pi$$

Thus

$$F_n = \frac{1}{2}\int_{-1}^{1} e^{-t}e^{-jn\omega_0 t}\, dt = \frac{1}{2}\left[\frac{e^{-(1+nj\pi)t}}{-(1 + nj\pi)}\right]_{-1}^{1}$$

$$= \frac{ee^{jn\pi} - e^{-1}e^{-jn\pi}}{2(1 + jn\pi)} \tag{5.38}$$

We can simplify the form of F_n by using the fact that

$$e^{j\pi} = -1$$

which implies that

$$e^{jn\pi} = (-1)^n \tag{5.39}$$

Therefore,

$$F_n = \frac{(-1)^n}{1 + jn\pi}\left[\frac{e - e^{-1}}{2}\right] = \frac{(-1)^n \sinh(1)}{1 + jn\pi}$$

$$= \frac{(-1)^n(1 - jn\pi)\sinh(1)}{1 + n^2\pi^2} \tag{5.40}$$

The expansion for $f(t)$ is therefore

$$f(t) = \sum_{n=-\infty}^{+\infty} F_n e^{jn\pi t} = \sum_{n=-\infty}^{\infty} \frac{(-1)^n(1 - jn\pi)\sinh(1)e^{jn\pi t}}{1 + n^2\pi^2} \tag{5.41}$$

■

Example 5.4 In Example 5.1, we expanded the rectangular waveform of Figure 5.2 in terms of the mutually orthogonal set of functions $\{\sin n\omega_0 t\}$. Suppose that instead of using the set $\{\sin n\omega_0 t\}$ we were to use the set $\{\cos n\omega_0 t\}$. It is easy to show (see Example 5.1) that this set is also mutually orthogonal on $[t_0, t_0 + (2\pi/\omega_0)]$. If we calculate the Fourier coefficients for the function $f(t)$ as given in Figure 5.2, we obtain

$$F_n = \frac{\int_0^{2\pi} f(t)\cos nt\, dt}{\int_0^{2\pi} \cos^2 nt\, dt} = \frac{1}{\pi}\left\{\int_0^{\pi} \cos nt\, dt - \int_{\pi}^{2\pi} \cos nt\, dt\right\}$$

$$= \frac{1}{\pi}\left\{-\left.\frac{\sin nt}{n}\right|_0^{\pi} + \left.\frac{\sin nt}{n}\right|_0^{2\pi}\right\} = 0 \qquad \text{for all } n \tag{5.42}$$

Equation 5.42 implies that $f(t)$ is orthogonal to all functions in the set $\cos nt$, $n = 0, 1, 2, \ldots$. ■

This example points out that in order to obtain an accurate representation of a function $f(t)$, one must use a complete set of orthogonal functions. The exponential set of functions $\{e^{jn\omega_0 t}\}$, $n = 0, \pm 1, \pm 2, \ldots$ is complete, whereas the

Table 5.2 Fourier Series Representation of a Periodic Function

Form	Series Representation	Fourier Coefficients	Conversion Formulas
Exponential	$f(t) = \sum_{n=-\infty}^{+\infty} F_n e^{jn\omega_0 t}$	$F_n = \frac{1}{T}\int_{-T/2}^{T/2} f(t)e^{-jn\omega_0 t}\, dt$	$F_0 = a_0$
			$F_n = \frac{1}{2}(a_n - jb_n)$
Trigonometric	$f(t) = \frac{a_0}{2} + \sum_{n=1}^{\infty}(a_n \cos n\omega_0 t + b_n \sin n\omega_0 t)$	$a_n = \frac{2}{T}\int_{-T/2}^{T/2} f(t)\cos n\omega_0 t\, dt$	$a_n = F_n + F_{-n}$
			$b_n = j(F_n - F_{-n})$
	$f(t) = \frac{a_0}{2} + \sum_{n=1}^{\infty} A_n \cos(n\omega_0 t + \theta_n)$	$b_n = \frac{2}{T}\int_{-T/2}^{T/2} f(t)\sin n\omega_0 t\, dt$	$A_n = (a_n^2 + b_n^2)^{1/2}$ $= 2\lvert F_n \rvert$
			$\theta_n = -\tan^{-1}\left(\frac{b_n}{a_n}\right)$

sets $\{\sin n\omega_0 t\}$ and $\{\cos n\omega_0 t\}$, $n = 0, 1, 2, \ldots$, are not individually complete. The exponential set is a convenient way of combining the sines and cosines into a complete set. One can, of course, use the trigonometric form of the exponential set $\{e^{jn\omega_0 t}\}$. In this case, the representation for $f(t)$ is given by

$$f(t) = \frac{a_0}{2} + \sum_{n=1}^{\infty} (a_n \cos n\omega_0 t + b_n \sin n\omega_0 t) \tag{5.43}$$

where

$$a_n = \frac{\omega_0}{\pi} \int_{t_0}^{t_0 + 2\pi/\omega_0} f(t) \cos n\omega_0 t \, dt \tag{5.44}$$

$$b_n = \frac{\omega_0}{\pi} \int_{t_0}^{t_0 + 2\pi/\omega_0} f(t) \sin n\omega_0 t \, dt \tag{5.45}$$

One can easily convert from the exponential to the trigonometric form (and vice versa) by using the conversion formulas given in Table 5.2.

Convergence of the Fourier Expansion

We must be careful at this point not to delude ourselves by thinking we have proved that a function $f(t)$ has a Fourier expansion that is a valid representation of it. What we have shown thus far is that *if* a function $f(t)$ can be represented by a Fourier expansion, *then* the coefficients are calculated according to (5.10). The question of whether the series converges to the function itself and the necessary and sufficient conditions under which the expansions will converge are known.† These conditions are somewhat technical. For engineering applications, the following sufficient conditions proposed by Dirichlet are usually all we need. The Dirichlet conditions are stated as follows: if $f(t)$ is bounded and of period T and if $f(t)$ has at most a finite number of maxima and minima in one period and a finite number of discontinuities, then the Fourier series for $f(t)$ converges to $f(t)$ at all points where $f(t)$ is continuous, and converges to the average of the right-hand and left-hand limits of $f(t)$ at each point where $f(t)$ is discontinuous.

This theorem of Dirichlet indicates that the function need not be continuous in order to possess a valid Fourier expansion. If, as in Figure 5.6, $f(t)$ is piecewise continuous, then at points of discontinuity, the value the Fourier series expansion takes on is the value halfway between the right- and left-hand limits of $f(t)$.

† L. Carleson, "On Convergence and Growth of Partial Sums by Fourier Series," *Acta Mathematica*, June 1966, 135–157.

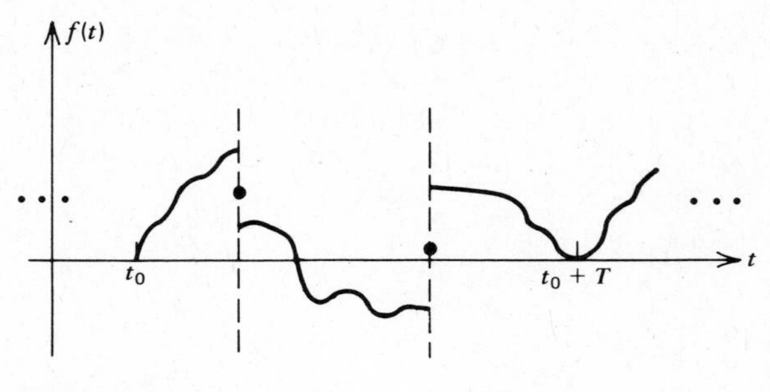

Figure 5.6

5.4 THE COMPLEX FOURIER SPECTRUM

The Fourier series expansion of a function $f(t)$ can be used for two classes of functions. One can represent an aperiodic function on a finite interval, say $[0, T]$. In this case, the Fourier series converges to the periodic extension of $f(t)$ outside of $[0, T]$: that is, to $f(t + nT)$, $n = \pm 1, \pm 2, \ldots$. One also uses the Fourier series expansion to represent a periodic function $f(t)$ over any interval of interest. Let us, for the moment, consider this latter use of the Fourier series expansion.

One interpretation of the Fourier series expansion for a periodic function is that we are decomposing the function $f(t)$ in terms of its harmonics: that is, its various frequency components. If $f(t)$ is periodic of period T, then it has frequency components at the radian frequencies $n\omega_0$, $n = 1, 2, \ldots$ where $\omega_0 = 2\pi/T$. The set or collection of these frequency components that make up $f(t)$ is called the *frequency spectrum* or *spectrum* of $f(t)$. In the case of a periodic function, this spectrum is discrete; that is, the spectrum has nonzero value only for the frequencies $n\omega_0$, $n = 1, 2, \ldots$. The spectrum is an alternate representation for $f(t)$ and given the spectrum, we can specify $f(t)$. Thus we have two methods of specifying a periodic function $f(t)$. We can define $f(t)$ in the time domain by some kind of time waveform description or we can specify $f(t)$ in the frequency domain using its frequency spectrum. The frequency spectrum is often displayed graphically by a so-called *line spectrum*. The amplitude of each harmonic is represented by a vertical line proportional to the amplitude of the harmonic. The discrete frequency spectrum is thus a graph of equally spaced lines with heights proportional to the amplitudes of the component frequencies contained in $f(t)$. Note that the phase of each harmonic must also be specified.

We shall use the exponential form of the Fourier series to represent the discrete spectrum. In the exponential form, the basis functions are $e^{jn\omega_0 t}$, $n = 0, \pm 1, \pm 2, \ldots$. Thus the discrete spectrum exists at the frequencies $0, \pm\omega_0, \pm 2\omega_0, \ldots$. The lines at negative frequencies are a necessary part of the spectrum because it takes a negative and a corresponding positive exponent to create a real-valued function. That is,

$$e^{jn\omega_0 t} + e^{-jn\omega_0 t} = 2\cos n\omega_0 t \tag{5.46}$$

The negative and corresponding positive frequency components in the exponential series expansion constitute a mathematical description of a real sinusoid. Each by itself is nonphysical in the sense that neither can be physically measured—only the combination $(e^{jn\omega_0 t} + e^{-jn\omega_0 t})$ has a physical interpretation.

For a periodic function of period T, the exponential Fourier series is

$$f(t) = \sum_{n=-\infty}^{\infty} F_n e^{jn\omega_0 t} \tag{5.47}$$

where

$$F_n = \frac{1}{T}\int_{-T/2}^{T/2} f(t)e^{-jn\omega_0 t}\,dt \tag{5.48}$$

In general, the F_n, $n = 0, \pm 1, \pm 2, \ldots$ are complex valued. Thus, we write $F_n = |F_n|e^{j\theta_n}$ where $|F_n|$ is the magnitude of F_n and θ_n is the angle of F_n. Therefore, in general, we need two discrete or line spectra to represent $f(t)$ in the frequency domain. The *amplitude spectrum* of $f(t)$ is a specification of $|F_n|$ as a function of $n\omega_0$. Similarly, the *phase spectrum* is a specification of θ_n as a function of $n\omega_0$. In many cases, F_n is not complex, and so we can use a single spectrum to represent $f(t)$.

Example 5.5 Find the frequency spectrum of the periodic function in Figure 5.7. This periodic function is called the periodic gate function. To find the spectrum, we calculate F_n from

$$F_n = \frac{1}{T}\int_{-T/2}^{T/2} f(t)e^{-jn\omega_0 t}\,dt, \qquad \omega_0 = \frac{2\pi}{T}$$

$$= \frac{1}{T}\int_{-d/2}^{d/2} Ae^{-jn\omega_0 t}\,dt$$

$$= \frac{-A}{jn\omega_0 T}\left[e^{-jn\omega_0 t}\right]_{-d/2}^{d/2}$$

$$= \frac{2A}{n\omega_0 T}\left[\frac{e^{jn\omega_0 d/2} - e^{-jn\omega_0 d/2}}{2j}\right] = \frac{2A}{n\omega_0 T}\sin\left(\frac{n\omega_0 d}{2}\right)$$

$$= \frac{Ad}{T}\left[\frac{\sin\left(\dfrac{n\omega_0 d}{2}\right)}{\dfrac{n\omega_0 d}{2}}\right] = \frac{Ad}{T}\operatorname{sinc}\left(\frac{n\omega_0 d}{2}\right) \tag{5.49}$$

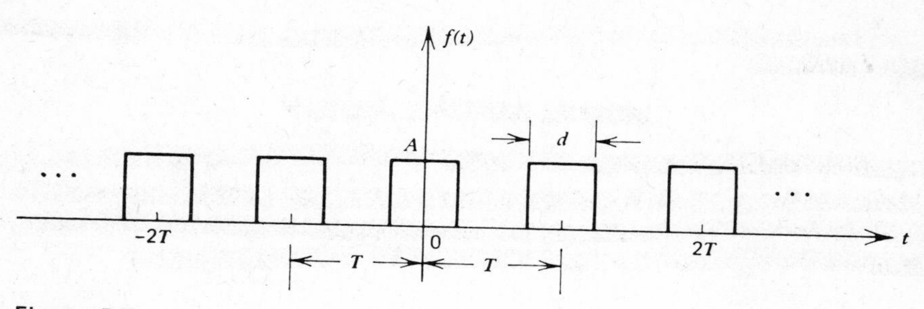

Figure 5.7

where we define sinc $x \triangleq (\sin x)/x$. This function plays an important role in signal representation. A plot of sinc x is shown in Figure 5.8. Appendix B tabulates values of the sinc function. The sinc function oscillates with period 2π and decays with increasing x. It has zeros at $n\pi$, $n = \pm 1, \pm 2, \ldots$ and is an even function of x. The spectrum of the gate function is thus

$$\begin{aligned} F_n &= \frac{Ad}{T} \operatorname{sinc}\left(\frac{n\omega_0 d}{2}\right) \\ &= \frac{Ad}{T} \operatorname{sinc}\left(\frac{n(2\pi/T)d}{2}\right) \\ &= \frac{Ad}{T} \operatorname{sinc}\left(\frac{n\pi d}{T}\right) \end{aligned} \tag{5.50}$$

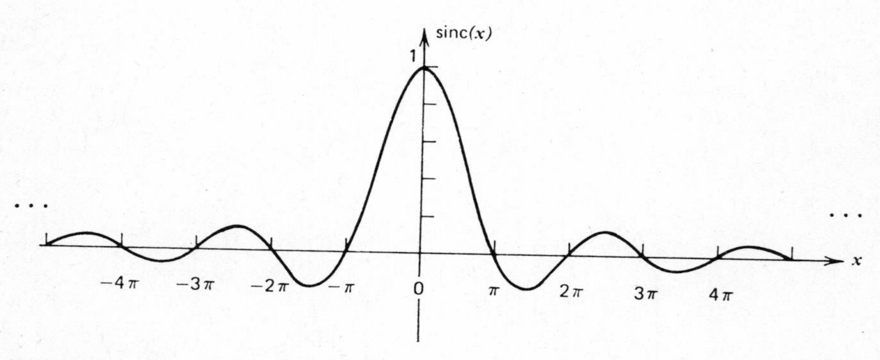

Figure 5.8

Thus we can represent $f(t)$ as

$$f(t) = \sum_{n=-\infty}^{\infty} \frac{Ad}{T} \operatorname{sinc}\left(\frac{n\pi d}{T}\right) e^{(jn2\pi/T)t} \tag{5.51}$$

From (5.50) we can see that the values F_n are real. Thus, we need only a single spectrum. To plot the spectrum, we consider three cases for d and T.

***Case 1*:** Suppose $d = \frac{1}{10}$ and $T = \frac{1}{2}$. In this case, the spectrum is $F_n = (A/5) \operatorname{sinc}(n\pi/5)$ and a plot is shown in Figure 5.9*a*. The fundamental frequency is $\omega_0 = 2\pi/T = 4\pi$, and so the harmonic spacing $\Delta\omega_0$ is also 4π.

***Case 2*:** Assume $d = \frac{1}{10}$ as before, and that we increase the period to 1. Now the spectrum is $F_n = (A/10) \operatorname{sinc}(n\pi/10)$ which is depicted in Figure 5.9*b*. The fundamental frequency is decreased to $\omega_0 = 2\pi$ and so also the harmonic spacing to 2π.

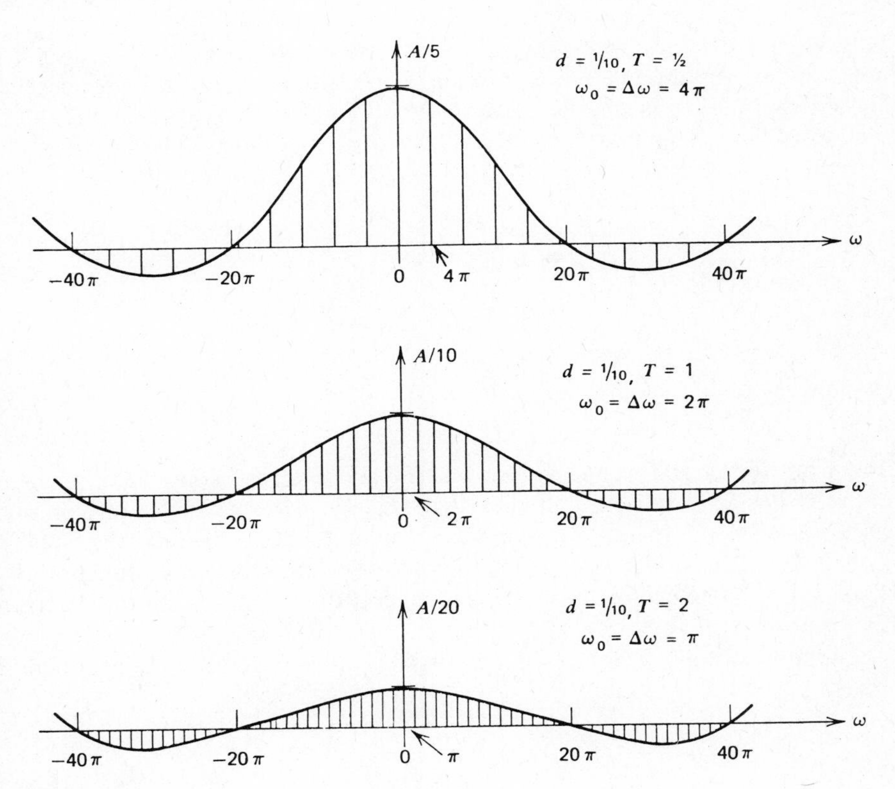

Figure 5.9

***Case 3*:** Assume $d = \frac{1}{10}$ and we further increase the period to 2. The spectrum is $F_n = (A/20)\,\text{sinc}\,(n\pi/20)$ as shown in Figure 5.9*c*. The spectrum now exists at $\omega = 0, \pm\pi, \pm 2\pi, \ldots$. ■

In this example, we observe that as the period T increases, two qualitative features of the spectrum are changed. The spacing or separation of the component frequencies of $f(t)$ decreases and the amplitudes of the frequency harmonics decrease. The spectrum thus becomes denser and the amplitudes smaller as T increases. Notice, however, that the *shape* of the frequency spectrum does not change with the period T. The shape of the envelope depends only on the periodic pulse shape that repeats every T seconds.

Properties of the Discrete Frequency Spectrum

The discrete frequency spectrum possesses certain properties that are useful in understanding the spectral representation. For example, the magnitude spectrum of every real function is an even function of $n\omega_0$ and the phase spectrum is an odd function of $n\omega_0$. This relationship can be shown as follows. The coefficient F_n is given by

$$F_n = \frac{1}{T}\int_{-T/2}^{T/2} f(t)e^{-jn\omega_0 t}\,dt \tag{5.52}$$

and

$$F_{-n} = \frac{1}{T}\int_{-T/2}^{T/2} f(t)e^{+jn\omega_0 t}\,dt \tag{5.53}$$

Thus, from (5.52) and (5.53) we see that F_n and F_{-n} are complex conjugates. That is, $F_n = F^*_{-n}$. This fact implies that $|F_n| = |F_{-n}|$, which in turn implies that $|F_n|$ is an even function of n (or $n\omega_0$). The phase of F_n is θ_n and the phase of F_{-n} is θ_{-n}, which equals $-\theta_n$. Thus, the phase spectrum is an odd function of n (or $n\omega_0$). These two symmetry properties suggest that we need plot the spectrum only for positive n.

Suppose we have a periodic function $f(t)$ representable by the exponential expansion

$$f(t) = \sum_{n=-\infty}^{\infty} F_n e^{jn\omega_0 t} \tag{5.54}$$

What happens to the spectrum of $f(t)$ if we shift the time origin of $f(t)$ by an amount τ? That is, what is the spectrum of $f(t \pm \tau)$? Intuitively, we might argue that the amplitude spectrum should not change because the amplitude spectrum depends on the waveform or shape of $f(t)$. However, the phase spectrum of the shifted version should change in some manner, because shifting the time origin changes the phase of all the harmonics that make up $f(t)$. From (5.54), we have

$$f(t \pm \tau) = \sum_{n=-\infty}^{\infty} F_n e^{jn\omega_0(t \pm \tau)} = \sum_{n=-\infty}^{\infty} F_n e^{\pm jn\omega_0 \tau} e^{jn\omega_0 t} = \sum_{n=-\infty}^{\infty} \hat{F}_n e^{jn\omega_0 t} \tag{5.55}$$

where $\hat{F}_n = F_n e^{\pm jn\omega_0 \tau}$. Represent the original spectrum in the form $F_n = |F_n| e^{j\theta_n}$. The spectrum of the shifted version is then

$$\hat{F}_n = |F_n| \, e^{j(\theta_n \pm n\omega_0 \tau)} \tag{5.56}$$

Equation 5.56 indicates that the amplitude spectrum of $f(t \pm \tau)$ is identical to the amplitude spectrum of $f(t)$: that is, $|F_n|$. The harmonic frequencies are identical, as can be seen from (5.55). The phase spectrum, however, is changed. The time shift of $\pm\tau$ causes a phase shift of $\pm n\omega_0 \tau$ radians in the nth harmonic: that is, the frequency component at $n\omega_0$. Thus, a time shift of $\pm\tau$ seconds does not affect the amplitude spectrum, but changes the phase spectrum by an amount of $\pm n\omega_0 \tau$ radians in the nth harmonic of $f(t)$.

The Power Spectrum of a Periodic Function

The power of a time signal waveform is often an important characterization of the signal. We define the power associated with the periodic signal $f(t)$ as the integral

$$P = \frac{1}{T} \int_{-T/2}^{T/2} f^2(t) \, dt \tag{5.57}$$

Equation 5.57 expresses the power of $f(t)$ in the time domain. We can also express the power in $f(t)$ in the frequency domain by calculating the power associated with each frequency component. This leads us to the idea of a *power spectrum* for $f(t)$ in which we calculate the power associated with each harmonic in $f(t)$.

Let the Fourier series for $f(t)$ be

$$f(t) = \sum_{n=-\infty}^{\infty} F_n e^{jn\omega_0 t} \tag{5.58}$$

Because the exponential set $\{e^{jn\omega_0 t}\}$, $n = 0, \pm 1, \pm 2, \ldots$ is a complete orthogonal set, Parseval's relation, (5.12), holds. That is,

$$\int_{-T/2}^{T/2} f^2(t)\, dt = \sum_{n=-\infty}^{\infty} |F_n|^2 \cdot T \tag{5.59}$$

or

$$\frac{1}{T}\int_{-T/2}^{T/2} f^2(t)\, dt = \sum_{n=-\infty}^{\infty} |F_n|^2 \tag{5.60}$$

The power in $f(t)$ can thus be calculated from

$$P = \cdots + |F_{-n}|^2 + \cdots + |F_{-1}|^2 + |F_0|^2 + |F_1|^2 + \cdots + |F_n|^2 + \cdots \tag{5.61}$$

Equation 5.61 indicates that the power in $f(t)$ can be calculated by adding together the powers associated with the frequency components in $f(t)$. The power associated with the frequency component at $n\omega_0$ radians is $|F_n|^2$ and that of $-n\omega_0$ is $|F_{-n}|^2$. Recall in this representation that it takes both frequency components at $\pm n\omega_0$ to form a single real harmonic. We know that

$$F_n = F^*_{-n}$$

which implies that

$$|F_n|^2 = |F_{-n}|^2$$

Thus the power in $f_n(t)$, the nth (real) harmonic of $f(t)$, is

$$p_n = |F_n|^2 + |F_{-n}|^2 = 2|F_n|^2 \tag{5.62}$$

Equation 5.62 is the power in the time function

$$\begin{aligned} f_n(t) &= F_n e^{+jn\omega_0 t} + F_{-n} e^{-jn\omega_0 t} \\ &= |F_n| e^{j(n\omega_0 t + \theta_n)} + |F_n| e^{-j(n\omega_0 t + \theta_n)} \\ &= 2|F_n| \cos(n\omega_0 t + \theta_n) \end{aligned} \tag{5.63}$$

Because the power in $A\cos(\omega t + \phi)$ is $A^2/2$, it follows that the power in $2|F_n|\cos(n\omega_0 t + \theta_n)$ is $2|F_n|^2$, as given by (5.62). To summarize: we can calculate the power in $f(t)$ in the frequency domain by calculating the power in each frequency component of $f(t)$. We then add these component powers to find the total power P. The collection of component powers, $|F_n|^2$, as a function of $n\omega_0$ is called the *power spectrum* of $f(t)$.

Example 5.6 What percentage of the total power is contained within the first zero crossing of the spectrum envelope for $f(t)$ as given in Figure 5.10? In this example, we have

$$f(t) = \sum_{n=-\infty}^{\infty} F_n e^{jn\omega_0 t}$$

where

$$\begin{aligned} F_n &= \frac{1}{T}\int_{-T/2}^{T/2} f(t)e^{-jn\omega_0 t}\,dt \\ &= \frac{Ad}{T}\operatorname{sinc}\left(\frac{n\pi d}{T}\right) \\ &= \frac{1}{5}\operatorname{sinc}\left(\frac{n\pi}{5}\right) \end{aligned} \tag{5.64}$$

The total power in $f(t)$ is

$$\begin{aligned} P &= \frac{1}{T}\int_{-T/2}^{T/2} f^2(t)\,dt \\ &= 4\int_{-1/40}^{1/40} (1)^2\,dt = 4\left[\frac{1}{40}+\frac{1}{40}\right] = 0.20 \end{aligned} \tag{5.65}$$

The spectrum of $f(t)$ is shown in Figure 5.11. The first zero crossing occurs at 40π rad/sec, and there are four harmonics plus the d.c. value within the first zero

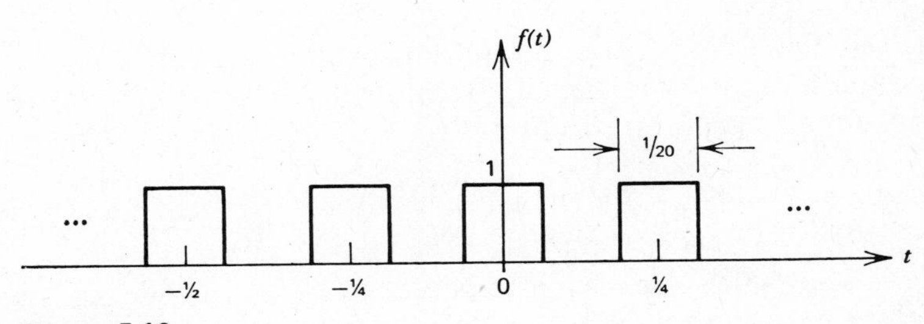

Figure 5.10

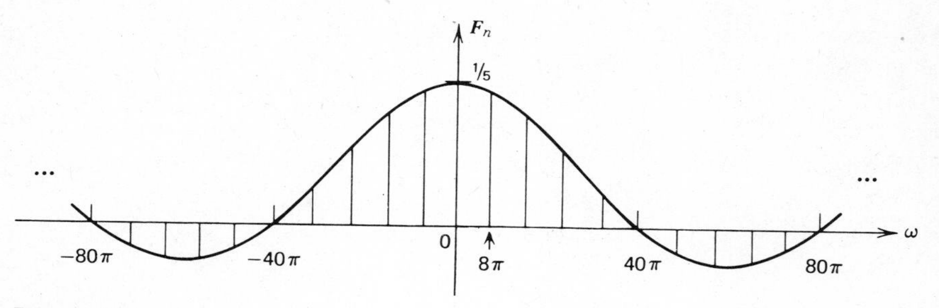

Figure 5.11

crossing. Thus, the power contained within the first zero crossing of the spectrum envelope is

$$\begin{aligned} P_{fzc} &= |F_0|^2 + 2\{|F_1|^2 + |F_2|^2 + |F_3|^2 + |F_4|^2\} \\ &= \left(\frac{1}{5}\right)^2 + \frac{2}{5^2}\left\{\text{sinc}^2\left(\frac{\pi}{5}\right) + \text{sinc}^2\left(\frac{2\pi}{5}\right)\right. \\ &\quad \left. + \text{sinc}^2\left(\frac{3\pi}{5}\right) + \text{sinc}^2\left(\frac{4\pi}{5}\right)\right\} \\ &= \left(\frac{1}{5}\right)^2 + \frac{2}{5^2}\{0.875 + 0.573 + 0.255 + 0.055\} \\ &\cong 0.04 + 0.141 = 0.181 \end{aligned} \tag{5.66}$$

Comparing (5.65) and (5.66), we see that 90.5% of the total power in $f(t)$ is contained within the first zero crossing of the spectrum for $f(t)$. For applications involving the power of $f(t)$, one might then reasonably represent $f(t)$ in terms of its first four harmonics and its average value. ■

The complex Fourier series furnishes us with a method of decomposing a signal in terms of a sum of elementary signals of the form $\{e^{jn\omega_0 t}\}$. We can use this representation for signals $f(t)$ that are:

1. Periodic, $f(t) = f(t + T)$, in which case the representation is valid on $(-\infty, \infty)$.
2. Aperiodic, in which case the representation is valid on a finite interval $[t_1, t_2]$. The periodic extension of $f(t)$ is obtained outside of $[t_1, t_2]$.

If we wish to extend the class of functions to include aperiodic functions on $(-\infty, \infty)$, we must use another kind of representation called a *Fourier integral* or *transform.* In this case we shall decompose $f(t)$ into a continuum of complex sinusoids of the form $e^{j\omega t}$. This representation is discussed in sections 5.8–5.14.

5.5 THE DISCRETE-TIME FOURIER TRANSFORM

We can also represent sequences in terms of complex exponentials as we shall next describe. In Chapter 4 we discussed the process of obtaining the frequency response function of a linear discrete-time system. If $H(z)$ is the transfer function, the frequency response is obtained by evaluating $H(z)$ around the unit circle, that is, for $z = e^{j\theta}$, in the z-plane. Thus,

$$H(e^{j\theta}) = H(z)|_{z=e^{j\theta}} \tag{5.67}$$

Since $H(z)$ is the Z-transform of the impulse response sequence, we also have

$$H(e^{j\theta}) = \sum_{k=-\infty}^{\infty} h_k e^{-jk\theta} \tag{5.68}$$

Equation 5.68 was obtained in Chapter 2 by using convolution. We shall term $H(e^{j\theta})$ the Discrete-Time Fourier Transform (DTFT) of the sequence h. We can obtain the coefficients h_k from $H(e^{j\theta})$ using the inverse transform

$$h_k = \frac{1}{2\pi} \int_{-\pi}^{\pi} H(e^{j\theta}) e^{jk\theta}\, d\theta \tag{5.69}$$

Equation 5.69 can be established by substituting for $H(e^{j\theta})$ from (5.68). This gives

$$h_k \stackrel{?}{=} \frac{1}{2\pi} \int_{-\pi}^{\pi} \sum_{n=-\infty}^{\infty} h_n e^{-jn\theta} e^{jk\theta}\, d\theta$$

Interchanging the sum and the integral gives

$$h_k \stackrel{?}{=} \frac{1}{2\pi} \sum_{n=-\infty}^{\infty} h_n \int_{-\pi}^{\pi} e^{j\theta(k-n)}\, d\theta$$

The integral on the right-hand side is nonzero for $k \neq n$ and is equal to 2π for $k = n$. Only one term of the sum on the right side is nonzero, namely, the term $k = n$. Thus the right side is h_k and the inversion integral is established.

Thus, given any sequence f, we can compute its DTFT as

$$F(e^{j\theta}) = \sum_{k=-\infty}^{\infty} f_k e^{-jk\theta} \tag{5.70}$$

Given the transform $F(e^{j\theta})$ we compute the sequence from

$$f_k = \frac{1}{2\pi} \int_{-\pi}^{\pi} F(e^{j\theta}) e^{jk\theta}\, d\theta \tag{5.71}$$

From our previous discussions on the complex Fourier series we can also argue that (5.71) follows directly. The transform $F(e^{j\theta})$ is periodic of period 2π. Therefore, it has a Fourier series. The coefficients are those given in (5.71). We have merely interchanged the variables corresponding to time and frequency in our previous discussion of Fourier series.

We interpret (5.71) as a decomposition of the sequence values f_k in terms of a continuum of complex sinusoids, each with amplitude and phase $F(e^{j\theta})/2\pi$. $F(e^{j\theta})$ is the Z-transform of the sequence f restricted to values of z on the unit circle in the z-plane. This spectrum of the sequence f is real-valued and measurable. Given the spectrum there corresponds a unique sequence f. Thus we have two physical and measurable descriptions for the same phenomenon. Engineers often specify systems in the transform domain. A description for the design of a low-pass filter with a certain bandwidth, transition region, and attenuation in the stopband is an example.

One word of caution. The description of a system in terms of $H(e^{j\theta})$ is essentially a steady-state kind of description of the system. If we wish to solve a transient problem, we should use the Z-transform analysis or one of the models dicussed in Chapter 2.

Example 5.7 What is the frequency response of a digital filter with impulse-response sequence h where

$$h_n = \begin{cases} 1, & 0 \leq n \leq N-1 \\ 0, & \text{otherwise} \end{cases}$$

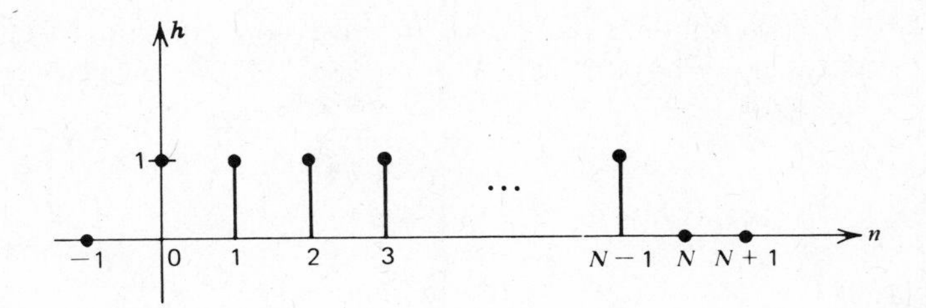

Figure 5.12

as shown in Figure 5.12? The corresponding spectrum is found using (5.68). The result is

$$H(e^{j\theta}) = \sum_{n=0}^{N-1} 1 \cdot e^{-jn\theta} = \frac{1 - e^{-jN\theta}}{1 - e^{-j\theta}} = \frac{e^{-j(N\theta/2)}}{e^{(-j\theta/2)}} \frac{(e^{j(N\theta/2)} - e^{-jN\theta/2)})}{(e^{(j\theta/2)} - e^{(-j\theta/2)})}$$

$$= e^{-j(N-1)(\theta/2)}\left[\frac{\sin(N\theta/2)}{\sin(\theta/2)}\right]$$

The magnitude response function is shown in Figure 5.13 for $N = 8$. The phase response is linear and equal to $-(N - 1)(\theta/2)$. Based on the amplitude response function in Figure 5.13 we would call this filter a crude lowpass filter (on the interval $[-\pi, \pi]$). Notice that we need to specify the response $H(e^{j\theta})$ only on the interval $[0, \pi]$ for real h_n. $H(e^{j\theta})$ is periodic of period 2π. However, for real h_n, $|H(e^{j\theta})|$ is an even function of θ about $\theta = 0$ and $\arg[H(e^{j\theta})]$ is an odd function of θ about $\theta = 0$.

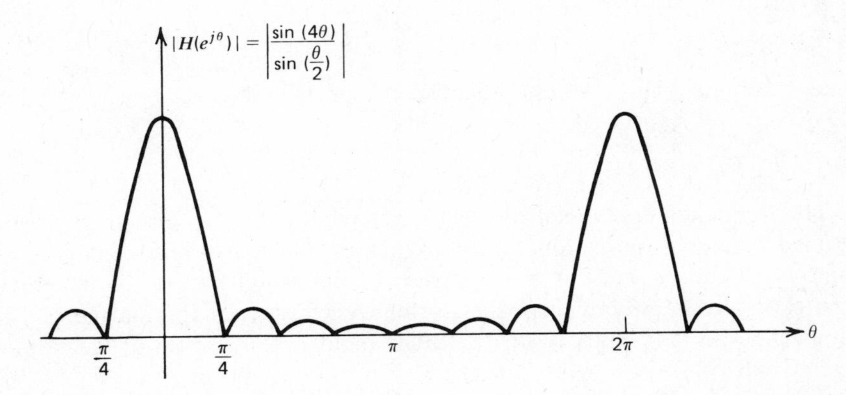

Figure 5.13

■

Example 5.8 As a simple example of the inversion integral (5.71) let us determine what sequence f has a DTFT $F(e^{j\theta}) = \cos\theta$. From (5.71) we have

$$f_n = \frac{1}{2\pi}\int_{-\pi}^{\pi} F(e^{j\theta})e^{jn\theta}\, d\theta$$

Substituting for $F(e^{j\theta})$ the exponential form of $\cos\theta$ gives us

$$f_n = \frac{1}{2\pi}\int_{-\pi}^{\pi}\left(\frac{e^{j\theta}+e^{-j\theta}}{2}\right)e^{jn\theta}\, d\theta$$

$$= \frac{1}{4\pi}\int_{-\pi}^{\pi}(e^{j(n+1)\theta}+e^{j(n-1)\theta})\, d\theta$$

The integral of $\exp(jk\theta)$ on $[-\pi, \pi]$ is zero for any integer k not zero. Thus, the only nonzero values for f_n occur for $n = -1$ and $n = 1$ in which case

$$f_{-1} = f_1 = \frac{1}{2}$$

Thus the sequence $\{\frac{1}{2}, 0, \frac{1}{2}\}$ has a spectrum $\cos\theta$.

Notice that this result is consistent with our previous discussion of Fourier series. The Fourier series representation of $\cos t$ is

$$\cos t = \sum_{n=-\infty}^{\infty} F_n e^{-jn\omega_0 t}$$

$$= \frac{e^{jt}+e^{-jt}}{2}$$

Hence the only nonzero values of the discrete spectrum F_n occur for $n = \pm 1$ with value $\frac{1}{2}$. ∎

The frequency content of discrete-time signals, that is, sequences f, can be described by a continuous function of frequency $F(e^{j\theta})$. For real sequences f, the symmetry and periodicity of the spectrum $F(e^{j\theta})$ imply that all the information in the spectrum is contained in the normalized frequency interval $[0, \pi]$. Continuous time periodic signals, on the other hand, are made up of harmonics at discrete values of frequency. In general, we must use harmonics specified in $[0, \infty]$ to completely describe continuous periodic signals.

We can use the spectrum of a sequence qualitatively to infer the smoothness or variability of the sequence values. Given two sequences f and g, the sequence with the larger value of amplitude spectrum at values of θ near π is the more rapidly changing sequence. The following example demonstrates this.

Example 5.9 Consider two sequences f and g defined as

$$f_n = \begin{cases} \left(\frac{1}{2}\right)^n, & n \geq 0 \\ 0, & n < 0, \end{cases} \qquad g_n = \begin{cases} \left(-\frac{1}{2}\right)^n, & n \geq 0 \\ 0, & n < 0 \end{cases}$$

The graphs of these sequences are shown in Figure 5.14. Because of the minus sign, the sequence g is more variable than the sequence f, as shown in Figure 5.14. Consider the spectrum of f:

$$\begin{aligned} F(e^{j\theta}) &= \sum_{n=0}^{\infty} \left(\frac{1}{2}\right)^n e^{-jn\theta} = \frac{1}{1 - \frac{1}{2}e^{-j\theta}} \\ &= \frac{1}{1 - \frac{1}{2}\cos\theta + \frac{1}{2}j\sin\theta} \\ &= \frac{1 - \frac{1}{2}\cos\theta - \frac{1}{2}j\sin\theta}{[(1 - \frac{1}{2}\cos\theta)^2 + (\frac{1}{2}\sin\theta)^2]} \end{aligned}$$

Thus,

$$|F(e^{j\theta})| = \frac{1}{[1 + \frac{1}{4}\cos^2\theta + \frac{1}{4}\sin^2\theta - \cos\theta]^{1/2}} = \frac{1}{[1.25 - \cos\theta]^{1/2}}$$

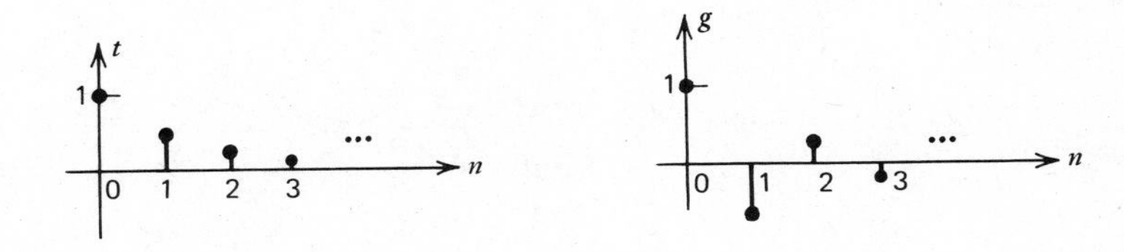

Figure 5.14

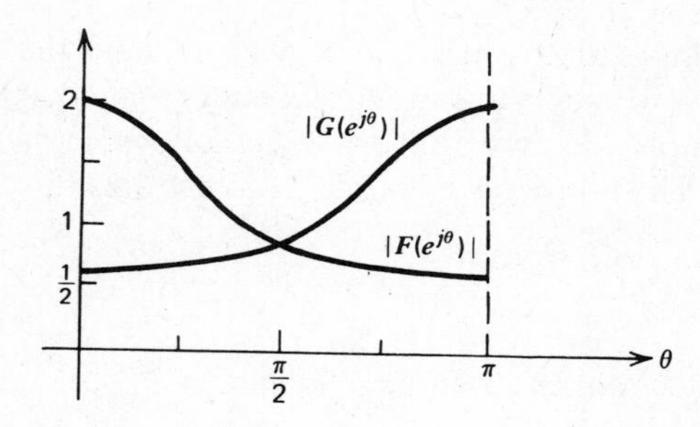

Figure 5.15

Similarly, we calculate the spectrum of g as

$$G(e^{j\theta}) = \sum_{n=0}^{\infty}\left(-\frac{1}{2}\right)^n e^{-jn\theta} = \sum_{n=0}^{\infty}\left(-\frac{e^{-j\theta}}{2}\right)^n = \frac{1}{1+\frac{1}{2}e^{-j\theta}}$$

$$= \frac{1}{1+\frac{1}{2}\cos\theta - \frac{1}{2}j\sin\theta}$$

$$= \frac{1+\frac{1}{2}\cos\theta + \frac{1}{2}j\sin\theta}{(1+\frac{1}{2}\cos\theta)^2 + (\frac{1}{2}\sin\theta)^2}$$

Thus

$$|G(e^{j\theta})| = \frac{1}{[1+\frac{1}{4}\cos^2\theta + \frac{1}{4}\sin^2\theta + \cos\theta]^{1/2}} = \frac{1}{[1.25+\cos\theta]^{1/2}}$$

The amplitude spectra are shown in Figure 5.15. The smoother sequence f results in an amplitude spectrum that is larger for small values of θ and smaller for large values of θ.

■

5.6 PROPERTIES OF THE DISCRETE-TIME FOURIER TRANSFORM

1. Convergence Properties

For a general sequence f, the spectrum $F(e^{j\theta})$ exists provided the series of (5.70) converges. If the sequence f is absolutely summable, that is,

$$\sum_{n=-\infty}^{\infty} |f_n| < \infty \tag{5.72}$$

the series is said to be absolutely convergent and the series in (5.70) converges uniformly to a continuous function of θ. This implies that the frequency response for a stable system always converges, since the impulse response sequence h of such a system is absolutely summable.

If a sequence is absolutely summable, then it also has finite energy, that is, $\sum |f_n|^2 < \infty$. The converse is not true. A sequence with finite energy may not be absolutely summable. However, in this latter case one can use other convergence methods so that series in (5.70) converges, not pointwise, but in a manner to minimize the mean square error. This nonuniform convergence leads to oscillations at points of discontinuity in the spectrum. These oscillations are known as Gibbs' phenomenon.† The oscillations can be smoothed using so-called window techniques. We shall discuss this in more detail in Chapter 7.

2. Linearity

The DTFT is a linear operation. If f and g are sequences with transform $F(e^{j\theta})$ and $G(e^{j\theta})$, respectively, then

$$\begin{aligned}\text{DTFT}\{\alpha f + \beta g\} &= \sum_n (\alpha f_n + \beta g_n)e^{-jn\theta} \\ &= \alpha \sum_n f_n e^{-jn\theta} + \beta \sum_n g_n e^{-jn\theta} \\ &= \alpha F(e^{j\theta}) + \beta G(e^{j\theta})\end{aligned}$$

3. Convolution

If $f \leftrightarrow F(e^{j\theta})$ and $g \leftrightarrow G(e^{j\theta})$, then $f * g \leftrightarrow F(e^{j\theta})G(e^{j\theta})$. Consider the DTFT of $f * g$.

$$\text{DTFT}\{f * g\} = \sum_{n=-\infty}^{+\infty} \sum_{k=-\infty}^{\infty} f_k g_{n-k} e^{-jn\theta}$$

Multiply on the right by $e^{-jk\theta}e^{jk\theta}$. We obtain

$$\text{DTFT}\{f * g\} = \sum_{n=-\infty}^{+\infty} \sum_{k=-\infty}^{\infty} f_k e^{-jk\theta} g_{n-k} e^{-j(n-k)\theta}$$

† See, for example, A. Papoulis, *The Fourier Integral and Its Applications*, McGraw-Hill, New York, 1962, p. 30.

Let $m = n - k$ and interchange the summations on k and m to yield

$$\text{DTFT}\{f * g\} = \sum_{k=-\infty}^{+\infty} f_k e^{-jk\theta} \sum_{m=-\infty}^{\infty} g_m e^{-jm\theta}$$

$$= F(e^{j\theta})G(e^{j\theta})$$

We can use this property, for example, to find the output spectrum of a digital filter given the input sequence u. If the output sequence is y and the impulse response sequence h, the output spectrum is

$$Y(e^{j\theta}) = H(e^{j\theta})U(e^{j\theta}) \tag{5.73}$$

where $h \leftrightarrow H(e^{j\theta})$ and $u \leftrightarrow U(e^{j\theta})$.

4. Time Shifting

If $\{f_k\} \leftrightarrow F(e^{j\theta})$, then $\{f_{k-n}\} \leftrightarrow e^{-jn\theta}F(e^{j\theta})$. By definition

$$\text{DTFT}\{f_{k-n}\} = \sum_{k=-\infty}^{\infty} f_{k-n} e^{-jk\theta}$$

Letting $m = k - n$,

$$\text{DTFT}\{f_{k-n}\} = \sum_{m=-\infty}^{\infty} f_m e^{-j(m+n)\theta}$$

$$= e^{-jn\theta} \sum_{m=-\infty}^{\infty} f_m e^{-jm\theta} = e^{-jn\theta}F(e^{j\theta}).$$

This property is similar to the corresponding time-shift property we discussed for periodic analog signals. Shifting the sequence in time multiplies the original spectrum by $e^{-jn\theta}$ which represents a linear phase shift, that is, a pure delay of n. The amplitude response of the original sequence is unchanged.

5. Parseval's Theorem

In the exponential Fourier series we were able to relate the total power in a continuous periodic signal to the power in each harmonic [see (5.12)]. We can develop the same kind of relationship for sequences and their spectra. Let f be a sequence with spectrum $F(e^{j\theta})$. Consider the sum

$$\sum_{n=-\infty}^{\infty} f_n f_n^* = \sum_{n=-\infty}^{\infty} |f_n|^2$$

Substituting for f_n^* we have

$$\sum_{n=-\infty}^{\infty} f_n f_n^* = \sum_{n=-\infty}^{\infty} f_n \frac{1}{2\pi} \int_{-\pi}^{\pi} F^*(e^{j\theta}) e^{-jn\theta} \, d\theta$$

$$= \frac{1}{2\pi} \int_{-\pi}^{\pi} F^*(e^{j\theta}) \sum_{n=-\infty}^{\infty} f_n e^{-jn\theta} \, d\theta$$

$$= \frac{1}{2\pi} \int_{-\pi}^{\pi} F^*(e^{j\theta}) F(e^{j\theta}) \, d\theta = \frac{1}{2\pi} \int_{-\pi}^{\pi} |F(e^{j\theta})|^2 \, d\theta$$

Thus

$$\sum_{n=-\infty}^{\infty} |f_n|^2 = \frac{1}{2\pi} \int_{-\pi}^{\pi} |F(e^{j\theta})|^2 \, d\theta \tag{5.74}$$

Parseval's theorem for continuous periodic signals relates power in the time domain to power in the harmonics. Parseval's theorem in the case of sequences relates energy in the sequence domain to energy in the spectrum. Power is energy divided by the length of some interval. In the case of periodic signals, the interval is the period. Note that the energy for periodic signals on $(-\infty, \infty)$ is infinite.

6. Frequency Convolution

If $f \leftrightarrow F(e^{j\theta})$ and $g \leftrightarrow G(e^{j\theta})$, then $f \cdot g \leftrightarrow F(e^{j\theta}) * G(e^{j\theta})/2\pi$. Consider the inverse transform of $F(e^{j\theta}) * G(e^{j\theta})$:

$$\text{IDTFT}\left\{\frac{F * G}{2\pi}\right\} = \left(\frac{1}{2\pi}\right)^2 \int_{-\pi}^{\pi} e^{jn\theta} \left[\int_{-\pi}^{\pi} F(e^{j\phi}) G(e^{j(\theta-\phi)}) \, d\phi\right] d\theta$$

$$= \left(\frac{1}{2\pi}\right)^2 \int_{-\pi}^{\pi} F(e^{j\phi}) \int_{-\pi}^{\pi} G(e^{j(\theta-\phi)}) e^{jn\theta} \, d\theta \, d\phi$$

Let $\xi = \theta - \phi$, then $d\xi = d\theta$ and $\theta = \xi + \phi$. Thus

$$\text{IDTFT}\left\{\frac{F * G}{2\pi}\right\} = \left(\frac{1}{2\pi}\right)^2 \int_{-\pi}^{\pi} F(e^{j\phi}) \int_{-\pi-\phi}^{\pi-\phi} G(e^{j\xi}) e^{jn(\xi+\phi)} \, d\xi \, d\phi$$

$$= \left(\frac{1}{2\pi}\right)^2 \int_{-\pi}^{\pi} F(e^{j\phi}) e^{jn\phi} \int_{-\pi-\phi}^{\pi-\phi} G(e^{j\xi}) e^{jn\xi} \, d\xi \, d\phi$$

$$= \frac{1}{2\pi} \int_{-\pi}^{\pi} F(e^{j\phi}) e^{jn\phi} g_n \, d\phi = f_n g_n$$

Recall that the $F(e^{j\theta})$ and $G(e^{j\theta})$ are periodic of period 2π. The convolution of F and G is only over a single period and is properly termed a *circular convolution* as opposed to the more usual aperiodic convolution that we have discussed in some detail.

Example 5.10 What is the spectrum of the sequence $y = fg$, where

$$f_n = \begin{cases} \frac{1}{2}, & n = \pm 1 \\ 0, & \text{otherwise,} \end{cases} \qquad g_n = \begin{cases} \frac{1}{2}, & n = \pm 1 \\ 0, & \text{otherwise} \end{cases}$$

Using the frequency convolution property we have

$$Y(e^{j\theta}) = \frac{F(e^{j\theta}) * G(e^{j\theta})}{2\pi}$$

where

$$F(e^{j\theta}) = \sum_{n=-\infty}^{\infty} f_n e^{-jn\theta} = \frac{1}{2} e^{j\theta} + \frac{1}{2} e^{-j\theta} = \cos\theta$$

and

$$G(e^{j\theta}) = \cos\theta$$

Thus

$$\begin{aligned} Y(e^{j\theta}) &= \frac{1}{2\pi} \int_{-\pi}^{\pi} \cos\phi \cos(\theta - \phi)\, d\phi \\ &= \frac{1}{2\pi} \int_{-\pi}^{\pi} \cos\phi[\cos\theta \cos\phi + \sin\theta \sin\phi]\, d\phi \\ &= \frac{1}{2\pi} \int_{-\pi}^{\pi} (\cos\theta \cos^2\phi + \sin\theta \sin\phi \cos\phi)\, d\phi \\ &= \frac{\cos\theta}{2\pi} \int_{-\pi}^{\pi} \cos^2\phi\, d\phi \\ &= \frac{\cos\theta}{2\pi}[\pi] = \frac{\cos\theta}{2} \end{aligned}$$

This result is easily checked by calculating the sequence values for y and then taking the DTFT. ■

5.7 FOURIER ANALYSIS AND THE DESIGN OF FIR FILTERS

The Fourier analysis we have discussed here for sequences can be used as the basis of a design technique for finite impulse-response (FIR) digital filters. Recall that FIR filters are characterized by a finite number of terms in the impulse-response sequence h.

The frequency response of a causal FIR filter is of the form

$$H(e^{j\theta}) = \sum_{n=0}^{N-1} h_n e^{-jn\theta} \tag{5.75}$$

where h_n, $n = 0, 1, \ldots, N-1$ is the impulse response sequence. Equation 5.75 follows from our previous discussions leading to (5.68). If we require that the coefficients h_n satisfy the conditions

$$h_n = h_{N-1-n} \tag{5.76}$$

then the filter will have linear phase (see Example 2.9). In what follows we shall assume that FIR filters are linear phase unless otherwise stated.

To explain the design of FIR filters using Fourier analysis we shall use an example to illustrate the steps. Suppose the desired frequency-response function $H_d(e^{j\theta})$ is an ideal lowpass filter with amplitude response shown in Figure 5.16. $H_d(e^{j\theta})$ is the DTFT of some impulse-response sequence h_d, that is,

$$H_d(e^{j\theta}) = \sum_{n=-\infty}^{\infty} h_d(n)e^{-jn\theta} \tag{5.77}$$

The coefficients h_d are given by the inversion integral

$$h_d(n) = \frac{1}{2\pi}\int_{-\pi}^{\pi} H_d(e^{j\theta})e^{jn\theta}\, d\theta \tag{5.78}$$

In general, $H_d(e^{j\theta})$ is piecewise constant with discontinuities at boundary points between bands, as in the present case. In such designs the sequence h_d

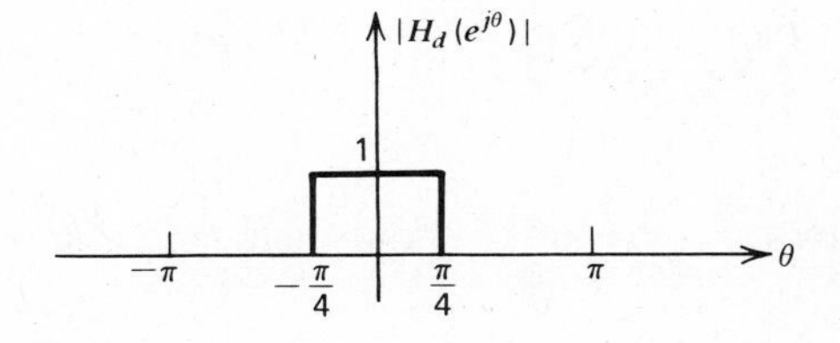

Figure 5.16

is of infinite duration. Thus we must truncate h_d to obtain an FIR design. This is a simple and direct design technique. Thus, if we wish N terms we simply choose the FIR impulse-response coefficients as

$$h(n) = \begin{cases} h_d(n), & -\left(\dfrac{N-1}{2}\right) \leq n \leq \dfrac{N-1}{2} \\ 0, & \text{otherwise} \end{cases} \tag{5.79}$$

where we have assumed N odd and have chosen the N taps weights $h(n)$ to be symmetric about $h(0)$, the middle term of the sequence h.

For the ideal lowpass filter, we find that the desired impulse-response coefficients are

$$h_d(n) = \frac{1}{2\pi} \int_{-\pi/4}^{\pi/4} 1 \cdot e^{jn\theta}\, d\theta = \frac{1}{2\pi} \left[\frac{e^{jn\theta}}{jn} \Bigg|_{-\pi/4}^{\pi/4} \right]$$

$$= \frac{1}{4} \operatorname{sinc}\left(\frac{n\pi}{4}\right), \qquad n = 0, \pm 1, \pm 2, \cdots$$

Suppose we restrict the FIR filter to $N = 9$ impulse-response coefficients. Then from (5.79) we choose these coefficients as

$$h(0) = \frac{1}{4}$$

$$h(1) = h(-1) = \frac{1}{4} \operatorname{sinc}\left(\frac{\pi}{4}\right)$$

$$h(2) = h(-2) = \frac{1}{4} \operatorname{sinc}\left(\frac{\pi}{2}\right)$$

$$h(3) = h(-3) = \frac{1}{4} \operatorname{sinc}\left(\frac{3\pi}{4}\right)$$

$$h(4) = h(-4) = \frac{1}{4} \operatorname{sinc}(\pi)$$

The first coefficient of the filter is $h(-4)$ and the final coefficient is $h(4)$ as shown in Figure 5.17.

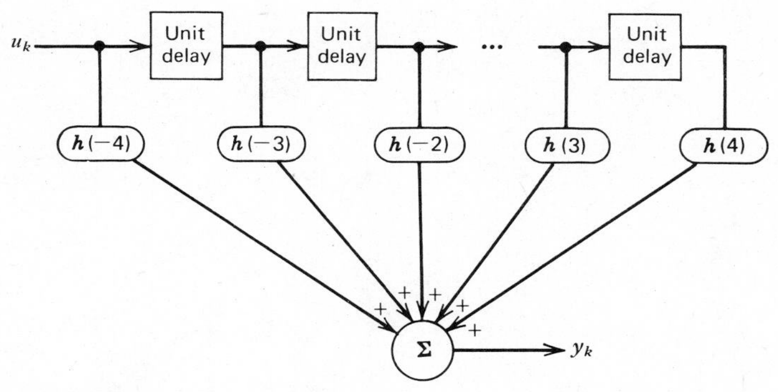

Figure 5.17

By choosing the coefficients h to be symmetric so that we satisfy (5.79), we essentially introduce a delay of one-half the length of the causal FIR filter. This is the linear phase term that defines the phase response of the filter.

The obvious question is how large should we choose N? As we choose N larger we expect our design to match the desired response more closely. In fact, since the $h(n)$'s are chosen as the Fourier series coefficients for the desired response $H_d(e^{j\theta})$, we know from the theory of Fourier series that our design is the best approximation in the sense of minimizing the mean square error

$$\frac{1}{2\pi}\int_{-\pi}^{\pi}[H_d(e^{j\theta}) - H(e^{j\theta})]^2\, d\theta$$

As comforting as this might seem, it still does not help us evaluate our design. We can obtain an intuitive and useful evaluation by use of the frequency convolution property. The actual response sequence is a truncated version h_d. Thus let

$$h(n) = h_d(n)w(n) \tag{5.80}$$

where w is the sequence

$$w(n) = \begin{cases} 1, & -\left(\dfrac{N-1}{2}\right) \le n \le \dfrac{N-1}{2} \\ 0, & \text{otherwise} \end{cases} \tag{5.81}$$

The w sequence merely truncates the h_d sequence. Using the frequency convolution property on (5.80) we obtain

$$H(e^{j\theta}) = \frac{1}{2\pi}\int_{-\pi}^{\pi} H_d(e^{j\phi})W(e^{j(\theta-\phi)})\, d\phi \tag{5.82}$$

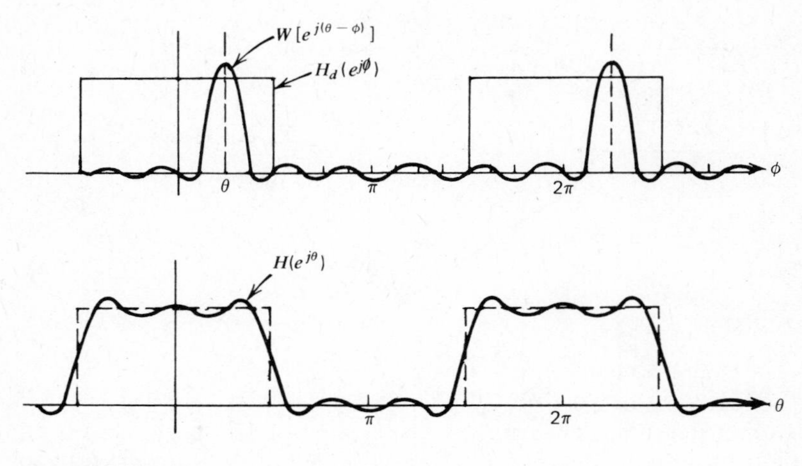

Figure 5.18

where $W(e^{j\theta})$ is the DTFT of the sequence $w(n)$. Equation 5.82 states that $H(e^{j\theta})$ is the desired response $H_d(e^{j\theta})$ convolved with a frequency function $W(e^{j\theta})$, called a window function. $W(e^{j\theta})$ has the effect of smoothing or smearing the desired response $H_d(e^{j\theta})$. Figure 5.18 depicts how the ideal low-pass response is shaped by the window function $W(e^{j\theta})$.

From (5.82) and Figure 5.18 it is clear that we want $W(e^{j\theta})$ to be narrow with respect to variations in $H_d(e^{j\theta})$. In this case $H(e^{j\theta})$ will approximate $H_d(e^{j\theta})$ closely. Thus one can study the approximation process by looking at the window $W(e^{j\theta})$ as a function of N. The rectangular window of (5.81) has the spectrum

$$W(e^{j\theta}) = \sum_{n=-(N-1)/2}^{(N-1)/2} 1 \cdot e^{-jn\theta}$$

$$= \frac{\sin(N\theta/2)}{\sin(\theta/2)}$$

which is sketched in Figure 5.13 for the case $N = 8$. The phase response of the window is zero. We shall discuss this design method in more detail in Chapter 7.

Fourier analysis is a decomposition of functions or sequences in terms of their harmonic content. Thus far we have considered the Fourier analysis of periodic functions of a continuous variable and of sequences defined on integer values. In the next section we shall enlarge our class of functions to include aperiodic functions of a continuous variable. The appropriate Fourier decomposition is in terms of the so-called *Fourier integral* or *transform.*

5.8 THE FOURIER TRANSFORM

The Fourier transform of a function $f(t)$ is defined as

$$F(j\omega) = \int_{-\infty}^{\infty} f(t)e^{-j\omega t}\,dt \tag{5.83}$$

with an inverse transform†

$$f(t) = \frac{1}{2\pi}\int_{-\infty}^{\infty} F(j\omega)e^{j\omega t}\,d\omega \tag{5.84}$$

As in the case of sequences (5.84) is interpreted as a decomposition of $f(t)$ in terms of a continuum of basis functions $e^{j\omega t}$. The function $F(j\omega)/2\pi$ plays the same role as the F_n's in the Fourier series representation. $F(j\omega)$ is called the frequency spectrum of $f(t)$. $F(j\omega)$ is, in general, a complex valued function of ω and so can be written in the form

$$F(j\omega) = |F(j\omega)|\,e^{j\theta(j\omega)} \tag{5.85}$$

where $|F(j\omega)|$ is the amplitude response and $\theta(j\omega)$ is the phase response.

Not all functions have a Fourier transform. One set of sufficient conditions for the existence of $F(j\omega)$ are the Dirichlet conditions:

1. $f(t)$ is absolutely integrable, that is, $\int_{-\infty}^{\infty} |f(t)|\,dt < \infty$,
2. $f(t)$ possesses a finite number of maxima and minima and finite number of discontinuities in any finite interval.

These conditions include all useful energy signals, that is, signals $f(t)$ for which $\int_{-\infty}^{\infty} f^2(t)\,dt < \infty$. However, there are a number of important signals like the unit step function that are not absolutely integrable. We can obtain Fourier integral representations for certain classes of these nonenergy signals by allowing

† We have chosen to use the notation $F(j\omega)$ for Fourier transform, in keeping with our earlier discussions of the frequency response $H(j\omega)$ in Chapter 3 and with the notation of Chapter 4. Other texts, notably those dealing with communication theory and not employing the Laplace transform, use $F(\omega)$ to denote the Fourier transform.

impulse functions as part of the Fourier representation. We shall classify signals $f(t)$ as being either

1. "Energy signals," that is,

$$\int_{-\infty}^{\infty} f^2(t)\,dt < \infty$$

2. "Power signals," that is, $f(t)$ has infinite energy but finite power. This means

$$\int_{-\infty}^{\infty} f^2(t)\,dt$$

does not exist, but $\lim_{T \to \infty} 1/T \int_{-T/2}^{T/2} f^2(t)\,dt$ is finite.

Transforms of Some Simple Energy Signals

To develop some insight into the continuous frequency spectrum for a function $f(t)$, we consider the Fourier transform for several simple energy functions.

Rectangular Pulse

Consider the rectangular pulse shown in Figure 5.19. This pulse is given by (5.86). We often call it a gate function.

$$g_T(t) = \begin{cases} 1, & |t| < \dfrac{T}{2} \\ 0, & \text{otherwise} \end{cases} \tag{5.86}$$

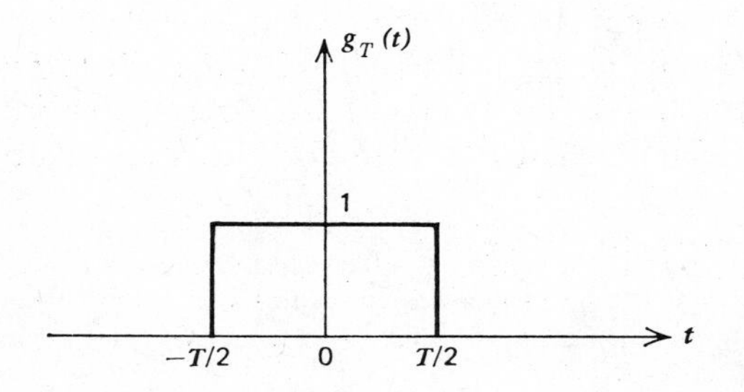

Figure 5.19

The transform of $g_T(t)$ is $G_T(j\omega)$ given by

$$G_T(j\omega) = \mathscr{F}\{g_T(t)\} = \int_{-\infty}^{\infty} g_T(t)e^{-j\omega t}\,dt$$

$$= \int_{-T/2}^{T/2} 1 \cdot e^{-j\omega t}\,dt$$

$$= \frac{e^{-j\omega t}}{-j\omega}\bigg|_{-T/2}^{T/2}$$

$$= \frac{1}{-j\omega}(e^{-j\omega T/2} - e^{+j\omega T/2})$$

$$= T \cdot \frac{\sin\left(\frac{\omega T}{2}\right)}{\left(\frac{\omega T}{2}\right)} = T \operatorname{sinc}\left(\frac{\omega T}{2}\right) \tag{5.87}$$

The amplitude spectrum is thus

$$|G_T(j\omega)| = T\left|\operatorname{sinc}\left(\frac{\omega T}{2}\right)\right| \tag{5.88}$$

and the phase spectrum is

$$\theta(j\omega) = \begin{cases} 0, & \operatorname{sinc}\left(\frac{\omega T}{2}\right) > 0 \\ \pi, & \operatorname{sinc}\left(\frac{\omega T}{2}\right) < 0 \end{cases} \tag{5.89}$$

In this case, the spectrum $G_T(j\omega)$ is a real number that is either positive or negative. The sign changes can be interpreted as phase changes of π radians. The plot of the spectrum of $G_T(j\omega)$ is shown in Figure 5.20. The shape of the spectrum in Figure 5.20 is dependent on the shape of $g_T(t)$. However, a very general property of transform pairs can be illustrated using $g_T(t)$ and $G_T(j\omega)$. We have shown for a periodic example (Example 5.6) that the majority of signal energy (power with periodic case) is contained between the first zero crossings of $G_T(j\omega)$. The first zero crossing of $G_T(j\omega)$ occurs in frequency at $f = 1/T$ Hz. As the pulse width T is decreased, this first zero crossing moves up in frequency. Conversely, as the pulse width T is increased, the first zero crossing moves toward the origin. This is a general property of all time-frequency transform pairs. The shorter the duration of a time signal, the more spread is its spectrum, and vice versa. We shall amplify this discussion when we consider the properties of the transform.

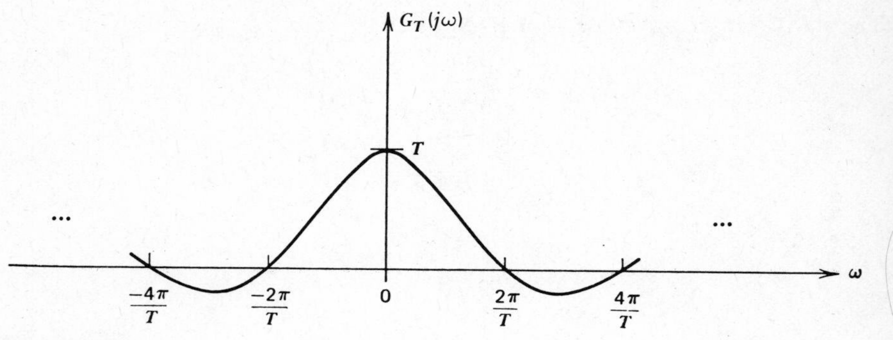

Figure 5.20

One-Sided Exponential Pulse

The exponential function is, as we have already seen, used again and again in the analysis of linear systems. Suppose we have the pulse shown in Figure 5.21.

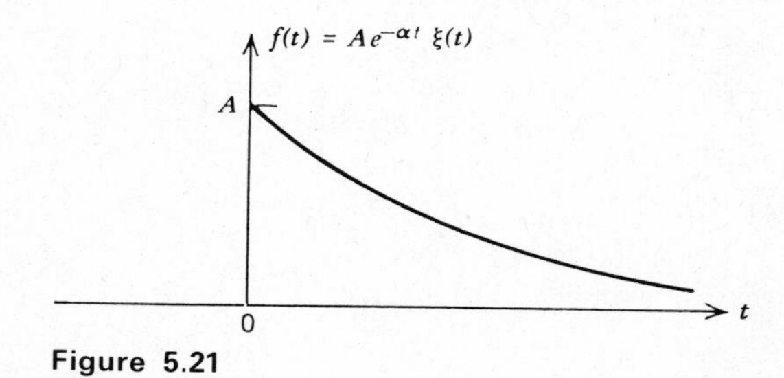

Figure 5.21

The Fourier transform of this function is

$$\begin{aligned} F(j\omega) &= \int_{-\infty}^{\infty} f(t)e^{-j\omega t}\,dt \\ &= \int_{-\infty}^{\infty} Ae^{-\alpha t}\xi(t)e^{-j\omega t}\,dt \\ &= A\int_{0}^{\infty} e^{-\alpha t}e^{-j\omega t}\,dt \\ &= \left.\frac{Ae^{-(\alpha+j\omega)t}}{-(\alpha+j\omega)}\right|_{0}^{\infty} \\ &= \frac{A}{\alpha+j\omega} \end{aligned} \tag{5.90}$$

The phase and amplitude spectra are shown in Figure 5.22. The analytic expressions for $|F(j\omega)|$ and $\theta(j\omega)$ are obtained by merely finding the magnitude and angle, respectively, of the complex function $F(j\omega)$ in (5.90). These expressions are

$$|F(j\omega)| = \frac{A}{(\alpha^2 + \omega^2)^{1/2}} \tag{5.91}$$

$$\theta(j\omega) = -\tan^{-1}\left(\frac{\omega}{\alpha}\right) \tag{5.92}$$

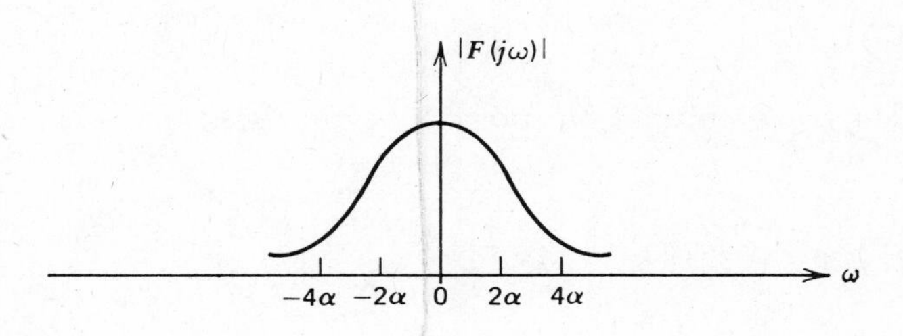

$$F(j\omega) = \frac{A(\alpha - j\omega)}{\alpha^2 + \omega^2}$$

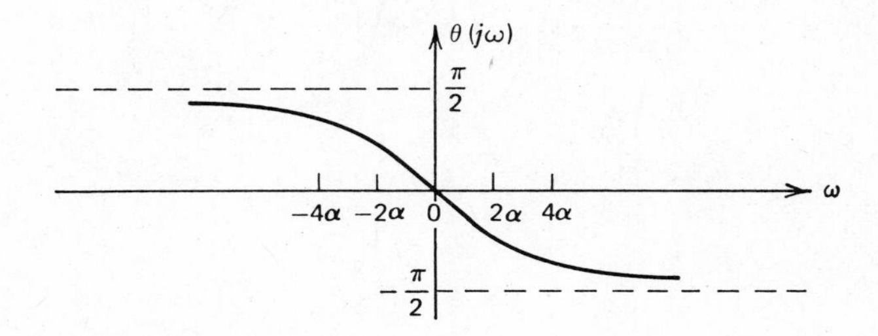

Figure 5.22

Triangular Rise

The triangular pulse of Figure 5.23 has the analytic expression given by

$$f(t) = \begin{cases} A\left(1 - \dfrac{|t|}{T}\right), & |t| < T \\ 0, & \text{otherwise} \end{cases} \tag{5.93}$$

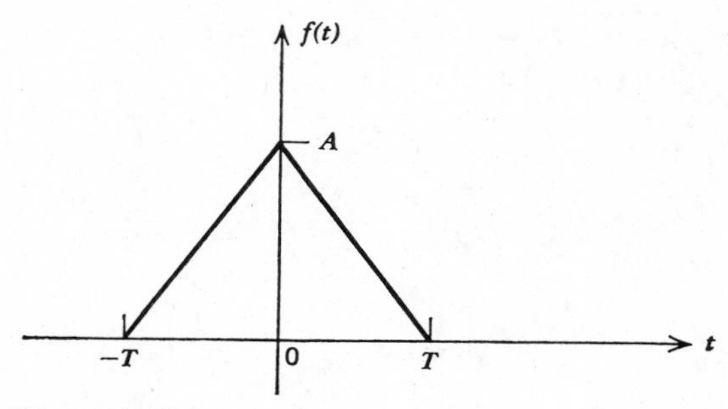

Figure 5.23

The Fourier integral of the time function can be calculated by use of the definition, as follows:

$$
\begin{aligned}
F(j\omega) &= \int_{-\infty}^{\infty} f(t)e^{-j\omega t}\,dt \\
&= \int_{-T}^{0}\left(\frac{At}{T}+A\right)e^{-j\omega t}\,dt + \int_{0}^{T}\left(\frac{-At}{T}+A\right)e^{-j\omega t}\,dt \\
&= \frac{A}{T}\left\{\left[\frac{te^{-j\omega t}}{-j\omega} - \frac{1}{(j\omega)^2}\,e^{-j\omega t}\right|_{-T}^{0}\right\} + \left.\frac{Ae^{-j\omega t}}{-j\omega}\right|_{-T}^{0} \\
&\quad + \frac{-A}{T}\left\{\left[\frac{te^{-j\omega t}}{-j\omega} - \frac{1}{(j\omega)^2}\,e^{-j\omega t}\right|_{0}^{T}\right\} + \left.\frac{Ae^{-j\omega t}}{-j\omega}\right|_{0}^{T}
\end{aligned}
\tag{5.94}
$$

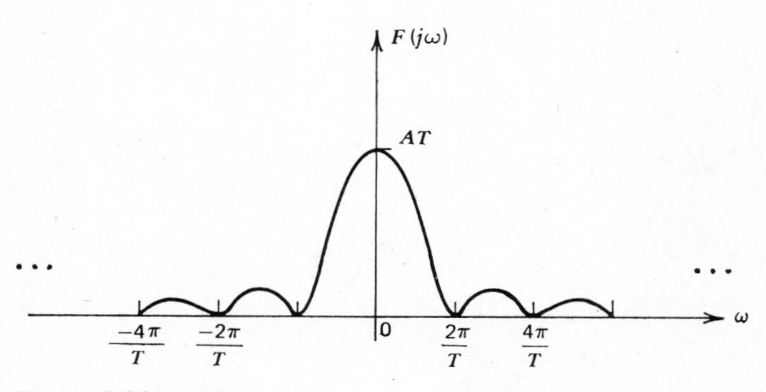

Figure 5.24

Table 5.3

Time Function, $f(t)$	Fourier Transform, $F(j\omega)$
1. $e^{-at}\xi(t)$	$\dfrac{1}{a + j\omega}$
2. $te^{-at}\xi(t)$	$\left(\dfrac{1}{a + j\omega}\right)^2$
3. $g_T(t) = \begin{cases} 1, & \lvert t \rvert < \dfrac{T}{2} \\ 0, & \text{otherwise} \end{cases}$	$T \operatorname{sinc}\left(\dfrac{\omega T}{2}\right)$
4. $\begin{cases} A\left(1 - \dfrac{\lvert t \rvert}{T}\right), & \lvert t \rvert < T \\ 0, & \lvert t \rvert > T \end{cases}$	$AT \operatorname{sinc}^2\left(\dfrac{\omega T}{2}\right)$
5. $e^{-a\lvert t \rvert}$	$\dfrac{2a}{a^2 + \omega^2}$
6. $e^{-at} \sin \omega_0 t \xi(t)$	$\dfrac{\omega_0}{(a + j\omega)^2 + \omega_0^2}$
7. $e^{-at} \cos \omega_0 t \xi(t)$	$\dfrac{a + j\omega}{(a + j\omega)^2 + \omega_0^2}$
8. e^{-at^2}	$\sqrt{\dfrac{\pi}{a}}\, e^{-\omega^2/4a}$
9. $\dfrac{t^{n-1}}{(n-1)!} e^{-at}\xi(t)$	$\dfrac{1}{(j\omega + a)^n}$
10. $\dfrac{1}{a^2 + t^2}$	$\dfrac{\pi}{a} e^{-a\lvert\omega\rvert}$
11. $\dfrac{\cos bt}{a^2 + t^2}$	$\dfrac{\pi}{2a} [e^{-a\lvert\omega - b\rvert} + e^{-a\lvert\omega + b\rvert}]$
12. $\dfrac{\sin bt}{a^2 + t^2}$	$\dfrac{\pi}{2aj} [e^{-a\lvert\omega - b\rvert} - e^{-a\lvert\omega + b\rvert}]$
13. $\cos \omega_0 t\left[\xi\left(t + \dfrac{T}{2}\right) - \xi\left(t - \dfrac{T}{2}\right)\right]$	$\dfrac{T}{2}\left[\operatorname{sinc}\left(\dfrac{(\omega - \omega_0)}{2} T\right) + \operatorname{sinc}\left(\dfrac{(\omega + \omega_0)}{2} T\right)\right]$

Simplifying the above expression (we leave the details to the reader) yields the transform as

$$F(j\omega) = AT \operatorname{sinc}^2\left(\frac{\omega T}{2}\right) \tag{5.95}$$

In this case, the frequency spectrum is a real nonnegative number for all ω and so can be plotted in a single graph, as shown in Figure 5.24. The transforms of other energy functions can be calculated in like manner. Table 5.3 summarizes some of the more useful transform pairs.

5.9 PROPERTIES OF THE FOURIER TRANSFORM

The Fourier transform is, as we have already pointed out, an alternate and equivalent method of representing a function $f(t)$. Having two descriptions is useful in engineering because often one description is easier to use in a particular application or one of the descriptions is more intuitive for a particular problem. This choice, of course, depends on one's own biases. Transforming between the two domains is relatively straightforward by means of the definitions. However, it is useful to study the effect in one domain caused by an operation in the other domain. Not only does this procedure allow one to transfer between the domains easily, but it also points out certain basic physical aspects of signals and systems that are not otherwise readily apparent (such as the inverse relationship between time duration and bandwidth). Thus one might ask, for example, what relationship exists between the transforms of a time function and the integral of this time function? What happens to the inverse of a frequency-domain function if the frequency-domain function is shifted in frequency? This section is concerned with answers to these kinds of questions.

The symmetry that exists between $f(t)$ and its transform $F(j\omega)$ is a particularly powerful property that we shall exploit time and again. The symmetry is readily apparent from the defining equations repeated here.

$$\begin{aligned} f(t) &= \frac{1}{2\pi}\int_{-\infty}^{\infty} F(j\omega)e^{j\omega t}\,d\omega \\ F(j\omega) &= \int_{-\infty}^{\infty} f(t)e^{-j\omega t}\,dt \end{aligned} \tag{5.96}$$

Actually, this symmetry could be made more complete by multiplying each integral by $1/\sqrt{2\pi}$ rather than multiplying by $1/2\pi$ and 1 as in (5.96). However, convention dictates that we use (5.96) to calculate $f(t)$ and $F(j\omega)$. We again use the notation $f(t) \leftrightarrow F(j\omega)$ to denote a transform pair. The reader should compare the properties of the Fourier transform with those of the Z-transform in Chapter 4. The properties are remarkably similar.

Symmetry (Duality)

One of the features of the transform equations in (5.96) is their symmetry in the variables t and ω. This symmetry can be used to great advantage in extending our table of transform pairs. If

$$f(t) \leftrightarrow F(j\omega)$$

then

$$F(jt) \leftrightarrow 2\pi f(-\omega) \tag{5.97}$$

Proof: Because

$$f(t) = \frac{1}{2\pi}\int_{-\infty}^{\infty} F(j\omega)e^{j\omega t}\, d\omega$$

then

$$2\pi f(-t) = \int_{-\infty}^{\infty} F(j\omega')e^{-j\omega' t}\, d\omega'$$

where we have replaced the dummy variable ω by ω'. Now if we replace t by ω, we have

$$2\pi f(-\omega) = \int_{-\infty}^{\infty} F(j\omega')e^{j\omega'\omega}\, d\omega'$$

Finally, to obtain a more recognizable form, we replace ω' by t. Thus

$$2\pi f(-\omega) = \int_{-\infty}^{\infty} F(jt)e^{-jt\omega}\, dt$$

$$= \mathscr{F}\{F(jt)\}$$

$$F(jt) \leftrightarrow 2\pi f(-\omega) \tag{5.98}$$

If $f(t)$ is an even function, $f(t) = f(-t)$, then (5.98) reduces to

$$\mathscr{F}\{F(jt)\} = 2\pi f(\omega)$$

Example 5.11 We can demonstrate the symmetry property using the triangular function

$$f(t) = \begin{cases} 1 - \dfrac{|t|}{A}, & |t| < A \\ 0, & |t| > A \end{cases}$$

The transform of this function is

$$F(j\omega) = A \operatorname{sinc}^2\left(\frac{\omega A}{2}\right)$$

The symmetry property states that the time function

$$F(jt) = A \operatorname{sinc}^2\left(\frac{tA}{2}\right)$$

has Fourier transform

$$2\pi f(-\omega) = \begin{cases} 2\pi\left(1 - \dfrac{|\omega|}{A}\right), & |\omega| < A \\ 0, & |\omega| > A \end{cases}$$

Figure 5.25 depicts the two transform pairs.

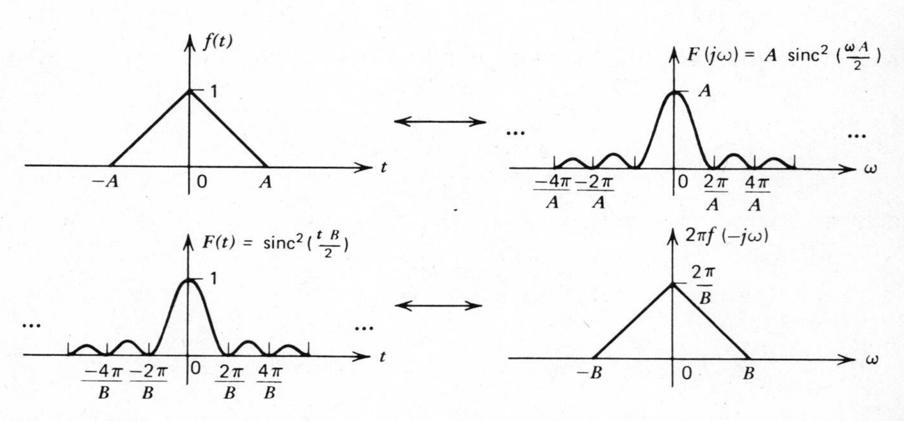

Figure 5.25

■

Linearity

The Fourier transform is a linear operation. That is, if

$$f_1(t) \leftrightarrow F_1(j\omega)$$
$$f_2(t) \leftrightarrow F_2(j\omega)$$

then

$$af_1(t) + bf_2(t) \leftrightarrow aF_1(j\omega) + bF_2(j\omega), \; a \text{ and } b \text{ are constants} \tag{5.99}$$

The proof of the above is immediately evident because the transforms are integrals of the time functions and integration is a linear operation.

Scaling Property

If

$$f(t) \leftrightarrow F(j\omega)$$

then for a real constant a,

$$f(at) \leftrightarrow \frac{1}{|a|} F\left(\frac{j\omega}{a}\right) \tag{5.100}$$

Proof: Assume $a > 0$. Then the transform of $f(at)$ is

$$\mathscr{F}\{f(at)\} = \int_{-\infty}^{\infty} f(at)e^{-j\omega t}\, dt$$

Let $x = at$, so that $dx = a\, dt$. Substituting in the above, we obtain

$$\mathscr{F}\{f(at)\} = \int_{-\infty}^{\infty} f(x)e^{-(j\omega x/a)} \frac{dx}{a}$$
$$= \frac{1}{a} F\left(\frac{\omega}{a}\right)$$

If $a < 0$, then one can show that

$$\mathscr{F}\{f(at)\} = \frac{-1}{a} F\left(\frac{j\omega}{a}\right)$$

Combining these two results yields

$$f(at) \leftrightarrow \frac{1}{|a|} F\left(\frac{j\omega}{a}\right)$$

The scaling property quantifies the time-duration to bandwidth relationship between a time function and its transform. If $|a| > 1$, then $f(at)$ as a function of t is the function $f(t)$ with a time scale compressed by a factor of a. Similarly, $F(j\omega/a)$ represents the function $F(j\omega)$ with a frequency scale, ω, expanded by the factor a. If $|a| < 1$, then $f(at)$ is an expansion of $f(t)$ and $F(j\omega/a)$ is a compression of $F(j\omega)$. Thus, as we compress the time duration of a signal, we expand the frequency spread of its spectrum. Compressing the time duration of a signal creates faster transitions, thereby requiring higher frequency components in the spectrum. Similarly, expanding the time duration of a signal means that transitions occur at more widely spaced intervals, which can be accomplished with lower frequency components in the spectrum.

Convolution

Convolution is a particularly powerful way of characterizing the input-output relationship of time-invariant linear systems. As we have seen in previous chapters, the convolution integral is not always simple to evaluate. The transform domain offers a convenient method of avoiding the convolution operation. There are two convolution theorems, one for the time domain and the other for the frequency domain.

Time Convolution. If

$$x(t) \leftrightarrow X(j\omega)$$
$$h(t) \leftrightarrow H(j\omega)$$

then

$$y(t) = \int_{-\infty}^{\infty} x(\tau)h(t-\tau)\,d\tau \leftrightarrow Y(j\omega) = X(j\omega)H(j\omega) \tag{5.101}$$

Proof:

$$\mathscr{F}\{y(t)\} = Y(j\omega) = \int_{-\infty}^{\infty} e^{-j\omega t}\left[\int_{-\infty}^{\infty} x(\tau)h(t-\tau)\,d\tau\right] dt$$
$$= \int_{-\infty}^{\infty} x(\tau)\left[\int_{-\infty}^{\infty} h(t-\tau)e^{-j\omega t}\,dt\right] d\tau$$

Now let $a = t - \tau$. Then $da = dt$ and $t = a + \tau$. Thus

$$Y(j\omega) = \int_{-\infty}^{\infty} x(\tau)\left[\int_{-\infty}^{\infty} h(a)e^{-j\omega(a+\tau)}\, da\right] d\tau$$

$$= \int_{-\infty}^{\infty} x(\tau)e^{-j\omega\tau}\, d\tau \int_{-\infty}^{\infty} h(a)e^{-j\omega a}\, da$$

$$= X(j\omega)H(j\omega)$$

The corresponding symmetric property involves the transform of a product of time functions.

Frequency Convolution. If

$$f(t) \leftrightarrow F(j\omega)$$

and

$$g(t) \leftrightarrow G(j\omega)$$

then

$$f(t)g(t) \leftrightarrow \frac{1}{2\pi} F(j\omega) * G(j\omega) \tag{5.102}$$

Proof: Consider the inverse transform of $[F(j\omega) * (G(j\omega)]/2\pi$. We have

$$\mathscr{F}^{-1}\left(\frac{F(j\omega) * G(j\omega)}{2\pi}\right) = \left(\frac{1}{2\pi}\right)^2 \int_{-\infty}^{\infty} e^{j\omega t} \int_{-\infty}^{\infty} F(ju)G(j\omega - ju)\, du\, d\omega$$

$$= \left(\frac{1}{2\pi}\right)^2 \int_{-\infty}^{\infty} F(ju) \int_{-\infty}^{\infty} G(j\omega - ju)e^{j\omega t}\, d\omega\, du$$

Let $x = \omega - u$: then $dx = d\omega$ and $\omega = x + u$. Thus

$$\mathscr{F}^{-1}\left\{\frac{F(j\omega) * G(j\omega)}{2\pi}\right\} = \left(\frac{1}{2\pi}\right)^2 \int_{-\infty}^{\infty} F(ju) \int_{-\infty}^{\infty} G(jx)e^{j(x+u)t}\, dx\, du$$

$$= \frac{1}{2\pi}\int_{-\infty}^{\infty} F(ju)e^{jut}\, du\, \frac{1}{2\pi}\int_{-\infty}^{\infty} G(jx)e^{jxt}\, dx$$

$$= f(t) \cdot g(t)$$

Thus the convolution operation in one domain is transformed into a product operation in the other domain. This is, in many cases, a sufficient simplification to justify taking transforms to circumvent the convolution operation. In the study of linear systems this simplicity is probably the primary reason for the widespread use of transform methods.

Example 5.12 In the following system, find the output voltage for an input voltage $u(t)$ of the form $te^{-at}\xi(t)$. Assume the constant $a = 1/RC$. The impulse response for the system is

$$h(t) = \frac{1}{RC} e^{-t/RC}\xi(t)$$

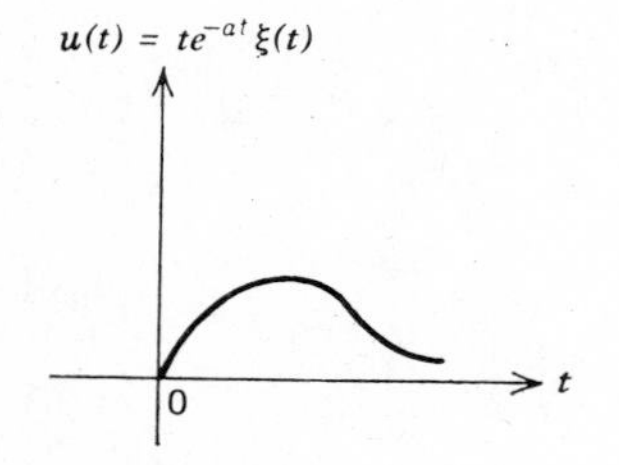

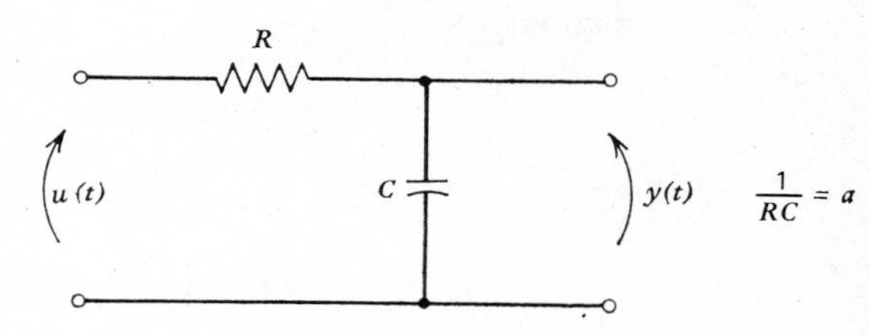

Figure 5.26

The output is given by

$$y(t) = u(t) * h(t)$$

In the transform domain, the above equation is

$$Y(j\omega) = U(j\omega)H(j\omega)$$

$H(j\omega) = \mathscr{F}[h(t)]$ is known as the system's transfer or system function. From Table 5.3, we see that

$$H(j\omega) = \mathscr{F}\left\{\frac{1}{RC} e^{-t/RC}\xi(t)\right\} = \frac{1}{RC}\left\{\frac{1}{\dfrac{1}{RC} + j\omega}\right\} = \frac{1}{1 + j\omega RC}$$

$$U(j\omega) = \mathscr{F}\{te^{-at}\xi(t)\} = \frac{1}{(a + j\omega)^2} = \frac{1}{\left(\dfrac{1}{RC} + j\omega\right)^2} = \frac{(RC)^2}{(1 + j\omega RC)^2}$$

Thus

$$Y(j\omega) = U(j\omega)H(j\omega)$$

$$= \left(\frac{1}{1+j\omega RC}\right)\left(\frac{RC}{1+j\omega RC}\right)^2 = \frac{(RC)^2}{(1+j\omega RC)^3}$$

$$\therefore\ y(t) = \mathscr{F}^{-1}\left\{\frac{(RC)^2}{(1+j\omega RC)^3}\right\} = \mathscr{F}^{-1}\left\{\frac{1}{RC}\frac{1}{\left(\frac{1}{RC}+j\omega\right)^3}\right\}.$$

$$= \frac{1}{RC}\frac{t^2e^{-t/RC}}{2}u(t)$$

■

Time Shifting

If

$$f(t) \leftrightarrow F(j\omega)$$

then

$$f(t-t_0) \leftrightarrow F(j\omega)e^{-j\omega t_0} \tag{5.103}$$

Proof. The Fourier transform of $f(t - t_0)$ is by definition

$$\mathscr{F}\{f(t-t_0)\} = \int_{-\infty}^{\infty} f(t-t_0)e^{-j\omega t}\,dt$$

Let $x = t - t_0$: then $dx = dt$ and $t = t_0 + x$. Thus,

$$\mathscr{F}\{f(t-t_0)\} = \int_{-\infty}^{\infty} f(x)e^{-j\omega(t_0+x)}\,dx$$

$$= e^{-j\omega t_0}F(j\omega)$$

The function $f(t - t_0)$ is $f(t)$ delayed by t_0 seconds. The theorem states that the original spectrum is multiplied by $e^{-j\omega t_0}$. This multiplication does not affect the amplitude spectrum. Each frequency component, however, is shifted in phase by an amount $-\omega t_0$. A little thought indicates that this result is reasonable, because a shift in time of t_0 corresponds to a phase shift of $-\omega t_0$ for a frequency component of ω rad/sec. For example, the double gate function of Fig. 5.27 has a spectrum

$$\mathscr{F}\{f(t)\} = e^{-j\omega D}T\,\text{sinc}\left(\frac{\omega T}{2}\right) + e^{j\omega D}T\,\text{sinc}\left(\frac{\omega T}{2}\right)$$

$$= 2T\cos(\omega D)\,\text{sinc}\left(\frac{\omega T}{2}\right)$$

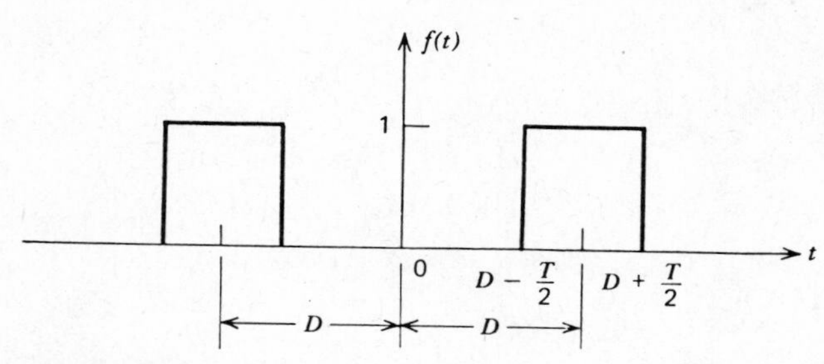

Figure 5.27

Frequency Shifting—Modulation

Frequency shifting or translation is an important operation in communication systems. This process is often known as modulation. If

$$f(t) \leftrightarrow F(j\omega)$$

then

$$f(t)e^{j\omega_0 t} \leftrightarrow F(j\omega - j\omega_0) \tag{5.104}$$

Proof: The transform of $f(t)e^{j\omega_0 t}$ is by definition

$$\begin{aligned}\mathscr{F}\{f(t)e^{j\omega_0 t}\} &= \int_{-\infty}^{\infty} f(t)e^{j\omega_0 t}e^{-j\omega t}\,dt \\ &= \int_{-\infty}^{\infty} f(t)e^{-j(\omega-\omega_0)t}\,dt \\ &= F(j\omega - j\omega_0)\end{aligned}$$

that is,

$$f(t)e^{j\omega_0 t} \leftrightarrow F(j\omega - j\omega_0)$$

The expression in (5.104) is the mathematical basis for understanding modulation. Consider, for example, the multiplication of a time function $f(t)$ by a sinusoid $\cos \omega_0 t$. The function $f(t)$ in this connection is known as the *modulating signal* and the sinusoid $\cos \omega_0 t$ is termed the *carrier* or *modulated signal*. Multiplying

$f(t)$ by $\cos \omega_0 t$ shifts the spectrum of $f(t)$ by an amount ω_0. In symbols, the spectrum of $f(t) \cos \omega_0 t$ is

$$\mathscr{F}\{f(t) \cos \omega_0 t\} = \mathscr{F}\left\{f(t)\left[\frac{e^{j\omega_0 t} + e^{-j\omega_0 t}}{2}\right]\right\}$$

$$= \frac{1}{2}\mathscr{F}\{f(t)e^{j\omega_0 t}\} + \frac{1}{2}\mathscr{F}\{f(t)e^{-j\omega_0 t}\}$$

$$= \frac{1}{2}[F(j\omega - j\omega_0) + F(j\omega + j\omega_0)]$$

Thus, multiplying a time function by $\cos \omega_0 t$ shifts the original spectrum so that half the original spectrum is centered about ω_0 and the other half is centered about $-\omega_0$. Figure 5.28 depicts the various time functions and their corresponding spectra. Notice that we can use frequency convolution to find the spectrum of $f(t) \cos \omega_0 t$, provided that we know the individual transforms. Thus

$$\mathscr{F}\{f(t) \cos \omega_0 t\} = \frac{1}{2\pi} \cdot \mathscr{F}\{f(t)\} * \mathscr{F}\{\cos \omega_0 t\} \tag{5.105}$$

We shall return to (5.105) when we have discussed the transform of $\cos \omega_0 t$.

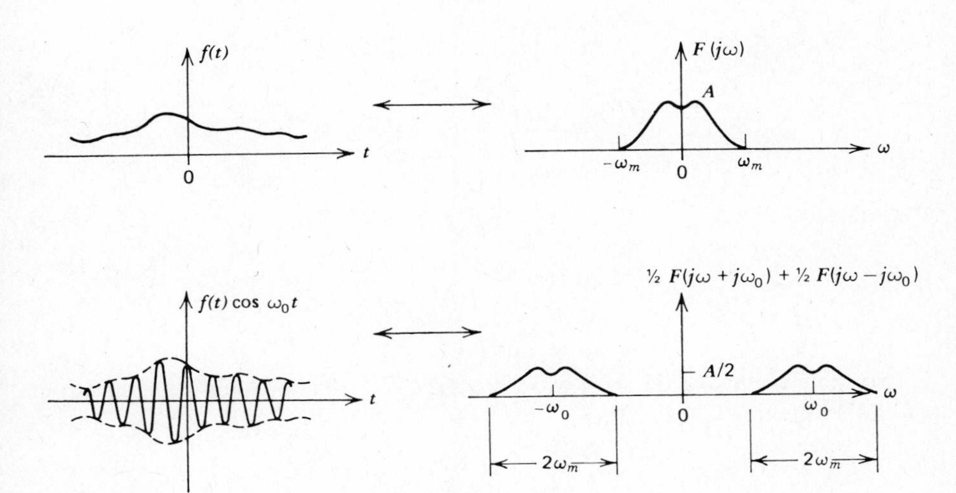

Figure 5.28

Time Differentiation and Integration

The Fourier transform can be used to solve linear differential equations. In this application, the transforms of time functions that are differentiated or integrated are important. If

$$f(t) \leftrightarrow F(j\omega)$$

then

$$\frac{df(t)}{dt} \leftrightarrow j\omega F(j\omega) \tag{5.106}$$

and

$$\int_{-\infty}^{t} f(t')\,dt' \leftrightarrow \frac{1}{j\omega} F(j\omega) \tag{5.107}$$

provided $F(0) = 0$.

Proof:

$$f(t) = \frac{1}{2\pi} \int_{-\infty}^{\infty} F(j\omega) e^{j\omega t}\,d\omega$$

Thus

$$\frac{df(t)}{dt} = \frac{1}{2\pi} \int_{-\infty}^{\infty} F(j\omega) j\omega e^{j\omega t}\,d\omega$$

which implies that

$$\frac{df(t)}{dt} \leftrightarrow (j\omega) F(j\omega)$$

We can extend this result to nth order derivatives by repeated differentiations within the integral. Thus

$$\frac{d^n f(t)}{dt^n} \leftrightarrow (j\omega)^n F(j\omega) \tag{5.108}$$

Consider the function $g(t)$ defined by

$$g(t) = \int_{-\infty}^{t} f(t')\, dt'$$

Let $g(t)$ have Fourier transform $G(\omega)$. Now

$$\frac{dg(t)}{dt} = f(t)$$

and so, using (5.106), we see that

$$j\omega G(j\omega) = F(j\omega)$$

or

$$G(j\omega) = \left(\frac{1}{j\omega}\right) F(j\omega)$$

However, for $g(t)$ to have a transform $G(j\omega)$, then, of course, $G(j\omega)$ must exist. One condition (that is somewhat more restrictive than absolute integrability) is that

$$\lim_{t \to \infty} g(t) = 0$$

This means

$$\int_{-\infty}^{+\infty} f(t)\, dt = 0$$

which is equivalent to $F(0) = 0$ because

$$F(j\omega)|_{\omega=0} = \int_{-\infty}^{\infty} f(t)\, dt$$

If $F(0) \neq 0$, then $g(t)$ is no longer an energy function and the transform of $g(t)$ includes an impulse function: that is,

$$\int_{-\infty}^{t} f(t')\, dt' \leftrightarrow \frac{1}{j\omega} F(\omega) + \pi F(0) \delta(\omega) \tag{5.109}$$

(see Section 5.11).

Table 5.4

1. Transformation	$f(t) \leftrightarrow F(j\omega)$
2. Linearity	$a_1 f_1(t) + a_2 f_2(t) \leftrightarrow a_1 F_1(j\omega) + a_2 F_2(j\omega)$
3. Symmetry	$F(jt) \leftrightarrow 2\pi f(-\omega)$
4. Scaling	$f(at) \leftrightarrow \frac{1}{\lvert a \rvert} F\left(j\frac{\omega}{a}\right)$
5. Delay	$f(t - t_0) \leftrightarrow e^{-j\omega t_0} F(j\omega)$
6. Modulation	$e^{j\omega_0 t} f(t) \leftrightarrow F(j\omega - j\omega_0)$
7. Convolution	$f_1(t) * f_2(t) \leftrightarrow F_1(j\omega) F_2(j\omega)$
8. Multiplication	$f_1(t) f_2(t) \leftrightarrow \frac{1}{2\pi} F_1(j\omega) * F_2(j\omega)$
9. Time Differentiation	$\frac{d^n}{dt^n} f(t) \leftrightarrow (j\omega)^n F(j\omega)$
10. Time Integration	$\int_{-\infty}^{t} f(\tau)\, d\tau \leftrightarrow \frac{F(j\omega)}{j\omega} + \pi F(0)\delta(\omega)$
11. Frequency Differentiation	$-jtf(t) \leftrightarrow \frac{dF(j\omega)}{d\omega}$
12. Frequency Integration	$\frac{f(t)}{-jt} \leftrightarrow \int F(j\omega')\, d\omega'$
13. Reversal	$f(-t) \leftrightarrow F(-j\omega)$

Thus we see that differentiation in the time domain corresponds to multiplication by $j\omega$ in the frequency domain. Similarly, integration in the time domain corresponds to division by $j\omega$ in the frequency domain.

Frequency Differentiation and Integration

If we differentiate the expression for $F(j\omega)$, we can show that if

$$f(t) \leftrightarrow F(j\omega)$$

then

$$-jtf(t) \leftrightarrow \frac{dF(j\omega)}{d\omega} \tag{5.110}$$

Similarly, within an additive constant

$$\frac{f(t)}{-jt} \leftrightarrow \int F(j\omega)\, d\omega \tag{5.111}$$

We leave the verification of these two properties as a problem for the reader.

Table 5.4 summarizes some of the important properties of the Fourier transform.

5.10 THE ENERGY SPECTRUM

For a periodic function, we showed that the power in a time waveform can be associated with the power contained in each harmonic component of the signal. The same kind of results apply to nonperiodic signals represented by Fourier transforms. For nonperiodic energy signals, the energy over the interval $(-\infty, \infty)$ is finite, whereas the power (energy per unit time) is zero. Thus the energy spectrum (rather than the power spectrum) is the more useful concept for nonperiodic energy signals. The energy associated with $f(t)$ is defined as

$$E = \int_{-\infty}^{\infty} f^2(t)\, dt \tag{5.112}$$

Using the Fourier integral representation for $f(t)$: that is,

$$f(t) = \frac{1}{2\pi} \int_{-\infty}^{\infty} F(j\omega) e^{j\omega t}\, d\omega$$

we can write the energy as

$$\begin{aligned} E &= \int_{-\infty}^{\infty} f^2(t)\, dt = \int_{-\infty}^{\infty} f(t) \left(\frac{1}{2\pi} \int_{-\infty}^{\infty} F(j\omega) e^{j\omega t}\, d\omega \right) dt \\ &= \frac{1}{2\pi} \int_{-\infty}^{\infty} F(j\omega) \left(\int_{-\infty}^{\infty} f(t) e^{j\omega t}\, dt \right) d\omega \\ &= \frac{1}{2\pi} \int_{-\infty}^{\infty} F(j\omega) F(-j\omega)\, d\omega \end{aligned} \tag{5.113}$$

For real $f(t)$, $F(-j\omega) = F^*(j\omega)$ so that (5.113) becomes

$$E = \frac{1}{2\pi} \int_{-\infty}^{\infty} F(j\omega) F^*(j\omega)\, d\omega = \frac{1}{2\pi} \int_{-\infty}^{\infty} |F(j\omega)|^2\, d\omega$$

That is,

$$\int_{-\infty}^{\infty} f^2(t)\, dt = \frac{1}{2\pi} \int_{-\infty}^{\infty} |F(j\omega)|^2\, d\omega \tag{5.114}$$

Equation 5.114 expresses the energy in $f(t)$ in terms of the continuous frequency spectrum of $f(t)$. The energy of $f(t)$ is given, as (5.110) indicates, by the area under the $|F(j\omega)|^2/2\pi$ curve. The function $|F(j\omega)|^2$ is a real and even function of ω, and so we can obtain the energy in $f(t)$ by integrating $|F(j\omega)|^2/\pi$ from zero to infinity. That is,

$$\begin{aligned} E &= \frac{1}{2\pi} \int_{-\infty}^{\infty} |F(j\omega)|^2 d\omega = \frac{1}{\pi} \int_{0}^{\infty} |F(j\omega)|^2\, d\omega \\ &= \int_{0}^{\infty} S(\omega)\, d\omega \end{aligned} \tag{5.115}$$

where $S(\omega) \triangleq |F(j\omega)|^2/\pi$ is called the *energy density spectrum* of $f(t)$. In our study of periodic functions, we associated certain quantities of power with each harmonic. In the case of energy signals, we associate energy with continuous bands of frequencies. The energy contained in the frequency band (ω_1, ω_2) is merely the area under $S(\omega)$ between ω_1 and ω_2. For example, if we have the gate function

$$g_T(t) = \begin{cases} 1, & |t| < \dfrac{T}{2} \\ 0, & |t| > \dfrac{T}{2} \end{cases} \tag{5.116}$$

then the energy spectrum is shown in Figure 5.29. The energy in $g_T(t)$ in the frequency band (ω_1, ω_2) is the shaded area.

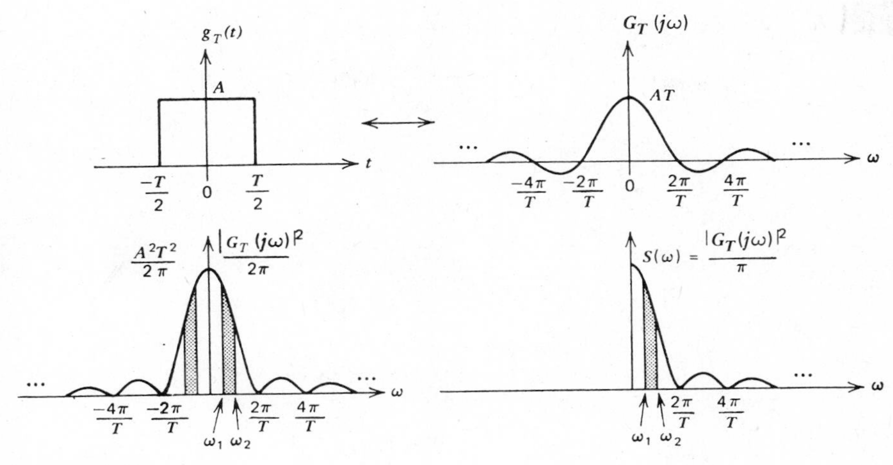

Figure 5.29

5.11 FOURIER TRANSFORMS OF POWER SIGNALS

We have thus far considered only energy signals: that is, functions that possess finite energy over the interval $(-\infty, \infty)$. These energy functions are absolutely integrable and so satisfy sufficient conditions for the existence of $F(\omega)$. However, there are several functions that are very useful and are not absolutely integrable: that is, they do not satisfy the condition

$$\int_{-\infty}^{\infty} |f(t)|\, dt < \infty \tag{5.117}$$

For example, sine waves or step functions do not satisfy (5.117). Many of these functions do, however, possess Fourier transforms if we allow the transforms to include impulse functions and, in some cases, higher-order singularity functions. The mathematical theory and justification for this process is known† but will not be covered here. We shall be content to obtain the transforms of these power signals by limiting processes. These limiting processes are straightforward, but they can be deceiving, because a wrong starting point often leads to an incorrect

† A. Erdélyi, *Operational Calculus and Generalized Functions*, Holt, Rinehart, and Winston, New York, 1962.

result. This class of signals is called power signals because the signal energy is infinite over $(-\infty, \infty)$, but the power is finite: that is,

$$P = \lim_{T \to \infty} \frac{1}{T} \int_{-T/2}^{T/2} f^2(t)\, dt < \infty \tag{5.118}$$

Impulse Function

The Fourier transform of the impulse function $\delta(t)$ is readily obtained by use of the defining relation for $\delta(t)$ given in (3.23):

$$F(j\omega) = \mathscr{F}\{\delta(t)\} = \int_{-\infty}^{\infty} \delta(t) e^{-j\omega t}\, dt = 1 \tag{5.119}$$

Thus we have the pair

$$\delta(t) \leftrightarrow 1 \tag{5.120}$$

shown in Figure 5.30. Using the time-shift theorem we can obtain the transform of the shifted impulse $\delta(t - t_0)$ as

$$\delta(t - t_0) \leftrightarrow e^{-j\omega t_0} \tag{5.121}$$

The transform of $\delta(t - t_0)$ has unit amplitude for all ω and a linear phase as shown in Figure 5.31. Using the symmetry property, we can derive the transform pairs

$$e^{j\omega_0 t} \leftrightarrow 2\pi\, \delta(\omega - \omega_0) \tag{5.122}$$

$$1 \leftrightarrow 2\pi\, \delta(\omega) \tag{5.123}$$

Thus the Fourier transform of the constant one is an impulse at the origin of area equal to 2π, as shown in Figure 5.32.

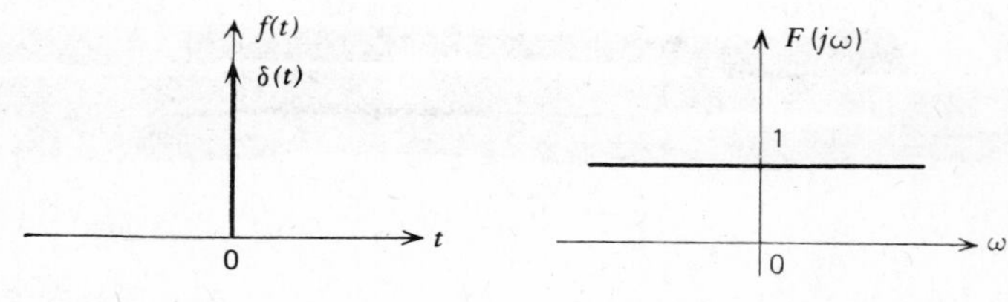

Figure 5.30

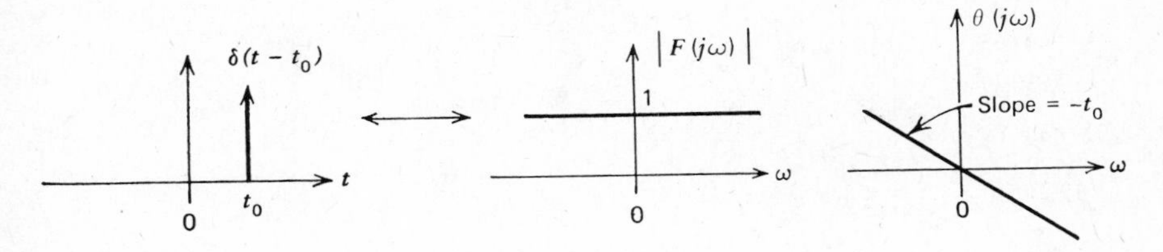

Figure 5.31

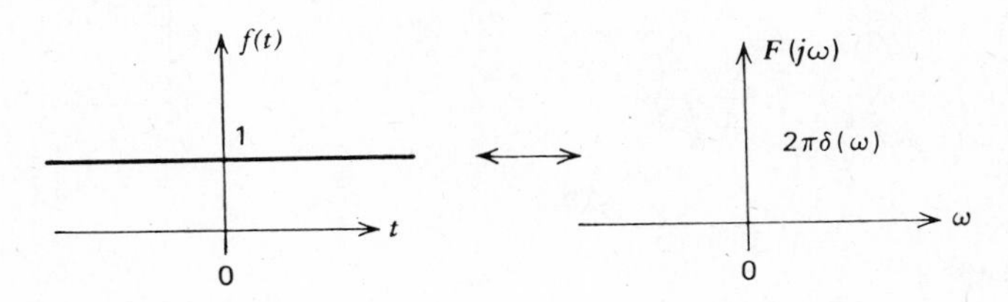

Figure 5.32

Sinusoidal Functions

We can use (5.122) to obtain the transforms of $\cos \omega_0 t$ and $\sin \omega_0 t$:

$$\cos \omega_0 t = \frac{e^{j\omega_0 t} + e^{-j\omega_0 t}}{2} \tag{5.124}$$

Using the transform pair $e^{j\omega_0 t} \leftrightarrow 2\pi\delta(\omega - \omega_0)$, we see that

$$\cos \omega_0 t \leftrightarrow \pi[\delta(\omega - \omega_0) + \delta(\omega + \omega_0)] \tag{5.125}$$

Similarly, we can write for the transform of $\sin \omega_0 t$

$$\sin \omega_0 t = \frac{e^{j\omega_0 t} - e^{-j\omega_0 t}}{2j} \leftrightarrow j\pi[\delta(\omega + \omega_0) - \delta(\omega - \omega_0)] \tag{5.126}$$

These transform pairs are depicted in Figure 5.33. Recall our discussion concerning frequency shifting. The calculation of the shifted spectrum involves a transform of the product $f(t) \cos \omega_0 t$. By the convolution theorem, we can write $\mathscr{F}\{f(t) \cos \omega_0 t\}$ using (5.125) as

$$\mathscr{F}\{f(t) \cos \omega_0 t\} = F(j\omega) * \frac{1}{2}[\delta(\omega + \omega_0) + \delta(\omega - \omega_0)]$$

$$= \frac{F(j\omega + j\omega_0) + F(j\omega - j\omega_0)}{2} \tag{5.127}$$

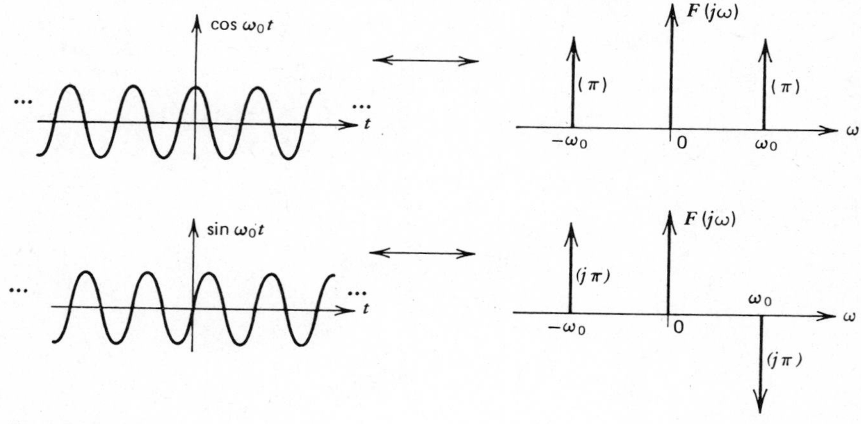

Figure 5.33

The above convolution is, of course, very easy to perform, because convolution with an impulse merely reproduces the function $F(j\omega)$ at the location of the impulse.

The Signum Function

The signum function, denoted by sgn (t), is defined as

$$\text{sgn}(t) = \begin{cases} -1, & t < 0 \\ 0, & t = 0 \\ 1, & t > 0 \end{cases} \tag{5.128}$$

Its graph is shown in Figure 5.34. The transform of the signum function can be

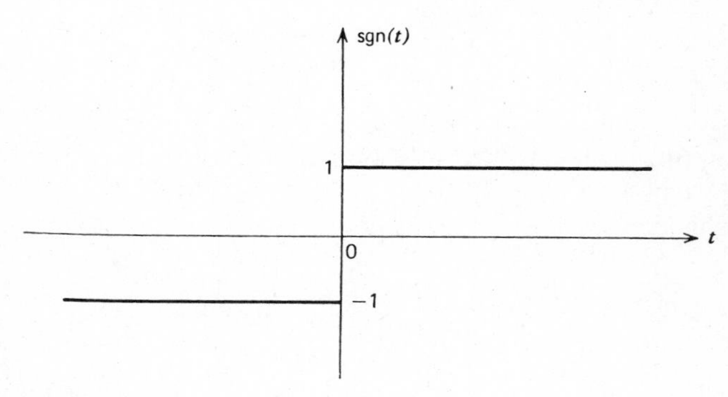

Figure 5.34

found by use of the time differentiation property. Recall that if

$$f(t) \leftrightarrow F(j\omega)$$

then

$$f^{(1)}(t) \leftrightarrow j\omega F(j\omega)$$

Suppose we differentiate the signum function. Its derivative is $2\delta(t)$: that is,

$$\frac{d}{dt}\,\text{sgn}\,(t) = 2\delta(t)$$

The transform of d/dt [sgn (t)] is

$$\frac{d}{dt}\,\text{sgn}\,(t) \leftrightarrow j\omega F(j\omega) \tag{5.129}$$

where

$$j\omega F(j\omega) = \mathscr{F}\{2\delta(t)\} = 2$$

Thus the transform of sgn (t) is

$$F(j\omega) = \frac{2}{j\omega}$$

and so we have the transform pair

$$\text{sgn}\,(t) \leftrightarrow \frac{2}{j\omega} \tag{5.130}$$

A sketch of the spectrum for sgn (t) is shown in Figure 5.35.

The Unit Step Function

The unit step function can be written in terms of the signum function as

$$\xi(t) = \frac{1}{2} + \frac{1}{2}\,\text{sgn}\,(t) \tag{5.131}$$

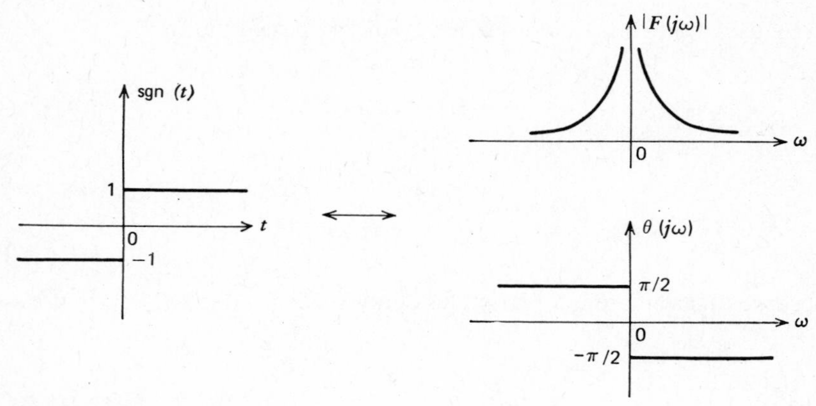

Figure 5.35

Thus, the transform of $\xi(t)$ is

$$\mathscr{F}\{\xi(t)\} = \mathscr{F}\left\{\frac{1}{2}\right\} + \frac{1}{2}\mathscr{F}\{\text{sgn}\,(t)\} = \pi\delta(\omega) + \frac{1}{j\omega} \tag{5.132}$$

In our discussion of the time integration property, we stated, in general, that the transform of integral of $f(t)$ includes an impulse in its transform: that is,

$$\int_{-\infty}^{t} f(t')\,dt' \leftrightarrow \frac{F(j\omega)}{j\omega} + \pi F(0)\delta(\omega) \tag{5.133}$$

We can use (5.132) to show the above. Let

$$g(t) = \int_{-\infty}^{t} f(t')\,dt'$$

We can write $g(t)$ as a convolution of $f(t)$ and $\xi(t)$: that is,

$$g(t) = f(t) * \xi(t) = \int_{-\infty}^{\infty} f(t')\xi(t - t')\,dt'$$

$$= \int_{-\infty}^{t} f(t')\,dt' \tag{5.134}$$

Using the convolution theorem for time functions, we see that

$$\begin{aligned} G(j\omega) &= \mathscr{F}\{f(t) * \xi(t)\} = \mathscr{F}\{f(t)\} \cdot \mathscr{F}\{\xi(t)\} \\ &= F(j\omega) \cdot \left(\pi\delta(\omega) + \frac{1}{j\omega}\right) \\ &= \pi F(j\omega)\delta(\omega) + \frac{F(j\omega)}{j\omega} \\ &= \pi F(0)\delta(\omega) + \frac{F(j\omega)}{j\omega} \end{aligned} \tag{5.135}$$

Thus we have the transform pair of (5.133).

Example 5.13 Once a few transform pairs are known, other pairs can be quickly derived through use of the properties of the Fourier transform. For example, consider the transform of the function $f(t)$ given by

$$f(t) = \cos \omega_0 t \xi(t) \tag{5.136}$$

The frequency convolution theorem can be used to obtain $F(j\omega)$ as

$$\begin{aligned} F(j\omega) &= \mathscr{F}\{\cos \omega_0 t \cdot \xi(t)\} = \frac{1}{2\pi} \mathscr{F}\{\cos \omega_0 t\} * \mathscr{F}\{\xi(t)\} \\ &= \frac{1}{2\pi} [\pi\delta(\omega - \omega_0) + \pi\delta(\omega + \omega_0)] * \left[\pi\delta(\omega) + \frac{1}{j\omega}\right] \\ &= \frac{\pi}{2} [\delta(\omega - \omega_0) + \delta(\omega + \omega_0)] + \frac{1}{2}\left[\frac{1}{j(\omega - \omega_0)} + \frac{1}{j(\omega + \omega_0)}\right] \\ &= \frac{\pi}{2} [\delta(\omega - \omega_0) + \delta(\omega + \omega_0)] + \frac{j\omega}{\omega_0^2 - \omega^2} \end{aligned} \tag{5.137}$$

$|F(j\omega)|$ is shown in Figure 5.36.

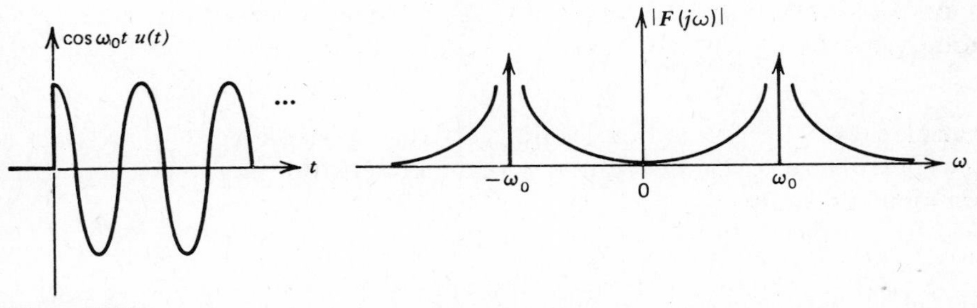

Figure 5.36 ■

Periodic Functions

Periodic functions can, of course, be represented as a sum of complex exponentials; thus, because we can transform complex exponentials by means of (5.122), we should be able to represent a periodic function using the Fourier integral. Assume that $f(t)$ is periodic of period T. We can express $f(t)$ in terms of a Fourier series as

$$f(t) = \sum_{n=-\infty}^{\infty} F_n e^{jn\omega_0 t} \tag{5.138}$$

The Fourier transform of $f(t)$ is therefore

$$\begin{aligned} F(j\omega) &= \mathscr{F}\{f(t)\} = \mathscr{F}\left\{\sum_{n=-\infty}^{\infty} F_n e^{jn\omega_0 t}\right\} \\ &= \sum_{n=-\infty}^{\infty} F_n \mathscr{F}\{e^{jn\omega_0 t}\} \\ &= 2\pi \sum_{n=-\infty}^{\infty} F_n \delta(\omega - n\omega_0) \end{aligned} \tag{5.139}$$

where the F_n's are the Fourier coefficients associated with $f(t)$ and are given by

$$F_n = \frac{1}{T}\int_{-T/2}^{T/2} f(t)e^{-jn\omega_0 t}\, dt \tag{5.140}$$

Equation 5.139 states that the Fourier transform of a periodic function consists of impulses located at the harmonic frequencies of $f(t)$. The area associated with each impulse is equal to 2π times the Fourier coefficient obtained via the exponential Fourier series. Equation 5.139 is really nothing more than an alternate representation of the information contained in the exponential Fourier series. This result should not really be a surprise to anyone familiar with the Fourier series representation.

Example 5.14 Find the Fourier transform of a periodic on-off pulse train as shown in Figure 5.37. The Fourier series representation for this function has been found previously as

$$f(t) = \sum_{n=-\infty}^{\infty} F_n e^{jn\omega_0 t}$$

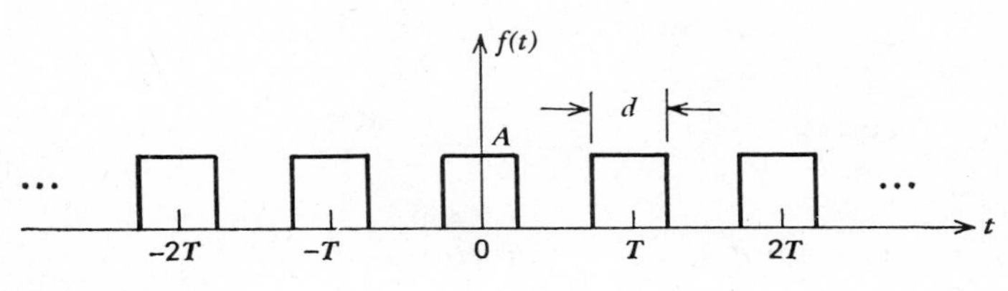

Figure 5.37

where the Fourier coefficients are

$$F_n = \frac{Ad}{T} \operatorname{sinc}\left(\frac{n\pi d}{T}\right)$$

The Fourier transform of $f(t)$ is therefore

$$\mathscr{F}\{f(t)\} = \frac{2\pi Ad}{T} \sum_{n=-\infty}^{\infty} \operatorname{sinc}\left(\frac{n\pi d}{T}\right)\delta(\omega - n\omega_0)$$

where $\omega_0 = 2\pi/T$. The transform of $f(t)$ consists of impulses located at $\omega = 0$, $\pm\omega_0, \pm 2\omega_0, \ldots$. Each impulse has an area associated with it of value $(2\pi Ad/T)$ sinc $(n\pi d/T)$, where n is the number of the harmonic. ■

Example 5.15 A special kind of periodic function is the unit impulse train shown in Figure 5.38. This function is useful in applications involving sampling of time waveforms. Because $f(t) = \sum_{k=-\infty}^{\infty} \delta(t - kT)$ is a periodic function, we can expand $f(t)$ in a Fourier series as

$$f(t) = \sum_{n=-\infty}^{\infty} F_n e^{jn\omega_0 t}$$

where

$$\begin{aligned} F_n &= \frac{1}{T} \int_{-T/2}^{T/2} f(t) e^{-jn\omega_0 t}\, dt = \frac{1}{T} \int_{-T/2}^{T/2} \delta(t) e^{-jn\omega_0 t}\, dt \\ &= \frac{1}{T} \end{aligned} \tag{5.141}$$

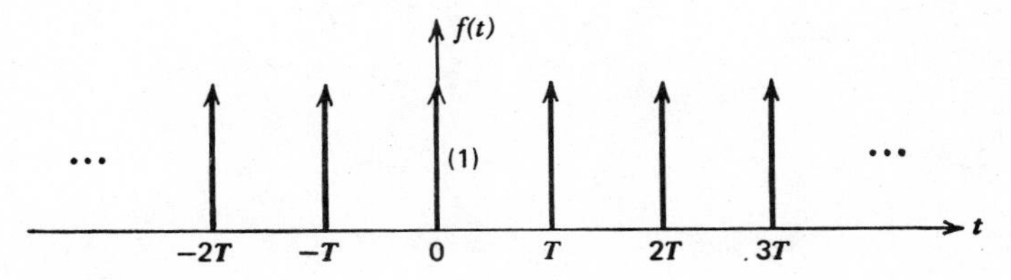

Figure 5.38

Each of the F_n's is the same constant, $1/T$. Thus, the Fourier series representation of the unit impulse train is

$$f(t) = \frac{1}{T} \sum_{n=-\infty}^{\infty} e^{jn\omega_0 t} \tag{5.142}$$

If we now take the Fourier transform on both sides of (5.142), we obtain

$$\begin{aligned} F(j\omega) &= \mathscr{F}\left\{\frac{1}{T} \sum_{n=-\infty}^{\infty} e^{jn\omega_0 t}\right\} \\ &= \frac{2\pi}{T} \sum_{n=-\infty}^{\infty} \delta(\omega - n\omega_0), \qquad \omega_0 = \frac{2\pi}{T} \end{aligned}$$

Table 5.5

Time Functions, $f(t)$	Fourier Transform, $F(j\omega)$
1. $k\delta(t)$	k
2. k	$2\pi k\delta(\omega)$
3. $\xi(t)$	$\pi\delta(\omega) + \dfrac{1}{j\omega}$
4. $\operatorname{sgn}(t)$	$2/j\omega$
5. $\cos \omega_0 t$	$\pi[\delta(\omega - \omega_0) + \delta(\omega + \omega_0)]$
6. $\sin \omega_0 t$	$j\pi[\delta(\omega + \omega_0) - \delta(\omega - \omega_0)]$
7. $e^{j\omega_0 t}$	$2\pi\delta(\omega - \omega_0)$
8. $t\,\xi(t)$	$j\pi\delta^{(1)}(\omega) - \dfrac{1}{\omega^2}$
9. $\sum_{k=-\infty}^{\infty} \delta(t - kT)$	$\omega_0 \sum_{n=-\infty}^{\infty} \delta(\omega - n\omega_0), \quad \omega_0 = \dfrac{2\pi}{T}$
10. $\sum_{n=-\infty}^{\infty} F_n e^{jn\omega_0 t}$	$2\pi \sum_{n=-\infty}^{\infty} F_n \delta(\omega - n\omega_0)$
11. $\dfrac{d^n\delta(t)}{dt^n}$	$(j\omega)^n$
12. $\lvert t \rvert$	$\dfrac{-2}{\omega^2}$
13. t^n	$2\pi j^n \dfrac{d^n\delta(\omega)}{d\omega^n}$

That is,

$$\sum_{k=-\infty}^{\infty} \delta(t - kT) \leftrightarrow \omega_0 \sum_{n=-\infty}^{\infty} \delta(\omega - n\omega_0) \tag{5.143}$$

A unit impulse train in the time domain has as a transform an impulse train in the frequency domain. The area associated with each impulse in the frequency domain is ω_0 and the impulses are located at the harmonic frequencies $n\omega_0 = n2\pi/T$, $n = 0, \pm 1, \pm 2, \ldots$. ■

Table 5.5 summarizes many of the useful Fourier transforms for power signals.

5.12 SAMPLING OF TIME SIGNALS

One of the useful applications of transform techniques concerns the sampling of time waveforms. Sampling is an important consideration in the transmission of information. The idea involved in sampling is, roughly speaking, the following. If values of a time waveform are taken close enough together, then these values can be used to recover all the intervening values of the waveform with complete accuracy. In other words, there is an interdependence between waveform values taken at neighboring time instants. The condition we need for this kind of behavior is that the function be bandlimited: that is, its Fourier transform is zero except for a finite band of frequencies. One form of the sampling theorem may be stated as

A bandlimited signal that has no spectral components at or above ω_m rad/sec can be uniquely represented by its sampled values spaced at uniform intervals that are not more than $1/2f_m$ seconds apart (where $\omega_m = 2\pi f_m$).

The above statement of the sampling theorem is not the most general statement, because we require uniform spacing of the sampled values. It is this form, however, that is probably most widely used. If the Fourier transform of $f(t)$ is zero beyond a certain frequency $\omega_m = 2\pi f_m$, then the samples of $f(t)$ spaced no further than $T = 1/2f_m$ seconds apart are equivalent to knowing $f(t)$ at all time instants. For example, if $f(t)$ as shown in Figure 5.39 has a transform $F(j\omega)$ as shown, then we need consider only the sampled values $f(nT)$, $n = 0, \pm 1, \pm 2, \ldots$ to have complete knowledge of $f(t)$ for every value of t.

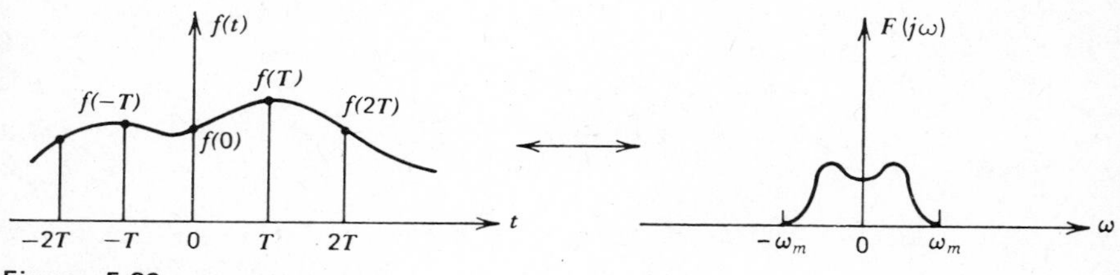

Figure 5.39

We can prove the sampling theorem with the help of the frequency convolution theorem. Consider a sampled version of $f(t)$ and denote it by $f_s(t)$. We can represent the sampled version of $f(t)$ by multiplying $f(t)$ by a unit impulse train with period T, equal to the sampling interval. That is,

$$f_s(t) = f(t) \cdot \sum_{n=-\infty}^{\infty} \delta(t - nT) \tag{5.144}$$

Now let us find the spectrum of $f(t)$. Using the frequency convolution theorem, we know that $F_s(j\omega)$ is the convolution of $F(j\omega)$ and the Fourier transform of the impulse train times $1/2\pi$. In symbols

$$F_s(j\omega) = \frac{F(j\omega) * \omega_0 \sum_{n=-\infty}^{\infty} \delta(\omega - n\omega_0)}{2\pi} \tag{5.145}$$

where $\omega_0 = 2\pi/T$. Thus, substituting for ω_0, we obtain the spectrum of the sampled values as

$$\begin{aligned} F_s(j\omega) &= \frac{1}{T} \sum_{n=-\infty}^{\infty} [F(j\omega) * \delta(\omega - n\omega_0)] \\ &= \frac{1}{T} \sum_{n=-\infty}^{\infty} F(j\omega - jn\omega_0) \end{aligned} \tag{5.146}$$

The right-hand side of (5.146) represents the function $F(j\omega)$ repeating itself every ω_0 rad/sec. If the width of the band of nonzero frequencies in $F(j\omega)$ is less than spacing between repeats of $F(j\omega)$, then the repeated versions of $F(j\omega)$ do not overlap. That is, $F(j\omega)$ will repeat periodically in the frequency domain without overlap, provided that $\omega_0 \geq 2\omega_m$, which implies that

$$\frac{2\pi}{T} \geq 2(2\pi f_m)$$

or

$$T \leq \frac{1}{2f_m} \tag{5.147}$$

As long as we sample $f(t)$ at intervals not more than $1/2f_m$ seconds apart $F_s(j\omega)$ will be a periodic replica of $F(j\omega)$. This result can be also shown graphically as in Figure 5.40. The spectrum of $F_s(j\omega)$ is the convolution of the impulse train $(\omega_0/2\pi)\sum_{n=-\infty}^{+\infty}\delta(\omega - n\omega_0)$ and $F(j\omega)$. Because the impulse train $(1/T)\sum \delta(\omega - n\omega_0)$ is an even function of ω, the mirror image is the same as the original function. To perform the convolution, we shift the impulse train past $F(j\omega)$. The impulses are separated by ω_0. Thus the reproduced versions of $F(j\omega)$ that make up $F_s(j\omega)$ are also spaced ω_0 rad/sec apart. This yields the $F_s(j\omega)$ spectrum shown in Figure 5.40. Notice that the repeated spectrum is multiplied by $1/T$.

We can easily recover $F(j\omega)$ [and so $f(t)$] from $F_s(j\omega)$ by merely filtering out all frequency components above ω_m rad/sec. This can be accomplished by a low-pass filter that allows transmission of all frequencies below ω_m and attenuates all frequencies above ω_m. Such a filter characteristic is shown dotted in Figure 5.40.

The interval $T = 1/2f_m$ is called the *Nyquist interval.* If we sample $f(t)$ with an interval between samples larger than $1/2f_m$ sec, then the repeated versions of $F(j\omega)$ that make up $F_s(j\omega)$ overlap and $F(j\omega)$ cannot be recovered from $F_s(j\omega)$ without some error. This, of course, is what our intuition would indicate. If samples are spaced too far apart, it seems logical that the signal $f(t)$ could not be recovered from its samples.

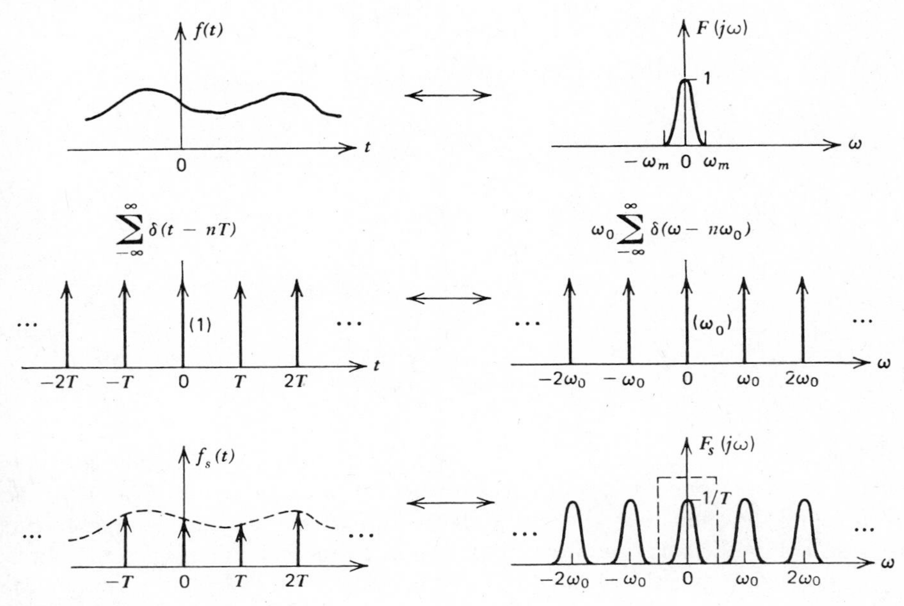

Figure 5.40

The process of recovering $f(t)$ from it sampled version $f_s(t)$ is accomplished by passing $f_s(t)$ through a lowpass filter. Mathematically, we can show this result using the theorem on time convolution. Let the sampling rate be $T = 1/2f_m$ and so $\omega_0 = 2\pi/T = 4\pi f_m = 2\omega_m$. The spectrum of $f_s(t)$ is given by (5.146).

$$F_s(j\omega) = \frac{1}{T} \sum_{n=-\infty}^{\infty} F(\omega - n\omega_0) = \frac{1}{T} \sum_{n=-\infty}^{\infty} F(\omega - 2n\omega_m) \tag{5.148}$$

The process of low-pass filtering is equivalent to multiplying $F_s(j\omega)$ by a function of ω which is 1 for $\omega < |\omega_m|$ and 0 otherwise: that is, a gate function $G_{2\omega_m}(j\omega)$.

$$G_{2\omega_m}(j\omega) = \begin{cases} 1, & |\omega| < \omega_m \\ 0, & |\omega| > \omega_m \end{cases} \tag{5.149}$$

Thus

$$\frac{F(j\omega)}{T} = F_s(j\omega) G_{2\omega_m}(j\omega)$$

or

$$F(j\omega) = T F_s(j\omega) G_{2\omega_m}(j\omega) \tag{5.150}$$

The time convolution theorem applied to (5.150) yields

$$\begin{aligned} f(t) &= T f_s(t) * \frac{\omega_m}{\pi} \operatorname{sinc} \omega_m t \\ &= f_s(t) * \operatorname{sinc} \omega_m t \\ &= \sum_{n=-\infty}^{\infty} f(nT)\delta(t - nT) * \operatorname{sinc} \omega_m t \\ &= \sum_{n=-\infty}^{\infty} f(nT) \operatorname{sinc} [\omega_m(t - nT)] \end{aligned} \tag{5.151}$$

The function sinc $[\omega_m(t - nT)]$ is often called an *interpolation function*, because it allows one to interpolate between the sampled values $f(nT)$ to find $f(t)$ for all t. Graphically, the result is shown in Figure 5.41. Each sampled value is multiplied by the sinc function centered on the sampled value. These functions are then added to give the original waveform. Notice that (5.151) is nothing more than an orthogonal expansion of $f(t)$ in terms of the basis set of functions

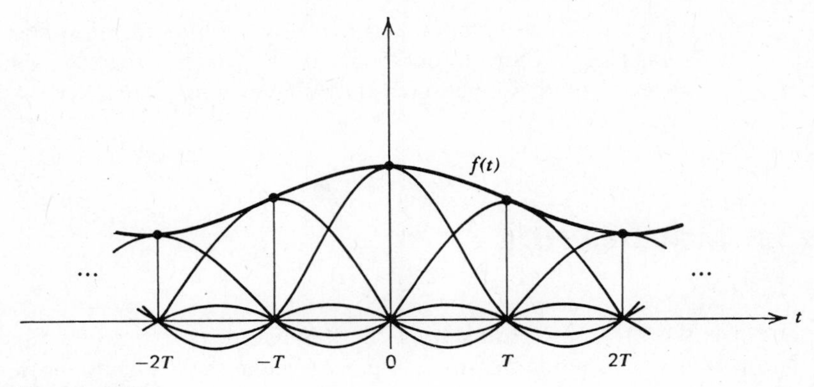

Figure 5.41

{sinc $[\omega_m(t - nT)]$}, $n = 0 \pm 1, \pm 2, \ldots$. In this representation, the generalized Fourier coefficients are merely the sample values of the original function.

The sampling theorem is often used in the representation of a finite-duration signal on the interval $[0, T]$. Sampling at the Nyquist rate implies that one needs $2f_m T$ samples to represent a signal that is band-limited to f_m Hz. This use of the sampling theorem is, strictly speaking, erroneous, because a signal cannot be simultaneously both of finite duration and also bandlimited. The discontinuities at each end of the time signal cause the spectrum to exist for all frequencies. One way of showing this fact is as follows. Let $f(t)$ be of finite time duration, for example $f(t) = 0$ for $|t| > T/2$. We can write $f(t)$ as

$$f(t) = g_T(t) f(t) \tag{5.152}$$

where $g_T(t)$ is the gate function

$$g_T(t) = \begin{cases} 1, & |t| < \dfrac{T}{2} \\ 0, & |t| > \dfrac{T}{2} \end{cases} \tag{5.153}$$

The spectrum of $f(t)$ is therefore

$$F(j\omega) = \frac{T}{2\pi} \operatorname{sinc}\left(\frac{\omega T}{2}\right) * F(j\omega) \tag{5.154}$$

No matter what we assume for the spectrum of $F(j\omega)$ on the right-hand side, $F(j\omega)$ on the left-hand side of (5.153) exists for all ω because of the convolution with sinc $(\omega T/2)$. Thus, we conclude that if $f(t)$ is of finite duration, its spectrum exists for all ω. However, the error in representing a finite-duration signal on $[0, T]$ by $2f_m T$ samples is often minor and the loss of information using the samples is usually more than compensated for by savings in complexity.

5.13 MODULATION

Modulation is the process of translating the frequency spectrum of a function. Modulation is used in communication systems to make the transmission process more efficient. For example, a time signal can be radiated effectively by an antenna only if the radiating antenna is of the order of one tenth or more of the wavelength of the frequencies comprising $f(t)$. If the signal to be transmitted is human speech, for instance, the lower limit on transmitted signal frequency is about 300 Hz. This frequency corresponds to a wavelength of 900,000 m. An efficient radiating antenna for human-speech signals would thus be on the order of 90,000 m. If, however, we modulate or shift the frequency spectrum to a higher band of frequencies, we can reduce the antenna size accordingly. In actual practice, all radio and television signals are modulated for efficient transmission.

The understanding of the basic principles in modulation follows directly from the frequency convolution theorem. Suppose $f(t)$ is a signal with spectrum $F(j\omega)$ centered about $\omega = 0$. If we want to shift the spectrum to a frequency ω_c, we need only to multiply $f(t)$ by $\cos \omega_c t$ That is, if

$$f(t) \leftrightarrow F(j\omega)$$

then

$$f(t) \cos \omega_c t \leftrightarrow \frac{F(j\omega + j\omega_c) + F(j\omega - j\omega_c)}{2} \tag{5.155}$$

The signal $f(t)$ is called the *modulating signal.* The function $\cos \omega_c t$ is called the *carrier* or the *modulated signal.* Because multiplying $\cos \omega_c t$ by $f(t)$ varies the amplitude of the carrier signal, this modulation process is called *amplitude modulation.* The process of recovering $f(t)$ from the translated version is called *demodulation* or *detection.* Demodulation amounts to the inverse operation of modulation: that is, a shift down in frequency rather than up.

Schematically, this process is shown in Figure 5.42. The spectrum of $f(t)$ is shown. The modulated signal has a spectrum given by (5.155). To demodulate, we multiply $f(t) \cos \omega_c t$ by $\cos \omega_c t$ and then lowpass filter. Now the spectrum

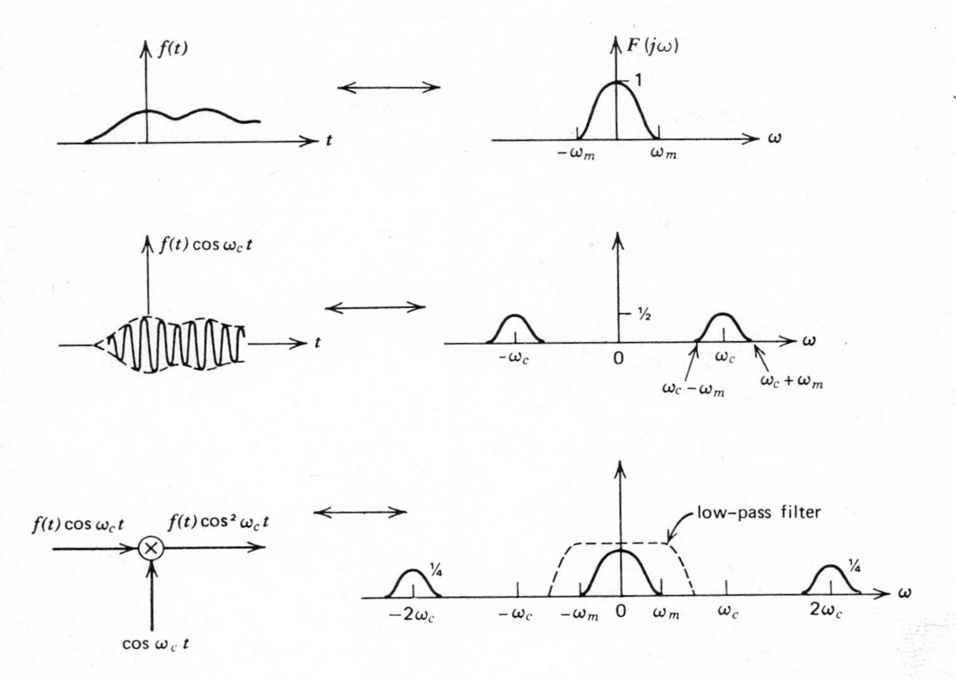

Figure 5.42

of $f(t) \cos \omega_c t \cdot \cos \omega_c t$ is the modulated spectrum convolved with the spectrum of $\cos \omega_c t$: that is,

$$\mathscr{F}\{f(t) \cos \omega_c t \cdot \cos \omega_c t\} = \frac{F(j\omega + j\omega_c) + F(j\omega - j\omega_c)}{2}$$

$$* \frac{\delta(\omega + \omega_c) + \delta(\omega - \omega_c)}{2}$$

$$= \frac{1}{2} F(j\omega) + \frac{1}{4} F(j\omega + j2\omega_c) + \frac{1}{4} F(j\omega - 2j\omega_c) \quad (5.156)$$

The low-pass filter attenuates the frequency components centered about $\pm 2\omega_c$ and thus permits recovery of the original spectrum. The process of multiplying by $\cos \omega_c t$ at the receiver to recover the original spectrum is known as *coherent* or *synchronous* demodulation. This demodulation process requires that one has a method of generating $\cos \omega_c t$ exactly. If the frequency and phase of the cosine function at the receiver are not perfect then one cannot recover the original signal without some error. Since it is expensive to build very stable oscillators for generating

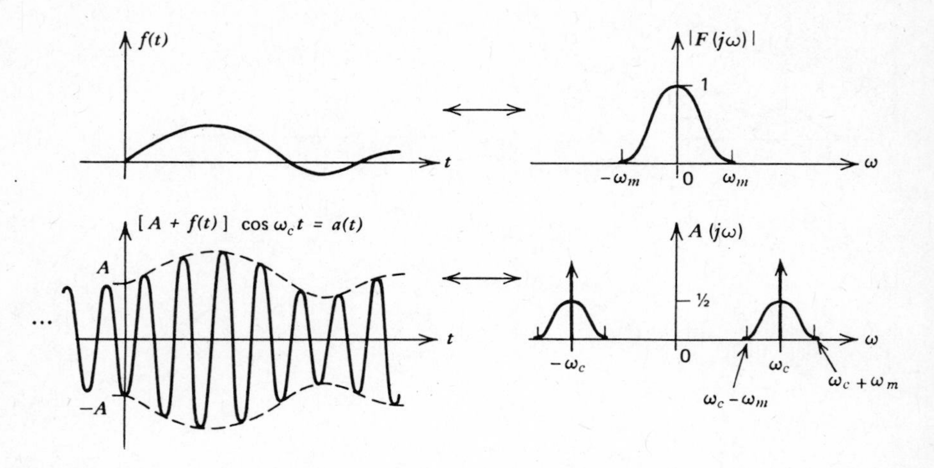

Figure 5.43

the receiver carrier, we often use other forms of demodulation that do not depend upon the generation of a local oscillator signal.

One method of circumventing the generation of a local oscillator is to transmit a large amount of a carrier signal. That is, we transmit $[A + f(t)]\cos \omega_c t$ instead of merely $f(t) \cos \omega_c t$. Here $A > f(t)|_{\max}$. The spectrum of a typical time function might appear as shown in Figure 5.43. The amplitude-modulated signal and its spectrum are also shown. The impulses in the spectrum represent the carrier $A \cos \omega_c t$. The usual method of demodulating $a(t)$ to recover $f(t)$ is to use an envelope detector, as shown in Figure 5.44. The envelope detector has an output that follows the envelope of the modulated signal. On a peak voltage cycle, the capacitor charges to the peak of the modulated signal. As the modulated signal drops, the diode is cut off and the capacitor discharges through the resistance R at a rate depending on the RC time constant. During the next cycle, the diode starts to conduct when the input voltage becomes greater than the capacitor voltage. The capacitor charges to the peak voltage of this new cycle and then discharges when the diode is cut off. During the cutoff period, the capacitor voltage

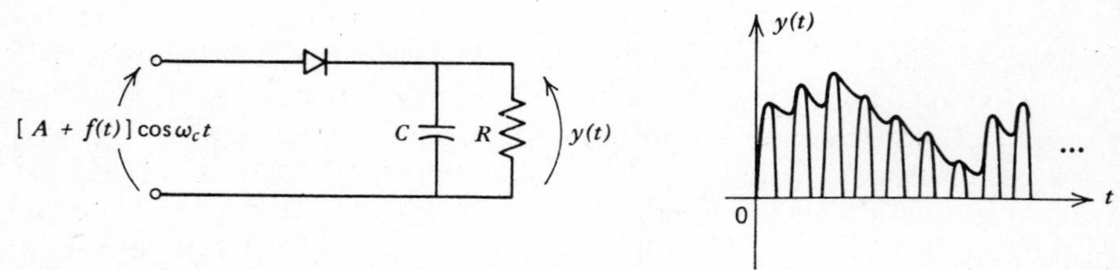

Figure 5.44

does not change appreciably. The RC time constant is adjusted to follow the envelope of the modulated signal. The output voltage $y(t)$ contains a ripple frequency ω_c that can be smoothed by another lowpass filter.

5.14 TRANSMISSION OF SIGNALS THROUGH LINEAR FILTERS

A system can be thought of as a method of processing an input signal $x(t)$ to form an output signal $y(t)$. The processing, of course, depends on the characteristic of the system, which we can characterize by $h(t)$ or $H(\omega)$, the impulse response and its Fourier transform, respectively. The output $y(t)$ for any input $x(t)$ can be found by convolution: that is,

$$y(t) = x(t) * h(t) \tag{5.157}$$

In the transform domain, the above is

$$Y(j\omega) = X(j\omega)H(j\omega) \tag{5.158}$$

The system thus modifies or filters the spectrum of the input. In other words, the transfer function $H(j\omega)$ changes the relative importance of the frequencies contained in the input signal both in amplitude and in phase. That is, $H(j\omega)$ weights the various frequency components of $x(t)$. This filtering pocess is most easily interpreted in the frequency domain.

For example, consider the lowpass filter approximation provided by the RC network of Figure 5.45. The impulse response of this system is (see Example 3.11)

$$h(t) = e^{-t}\xi(t) \tag{5.159}$$

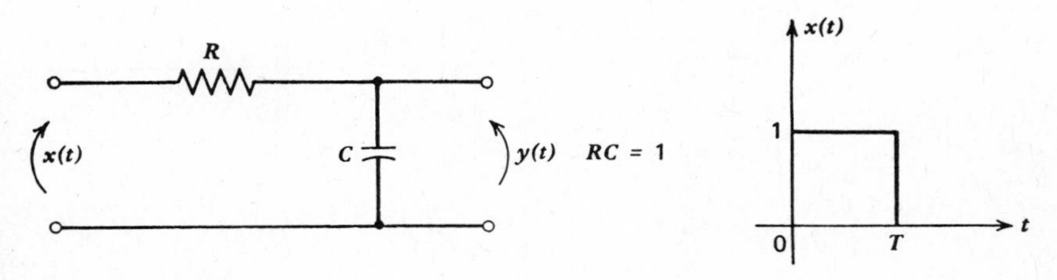

Figure 5.45

Thus the transfer function $H(j\omega)$ is

$$H(j\omega) = \mathscr{F}\{h(t)\} = \int_0^\infty e^{-t}e^{-j\omega t}\,dt = \frac{1}{1 + j\omega} \tag{5.160}$$

This system has a filter characteristic shown in Figure 5.46, where both $|H(j\omega)|$ and $\theta(j\omega)$ are plotted. If we excite this system with the pulse input shown in Figure 5.45, we can obtain the output spectrum of $y(t)$ from (5.158). Thus

$$|Y(j\omega)| = |X(j\omega)H(j\omega)| = |X(j\omega)|\,|H(j\omega)| \tag{5.161}$$

and

$$\theta_y(j\omega) = \theta_x(j\omega) + \theta_h(j\omega) \tag{5.162}$$

The input amplitude spectrum is thus filtered or shaped by the form of $|H(j\omega)|$. Likewise, the output phase is obtained by adding the phase characteristic of the system to the phase characteristic of the input. Figure 5.47 depicts the input and output amplitude spectra. The system in this example attenuates the higher-frequency components in $x(t)$ and thus approximates a low-pass filter. The output time function $y(t)$ is a distorted version of the input pulse, as shown in Figure 5.47. The attenuation of the high frequencies contained in $x(t)$ causes the output to have smoother corners, because it is the high frequencies in $x(t)$ that combine to produce its sharp transitions. Because the system attenuates all high-frequency components, the output voltage cannot change as rapidly as the input voltage.

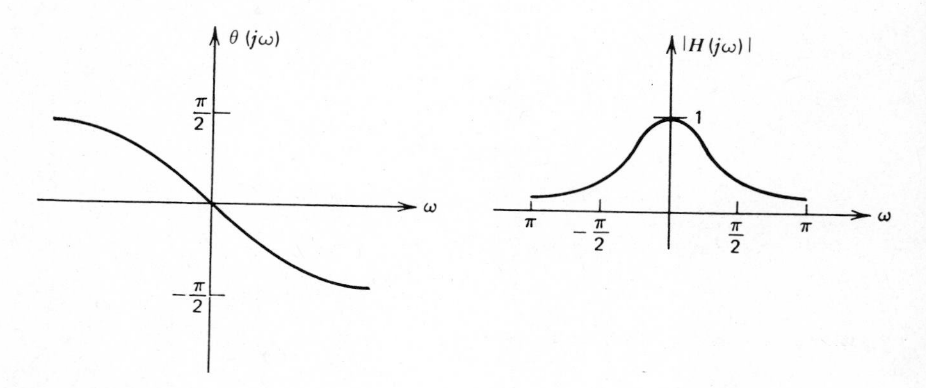

Figure 5.46

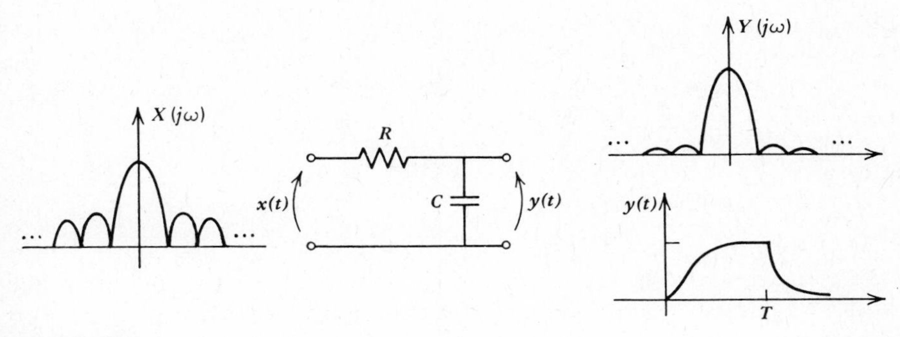

Figure 5.47

A Distortionless Filter

The preceding example was one in which the system distorted the input signal as it passed through the system. In certain applications, this effect is precisely what is desired. However, in other applications, we may desire distortionless transmission. In distortionless transmission, we permit a scale change of the input waveform (a gain change) and a time delay of the input signal, provided that the shape or form of the input signal is unchanged. Thus, if $x(t)$ is the input signal, a distortionless system has as an output $kx(t - \tau)$, where k is the gain change and τ the time delay. We can now deduce the frequency characteristic of a distortionless system. Let the Fourier transform of $x(t)$ be $X(j\omega)$. By the time-shifting theorem,

$$\mathscr{F}\{kx(t - \tau)\} = kX(j\omega)e^{-j\omega\tau} \tag{5.163}$$

we have the relationship between input and output as

$$X(j\omega)H(j\omega) = kX(j\omega)e^{-j\omega\tau} \tag{5.164}$$

As we compare the right-hand and left-hand sides of (5.164), it is clear that

$$H(j\omega) = ke^{-j\omega\tau} \tag{5.165}$$

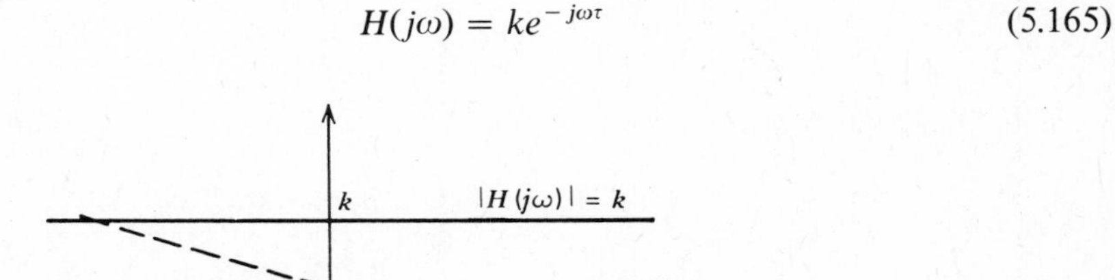

Figure 5.48

The magnitude of the transfer function is $|H(j\omega)| = k$, which is independent of ω. The phase shift of this transfer function is $-\omega\tau$. A plot of $H(j\omega)$ is shown in Figure 5.48. A filter is said to have *amplitude distortion* if $|H(j\omega)|$ is not constant, and *phase distortion* if $\theta(j\omega)$ is not linear through the origin.

Linear Phase Systems

Consider a system that has no phase distortion: that is, the phase shift of the filter is linear. A typical transfer function is shown in Figure 5.49. The analytic form of a linear phase system is

$$H(j\omega) = |H(j\omega)|e^{-j\omega\tau} \tag{5.166}$$

The impulse response of this system is

$$\begin{aligned} h(t) &= \mathcal{F}^{-1}\{|H(j\omega)|e^{-j\omega\tau}\} \\ &= \frac{1}{2\pi}\int_{-\infty}^{\infty} |H(j\omega)|e^{-j\omega\tau}e^{j\omega t}\,d\omega \\ &= \frac{1}{\pi}\int_{0}^{\infty} |H(j\omega)| \cos[\omega(t-\tau)]\,d\omega \end{aligned} \tag{5.167}$$

This impulse response is symmetrical about τ, because

$$h(t+\tau) = h(\tau - t)$$

Also, the maximum value of $h(t)$ is reached at $t = \tau$.

$$h_{\max} = h(\tau) = \frac{1}{\pi}\int_{0}^{\infty} |H(j\omega)|\,d\omega \tag{5.168}$$

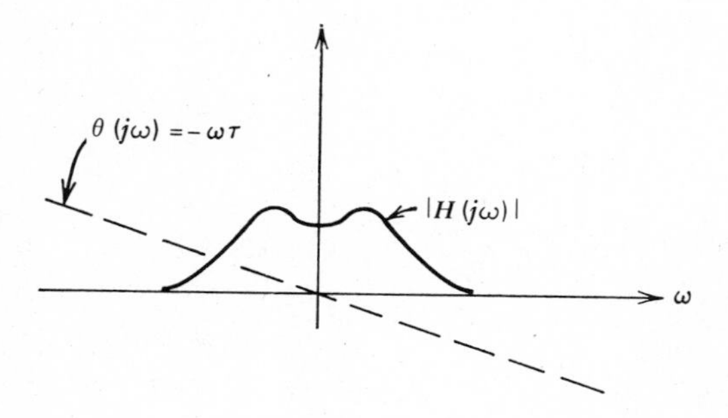

Figure 5.49

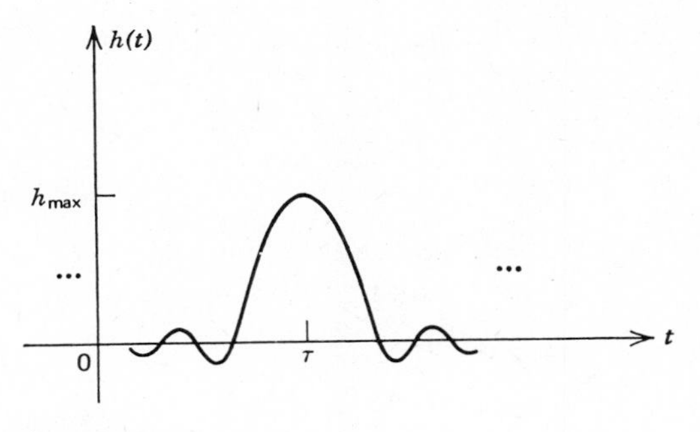

Figure 5.50

Any other value of t can only decrease the value of the integrand in (5.167), because $\cos \omega t$ takes on its maximum value at $t = 0$ for all ω. We conclude from these two properties that $h(t)$ has the general form sketched in Figure 5.50. The impulse response is, in general, nonzero for $t < 0$.

Ideal Filters

An ideal filter is a system that transmits without distortion all the frequencies in a certain band. The amplitude spectrum over the band is a constant and the phase spectrum over the band is linear. Figure 5.51 depicts the filter characteristics for the ideal lowpass and bandpass filters.

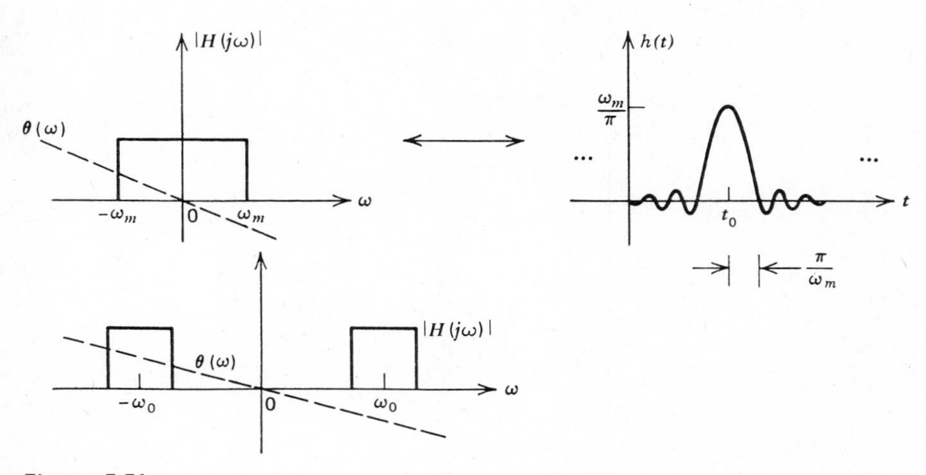

Figure 5.51

The ideal low-pass filter is defined by the transfer $H(j\omega)$ given by

$$H(j\omega) = \begin{cases} e^{-j\omega t_0}, & \omega < |\omega_m| \\ 0, & \omega > |\omega_m| \end{cases} \tag{5.169}$$

The frequency ω_m is often called the cutoff frequency of the filter. We can find the impulse response of this system by taking the inverse Fourier transform of (5.169). Thus

$$\begin{aligned} h(t) &= \mathscr{F}^{-1}\{H(j\omega)\} \\ &= \frac{1}{2\pi}\int_{-\infty}^{\infty} H(j\omega)e^{j\omega t}\,d\omega \\ &= \frac{1}{2\pi}\int_{-\omega_m}^{\omega_m} e^{j\omega(t-t_0)}\,d\omega = \frac{1}{2\pi j(t-t_0)}\,e^{j\omega(t-t_0)}\Big|_{-\omega_m}^{\omega_m} \\ &= \frac{\omega_m}{\pi}\operatorname{sinc}\,[\omega_m(t-t_0)] \end{aligned} \tag{5.170}$$

A sketch of the impulse response is shown in Figure 5.51. We conclude from the impulse-response function that the peak value of the response, ω_m/π, is proportional to the cut-off frequency. The width of the main pulse is $2\pi/\omega_m$. It is convenient to characterize this width as the *effective duration* of the output pulse T_d. Notice that as $\omega_m \to \infty$ the filter becomes an all-pass filter, $T_d \to 0$, and the output response peak $\to \infty$. In other words, the output response approaches the input, an impulse. Again, the impulse response is not causal, of course, because of the ideal characteristic of the amplitude spectrum. If we limit the frequency spectrum of $h(t)$ to be nonzero for only a finite band of frequencies, then $h(t)$ must exist for all time.

The step response $g(t)$ of this ideal lowpass filter can be obtained by integrating the impulse response. Thus,

$$g(t) = \int_{-\infty}^{t} h(t')\,dt' = \frac{1}{\pi}\int_{-\infty}^{t} \frac{\sin[\omega_m(t'-t_0)]}{t'-t_0}\,dt'$$

In the above integral, let $x = \omega_m(t' - t_0)$. Then we have

$$\begin{aligned} g(t) &= \frac{1}{\pi}\int_{-\infty}^{\omega_m(t-t_0)} \frac{\sin x}{x}\,dx \\ &= \frac{1}{\pi}\int_{-\infty}^{0} \frac{\sin x}{x}\,dx + \frac{1}{\pi}\int_{0}^{\omega_m(t-t_0)} \frac{\sin x}{x}\,dx \end{aligned} \tag{5.171}$$

The integral $\int_0^y (\sin x)/x \, dx$ is a tabulated function known as the sine-integral function and denoted by

$$Si(y) \triangleq \int_0^y \text{sinc } x \, dx$$

From the properties of the sinc function, we can deduce the following:

1. $Si(y)$ is an odd function: that is, $Si(y) = -Si(-y)$
2. $Si(0) = 0$
3. $Si(\infty) = \pi/2$, $Si(-\infty) = -\pi/2$

A sketch of $Si(y)$ is shown in Figure 5.52. Using the sine-integral function, the step response can be expressed as

$$g(t) = \frac{1}{2} + \frac{1}{\pi} Si[\omega_m(t - t_0)] \tag{5.172}$$

A sketch of the step response $g(t)$ is shown in Figure 5.53.

We again observed the distortion in the step input as it passes through this system, resulting from the limited passband of the filter. We also note that the response is nonzero before $t = 0$. As $\omega_n \to \infty$, the response $g(t)$ becomes

$$g(t) = \begin{cases} \frac{1}{2} + Si(-\infty) = \frac{1}{2} - \frac{1}{2} = 0, & t < t_0 \\ \frac{1}{2} + Si(\infty) = \frac{1}{2} + \frac{1}{2} = 1, & t > t_0 \end{cases} \tag{5.173}$$

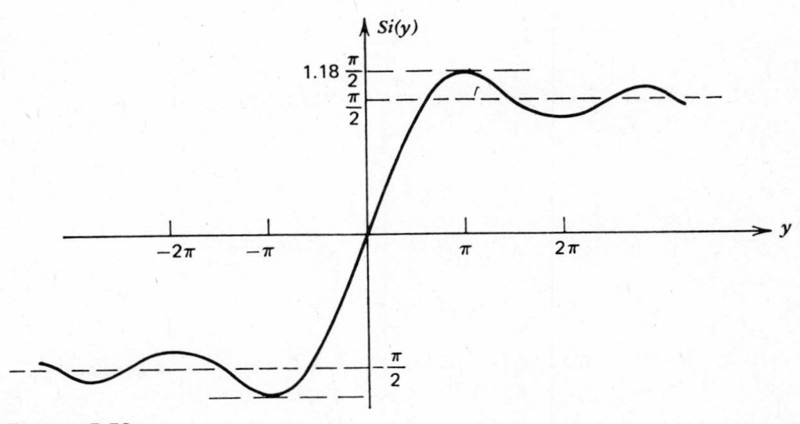

Figure 5.52

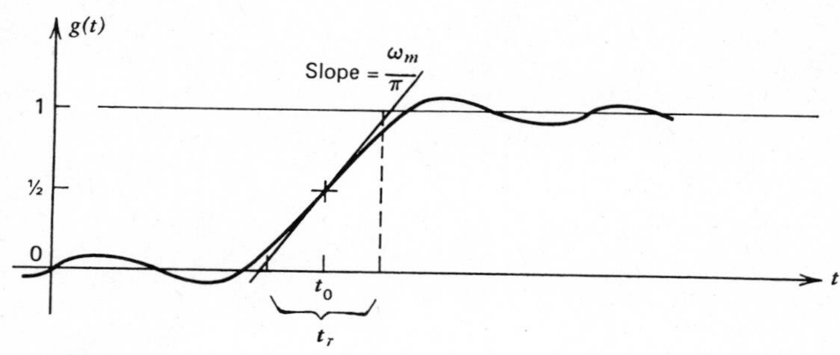

Figure 5.53

that is, $g(t)$ approaches a delayed unit step $\xi(t - t_0)$, as it should. Notice that the abrupt rise of the input step corresponds to the more gradual rise of the output in the region denoted by t_r. The rise time or buildup of the response of $g(t)$ can be related to the cut-off frequency ω_m. Define the rise time t_r as the interval between the intercepts of the tangent at $t = 0$ with the lines $g(t) = 0$ and $g(t) = 1$. From Figure 5.53, we conclude that

$$\left.\frac{dg(t)}{dt}\right|_{t=t_0} = \frac{1}{t_r} = \frac{d}{dt}\left(\left.\frac{1}{\pi}\, Si[\omega_m(t - t_0)]\right|_{t=t_0}\right) = \frac{\omega_m}{\pi}$$

Hence

$$t_r = \frac{\pi}{\omega_m} \tag{5.174}$$

or

$$\omega_m t_r = \pi \tag{5.175}$$

The rise time t_r is inversely proportional to the filter bandwidth. This statement is a paraphrasing of the scaling property. Equation 5.175 indicates that

$$\text{(bandwidth)} \times \text{(rise time)} = \text{constant} \tag{5.176}$$

Bandwidth

Bandwidth is a number we use to measure the extent of the significant band of frequencies in the spectrum of a time function. There are many ways to define the bandwidth of a time function. For example, if $H(j\omega)$ is the transfer function of a

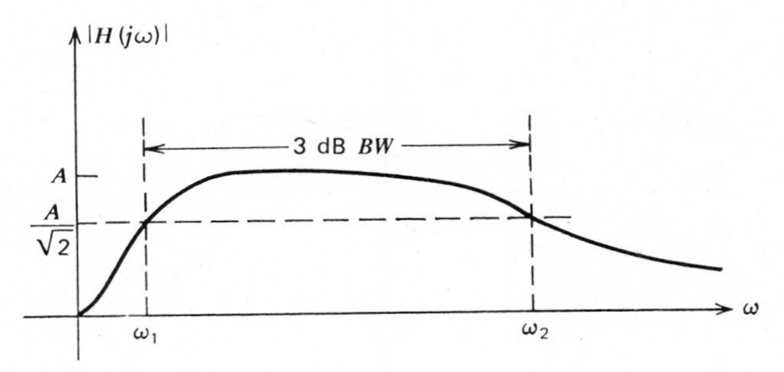

Figure 5.54

system, the bandwidth of the system can be defined as that interval of frequencies for which $|H(j\omega)|$ remains within $1/\sqrt{2}$ (3 dB) of its maximum value. (This assumes that $|H(j\omega)|$ is unimodal.) If the amplitude spectrum for $H(j\omega)$ is as shown in Figure 5.54, the bandwidth by this definition is $(\omega_2 - \omega_1)$ radians/second.

Another common definition of bandwidth is the *equivalent rectangular bandwidth*, denoted by BW_e. In this definition we match the area under the energy-density spectrum $|H(j\omega)|^2$ of the function $h(t)$ to the area of a rectangle with height equal to the maximum value of $|H(j\omega)|^2$ and width $2BW_e$.

$$BW_e = \frac{\int_{-\infty}^{\infty} |H(j\omega)|^2/d\omega}{|H(j\omega)|^2_{\max}} \tag{5.177}$$

This definition is usually applied to lowpass spectra as shown in Figure 5.55.

A similar definition can be used to define the *effective duration*† of a time signal $h(t)$.

$$T_e = \frac{\left(\int_{-\infty}^{\infty} |h(t)|dt\right)^2}{\int_{-\infty}^{\infty} |h(t)|^2 dt}$$

Using this definition of duration we can show (see Problem 5.37) that

$$T_e \cdot BW_e \geq \frac{1}{2}$$

which states that the bandwidth of a pulselike signal is inversely proportional to the effective duration of the signal. This again is a restatement of the time-scaling property of the Fourier transform.

† This definition of effective duration for a time pulse is the ratio of the total area squared or "content" of the function (with area above and below the axis counted as positive) divided by the energy of the function.

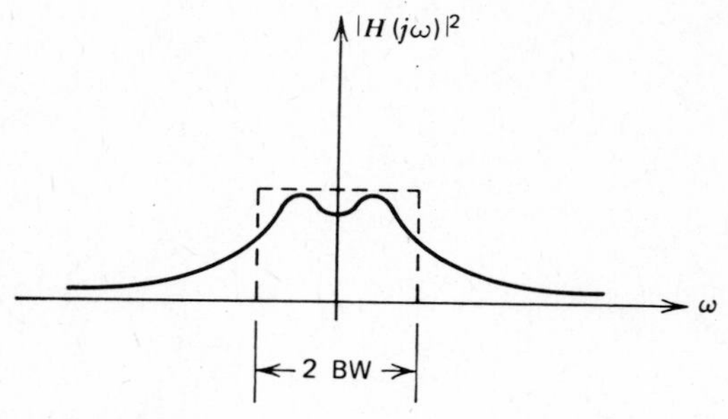

Figure 5.55

Energy Spectra Through Linear Systems

The energy spectrum of a signal $f(t)$ is

$$S(\omega) = \frac{1}{\pi} |F(j\omega)|^2$$

The energy spectrum of the output $y(t)$ of a linear system with input $f(t)$ is therefore

$$\begin{aligned} S_{\text{out}}(\omega) &= \frac{1}{\pi} | Y(j\omega)|^2 \\ &= \frac{1}{\pi} |F(j\omega)H(j\omega)|^2 = \frac{1}{\pi} |F(j\omega)|^2 |H(j\omega)|^2 \\ &= S_{\text{in}}(\omega) |H(j\omega)|^2 \end{aligned} \tag{5.178}$$

where $H(j\omega)$ is the transfer function of the linear system. Notice that the output energy spectrum is independent of the phase characteristics of both the input and the system.

5.15 NUMERICAL CALCULATION OF FOURIER TRANSFORMS—THE DISCRETE-FOURIER TRANSFORM

The numerical calculation of the frequency spectrum of a time waveform $f(t)$ is of great practical importance. For example, if we multiply a signal spectrum by the spectrum of a system's impulse response, we have the output spectrum of the system. If one can numerically calculate Fourier transforms efficiently, this method of computation is useful in the analysis of systems. There are other applications in which the spectrum is more informative than the time waveform. If one has data with certain periodicities contained in the data, it is often easier to find these periodic components using spectral analysis rather than using time domain methods.

The numerical calculation of the Fourier transform of a continuous-time waveform $f(t)$ involves two sampling processes. To calculate the Fourier integral of (5.179)

$$F(j\omega) = \int_{-\infty}^{\infty} f(t)e^{-j\omega t}\,dt \tag{5.179}$$

we must approximate the integral by a sum. This implies that $f(t)$ is represented by a set of samples $f(kT)$. Furthermore, we cannot calculate $F(j\omega)$ for the continuum of all ω. We must represent $F(j\omega)$ by a set of frequency samples $\{F(jm\Omega)\}$. These two sampling processes that are essential for numerical calculation are the basis for considering the Fourier analysis of sequences in which the spectrum is itself a sequence of complex values representing the sampled spectrum.

We have seen that the process of time sampling creates in the frequency domain a periodicity in the spectrum. For example, the spectrum of the sequence f is

$$F(e^{j\theta}) = \sum_{k=-\infty}^{\infty} f_k e^{-jk\theta} \tag{5.180}$$

The spectrum $F(e^{j\theta})$ is always periodic with period 2π. We have seen this same kind of periodicity in our discussion of the sampling theorem. A continuous-time signal $f(t)$ with spectrum $F(j\omega)$, if time sampled, yields a new spectrum $1/T \sum F(j\omega - n(2\pi/T))$. This new spectrum is periodic with period $1/T$ Hertz. The Fourier series representation of a periodic waveform $f(t)$ is precisely the same idea with the two domains interchanged. A sampled spectrum, that is, the Fourier coefficients F_n, correspond to a periodic waveform f in the other domain.

Thus we should anticipate that relating time samples of a signal $f(kt)$ to frequency spectrum samples $\{F(jm\Omega)\}$ causes *both sequences to be periodic*. In this case both domains will consist of periodic sequences. This periodicity creates certain problems but it is impossible to avoid. Any numerical calculation of the spectrum and the corresponding inverse must be in terms of these periodic sequences.

Consider a sequence of N samples $\{f(kt)\}_0^{N-1} = f$. We have previously obtained the spectrum of such a sequence as

$$F(e^{j\theta}) = \sum_{k=0}^{N-1} f_k e^{-jk\theta} \tag{5.181}$$

In terms of the unnormalized frequency variable ω, where $\theta = \omega T$, we have

$$F(e^{j\omega T}) = \sum_{k=0}^{N-1} f_k e^{-jk\omega T} \tag{5.182}$$

The sequence f is aperiodic as originally defined, while the spectrum $F(e^{j\omega t})$ is periodic. If we now sample this spectrum at intervals $\Delta\omega = \Omega$, the resulting spectrum is

$$F(e^{jn\Omega T}) = \sum_{k=0}^{N-1} f_k e^{-jkn\Omega T} \tag{5.183}$$

For notational simplicity we shall denote the sample values $F(e^{jn\Omega T})$ as F_n. We choose the frequency spacing Ω to satisfy

$$\Omega = \frac{2\pi}{NT} \tag{5.184}$$

This particular choice of Ω yields exactly N distinct values of $F(e^{jn\Omega T}) = F_n$. The sample values F_n are also periodic since they resulted from sampling a periodic function. Their period is N.

This choice of Ω permits us to invert the discrete Fourier transform of (5.183) precisely. We can recover the time sequence f from the complex frequency spectrum sequence F using

$$f_k = \frac{1}{N} \sum_{n=0}^{N-1} F_n e^{jkn\Omega T} \tag{5.185}$$

Equation 5.185 can be verified by substituting the expression for F_n from 5.182. This gives us

$$f_k = \frac{1}{N} \sum_{n=0}^{N-1} \sum_{m=0}^{N-1} f_m e^{-jmn(2\pi/NT)\cdot T} e^{jkn(2\pi/NT)\cdot T}$$

$$= \frac{1}{N} \sum_{n=0}^{N-1} \sum_{m=0}^{N-1} f_m e^{jn(2\pi/N)(k-m)}$$

Interchanging the summations gives

$$f_k = \frac{1}{N} \sum_{m=0}^{N-1} f_m \sum_{n=0}^{N-1} e^{jn(2\pi/N)(k-m)} \tag{5.186}$$

Evaluating the inner sum using the formula for the partial sum of a geometric sequence,

$$\sum_{n=0}^{N-1} e^{jn(2\pi/N)(k-m)} = \frac{1 - e^{jN(2\pi/N)(k-m)}}{1 - e^{j(2\pi/N)(k-m)}} = \begin{cases} 0, & k \neq m \\ N, & k = m \end{cases}$$

Thus the right-hand side of (5.186) is zero except for $k = m$. There is, therefore, only one nonzero term that occurs for $m = k$ and has value f_k. Thus we have the discrete Fourier transform (DFT) pair

$$F_n = \sum_{k=0}^{N-1} f_k e^{-jkn(2\pi/N)} = \sum_{k=0}^{N-1} f_k W^{-kn} \tag{5.187}$$

$$f_k = \frac{1}{N} \sum_{n=0}^{N-1} F_n e^{jkn(2\pi/N)} = \frac{1}{N} \sum_{n=0}^{N-1} F_n W^{kn} \tag{5.188}$$

where $W = e^{j2\pi/N}$.

This transform pair is an *exact and faithful relationship.* There is no error in going from sequence f to sequence F and visa versa. There is still a question as to how well these two sequences approximate the continuous Fourier transform pair in (5.187) and (5.188). The sequence F is periodic of period N as we have previously explained and as one can easily verify. The process of sampling the continuous transform $F(e^{j\omega T})$ in ω causes the sequence f also to be periodic of period N *whether or not the original sequence f is periodic.* This is easily verified by calculating f_{k+N} as shown below.

$$f_{k+N} = \frac{1}{N} \sum_{n=0}^{N-1} F_n e^{j(k+N)n(2\pi/N)}$$

$$= \frac{1}{N} \sum_{n=0}^{N-1} F_n e^{jkn(2\pi/N)} e^{jN(2\pi/N)} = f_k$$

In using the DFT to approximate the Fourier transform $F(j\omega)$ of an aperiodic time function f, we shall always have to approximate the aperiodic spectrum and time function by periodic sequences representing samples of the spectrum and the time function.

If, as shown in Figure 5.56, $f(t)$ is represented by N samples spaced T seconds apart, the total duration is $R = NT$ seconds. The corresponding continuous spectrum is

$$F(j\omega) = \frac{1}{T} \sum_{k=-\infty}^{\infty} F\left(j\omega - jk\frac{2\pi}{T}\right) \tag{5.189}$$

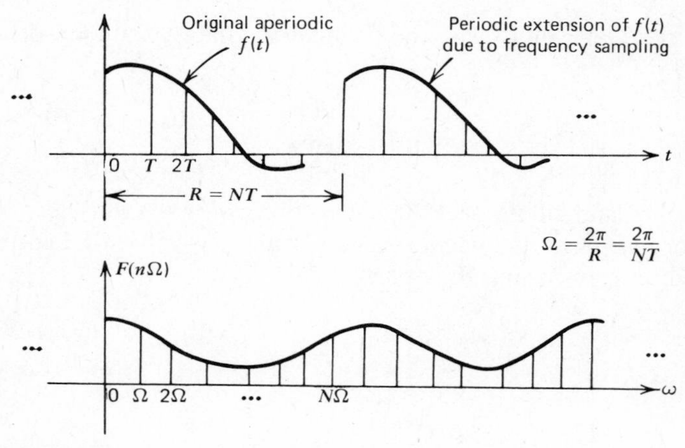

Figure 5.56

as we have previously derived in our discussion on sampling in Section 5.12. If we now sample $F(j\omega)N$ times with a spacing between samples of $\Omega = 2\pi/R = 2\pi/NT$, the spectral values are

$$F(jn\Omega) = \frac{1}{T} \sum_{k=-\infty}^{\infty} F\left(jn\Omega - jk\frac{2\pi}{T}\right) \tag{5.190}$$

The corresponding time function is found by convolving, in the time domain, the original sampled waveform with a periodic impulse waveform with time spacing $R = 2\pi/\Omega = NT$. We thus obtain an aliased version of the original time waveform that is periodic of period NT, as shown in Figure 5.56. Thus when we approximate the spectrum of an aperiodic function $f(t)$, we must choose NT to be large enough so that the aliased version $\sum f(t - mNT)$ is a good approximation to $f(t)$. The duration NT also defines the frequency resolution of the spectrum, that is, the spacing between frequency samples.

Example 5.16 To illustrate the use of the *DFT* in the numerical calculation of the Fourier integral, consider the calculation of the spectrum for the pulse function shown in Figure 5.57. The Fourier transform for this $f(t)$ is easily calculated as

$$F(j\omega) = \int_0^1 1 \cdot e^{-j\omega t}\, dt = e^{-j\omega/2} \operatorname{sinc}\left(\frac{\omega}{2}\right) \tag{5.191}$$

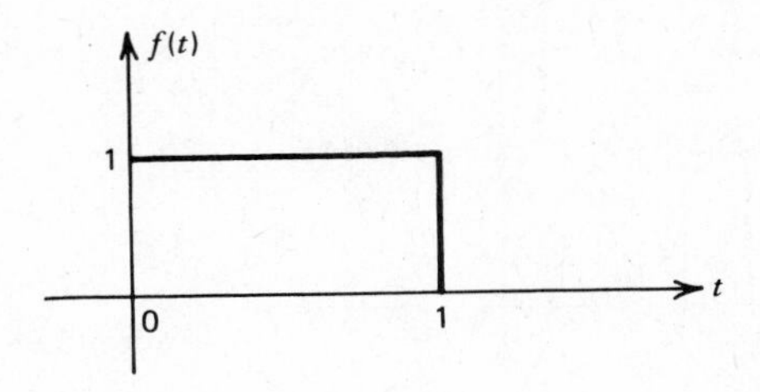

Figure 5.57

To use a *DFT* to approximate this calculation we must first represent $f(t)$ by a sequence of its sampled values. Suppose we choose $N = 10$ sample values, 5 of which fall within the pulse, 5 of which are zero. We can view this approximation as a series of rectangles each with an area of $f(kT) \cdot T$ as shown in Figure 5.58. Thus we approximate $f(t)$ as

$$f(t) \approx \begin{cases} f(0), & 0 \le t < T \\ f(T), & T \le t < 2T \\ \vdots & \vdots \\ f(4T), & 4T \le t < 5T \end{cases} \tag{5.192}$$

Now replace each rectangle by an impulse of the same area to obtain

$$f_s(t) = \sum_{k=0}^{4} T f(kT)\delta(t - kT) \tag{5.193}$$

as we have previously discussed in Section 5.12. The corresponding spectrum is

$$\begin{aligned} \hat{F}(j\omega) &= \mathscr{F}\left\{T \sum_{k=0}^{4} f(kT)\delta(t - kT)\right\} \\ &= T \sum_{k=0}^{4} f(kT)e^{-j\omega kT} \end{aligned}$$

$\hat{F}(j\omega)$ evaluated for $\omega = n\Omega$ is

$$\begin{aligned} \hat{F}(jn\Omega) &= T \sum_{k=0}^{4} f(kT)e^{-jnk\Omega T} \\ &= T \cdot DFT\{f_k\} \end{aligned} \tag{5.194}$$

That is, our sampled approximation to the continuous spectrum is T times the *DFT* of the time samples, $\{f(kT)\}$.

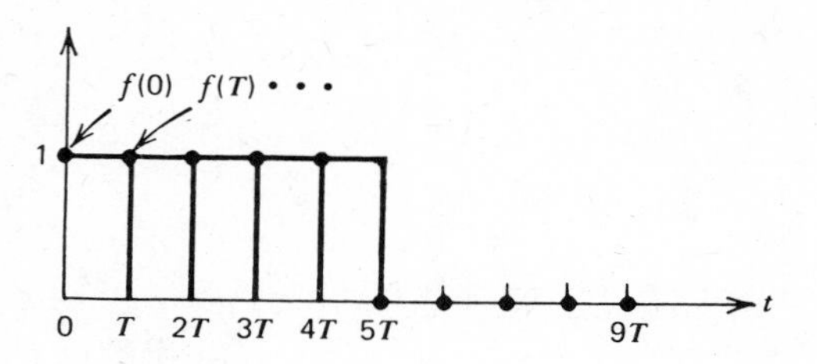

Figure 5.58

For the example here $T = \frac{1}{5}$ and $NT = (10)(\frac{1}{5}) = 2$. The frequency spacing is $\Omega = 2\pi/NT = \pi$ radians. The DFT of the samples f is

$$\frac{1}{T}\hat{F}(jn\Omega) = \sum_{k=0}^{9} f(kT)e^{-jkn\Omega T}$$

$$= \sum_{k=0}^{4} f\left(\frac{k}{5}\right)e^{-jkn\pi/5} \tag{5.195}$$

If we evaluate $(\hat{F}(jn\Omega)/T)$ for $n = 0, 1, 2, \ldots, 9$, we obtain 10 values of the sampled spectrum corresponding to frequency values $\omega = 0, \pi, 2\pi, \ldots, 9\pi$ radians/second. We must multiply by T to rescale each value as in (5.194).

For $n = 1$, corresponding to a frequency π radians/second, we have

$$F_1 = \frac{\hat{F}(j\Omega)}{T} = \sum_{k=0}^{4} f\left(\frac{k}{5}\right)e^{-jk\pi/5}$$

$$= 1.0 + (0.809017 - j0.587785) + (0.309017 - j0.951057)$$

$$+ (-0.309017 - j0.951057) + (-0.809017 - j0.587785)$$

$$= 1.0 - j3.077684$$

Thus $|F_1| = 3.236068$ and $\arg[F_1] = -72.0°$. The approximation to $F(j\omega)$ at $\omega = \pi$ radians/seconds is $|F_1|T$. That is,

$$|F(j\omega)||_{\omega=\pi} \cong |F_1| \cdot T = (3.236068)\left(\frac{1}{5}\right) = 0.647214$$

The actual value of the magnitude spectrum from (5.191) is

$$|F(j\omega)|\Big|_{\omega=\pi} = \left|e^{-j\omega/2}\operatorname{sinc}\left(\frac{\omega}{2}\right)\right|\Bigg|_{\omega=\pi} = \operatorname{sinc}\left(\frac{\pi}{2}\right) \cong 0.636620$$

The error in the DFT approximation of the magnitude spectrum is thus 1.7%. For $n = 0, 1, 2, \ldots, 9$, we obtain DFT approximations of the true spectrum for frequency values $\omega = 0, \pi, 2\pi, 3\pi, \ldots, 9\pi$. For real sequence values f, the DFT values F_n are symmetric with respect to $N/2$. Essentially, N real values of f are equivalent to $N/2$ complex values of F.

To increase the accuracy of the DFT we can decrease T and increase N. Suppose we halve T and double N so that $T = \frac{1}{10}$, $N = 20$. The frequency spacing is unchanged since $NT = 2$ as before. Calculation of the approximate spectrum at π radians/second is again based on F_1, where

$$F_1 = \sum_{k=0}^{19} f\left(\frac{k}{10}\right) e^{-jk2\pi/20} = \sum_{k=0}^{9} 1 \cdot e^{-jk\pi/10}$$

The approximation to $F(j\pi)$ is

$$F(j\omega)|_{\omega=\pi} \cong T \cdot F_1 = \frac{1}{10} \sum_{k=0}^{9} e^{-jk\pi/10} = 0.1 - j0.63138$$

The magnitude approximation is

$$|F(j\omega)||_{\omega=\pi} \cong \sqrt{0.1^2 + 0.63138^2} = 0.639245$$

The error has been reduced to 0.41 %. Decreasing T means that the aliased versions of $F(jn\Omega - jk2\pi/T)$ in (5.190) are separated further in frequency, thus decreasing the error in the approximation.

Table 5.6 and Figure 5.59 compare the DFT approximations for $N = 10$ and $N = 20$ and $\Omega = \pi$ with the true spectrum $e^{-j\omega/2}$ sinc $(\omega/2)$. Notice that increasing N (for the same frequency spacing Ω) increases the number and accuracy of the frequency samples.

If we wish to obtain a different harmonic spacing, we can do so by changing the duration NT. Suppose we choose $N = 12$ and $T = \frac{1}{3}$ so that $NT = 4$. Then the frequency spacing is $\Omega = 2\pi/NT = 2\pi/4 = \pi/2$. The DFT values corresponding to this approximation are shown in Figure 5.59. In this last case the error is larger than previously because the value of T is large, which reduces the spacing between aliased versions of $F(jn\Omega - jk2\pi/T)$ in (5.190).

We can summarize the closeness of the approximation $\hat{F}(jn\Omega)$ to the true sampled spectrum $F(jn\Omega)$ as follows. We have that

$$\hat{F}(jn\Omega) = T \cdot DFT\{f_k\}$$

We wish to determine how well $\hat{F}(jn\Omega)$ approximates the true sampled spectrum $F(j\omega) = F(jn\Omega)$ as a function of T, the spacing between time samples, and N, the

Table 5.6

n	$\omega = n\Omega$	$\lvert F(j\omega)\rvert$	$T \cdot \lvert F_n\rvert$ $N = 10,\ T = \frac{1}{5}$	Percent Error	$T \cdot \lvert F_n\rvert$ $N = 20,\ T = \frac{1}{10}$	Percent Error
0	0	1	1	0	1	0
1	π	0.636620	0.647214	1.7%	0.639245	0.41%
2	2π	0	0	0	0	0
3	3π	0.212206	0.247212	16.5%	0.2202689	3.8%
4	4π	0	0	0	0	0
5	5π	0.12732	0.20	57.1%	0.1414214	11.1%
6	6π	0	0	0	0	0
7	7π	0.0909457	0.247212	[a]—	0.112232	23.4%
8	8π	0	0	0	0	0
9	9π	0.070735	0.647214	—	0.1012465	43.1%
10	10π	0	1	—	0	0
11	11π	0.057874	0.647214	—	0.1012465	[a]—
12	12π	0	0	0	0	0

[a] Foldover frequency of spectrum. See Figure 5.59.

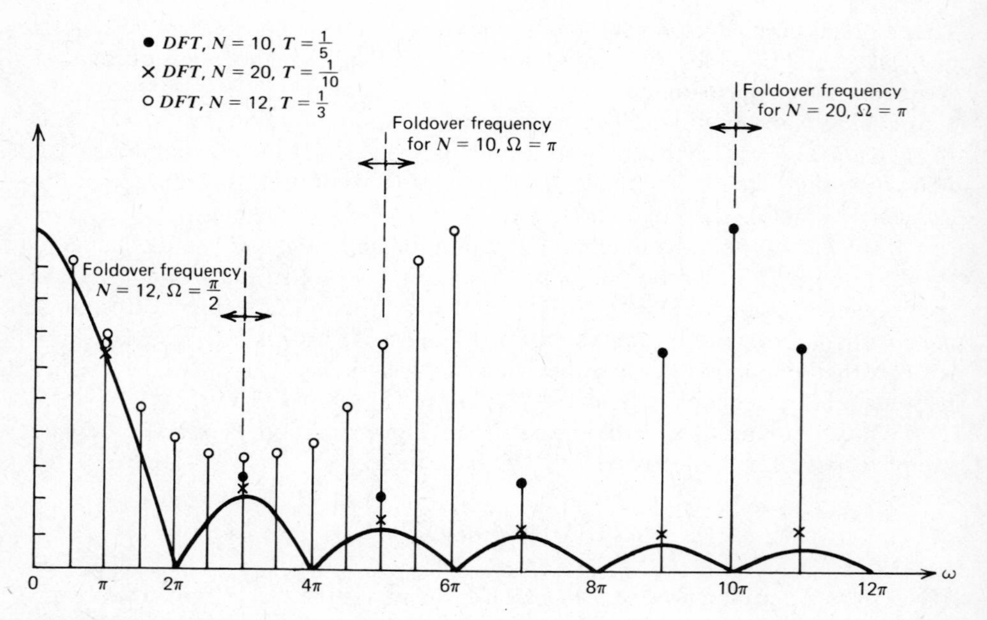

Figure 5.59

total number of samples. Because $f(t)$ is zero except on the interval [0, 1], the number of samples in [0, 1] is related to the spacing T between time samples as

$$T = \frac{1}{L}, \qquad L = \text{number of samples in } [0, 1]$$

The spacing of the frequency harmonics is

$$\Omega = \frac{2\pi}{NT} = \frac{2\pi L}{N}$$

where N is the total number of samples as before. The estimate of the sampled spectrum is thus

$$\begin{aligned}
\hat{F}(jn\Omega) &= T \cdot \sum_{k=0}^{L-1} W^{-nk}, \qquad W = e^{j2\pi/N} \\
&= \frac{1}{L} \sum_{k=0}^{L-1} W^{-nk} = \frac{1}{L} \frac{1 - W^{-nL}}{1 - W^{-n}} \\
&= \frac{1}{L} \frac{W^{-nL/2}}{W^{-n/2}} \cdot \frac{W^{nL/2} - W^{-nL/2}}{W^{n/2} - W^{-n/2}} \\
&= \frac{1}{L} e^{-j(2\pi/N)(nL/2 - n/2)} \frac{\sin(2\pi/N \cdot nL/2)}{\sin(2\pi/N \cdot n/2)} \\
&= \frac{e^{-j(n\pi L/N)} \operatorname{sinc}(n\pi L/N)}{e^{-jn\pi/N} \operatorname{sinc}(n\pi/N)}
\end{aligned}$$

Thus

$$\hat{F}(jn\Omega) = F(jn\Omega)\{F(jn\Omega T)\}^{-1} \tag{5.196}$$

Equation 5.196 states that the approximation to the sampled spectrum $\hat{F}(jn\Omega)$ is the true sampled spectrum $F(jn\Omega)$ divided by $F(jn\Omega T)$. Thus we can easily calculate how well $\hat{F}(jn\Omega)$ approximates $F(jn\Omega)$ for this particular example. For example, for N large we have $n\Omega T = n\pi/N \cong 0$. This implies that $F(jn\Omega T) = 1$ and so $\hat{F}(jn\Omega) = F(jn\Omega)$. At the foldover frequency that occurs for $n = N/2$ we have $n\Omega T = n\pi/2$. This means $F(jn\Omega T) = F(j\pi/2) = e^{-j\pi/2} \operatorname{sinc}(\pi/2)$. The error is thus $\pi/2$ radians in phase and $[\operatorname{sinc}(\pi/2)]^{-1} = \pi/2 \cong 1.571$ in magnitude. The error in magnitude is thus 57.1%. ■

5.16 PROPERTIES OF THE DISCRETE-FOURIER TRANSFORM

The DFT is useful as a numerical method for calculating the Fourier transform of a continuous function. In terms of the time sequence f and the sequence of frequency samples F, the DFT is an exact relationship. Outside of the range $0 \leq n \leq N - 1$, the time and frequency samples sequences are periodic. We can picture each of these N-point sequences arranged uniformly around a circle.

The properties of the DFT are similar to those of the continuous Fourier transform. The periodicity of the sequences, however, introduces certain basic differences.

1. *Linearity*. The DFT is a linear operation. Thus,

$$DFT\{\alpha f + \beta g\} = \alpha F + \beta G$$

where

$$DFT\{f\} = F \qquad \text{and} \qquad DFT\{g\} = G \tag{5.197}$$

2. *Time Shifting*. Let $DFT\{f\} = F$. The DFT of a shifted version of the sequence f is

$$DFT\{f[(n - i) \text{ Mod } N]\} = F(k\Omega)e^{-j\Omega Tki} \tag{5.198}$$

The notation $(n - i)$ Mod N means to divide $(n - i)$ by N and retain the remainder only. This is necessary because a shift of CN, C an integer, leaves the sequence unchanged because f is periodic of period N.

3. *Convolution*. If $DFT(f) = F$ and $DFT(g) = G$ assuming all sequence N long, then

$$IDFT\{FG\} = \sum_{i=0}^{N-1} f[(n - i) \text{ Mod } N]g(i) \tag{5.199}$$

Equation 5.199 states that the product of two DFTs is the *circular convolution* of the corresponding sequences. The easiest way to calculate (5.199) is to periodically extend both sequences and then compute the indicated convolution.

One of the reasons we use the transform domain is to convert aperiodic convolutions into a product of the transforms. Unfortunately, using DFTs we cannot obtain aperiodic convolutions of sequences without some additional thought. One easy method of obtaining aperiodic convolutions using the *DFT* is to redefine the original sequences by adding zeros to the sequence values to introduce a "guard band." By performing a circular convolution with enough zeros added to the original sequences, one can obtain the aperiodic convolution of the sequences (which repeats periodically).

Example 5.17 Suppose we wish to obtain the aperiodic convolution of the two sequences f and g defined below using the *DFT*. These sequences periodically extended are shown in Figure 5.60. If we were to take the *IDFT* of the sequence $F \cdot G$, the result would be the *circular* convolution of f and g. This circular convolution is:

$$y(n) = IDFT\{FG\} = \sum_{i=0}^{3} f[(n-i) \text{ Mod } 3]g(i)$$

$$y(0) = \sum_{i=0}^{3} f[(-i) \text{ Mod } 3]g(i) = 1 \cdot 1 + 4 \cdot 3 + 3 \cdot 2 + 2 \cdot 1 = 21$$

$$y(1) = 2 \cdot 1 + 1 \cdot 3 + 4 \cdot 2 + 3 \cdot 1 = 16$$

$$y(2) = 3 \cdot 1 + 2 \cdot 3 + 1 \cdot 2 + 4 \cdot 1 = 15$$

$$y(3) = 4 \cdot 1 + 3 \cdot 3 + 2 \cdot 2 + 1 \cdot 1 = 18$$

etc.

To obtain the *aperiodic* or *linear* convolution of f and g, we must first append enough zeros to the ends of these sequences so that successive periods cannot overlap in our convolution sum. Suppose in general that f is of length L_1 and g is of

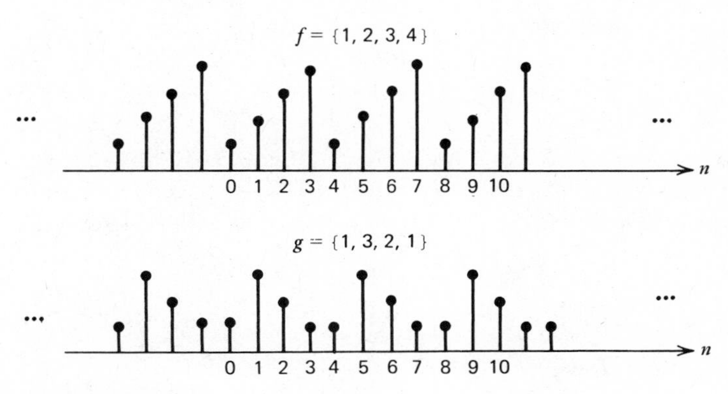

Figure 5.60

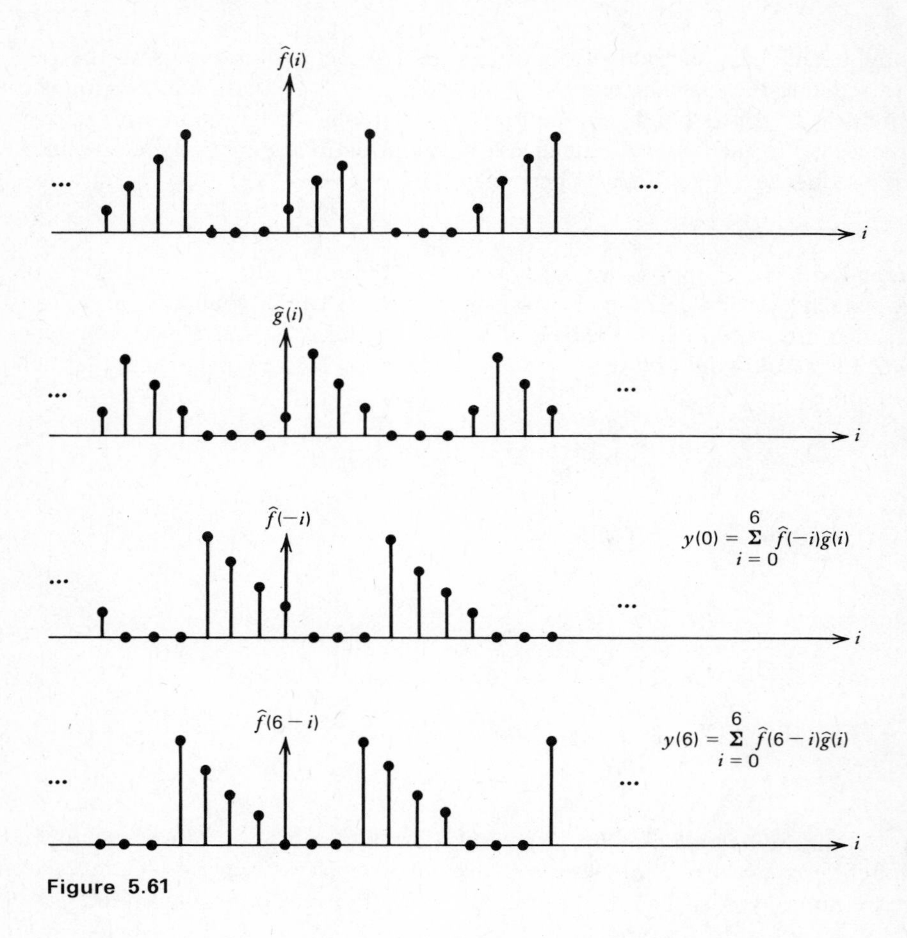

Figure 5.61

length L_2. Then we must add $L_2 - 1$ zeros to f and $L_1 - 1$ zeros to g, yielding the augmented sequences of f and g, both of length $N = L_1 + L_2 - 1$. Now the circular convolution of f and g will yield the linear convolution of f and g as desired.

In our example $L_1 = L_2 = 4$ and $N = 7$. Figure 5.61 shows f_1, g_1 and the time-reversed and shifted sequences $f(-n)$ and $f(6 - n)$ needed to compute $y(0)$ and $y(6)$, respectively. The sequence $y(0) \cdots y(6)$ is exactly the 7-length linear convolution we desire:

$$
\begin{aligned}
y &= f * g \\
&= IDFT\{\hat{F}\hat{G}\} = IDFT\{DFT(\hat{f}) \cdot DFT(\hat{g})\}
\end{aligned}
$$

5.17 THE FAST FOURIER TRANSFORM (FFT)

The fast Fourier transform is an algorithm used to compare the *DFT* and its inverse. By using the *FFT* one can reduce the number of computations dramatically. The discovery of the *FFT* has made the digital computation of frequency spectra a practical reality. By choosing the length of the data sequence N to be a power of 2, it is possible to reduce the number of multiplications in the calculation of the *DFT* from N^2 to approximately $N \log_2 N$. For example, a 4096-length sequence, $N = 2^{12}$, requires 16,777,216 multiplications using the *DFT* directly. Using the *FFT* algorithm this can be reduced by a factor of 341 to 49,152 multiplications.

Any algorithm for computing F from the sequence values f as in (5.187) can be used to compute f from the sequence F. Referring to (5.188) it is necessary only to divide by $1/N$ and change the sign of the exponent in the summation. See Appendix D for an *FFT* algorithm written in FORTRAN.

5.18 SUMMARY

Fourier analysis is an analysis of linear systems based on the representation of sequences and continuous-time functions in terms of sinusoids or harmonics. It is a transform domain analysis that engineers find very useful because many of the properties and results of the transform have physical interpretation in both the time domain and the transform domain. Unless transient effects of a linear system are needed explicitly, Fourier methods are generally preferred to either Laplace or Z transforms (depending on whether the system is continuous or discrete-time).

We have considered in detail four methods of analysis as summarized in Table 5.1. The physical interpretation of the transform is essentially identical for all four cases. The mathematical details, however, depend heavily on the characteristics of the time-domain description. Transform methods provide a unifying mathematical approach to the study of a variety of physical phenomena. The same concepts apply to problems in communication theory, control, information processing, and other linear analysis. Waveforms, either continuous-time or sampled, and their corresponding spectra arise in a variety of applications not only as electrical signals but also as optical or acoustical waveforms.

The great generality of Fourier methods is not always appreciated. We hope this brief introduction will stimulate the reader to further study.

PROBLEMS

5.1. Find the frequency spectrum of the following waveforms:

(a)

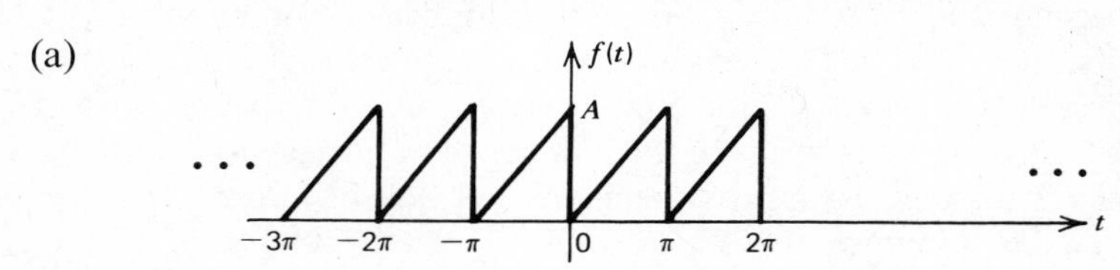

(b)

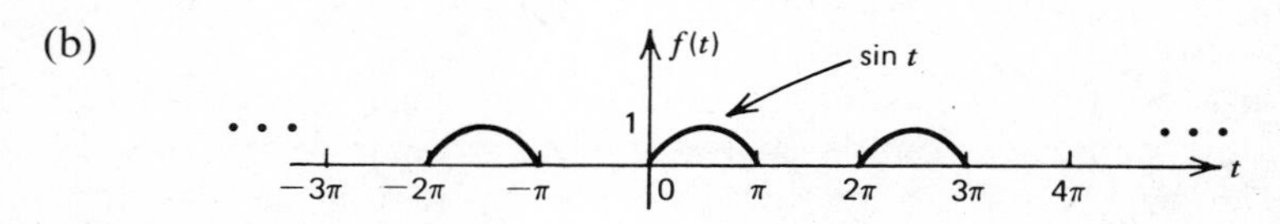

5.2 Define the energy of a signal $f(t)$ on the interval $[t_1, t_2]$ as

$$\text{Energy} = \int_{t_1}^{t_2} f^2(t')\, dt'$$

(a) What is the energy of the sum of two functions $f_1(t)$ and $f_2(t)$ on $[t_1, t_2]$?
(b) What is the energy of $f_1(t) + f_2(t)$ if $f_1(t)$ and $f_2(t)$ are orthogonal on $[t_1, t_2]$?

5.3. Suppose an electric circuit is excited by a voltage $v(t)$ given by

$$v(t) = V_0 + \sum_{n=1}^{\infty} V_n \cos(n\omega_0 t + \theta_n)$$

The corresponding steady-state current is $i(t)$, given by

$$i(t) = I_0 + \sum_{n=1}^{\infty} I_n \cos(n\omega_0 t + \phi_n)$$

as shown. Define the input power at the input terminals as

$$P = \frac{1}{T}\int_{-T/2}^{T/2} v(t)i(t)\, dt, \qquad T = \frac{2\pi}{\omega_0}$$

Show that the input power can also be written as

$$P = V_0 I_0 + \sum_{n=1}^{\infty} \frac{V_n I_n}{2} \cos(\theta_n - \phi_n)$$

5.4. A square-wave voltage $v(t)$ whose waveform is shown below is applied to a series RL circuit. Find the first five harmonics in the response current $i(t)$.

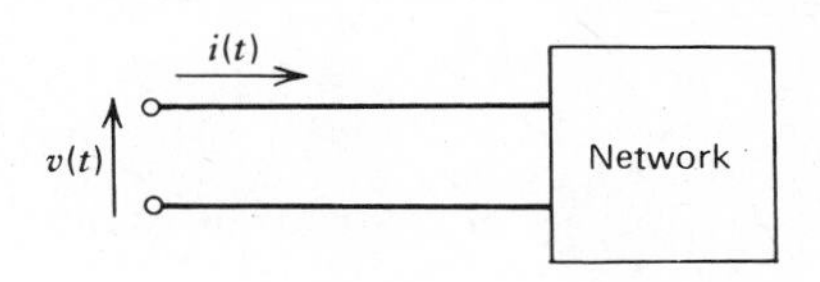

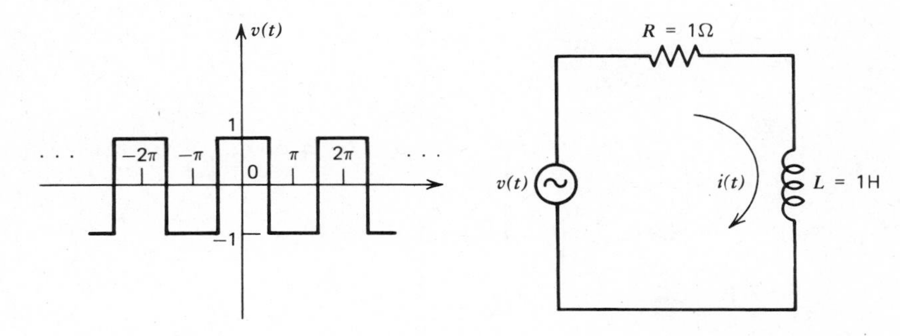

5.5. A periodic function $f(t)$ is known to have only the first ten harmonics (the remaining harmonics all have zero coefficients). Show that such a signal is uniquely specified by any $(2 \cdot 10 + 1) = 21$ sample values in a single period. In other words, show that the set of samples $\{f(t_1), f(t_2), \ldots, f(t_{21})\}$, $t_i \in [t_0, t_0 + T]$, can precisely specify $f(t)$ on $[t_0, t_0 + T]$.

5.6. A periodic waveform as shown below is to be approximated by sinusoids. How many sinusoids (added together) are needed to approximate $f(t)$ so that the approximating series is within 1% of the total energy of $f(t)$ over a single period?

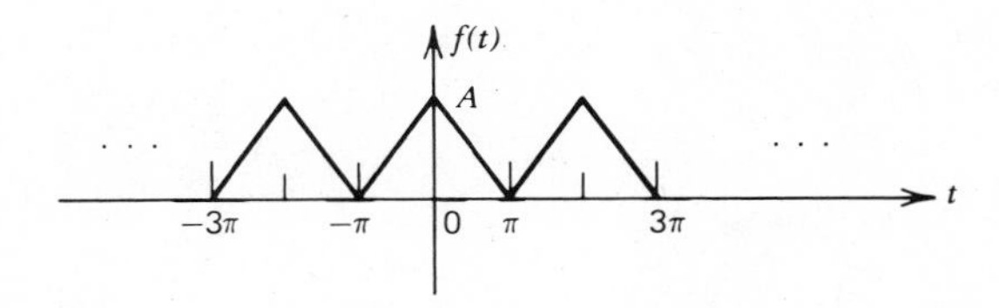

5.7. Sam Mooge has been experimenting with various waveforms for use in his electronic music machine. He is now searching for a signal that is amplitude limited between the values A and $-A$ and has maximum power at a particular frequency ω_0. What waveform should Sam use?

(a) Prove your answer for the class of all even or odd functions.

(b) Prove your answer in general.

5.8. The following differential equation is a model for a linear system with input $u(t)$ and output $y(t)$.

$$\frac{dy(t)}{dt} + y(t) = u(t)$$

If the input $u(t)$ is the square waveform shown, then find the third harmonic in the output.

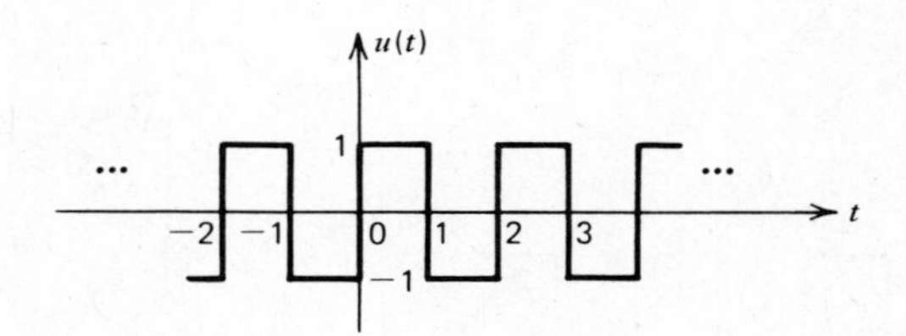

5.9. Fill in the blanks in the following table:

f_k	$F(e^{j\theta})$
	$\dfrac{e^{j2\theta} - 1}{e^{j2\theta} + 4}$
$e^{-3k} \sin(4k)$	
	$\dfrac{e^{jn\theta} - 1}{e^{j\theta} - 1}$

5.10. The following block diagram represents a connection of three subsystems in terms of their frequency-response functions. Find the impulse response of the entire system.

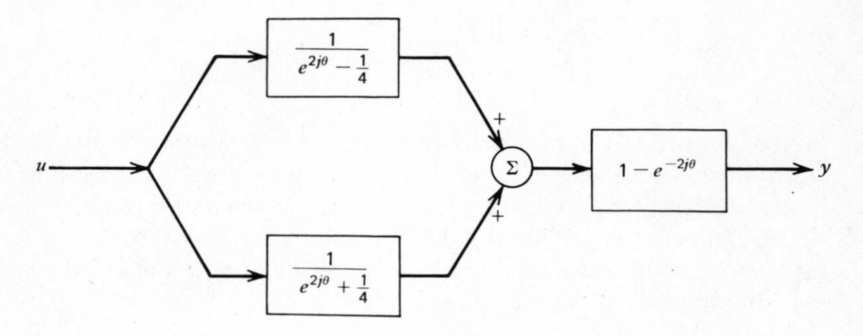

5.11. A high-pass digital filter (ideal) has the frequency-response function $H(e^{j\theta})$ given by

$$H(e^{j\theta}) = \begin{cases} 0, & -\dfrac{\pi}{2} < \theta < \dfrac{\pi}{2} \\ 1, & \text{otherwise} \end{cases}$$

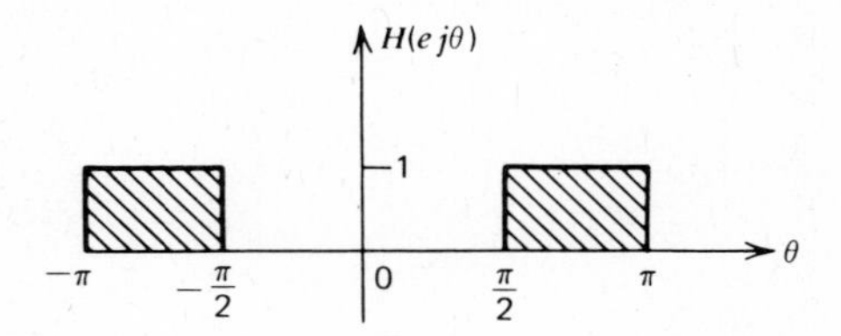

(a) Find the impulse-response sequence h_k, $-\infty < k < \infty$.

(b) Consider an FIR digital filter implentation of $H(e^{j\theta})$ whose impulse response g_k is limited to just 9 terms where

$$g_k = \begin{cases} 0, & k < 0 \\ h_{k-4}, & 0 \le k \le 8 \\ 0, & k > 8 \end{cases}$$

Plot the magnitude and phase of the FIR digital filter's frequency response.

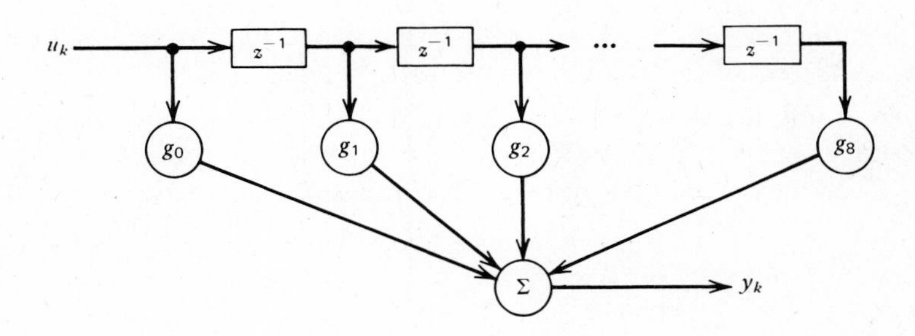

5.12. A discrete-time circuit has the frequency-response function

$$H(e^{j\theta}) = \frac{e^{2j\theta} - \frac{1}{2}e^{j\theta} - \frac{3}{4}}{e^{2j\theta} - \frac{1}{4}e^{j\theta} - \frac{1}{8}}.$$

What is the impulse-response sequence?

5.13. The signals $u_1(t)$ and $u_2(t)$ shown in the sketch below are obtained from adjacent pins on a counter chip.

(a) Sketch and label the frequency spectra of these signals.

(b) A "tone detector" is essentially a very narrow band filter as shown below. Can you choose ω_0 to detect $u_2(t)$ but not $u_1(t)$?

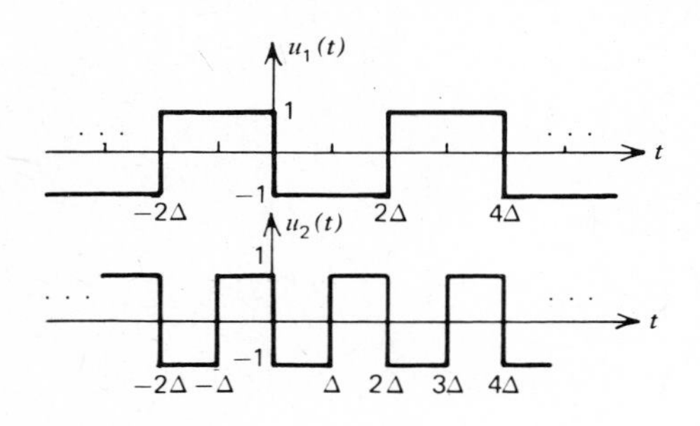

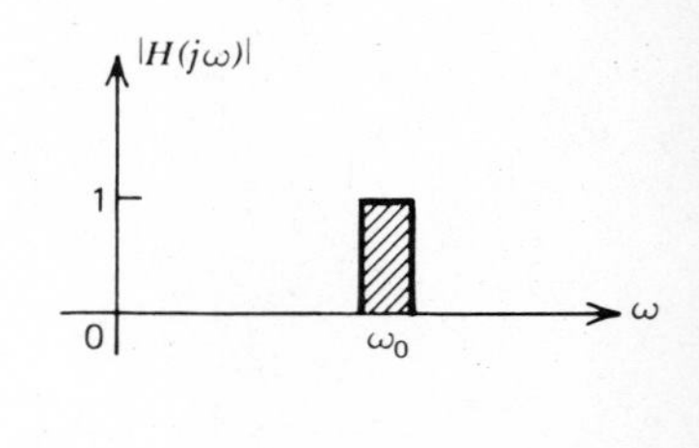

5.14. A digital filter has an output sequence given by

$$y_k = \sum_{n=-\infty}^{\infty} h_n u_{k-n} \qquad \text{where} \qquad h_n = \begin{cases} 0, & n < 0 \\ (\frac{1}{2})^n, & n \geq 0 \end{cases}$$

What is the frequency-response function $H(e^{j\theta})$ for this filter?

5.15. The input

$$u_n = \begin{cases} (\frac{1}{2})^{-n}, & n \leq 0 \\ 0, & n > 0 \end{cases}$$

produces an output whose transform is $2 \cos \theta/(2 - \cos \theta - j \sin \theta)$. What is the frequency-response function $H(e^{j\theta})$ for this filter?

5.16. A stable linear filter has an amplitude response shown below. Determine the transfer function $H(j\omega)$ or the differential equation for this system. Keep the order of the system to a minimum.

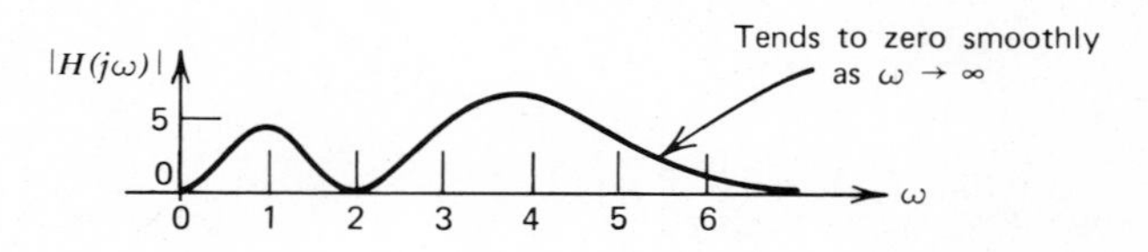

5.17. Sketch $f(t)$ and $F(j\omega)$ and obtain $F(j\omega)$ in closed form if $f(t)$ is given by

$$f(t) = \frac{1}{\sqrt{2\pi}} e^{-t^2/2} \cos \omega_0 t$$

5.18. (a) How fast should $f(t)$ (see sketch below) be sampled so that perfect reconstruction is possible?

(b) Suppose $f(t)$ is sampled every π/ω_0 seconds. Carefully sketch the spectrum of the sampled waveform.

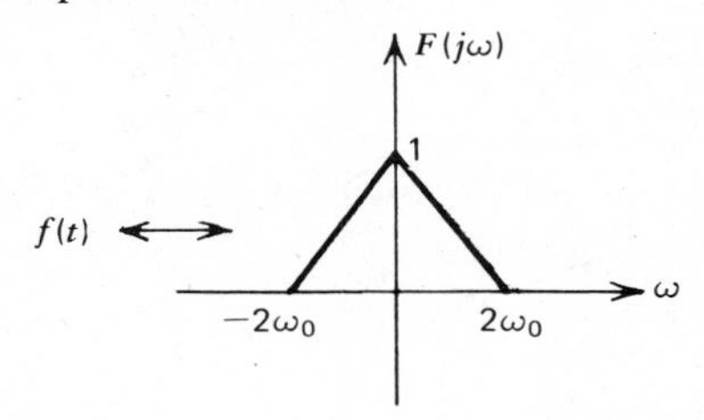

5.19. The Fourier transform of $f(t)$ and $g(t)$ are defined below.

$$G(j\omega) = \begin{cases} \cos \omega, & |\omega| < \pi/2 \\ 0, & \text{otherwise} \end{cases}$$

$$F(j\omega) = G(j\omega - j\omega_0) + G(j\omega + j\omega_0)$$

(a) Find $g(t)$ in closed form.
(b) Find $f(t)$ in closed form.
(c) How fast should $g(t)$ be sampled for perfect reconstruction?
(d) In the demodulation scheme shown below, can you find A, ω_1, ω_2 so that $y(t) = g(t)$?

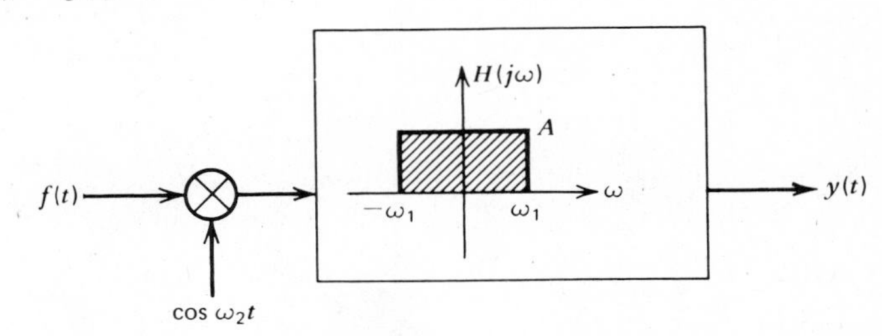

5.20. The following set of samples taken from the waveform $f(t)$ at the Nyquist interval defines a bandlimited function, $\{\ldots, 0, 0, 10, 30, 50, 50, -40, -10, 0, \ldots\}$. Give an exact expression for $f(t)$. Estimate the maximum value of this waveform as best you can.

5.21. Suppose a function $f(t)$ is continuous and bandlimited to W rad/sec. Show that $f(t)$ can be represented as

$$f(t) = \frac{\omega}{\pi} [f(t) * \text{sinc}(\omega t)] \qquad \text{for all } \omega > W$$

5.22. A communications channel has a bandwidth of 200 kHz. Pulses are to be transmitted over this channel with information coded by varying the amplitude of the pulses. If the pulses all have width δ sec and separation between centers of T sec, what values of δ and T should be chosen to transmit as much information per unit time as possible?

5.23. Sketch the output waveform frequency spectrum for the following system. Is this system linear?

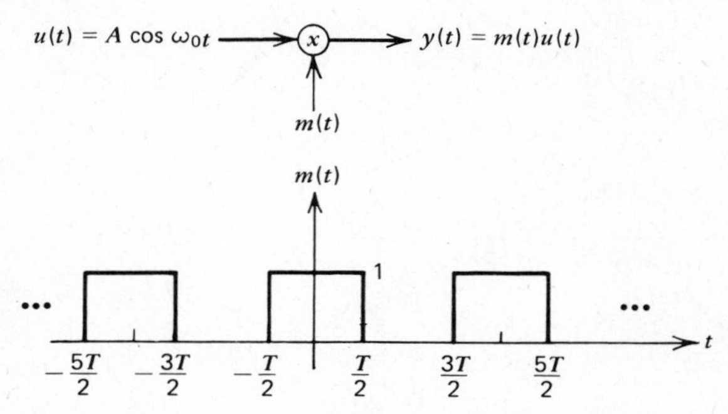

5.24. Sketch the phase spectrum between $[-2\pi, 2\pi]$ rad/sec of the time function shown, consisting of 2 pulses whose center values are separated by D seconds.

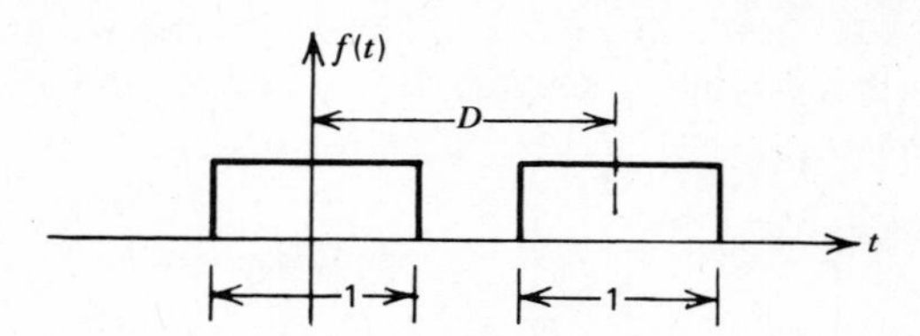

5.25. Find the time function $g(t)$ for the frequency spectrum $G(j\omega)$ as shown. Use properties of the Fourier transform to help determine the time function.

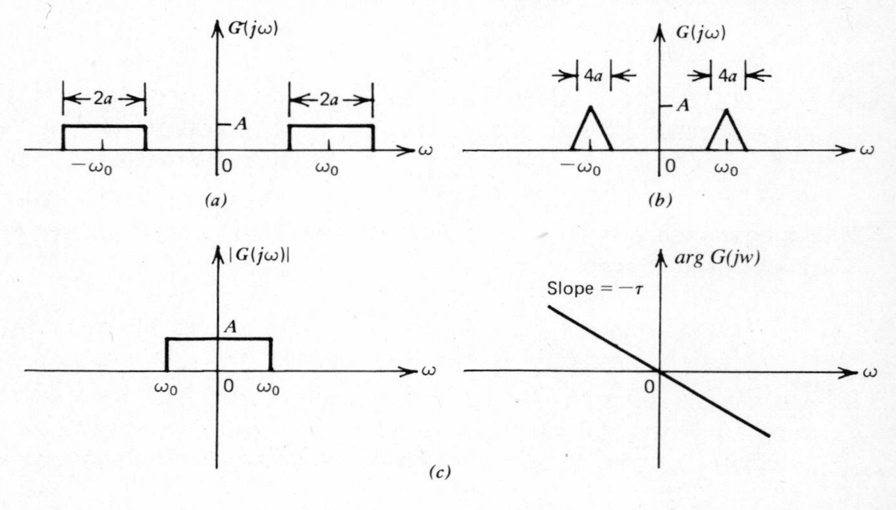

5.26. In the following system, the output is obtained by squaring the input. Sketch, in detail, the output frequency spectrum for an input $x(t)$ given by

(a) $x(t) = \text{sinc}\,(Wt)$

(b) $x(t) = \sum_{n=80}^{90} \cos(n\omega_0 t), \qquad \omega_0 = 10^3$ rad/sec

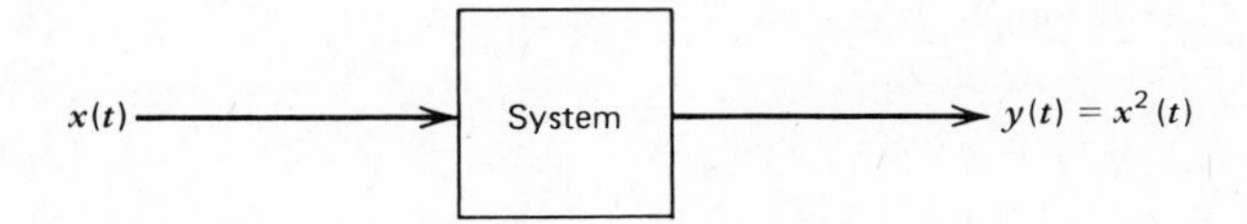

5.27. A pulse code modulation (PCM) system is a system that takes input analog information $u(t)$, samples the waveform at the Nyquist rate, quantizes the samples, and then codes the quantized sampled values by a binary code of 0s and 1s. These coded samples are then transmitted. For example, if the quantization operation performed on the sampled values divides the possible sampled values into 8 regions, the binary code for each sample would be chosen from one of the following: {000, 001, 010, 011, 100, 101, 110, 111}. Suppose the transmission channel has a bandwidth of W Hz and the input signal is bandlimited to f_m Hz, where $W > f_m$. What is the finest quantization one can obtain if the input signal is to be transmitted in the actual duration of the signal?

5.28. Sketch the frequency spectrum of $y(t)$ for the following modulation system where $f(t) = 2\cos 10t + 4\cos 20t$ and $m(t) = \cos 200t$.

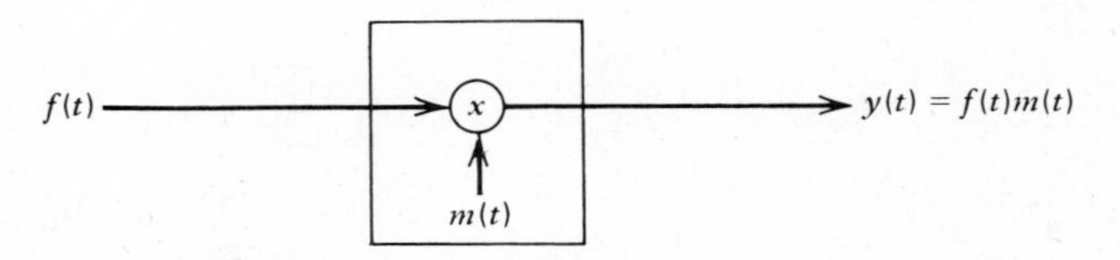

5.29. The impulse response of the following system is desired. The only equipment available to obtain this response is an oscilloscope and a pulse generator with a pulse width that can be varied between 1 msec and 1 sec. Can the system's impulse response be found by use of this equipment? Explain.

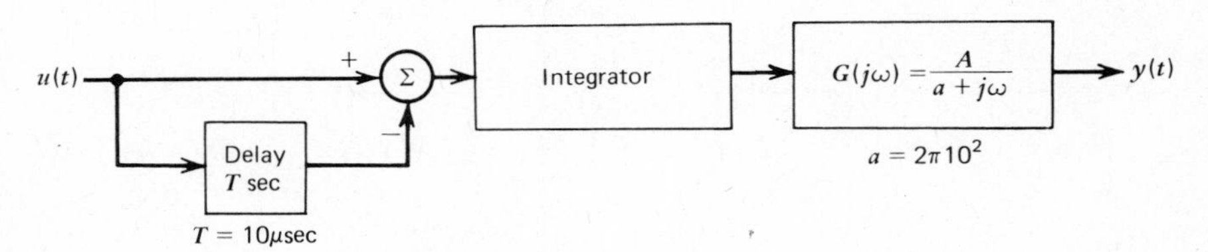

5.30. The waveform $f(t)$ modulates the pulse train $p_T(t)$ as shown below. Consider two cases. In case a, the modulated pulses in $m_a(t)$ have constant amplitudes. In case b, the pulses in $m_b(t)$ follow $f(t)$.

(1) Express the time waveforms $m_a(t)$ and $m_b(t)$ in terms of $f(t)$, $p(t)$, and the time impulse train $\delta_T(t) = \sum_{n=-\infty}^{\infty} \delta(t - nT)$, using the operations of multiplication and convolution.

(2) Find the spectra of $m_a(t)$ and $m_b(t)$ in terms of the transforms $F(j\omega)$ and $P(j\omega)$. Sketch $M_a(j\omega)$ and $M_b(j\omega)$ for $F(j\omega)$ as shown and $\tau = T/4$, and compare.

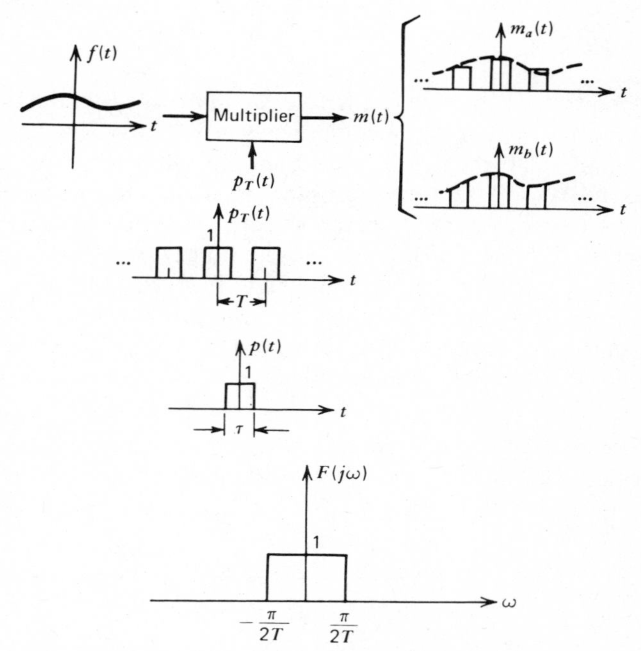

5.31. A function $f(t)$ has the Fourier transform shown below. Carefully sketch the Fourier transform of $f^2(t)$.

$$F(j\omega) = \delta(\omega) + \begin{cases} 1, & 2 \le |\omega| \le 4 \\ 0, & \text{otherwise} \end{cases}$$

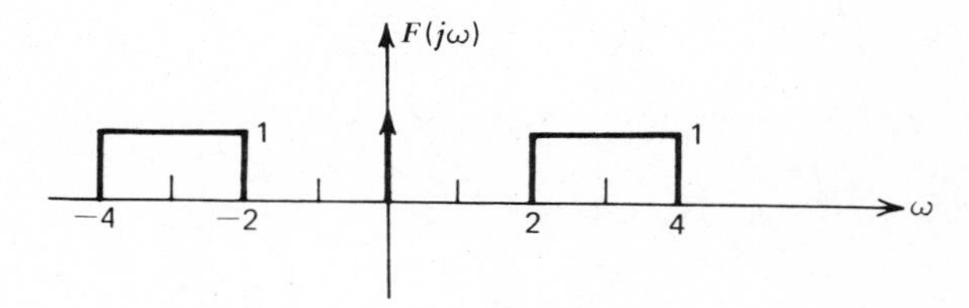

5.32. Use the energy theorem: that is,

$$\int_{-\infty}^{\infty} f^2(t)\, dt = \frac{1}{2\pi} \int_{-\infty}^{\infty} |F(j\omega)|^2\, d\omega$$

to find the value of

(a)
$$\int_{-\infty}^{\infty} \text{sinc}^2\,(t)\, dt$$

(b)
$$\int_{-\infty}^{\infty} \frac{dx}{(1 + x^2)^2}$$

5.33. A full-wave rectifier is a system whose output $y(t)$ to an input $u(t)$ is $y(t) = |u(t)|$. Suppose $a(t) \sin \omega_0 t$, $a(t) > 0$, is the input to a full-wave rectifier. Show that the output spectrum is

$$Y(j\omega) = -\frac{2}{\pi} \sum_{n=-\infty}^{\infty} \frac{A(j\omega - 2\,jn\omega_0)}{4n^2 - 1}, \qquad A(j\omega) = \mathscr{F}\{a(t)\}.$$

5.34. The operation of a sampling oscilloscope can be described (in a simplified manner) as follows. The input waveform $u(t)$ is multiplied by $g(t)$, a periodic waveform of unit impulses spaced $(1 + \alpha)T$ seconds apart. The waveform $u(t)g(t)$ is then passed through a a lowpass filter $H(j\omega)$ where

$$H(j\omega) = \begin{cases} 1, & |\omega| \le \dfrac{\pi}{(1 + \alpha)T} \\[2ex] 0, & |\omega| > \dfrac{\pi}{(1 + \alpha)T} \end{cases}$$

Suppose $u(t)$ is periodic of period T and is bandlimited, that is, $U(j\omega) = 0$, $|\omega| > \pi/\alpha T$.

(a) Sketch the Fourier transform of a typical $u(t)$.
(b) Sketch the Fourier transform of $g(t)$.
(c) Sketch the Fourier transform of $u(t)g(t)$.
(d) Show that $y(t) = \text{constant} \cdot u(\beta t)$ and find β in terms of α. [This means $y(t)$ is proportional to a "time stretched" version of $u(t)$.]

5.35. A time function $f(t) = \text{sinc}\ 4t$ is sampled every $2\pi/3$ seconds. Find and sketch (to scale) the frequency spectrum of the sampled time function $f_s(t)$.

5.36. Sketch the output waveform (carefully) for an input waveform $u(t)$ as shown. Assume the sampler is ideal with an output $u_s(t)$ given by

$$u_s(t) = u(t) \sum_{n=-\infty}^{+\infty} \delta(t - nT) = \sum_{n=-\infty}^{\infty} u(nT)\delta(t - nT)$$

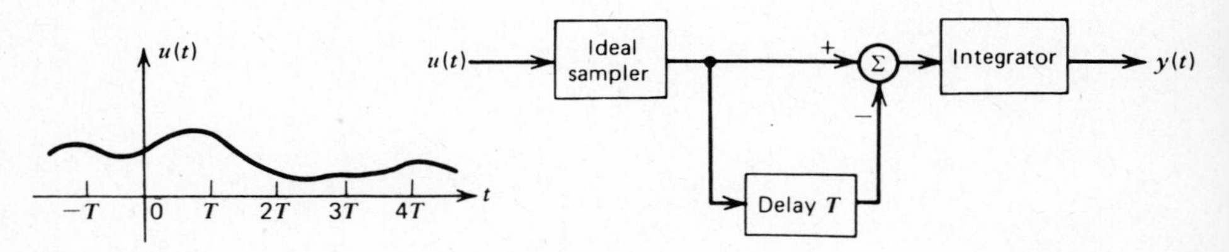

5.37. Using a special form of Schwarz's inequality,

$$\left| \int_{-\infty}^{\infty} g(t)\, dt \right|^2 \leq \int_{-\infty}^{+\infty} |g(t)|^2\, dt,$$

show that $T_e \cdot BW_e \geq 2\pi$ where

$$BW_e = \frac{\int_{-\infty}^{+\infty} |H(j\omega)|^2\, d\omega}{|H(0)|^2} \quad \text{and} \quad T_e = \frac{\left(\int_{-\infty}^{+\infty} |h(t)|\, dt\right)^2}{\int_{-\infty}^{+\infty} |h(t)|^2\, dt}$$

[Assume $|H(0)|$ is $\max_{\omega} |H(j\omega)|$.]

5.38. Define the duration of a time signal by the second moment of $|f(t)|^2$ about the origin, that is, define the duration D_t of $f(t)$ as

$$D_t^2 = \int_{-\infty}^{+\infty} t^2 |f(t)|^2\, dt$$

Similarly define the duration of its Fourier transform $F(j\omega)$ as D_ω where

$$D_\omega^2 = \int_{-\infty}^{\infty} \omega^2 |F(j\omega)|^2\, d\omega$$

[D_t and D_ω are a measure of the spread of the functions $|f(t)|$ and $|F(j\omega)|$, respectively. In statistical terms they correspond to the standard deviation of $|f(t)|$ and $|F(j\omega)|$.]

Assume further that

$$\int_{-\infty}^{+\infty} |f(t)|^2\, dt = \frac{1}{2\pi} \int_{-\infty}^{\infty} |F(j\omega)|^2\, d\omega = 1$$

Using Schwarz's inequality,

$$\left| \int_a^b f(t)g(t)\, dt \right|^2 \leq \int_a^b |f(t)|^2\, dt \int_a^b |g(t)|^2\, dt$$

with equality if and only if $f(t) = kg^*(t)$, and Parseval's theorem, show that

$$D_t D_\omega \geq \sqrt{\frac{\pi}{2}}$$

with equality only for Gaussian signals,

$$f(t) = \sqrt{(\beta/\pi)}e^{-\beta t^2}.$$

This result is similar to the result of Problem 5.37. The measures of duration and bandwidth are not the same but the essential result is identical.

5.39. A continuous-time pulse $u_a(t)$ of duration 50 msec and approximately band limited to 1 kHz is sampled and transformed by the three systems shown below. Compare quantitatively the three *DFT*s U_1, U_2, and U_3.

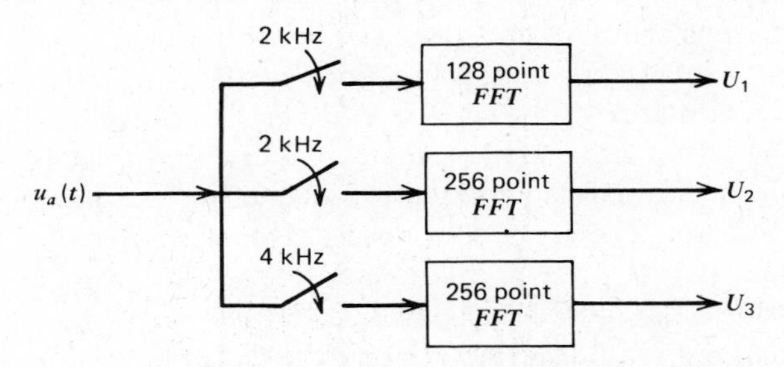

5.40. Design a simple FIR filter consisting of two coefficients that can be used to find changes in a rapidly varying signal that is combined with a slowly varying disturbance. (This kind of structure would be useful in detecting abrupt changes in a waveform.) In other words, we wish to obtain an FIR filter that emphasizes high frequencies and suppresses low frequencies. Plot the phase and amplitude response of your design.

CHAPTER 6
THE LAPLACE TRANSFORM

The Fourier transform of a function $f(t)$ is a representation of the function as a continuous sum of exponential functions of the form $e^{j\omega t}$, where ω is a real frequency. This interpretation of the Fourier transform gives one a physical feeling for an otherwise abstract mathematical transformation. The Laplace transform can be viewed as an extension of the Fourier transform in which we represent a function $f(t)$ by a continuous sum of exponential functions of the form e^{st}, where $s = \sigma + j\omega$ is a complex frequency. From this viewpoint, the Fourier transform is a special case in which $s = j\omega$.

Recall that the Fourier transform pair for $f(t)$ is

$$F(j\omega) = \int_{-\infty}^{\infty} f(t)e^{-j\omega t}\, dt \tag{6.1}$$

$$f(t) = \frac{1}{2\pi}\int_{-\infty}^{\infty} F(j\omega)e^{j\omega t}\, d\omega \tag{6.2}$$

Now suppose we define a function $a(t)$ as $f(t)$ times the function $e^{-\sigma t}$, where σ is a real number: that is, $a(t) = f(t)e^{-\sigma t}$. The Fourier transform of $a(t)$ is

$$\begin{aligned}\mathscr{F}\{a(t)\} &= \int_{-\infty}^{\infty} a(t)e^{-j\omega t}\, dt = \int_{-\infty}^{\infty} f(t)e^{-\sigma t}e^{-j\omega t}\, dt \\ &= \int_{-\infty}^{\infty} f(t)e^{-(\sigma + j\omega)t}\, dt \end{aligned} \tag{6.3}$$

Equation 6.3 can thus be written as

$$F(\sigma + j\omega) = \int_{-\infty}^{\infty} f(t)e^{-(\sigma + j\omega)t}\, dt \tag{6.4}$$

and $f(t)$ can be expressed as

$$f(t) = \frac{1}{2\pi}\int_{-\infty}^{\infty} F(\sigma + j\omega)e^{(\sigma + j\omega)t}\, d\omega \tag{6.5}$$

Now let s be the complex frequency $\sigma + j\omega$: that is, $s = \sigma + j\omega$. Then $d\omega = (1/j)\, ds$ and (6.4) and (6.5) become

$$F(s) = \int_{-\infty}^{\infty} f(t)e^{-st}\, dt \tag{6.6}$$

$$f(t) = \frac{1}{2\pi j}\int_{\sigma - j\infty}^{\sigma + j\infty} F(s)e^{st}\, ds \tag{6.7}$$

The limits on the integral of (6.7) result because of the substitution $s = \sigma + j\omega$. Equations (6.6) and (6.7) constitute the *complex Fourier transform pair* or the *bilateral Laplace transform pair* or the *two-sided Laplace transform pair*. We shall use the latter designation with the notation

$$F(s) = \mathscr{L}_b\{f(t)\}$$
$$f(t) = \mathscr{L}_b^{-1}\{F(s)\}$$

The two-sided Laplace transform of $f(t)$ can be obtained from the Fourier transform of $f(t)$ by merely substituting s for $j\omega$ provided $f(t)$ is absolutely integrable. For functions that possess Fourier transforms only in the limit, that is, power functions, one must evaluate the Laplace transform directly using (6.6). Notice that in this development we have multiplied a given function $f(t)$ by $e^{-\sigma t}$, where σ is any real number. Thus the convergence of the resulting Fourier integral is greatly enhanced by the so-called convergence factor $e^{-\sigma t}$. This means that the Laplace transform exists for many functions for which there is no Fourier transform. This is one of the principal advantages of the Laplace transform: the ability to transform functions that are not absolutely integrable.

6.1 CONVERGENCE OF THE LAPLACE TRANSFORM

To appreciate the generality possessed by the two-sided Laplace transform, one must understand the convergence properties of the integral (6.6). The two-sided Laplace transform exists if

$$F(s) = \int_{-\infty}^{\infty} f(t)e^{-st}\, dt \tag{6.8}$$

is finite. Therefore $F(s)$ is guaranteed to exist if

$$\int_{-\infty}^{\infty} |f(t)e^{-st}|\, dt = \int_{-\infty}^{\infty} |f(t)|e^{-\sigma t}\, dt$$

is finite. Suppose there exists a real positive number $\mathscr{R}$ so that for some real α and β we know that

$$|f(t)| < \begin{cases} \mathscr{R}e^{\alpha t}, & t > 0 \\ \mathscr{R}e^{\beta t}, & t < 0 \end{cases} \tag{6.9}$$

Then $F(s)$ converges for

$$\alpha < \text{Re}\,(s) < \beta$$

To verify this conclusion, write (6.8) as

$$|F(s)| \leq \int_{-\infty}^{0} |f(t)e^{-st}|dt + \int_{0}^{\infty} |f(t)e^{-st}|dt$$

Using the inequalities of (6.9), we see that

$$\begin{aligned} |F(s)| &< \int_{-\infty}^{0} \mathscr{R}e^{(\beta-\sigma)t}\, dt + \int_{0}^{\infty} \mathscr{R}e^{(\alpha-\sigma)t}\, dt \\ &< \mathscr{R}\left[\frac{1}{\beta - s}\, e^{(\beta-\sigma)t}\Big|_{-\infty}^{0} + \frac{1}{\alpha - s}\, e^{(\alpha-\sigma)t}\Big|_{0}^{\infty}\right] \end{aligned} \tag{6.10}$$

In (6.10) the first integral [negative time portion of $f(t)$] converges for Re $(s) < \beta$. Similarly, the second integral [positive time portion of $f(t)$] converges for Re $(s) > \alpha$. Thus both integrals, and therefore $F(s)$, converge in the common region, $\alpha <$ Re $(s) < \beta$. Notice that the left-hand limit α of the convergence region is governed by the behavior of the positive-time portion of $f(t)$, while the right-hand limit β depends on the behavior of the negative-time portion of $f(t)$.

Because $F(s)$ converges in the strip $\alpha < \sigma < \beta$, there are no singularities or poles† of $F(s)$ in this region. Now if $f(t)$ is nonzero for $(0, \infty)$ only, then the left-hand limit α alone defines the convergence region, and all the poles of $F(s)$ must be to the left of α. Likewise, if $f(t)$ is nonzero for $(-\infty, 0)$ only, then β alone defines the convergence region. In this case, all poles of $F(s)$ must lie to the right of β. In other words, poles to the left of the convergence region arise from the positive time portion of $f(t)$ and poles to the right of the convergence region arise from the negative time portion of $f(t)$. This simple fact is the key to finding inverse transforms for two-sided Laplace transforms. Whenever the region of convergence of $F(s)$ includes the $j\omega$ axis, then one can obtain the Fourier transform from $F(s)$ simply by substituting $j\omega$ for s and vice versa. If the region of convergence does not include the $j\omega$ axis, then the function $f(t)$ is not absolutely integrable and so does not possess a Fourier transform in the strict sense. Figure 6.1 depicts several time functions and their corresponding regions of convergence.

In summary, the two-sided Laplace transform exists, provided that $\int_{-\infty}^{\infty} |f(t)|e^{-\sigma t}\,dt$ is finite. This is equivalent to requiring the existence of real numbers, $\mathscr{R}$, α, and β such that

$$|f(t)| < \begin{cases} \mathscr{R}e^{\alpha t}, & t > 0 \\ \mathscr{R}e^{\beta t}, & t < 0 \end{cases}$$

A function that satisfies the above inequalities is said to be of exponential order. We can also define an exponential-order function as one that satisfies

$$\lim_{t \to \infty} e^{\alpha t} f(t) = 0, \qquad \lim_{t \to -\infty} e^{\beta t} f(t) = 0$$

for some α and β. If a signal $f(t)$ is of exponential order, $f(t)$ does not grow more rapidly than $\mathscr{R}e^{\alpha t}(t > 0)$ and $\mathscr{R}e^{\beta t}(t < 0)$. There are functions that are not of exponential order, such as e^{t^3} or t^t. However, it is difficult to envision a physical problem where they might arise.

† A pole of $F(s)$ is a value of s such that $|F(s)| \to \infty$. For example, if $F(s) = k(s + c)/(s - a)(s + b)$, then a and $-b$ are poles of $F(s)$.

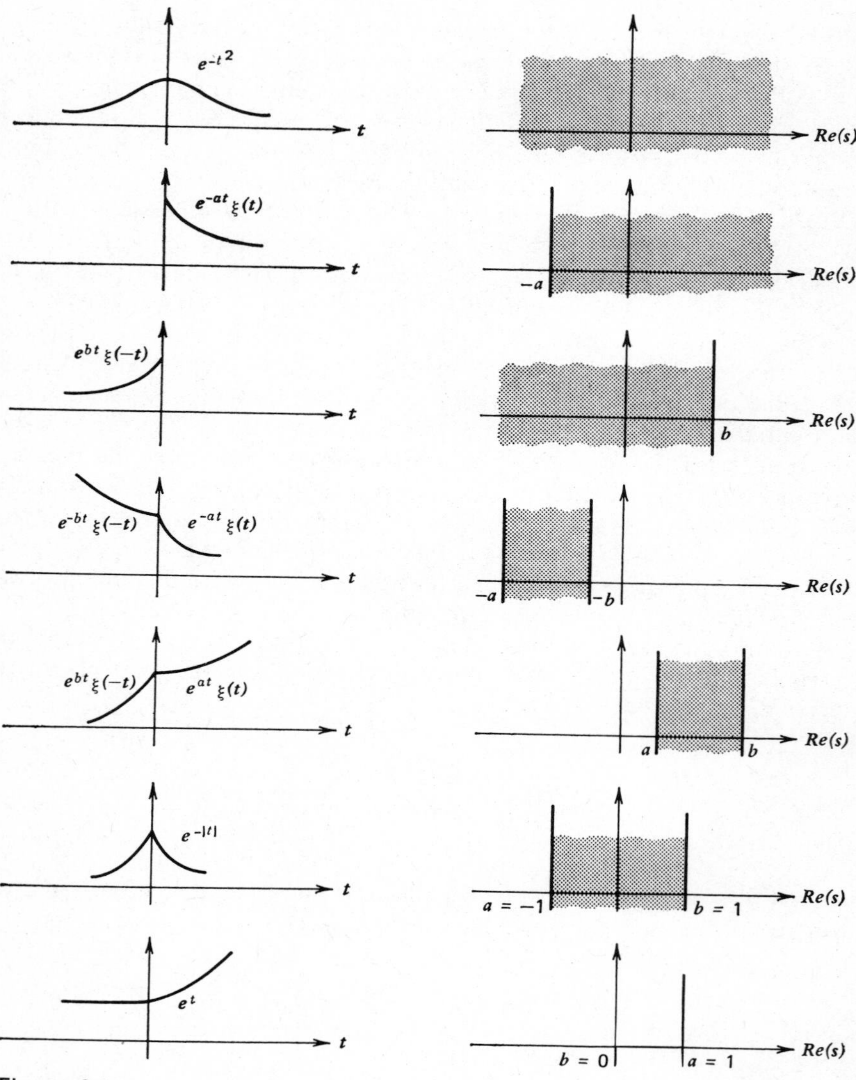

e^{-t^2}
t
Re(s)
$e^{-at}\xi(t)$
t
−a
Re(s)
$e^{bt}\xi(-t)$
t
b
Re(s)
$e^{-bt}\xi(-t)$
$e^{-at}\xi(t)$
t
−a
−b
Re(s)
$e^{bt}\xi(-t)$
$e^{at}\xi(t)$
t
a
b
Re(s)
$e^{-|t|}$
t
a = −1
b = 1
Re(s)
e^{t}
t
b = 0
a = 1
Re(s)

Figure 6.1

6.2 THE ONE-SIDED OR UNILATERAL LAPLACE TRANSFORM

In the majority of applications involving Laplace transforms, the time functions of interest are causal: that is, $f(t) = 0$ for $t < 0$. For example, if the forcing function applied to a causal system is zero for negative values of t, the response of the system is also causal. Generally, in the analysis of physical systems, one can assume an excitation that begins at $t = 0$. Thus, one often restricts the class of functions for use in the Laplace transform to functions nonzero only for $0 < t < \infty$. The resulting Laplace transform is termed the *unilateral* or *one-sided Laplace transform.* In this case, the transform equations (6.6) and (6.7) reduce to

$$F(s) = \int_0^{\infty} f(t)e^{-st}\, dt \tag{6.11}$$

$$f(t) = \frac{1}{2\pi j} \int_{\sigma - j\infty}^{\sigma + j\infty} F(s)e^{st}\, ds \tag{6.12}$$

We shall use the shorthand notation

$$F(s) = \mathscr{L}\{f(t)\}$$
$$f(t) = \mathscr{L}^{-1}\{F(s)\}$$

for the one-sided Laplace transform pair. The relationship of the two-sided to the one-sided Laplace transform is analogous to what occurs in the one- and two-sided Z-transforms. For example, if one is dealing with time functions nonzero on $(-\infty, \infty)$, then the region of convergence must be known in order to obtain a unique inverse. This same information is needed to obtain unique inverses in the case of two-sided Z-transforms (see Section 4.5). The properties of the one-sided and two-sided Laplace transforms differ only slightly, in much the same way as we have already seen for the case of Z-transforms.

6.3 PROPERTIES OF THE LAPLACE TRANSFORM

The properties of the two-sided Laplace transform are so similar to those of the Fourier transform of Chapter 5 that we merely summarize them here. Notice that the region of convergence is also added so as to obtain a unique transform pair.

1. *Linearity*

$$af_1(t) + bf_2(t) \leftrightarrow aF_1(s) + bF_2(s), \qquad \max(\alpha_1, \alpha_2) < \sigma < \min(\beta_1, \beta_2)$$

2. *Scaling*

$$f(at) \leftrightarrow \frac{1}{|a|} F\left(\frac{s}{a}\right), \qquad |a|\alpha < \sigma < |a|\beta$$

3. *Time Shift*

$$f(t - \tau) \leftrightarrow F(s)e^{-\tau s}, \qquad \alpha < \sigma < \beta$$

4. *Frequency Shift*

$$e^{-at} f(t) \leftrightarrow F(s + a), \qquad \alpha - \mathrm{Re}(a) < \sigma < \beta - \mathrm{Re}(a)$$

5. *Time Convolution*

$$f_1(t) * f_2(t) \leftrightarrow F_1(s)F_2(s), \qquad \text{same as (1)}$$

6. *Frequency Convolution*

$$f_1(t)f_2(t) \leftrightarrow \frac{1}{2\pi j} \int_{c-j\infty}^{c+j\infty} F_1(u)F_2(s-u)\,du, \qquad \begin{cases} \alpha_1 + \alpha_2 < \sigma < \beta_1 + \beta_2 \\ \alpha_1 < c < \beta_1 \end{cases}$$

7. *Time Differentiation*

$$\frac{df(t)}{dt} \leftrightarrow sF(s), \qquad \text{same as (3)}$$

8. *Time Integration*

$$\int_{-\infty}^{t} f(u)\,du \leftrightarrow \frac{F(s)}{s}, \qquad \max(\alpha, 0) < \sigma < \beta$$

$$\int_{t}^{\infty} f(u)\,du \leftrightarrow \frac{F(s)}{s}, \qquad \alpha < \sigma < \min(\beta, 0)$$

9. Frequency Differentiation

$$(-t)^n f(t) \leftrightarrow \frac{d^n F(s)}{ds^n}, \qquad \text{same as (3)}$$

These properties for the two-sided Laplace transform hold for the one-sided transform with the exception of properties (7) and (8). There are also initial and final value theorems for one-sided transforms. The time differentiation and time integration properties are very important properties for the one-sided transform for the analysis of transient phenomena in linear systems. The time-convolution theorem again plays a central role in the transform analysis of linear systems.

10. One-Sided Time Differentiation

If

$$f(t) \leftrightarrow F(s)$$

then

$$f^{(1)}(t) \leftrightarrow sF(s) - f(0) \tag{6.13}$$

and, in general, we have

$$\frac{d^n f(t)}{dt^n} \leftrightarrow s^n F(s) - s^{n-1} f(0) - s^{n-2} f^{(1)}(0) - \cdots - f^{(n-1)}(0) \tag{6.14}$$

Proof. By definition,

$$\mathscr{L}\left\{\frac{df(t)}{dt}\right\} = \int_0^\infty \frac{df(t)}{dt} e^{-st}\, dt$$

Integrate by parts to obtain

$$\mathscr{L}\{f^{(1)}(t)\} = f(t)e^{-st}\big|_0^\infty + s\int_0^\infty f(t)e^{-st}\, dt$$

$$= f(t)e^{-st}\big|_{t=\infty} - f(0) + sF(s)$$

Now because $F(s)$ exists, it follows that $f(t)e^{-st}$ evaluated at $t = \infty$ is zero. Thus

$$\mathscr{L}\{f^{(1)}(t)\} = sF(s) - f(0)$$

The general case is found by repeated application of the above process. The value of $f(0)$, $f^{(1)}(0)$, ... that is used must be consistent with the definition of the integral in (6.11). If $f(t)$ is discontinuous at the origin, $f^{(1)}(t)$ will possess an impulse equal in area to the height of the discontinuity. If this impulse is to be included in the Laplace transform of $f(t)$, then $f(0)$, $f^{(1)}(0)$, ... are evaluated at 0^+, and the one-sided Laplace integral of (6.11) has a lower limit of 0^+. One can also exclude these discontinuities at the origin provided $f(0)$, $f^{(1)}(0)$, ... are evaluated at 0^- and the one-sided Laplace integral has a lower limit of 0^-. In other words, we can use either 0^- or 0^+, in (6.11) and (6.14), provided that we are consistent in our choice of "$t = 0$." To illustrate, consider the following example.

Example 6.1. Find the transform of $f^{(1)}(t)$, where $f(t) = e^{-at}\xi(t)$ as shown in Figure 6.2.

a. Suppose we consider $t = 0$ as being $t = 0^-$. Then, using the defining integral, we have

$$\mathscr{L}\{f^{(1)}(t)\} = \int_{0^-}^{\infty} f^{(1)}(t)e^{-st}\,dt = \int_{0^-}^{\infty} [e^{-at}\,\delta(t) - ae^{-at}\xi(t)]e^{-st}\,dt$$

$$= \int_{0^-}^{\infty} (e^{-at}e^{-st}\,\delta(t) - ae^{-(a+s)t})\,dt$$

$$= 1 + \left.\frac{ae^{-(a+s)t}}{a+s}\right|_{0^-}^{\infty} = 1 - \frac{a}{a+s} = \frac{s}{s+a}$$

Because we are using 0^- as $t = 0$, $f(0^-) = 0$. Thus, (6.13) yields

$$\mathscr{L}\{f^{(1)}(t)\} = sF(s) - f(0^-) = sF(s)$$

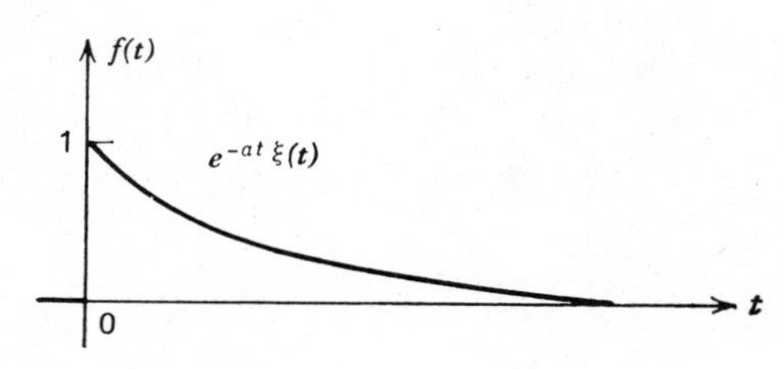

Figure 6.2

where

$$F(s) = \int_0^\infty e^{-at}e^{-st}\,dt = \frac{e^{-(a+s)t}}{-(s+a)}\bigg|_0^\infty = \frac{1}{s+a}$$

$$\therefore\ \mathscr{L}\{f^{(1)}(t)\} = sF(s) = \frac{s}{s+a}$$

which agrees with the expression obtained by direct calculation.

b. Suppose we consider $t = 0$ as being $t = 0^+$. The defining integral is therefore

$$\mathscr{L}\{f^{(1)}(t)\} = \int_{0^+}^\infty f^{(1)}(t)e^{-st}\,dt = \int_{0^+}^\infty [e^{-(a+s)t}\,\delta(t) - ae^{-(a+s)t}]\,dt$$

$$= 0 + \frac{ae^{-(a+s)t}}{s+a}\bigg|_{0^+}^\infty = \frac{-a}{s+a}$$

If the lower limit is 0^+ in (6.11), then $f(0^+) = 1$. Thus we have

$$\mathscr{L}\{f^{(1)}(t)\} = sF(s) - f(0^+) = s\left(\frac{1}{s+a}\right) - 1 = \frac{-a}{s+a}$$

which again agrees with the direct calculation. Thus, the Laplace transform depends on whether we use 0^- or 0^+, but (6.13) and (6.14) give results consistent with the defining integral. ■

This property of the one-sided Laplace transform furnishes one with an extremely effective method of solving transient problems in linear systems. The initial conditions of a system can easily be incorporated into the solution by use of (6.13) or (6.14).

11. One-Sided Time Integration

If

$$f(t) \leftrightarrow F(s)$$

then

$$\int_0^t f(u)\,du \leftrightarrow \frac{F(s)}{s} \tag{6.15}$$

and

$$\int_{-\infty}^{t} f(u)\,du \leftrightarrow \frac{F(s)}{s} + \frac{f^{(-1)}(0)}{s} \tag{6.16}$$

where

$$f^{(-1)}(0) \triangleq \int_{-\infty}^{t} f(u)\,du \bigg|_{t=0}$$

Proof: By definition,

$$\mathscr{L}\left\{\int_0^t f(u)\,du\right\} = \int_0^\infty \left(\int_0^t f(u)\,du\right) e^{-st}\,dt$$

Integrating by parts we obtain

$$\mathscr{L}\left\{\int_0^t f(u)\,du\right\} = \frac{-e^{-st}}{s} \int_0^t f(u)\,du \bigg|_0^\infty + \frac{1}{s}\int_0^\infty f(t)e^{-st}\,dt$$

If $f(t)$ is of exponential order, so then is its integral. Thus the term $e^{-st}\int_0^t f(u)\,du$ evaluated at $t = \infty$ is zero for values of s in the convergence region. Thus

$$\mathscr{L}\left\{\int_0^t f(u)\,du\right\} = \frac{F(s)}{s}$$

If the lower limit is $-\infty$, then we obtain

$$\begin{aligned}\int_{-\infty}^{t} f(u)\,du &= \int_{-\infty}^{0} f(u)\,du + \int_0^t f(u)\,du \\ &= f^{(-1)}(0)\xi(t) + \int_0^t f(u)\,du\end{aligned}$$

Now

$$\mathscr{L}\{\xi(t)\} = \int_0^\infty e^{-st}\,dt = \frac{e^{-st}}{-s}\bigg|_0^\infty = \frac{1}{s}$$

Thus

$$\mathscr{L}\left\{\int_{-\infty}^{t} f(u)\,du\right\} = \frac{f^{(-1)}(0)}{s} + \frac{F(s)}{s}$$

This property of the Laplace transform is useful in transforming equations containing integrals and derivatives, as the next example demonstrates.

Example 6.2 In the following RLC network, find the current resulting from the voltage source $v(t)$. The switch closes at $t = 0$. The inductor has an initial current of $i(0)$ and the capacitor an initial charge of $q(0)$. Using Kirchoff's voltage law, we see that

$$v(t) = Ri(t) + L\frac{di(t)}{dt} + \frac{1}{C}\int_{-\infty}^{t} i(t')\,dt' \tag{6.17}$$

Usually, we solve (6.17) by differentiating it to remove the integral and obtain a differential equation: however, using (6.16) directly, we can take Laplace transforms on both sides of (6.17) to obtain

$$V(s) = RI(s) + L[sI(s) - i(0)] + \frac{1}{C}\left[\frac{I(s)}{s} + \frac{i^{(-1)}(0)}{s}\right]$$

Now $i(0)$ is merely the current in the inductor at $t = 0$. The term

$$\frac{i^{(-1)}(0)}{C} = \frac{\int_{-\infty}^{0} i(t')\,dt'}{C} = \frac{q(0)}{C} = v(0)$$

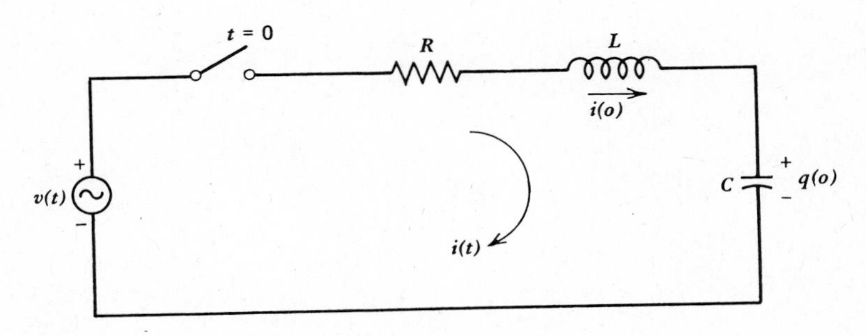

Figure 6.3

is the voltage across the capacitor at $t = 0$. Thus, solving for $I(s)$, we have

$$I(s) = \frac{V(s) - Li(0) + \dfrac{v(0)}{s}}{R + Ls + \dfrac{1}{Cs}}$$

Given the parameter values, we can take the inverse transform of $I(s)$ to obtain $i(t)$. The meaning of $t = 0$ in this problem is just after the switch closes at $t = 0^+$. ■

12. *Initial Value*

The initial-value property permits one to calculate $f(0)$ directly from the transform $F(s)$ without the need of inverting the transform. It is important to understand which value of $f(0)$ this theorem gives when $f(t)$ is discontinuous at the origin. Now (6.13) states

$$\mathscr{L}\{f^{(1)}(t)\} = \int_0^\infty f^{(1)}(t)e^{-st}\,dt = sF(s) - f(0) \tag{6.18}$$

If $f(t)$ is continuous at the origin, $f^{(1)}(t)$ is finite and as $s \to \infty$, the integral goes to zero. Thus

$$\lim_{s \to \infty} sF(s) = f(0) \tag{6.19}$$

If $f(t)$ is discontinuous at the origin, then $f^{(1)}(t)$ contains an impulse term, namely $[f(0^+) - f(0^-)]\,\delta(t)$. Thus, if 0^- is taken as the lower limit in (6.18), we obtain

$$\lim_{s \to \infty} \int_0^\infty f^{(1)}(t)e^{-st}\,dt = f(0^+) - f(0^-) = \lim_{s \to \infty} [sF(s) - f(0^-)]$$

Therefore

$$\begin{aligned} \lim_{s \to \infty} sF(s) &= f(0^+) - f(0^-) + f(0^-) \\ &= f(0^+) \end{aligned} \tag{6.20}$$

If 0^+ is taken as the lower limit, then as $s \to \infty$ (6.18) is

$$\lim_{s \to \infty} \int_{0^+}^\infty f^{(1)}(t)e^{-st}\,dt = 0$$

so that

$$0 = \lim_{s \to \infty} [sF(s) - f(0^+)]$$

which again yields

$$\lim_{s \to \infty} sF(s) = f(0^+) \tag{6.21}$$

Thus the initial value is always the limit of $f(t)$ as $t \to 0^+$, provided that it exists, independent of the lower limit, 0^- or 0^+, used in the defining equation for the Laplace transform.

Example 6.3 In the proof of the property (11), we calculated the Laplace transform of the unit step function $\xi(t)$ as $1/s$. The initial value of the unit step by (6.21) is therefore

$$\lim_{s \to \infty} sF(s) = \lim_{s \to \infty} s\left(\frac{1}{s}\right) = 1$$

which agrees with the correct value of $\xi(0^+)$. ■

13. Final Value

The value which $f(t)$ approaches as t becomes large may also be found directly from its transform $F(s)$. Using (6.13) again, we take the limit as $s \to 0$.

$$\lim_{s \to 0} \int_0^{\infty} f^{(1)}(t)e^{-st}\, dt = \lim_{s \to 0} [sF(s) - f(0)] \tag{6.22}$$

If $sF(s)$ has all its singularities in the left half of the s plane, then $\lim_{s \to 0} sF(s)$ exists. Then (6.22) can be written

$$\int_0^{\infty} f^{(1)}(t)\, dt = \lim_{s \to 0} [sF(s) - f(0)]$$

or

$$\lim_{t \to \infty} f(t) - f(0) = \lim_{s \to 0} sF(s) - f(0)$$

so that

$$\lim_{t \to \infty} f(t) = \lim_{s \to 0} sF(s)$$

A simple pole in $F(s)$ at the origin is permitted, but otherwise all other poles of $F(s)$ must be strictly in the left half of the s plane.

Example 6.4 Find the final value of the time function that corresponds to the transform (one-sided)

$$F(s) = \frac{1}{s + a}, \qquad a > 0$$

The singularity of $sF(s)$ is a pole at $s = -a$ (which is in the left half of the s plane). Thus

$$\lim_{s \to 0} sF(s) = \lim_{s \to 0} \left(\frac{s}{s + a} \right) = 0$$

This expression is correct from our knowledge of the time function $f(t) = e^{-at}\xi(t)$, which approaches zero as $t \to \infty$. ■

6.4 LAPLACE TRANSFORMS OF SIMPLE FUNCTIONS

It is seldom necessary to evaluate a Laplace transform by integration. Usually with systems that have a small number of natural modes, one need employ only a small number of basic transforms and knowledge of their properties. We have already calculated the key pair, namely

$$e^{-at}\xi(t) \leftrightarrow \frac{1}{s + a} \tag{6.23}$$

For $a = 0$, we have the unit step transform pair

$$\xi(t) \leftrightarrow \frac{1}{s} \tag{6.24}$$

If $a = +j\omega$ or $-j\omega$, then

$$e^{j\omega t}\xi(t) \leftrightarrow \frac{1}{s + j\omega} \tag{6.25}$$

$$e^{-j\omega t}\xi(t) \leftrightarrow \frac{1}{s - j\omega} \tag{6.26}$$

Adding the last two pairs yields a transform pair for the truncated cosine function:

$$\cos \omega t\xi(t) \leftrightarrow \frac{1}{2}\frac{1}{s + j\omega} + \frac{1}{2}\frac{1}{s - j\omega}$$

$$\leftrightarrow \frac{s}{s^2 + \omega^2} \tag{6.27}$$

Subtracting (6.26) from (6.25) yields a transform pair for the truncated sine function.

$$\sin \omega t\xi(t) \leftrightarrow \frac{1}{2j}\frac{1}{s - j\omega} + \frac{1}{2j}\frac{1}{s + j\omega}$$

$$\leftrightarrow \frac{\omega}{s^2 + \omega^2} \tag{6.28}$$

The time-integration property applied to the unit step function $\xi(t)$ yields a ramp function $t\xi(t)$ with the transform

$$t\xi(t) \leftrightarrow \frac{1}{s^2} \tag{6.29}$$

The time-derivative property applied to the unit step function $\xi(t)$ yields an impulse function $\delta(t)$ with the transform

$$\delta(t) \leftrightarrow 1 \tag{6.30}$$

One can apply the time-derivative property to the impulse function $\delta(t)$ to obtain $\delta^{(1)}(t)$ with the transform

$$\delta^{(1)}(t) \leftrightarrow s \tag{6.31}$$

Suppose that $a = \sigma \pm j\omega$ in (6.23). The resulting transform pair is

$$e^{-(\sigma \pm j\omega)t}\xi(t) \leftrightarrow \frac{1}{s + (\sigma \pm j\omega)} \tag{6.32}$$

If we break (6.32) into real and imaginary parts, we obtain two transform pairs.

$$e^{-\sigma t} \cos \omega t\xi(t) \leftrightarrow \frac{s + \sigma}{(s + \sigma)^2 + \omega^2} \tag{6.33}$$

$$e^{-\sigma t} \sin \omega t\xi(t) \leftrightarrow \frac{\omega}{(s + \sigma)^2 + \omega^2} \tag{6.34}$$

Table 6.1 summarizes some of the more useful transform pairs primarily for one-sided transforms.

Table 6.1 Useful Transform Pairs

$f(t)$	$F(s)$	Convergence Region
1. $e^{-at}\xi(t)$	$\dfrac{1}{s+a}$	$-\text{Re}(a) < \text{Re}(s)$
2. $\xi(t)$	$\dfrac{1}{s}$	$0 < \text{Re}(s)$
3. $t\xi(t)$	$\dfrac{1}{s^2}$	$0 < \text{Re}(s)$
4. $t^n\xi(t)$	$n!/s^{n+1}$	$0 < \text{Re}(s)$
5. $\delta(t)$	1	all s
6. $\delta^{(1)}(t)$	s	all s
7. $\text{sgn}\, t$	$\dfrac{2}{s}$	$\text{Re}(s) = 0$
8. $-\xi(-t)$	$\dfrac{1}{s}$	$\text{Re}(s) < 0$
9. $te^{-at}\xi(t)$	$\dfrac{1}{(s+a)^2}$	$-\text{Re}(a) < \text{Re}(s)$
10. $t^n e^{-at}\xi(t)$	$\dfrac{n!}{(s+a)^{n+1}}$	$-\text{Re}(a) < \text{Re}(s)$
11. $e^{-a\|t\|}\xi(t)$	$\dfrac{2a}{a^2 - s^2}$	$-\text{Re}(a) < \text{Re}(s) < \text{Re}(a)$
12. $(1 - e^{-at})\xi(t)$	$\dfrac{a}{s(s+a)}$	$\max(0, -\text{Re}(a)) < \text{Re}(s)$
13. $\cos\omega t\xi(t)$	$\dfrac{s}{s^2+\omega^2}$	$0 < \text{Re}(s)$
14. $\sin\omega t\xi(t)$	$\dfrac{\omega}{s^2+\omega^2}$	$0 < \text{Re}(s)$
15. $e^{-\sigma t}\cos\omega t\xi(t)$	$\dfrac{s+\sigma}{(s+\sigma)^2+\omega^2}$	$-\sigma < \text{Re}(s)$
16. $e^{-\sigma t}\sin\omega t\xi(t)$	$\dfrac{\omega}{(s+\sigma)^2+\omega^2}$	$-\sigma < \text{Re}(s)$
17. $\begin{cases} 1 - \|t\|, & \|t\| < 1 \\ 0, & \|t\| > 1 \end{cases}$	$\left(\dfrac{\sinh s/2}{s/2}\right)^2$	all s
18. $\sum_{n=0}^{\infty} \delta(t - nT)$	$\dfrac{1}{1 - e^{-sT}}$	all s

6.5 INVERSION OF THE LAPLACE TRANSFORM

In order to use the Laplace transform effectively in system analysis, one must be able to carry out the transformation from the s-domain to the t domain easily. This inverse transformation can be accomplished in several ways. The most direct method is to use the defining equation for $f(t)$: that is,

$$f(t) = \frac{1}{2\pi j} \int_{\sigma - j\infty}^{\sigma + j\infty} F(s)e^{st}\, ds \tag{6.35}$$

Evaluation of (6.35) requires an understanding of complex variables. The integration is usually calculated by means of a line integral in the s-plane. It is not always necessary to use (6.35) to find $f(t)$, and for certain classes of systems, other methods are easier.

As with the Z-transform, if one has $F(s)$ and its region of convergence, then the corresponding time function $f(t)$ is unique. If $F(s)$ is a one-sided transform, then knowledge of the region of convergence is not needed to obtain a unique inverse. This situation is analogous to the one that occurs for one-sided and two-sided Z-transforms. The following example demonstrates these ideas.

Example 6.5 Consider the two time functions $f_1(t) = e^{at}\xi(t)$ and $f_2(t) = -e^{at}\xi(-t)$, shown in Figure 6.4. The corresponding Laplace transforms are

$$F_1(s) = \int_{-\infty}^{\infty} e^{at}\xi(t)e^{-st}\, dt = \int_{0}^{\infty} e^{at}e^{-st}\, dt$$

$$= \frac{-1}{s-a} e^{-(s-a)t}\Bigg|_0^{\infty} = \frac{1}{s-a}, \qquad \sigma > a$$

$$F_2(s) = \int_{-\infty}^{\infty} -e^{at}\xi(-t)e^{-st}\, dt = \int_{-\infty}^{0} -e^{at}e^{-st}\, dt$$

$$= \frac{1}{s-a} e^{-(s-a)t}\Bigg|_{-\infty}^{0} = \frac{1}{s-a}, \qquad \sigma < a$$

The Laplace transform of $f_1(t)$ and $f_2(t)$ is $1/(s-a)$. However, the regions of convergence are different. If one allows time functions to be nonzero for both positive and negative t, then one must specify the region of convergence of $F(s)$ in order to specify a unique time function.

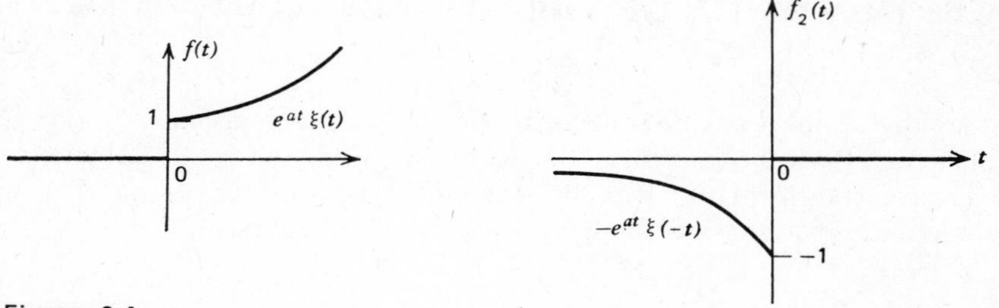

Figure 6.4

Partial Fraction Expansions

Linear time-invariant systems with lumped parameters generally lead to transforms that are rational functions of s: that is, transforms that are ratios of polynomials in s. Rational functions of this form can be represented as a sum of simpler fractions whose inverse transforms are tabulated. Assume for the present that we are concerned only with causal time functions, so that the transforms $F(s)$ are one sided. We can thus dispense with considering the convergence region in taking inverse transforms. We assume that $F(s)$ is of the form

$$F(s) = \frac{a_0 + a_1 s + a_2 s^2 + \cdots + a_m s^m}{b_0 + b_1 s + b_2 s^2 + \cdots + b_n s^n} \tag{6.36}$$

Without loss of generality, we assume $m < n$. If this is not the case, then we can always write $F(s)$ as the sum of a polynomial $Q(s)$ of degree $m - n$ plus a ratio of polynomials with the numerator degree one less than the denominator degree.

In (6.36), to find a partial fraction expansion, we first find the roots of the denominator polynomial, $p_1, p_2, \ldots, p_n$. These roots are the poles or singularities of $F(s)$. We write $F(s)$ as

$$\begin{aligned} F(s) &= \frac{a_0 + a_1 s + a_2 s^2 + \cdots + a_m s^m}{b_n(s - p_1)(s - p_2) \cdots (s - p_n)} \\ &= \frac{c_1}{s - p_1} + \frac{c_2}{s - p_2} + \cdots + \frac{c_n}{s - p_n} \end{aligned} \tag{6.37}$$

The coefficients c_i in (6.37) can be found for the case of nonrepeated poles as

$$c_i = (s - p_i)F(s)|_{s = p_i} \tag{6.38}$$

Equation 6.38 is obtained by multiplying both sides of (6.37) by $(s - p_i)$ and then evaluating the resulting equation at $s = p_i$. In the case of multiple roots, say p_i repeated r times, the expansion of (6.37) must include terms

$$\frac{c_{i_1}}{s - p_i} + \frac{c_{i_2}}{(s - p_i)^2} + \cdots + \frac{c_{i_r}}{(s - p_i)^r}$$

The coefficients c_{i_k} are evaluated by multiplying both sides of (6.37) by $(s - p_i)^r$, differentiating $(r - k)$ times, and then evaluating the resultant equation at $s = p_i$. Thus

$$\begin{aligned}
c_{i_r} &= (s - p_i)^r F(s)\big|_{s=p_i} \\
c_{i_{r-1}} &= \frac{d}{ds}\{(s - p_i)^r F(s)\}\Big|_{s=p_i} \\
&\cdots \\
c_{i_{r-k}} &= \frac{1}{k!}\frac{d^k}{ds^k}\{(s - p_i)^r F(s)\}\Big|_{s=p_i} \\
&\cdots \\
c_{i_1} &= \frac{1}{(r-1)!}\frac{d^{r-1}}{ds^{r-1}}\{(s - p_i)^r F(s)\}\Big|_{s=p_i}
\end{aligned} \tag{6.39}$$

The general form, assuming that the pole p_i is repeated r_i times, for $i = 1, 2, \ldots, k$ and $r_1 + r_2 + \cdots + r_k = n$, is

$$\begin{aligned}
F(s) = {} & \frac{c_1}{s - p_1} + \cdots + \frac{c_{i_1}}{s - p_i} \\
& + \frac{c_{i_2}}{(s - p_i)^2} + \cdots + \frac{c_{i_r}}{(s - p_i)^{r_i}} + \cdots + \frac{c_k}{(s - p_k)^{r_k}}
\end{aligned} \tag{6.40}$$

The above expansions can always be checked by recombining the terms in the partial fraction expansion. These expansions are valid for complex roots also. However, because complex roots always occur as a complex conjugate pair [for real $f(t)$], they can be expressed as a single quadratic factor. Quadratic factors of the denominator polynomial can be separated either as a single entity or as two simple poles. Suppose there is a complex conjugate pair of poles in the denominator polynomial. Three alternate representations for these roots are

$$\begin{aligned}
(s + \alpha + j\omega_0)(s + \alpha - j\omega_0) &= (s + \alpha)^2 + \omega_0^2 \\
&= s^2 + 2\alpha s + (\alpha^2 + \omega_0^2) \\
&= s^2 + as + b
\end{aligned} \tag{6.41}$$

The corresponding partial fraction expansion can take any of the following forms, depending on the representation used in (6.41).

$$\begin{aligned} F(s) &= \frac{A(s)}{B(s)} = \frac{a_0 + a_1 s + a_2 s^2 + \cdots + a_m s^m}{b_0 + b_1 s + b_2 s^2 + \cdots + b_n s^n} = \frac{A(s)}{B_1(s)(s^2 + as + b)} \\ &= \frac{\tilde{a}_1 s + \tilde{a}_2}{s^2 + as + b} + \frac{A_1(s)}{B_1(s)} \\ &= \frac{\tilde{b}_1 s + \tilde{b}_2}{(s + \alpha)^2 + \omega_0^2} + \frac{A_1(s)}{B_1(s)} \\ &= \frac{\tilde{c}_1 + j\tilde{c}_2}{s + \alpha + j\omega_0} + \frac{\tilde{c}_1 + j\tilde{c}_2}{s + \alpha - j\omega_0} + \frac{A_1(s)}{B_1(s)} \end{aligned} \tag{6.42}$$

The constants $\tilde{a}$, $\tilde{b}$, and $\tilde{c}$ in (6.42) are all real numbers. Usually, the latter two forms in (6.42) are the easiest to use. Notice that in the last form in (6.42), the two factors are complex conjugates, and so only one factor need be evaluated. The other can then be written down immediately. The following examples demonstrate finding inverse transforms by partial fraction expansions.

Example 6.6 Assume that $f(t)$ is a causal function and that $F(s)$ is given by $F(s) = (s + 3)/(s^3 + 3s^2 + 6s + 4)$. Because $f(t)$ is causal, we need not worry about the convergence region of $F(s)$ to obtain a unique inverse. We factor the denominator polynomial by guessing at a root, say a, and then dividing $s^3 + 3s^2 + 6s + 4$ by $s - a$. We find quickly that $s = -1$ is a root, and so

$$F(s) = \frac{s + 3}{[(s + 1)^2 + 3](s + 1)} = \frac{A}{s + 1} + \frac{Bs + C}{s^2 + 2s + 4} \tag{6.43}$$

We evaluate A by multiplying both sides of (6.43) by $(s + 1)$, then setting $s = -1$:

$$A = (s + 1)F(s)|_{s=-1} = \frac{(-1) + 3}{[(-1) + 1]^2 + 3} = \frac{2}{3}$$

The values of B and C can now be found by clearing (6.43) of fractions and equating the coefficients of powers of s.

$$s + 3 = \frac{2}{3}s^2 + \frac{4}{3}s + \frac{8}{3} + Bs^2 + Bs + Cs + C$$

Now we set

$$\frac{2}{3} + B = 0, \qquad B = -\frac{2}{3}$$

and

$$\frac{4}{3} + B + C = 1, \qquad C = \frac{1}{3}$$

Therefore we have that

$$\begin{aligned} F(s) &= \frac{2}{3}\frac{1}{s+1} + \frac{-\frac{2}{3}s + \frac{1}{3}}{s^2 + 2s + 4} \\ &= \frac{2}{3}\frac{1}{s+1} - \frac{2}{3}\frac{s - \frac{1}{2}}{(s+1)^2 + 3} \\ &= \frac{2}{3}\frac{1}{s+1} - \frac{2}{3}\frac{s+1}{(s+1)^2 + 3} + \frac{1}{\sqrt{3}}\frac{\sqrt{3}}{(s+1)^2 + 3} \end{aligned} \tag{6.44}$$

Equation 6.44 is now in a form that can be found immediately in Table 6.1. Thus

$$f(t) = \left(\frac{2}{3}e^{-t} - \frac{2}{3}e^{-t}\cos\sqrt{3}t + \frac{1}{\sqrt{3}}e^{-t}\sin\sqrt{3}t\right)\xi(t) \qquad \blacksquare$$

Example 6.7 Assume again that $f(t)$ is causal and that its transform is given by

$$F(s) = \frac{2s^2 + 3s + 3}{(s+1)(s+3)^3}$$

In this case, $F(s)$ has a first-order pole at $s = -1$ and third-order pole at $s = -3$. We thus expand $F(s)$ as

$$F(s) = \frac{2s^2 + 3s + 3}{(s+1)(s+3)^3} = \frac{A}{s+1} + \frac{C_1}{s+3} + \frac{C_2}{(s+3)^2} + \frac{C_3}{(s+3)^3} \tag{6.45}$$

We can evaluate coefficient A as in Example 6.6:

$$A = (s+1)F(s)\big|_{s=-1} = \frac{1}{4}$$

Coefficient C_3 is found by multiplying (6.45) by $(s + 3)^3$ and then evaluating at $s = -3$,

$$C_3 = (s + 3)^3 F(s)|_{s=-3} = -6$$

To find C_2 and C_1, we have to differentiate. Thus

$$\begin{aligned} C_2 &= \frac{d}{ds}\{(s + 3)^3 F(s)\}\bigg|_{s=-3} = \frac{d}{ds}\left\{\frac{2s^2 + 3s + 3}{s + 1}\right\}\bigg|_{s=-3} \\ &= \left(\frac{4s + 3}{s + 1} - \frac{2s^2 + 3s + 3}{(s + 1)^2}\right)\bigg|_{s=-3} = \frac{3}{2} \end{aligned}$$

Also,

$$\begin{aligned} C_1 &= \frac{1}{2!}\frac{d^2}{ds^2}\{(s + 3)^3 F(s)\}\bigg|_{s=-3} = \frac{1}{2}\frac{d}{ds}\left\{\frac{4s + 3}{s + 1} - \frac{2s^2 + 3s + 3}{(s + 1)^2}\right\}\bigg|_{s=-3} \\ &= \frac{1}{2}\left\{\frac{4}{s + 1} - \frac{4s + 3}{(s + 1)^2} - \frac{4s + 3}{(s + 1)^2} + \frac{2(2s^2 + 3s + 3)}{(s + 1)^3}\right\}\bigg|_{s=-3} = -\frac{1}{8} \end{aligned}$$

and so

$$F(s) = \frac{\frac{1}{4}}{s + 1} + \frac{-\frac{1}{8}}{s + 3} + \frac{\frac{2}{3}}{(s + 3)^2} + \frac{-6}{(s + 3)^3}$$

Using Table 6.1, we obtain the inverse transform as

$$f(t) = \left(\frac{1}{4}e^{-t} - \frac{1}{8}e^{-3t} + \frac{3}{2}te^{-3t} - 3t^2e^{-3t}\right)\xi(t)$$ ■

Inversion of Two-Sided Laplace Transforms

We have to this point considered finding inverse transforms only for causal functions. Suppose now that $f(t)$ is noncausal. In order to find $f(t)$ from $F(s)$ we must know the region of convergence for $F(s)$. The location of the poles of $F(s)$ with respect to the convergence region determines whether a given singularity refers to a positive or negative time portion of $f(t)$. If, for example, a pole of $F(s)$ lies to the right of the convergence region, then by our discussion in Section 6.1, this pole gives rise to a nonzero time function for $t < 0$. Similarly, poles to the left of the convergence region give rise to a positive time portion of $f(t)$.

To illustrate the procedure suppose that $F(s)$ is given by

$$F(s) = \frac{2s}{(s + 1)(s + 2)}, \qquad -2 < \text{Re}(s) < -1$$

In this case, we see that the pole at $s = -1$ lies to the right of the convergence region and the pole at $s = -2$ lies to the left of the convergence region. Thus, the pole at $s = -1$ gives rise to a time function nonzero for $t < 0$ and the pole at $s = -2$ gives rise to a time function nonzero for $t > 0$. We now expand $F(s)$ in partial fractions to obtain

$$F(s) = \frac{2s}{(s + 1)(s + 2)} = \frac{-2}{s + 1} + \frac{4}{s + 2}, \qquad -2 < \text{Re}(s) < -1$$

Having identified the poles that correspond to negative time portions and positive time portions of $f(t)$, we can proceed with inverting $F(s)$. The term $4/(s + 2)$ yields, from Table 6.1,

$$\frac{4}{s + 2} \leftrightarrow 4e^{-2t}\xi(t)$$

The term $-2/(s + 1)$ corresponds to a negative time portion of $f(t)$. Using the reasoning of Example 6.1, we find that $f(t) = e^{at}\xi(-t)$ has transform $-1/(s - a)$. Thus,

$$\frac{-2}{s + 1} \leftrightarrow 2e^{-t}\xi(-t)$$

The inverse transform of $F(s)$ is thus

$$f(t) = 2e^{-t}\xi(-t) + 4e^{-2t}\xi(t)$$

The inverse transform of many two-sided Laplace transforms can be found using the procedure outlined above. First, expand $F(s)$ in a partial fraction expansion. Identify which poles lie to the right and the left of the convergence region. For poles to the left of the convergence region, invert each term to obtain a time function that is nonzero for $t > 0$. For poles to the right of the convergence region, invert each term to obtain a time function that is nonzero for $t < 0$. The key transform pairs are

$$\frac{t^n e^{-at}\xi(t)}{n!} \leftrightarrow \frac{1}{(s + a)^{(n+1)}}, \qquad \text{Re}(s) > -a \tag{6.46}$$

$$\frac{(-t)^n e^{-at}\xi(-t)}{n!} \leftrightarrow \frac{1}{(s + a)^{(n+1)}}, \qquad \text{Re}(s) < -a \tag{6.47}$$

Another method of finding the inverse of two-sided Laplace transforms makes use of tables for one-sided transforms. The underlying idea is to express the Laplace transfer form as the sum of two one-sided transforms.

Consider a function $f(t)$ nonzero for both positive and negative t, as shown in Figure 6.5. We can always decompose $f(t)$ into a sum of two functions, $f_1(t)$ nonzero for $t > 0$ and $f_2(t)$ nonzero for $t < 0$. Thus

$$f(t) = f_1(t)\xi(t) + f_2(t)\xi(-t)$$

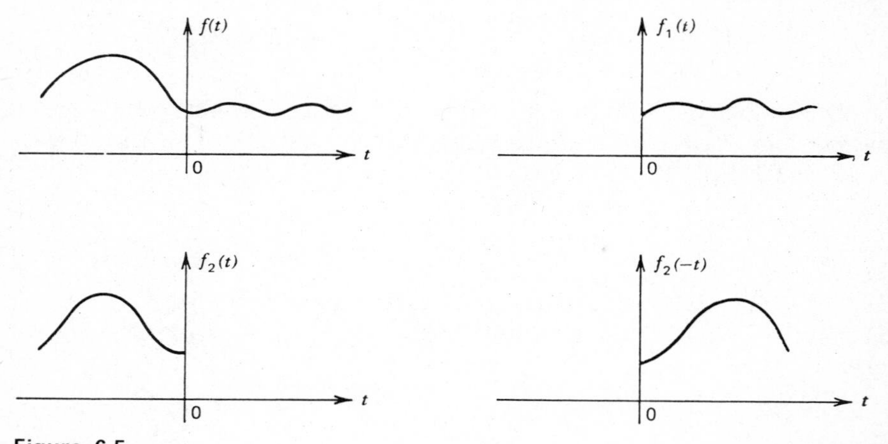

Figure 6.5

Suppose we now write the Laplace transform of $f(t)$:

$$\mathscr{L}_b\{f(t)\} = \int_{-\infty}^{\infty} f(t)e^{-st}\,dt = \int_{-\infty}^{0} f_2(t)e^{-st}\,dt + \int_{0}^{\infty} f_1(t)e^{-st}\,dt \tag{6.48}$$

The first integral in (6.48) can be put in standard one-sided form by substituting $t' = -t$. Thus

$$\begin{aligned} F(s) &= \int_{0}^{\infty} f_2(-t')e^{st'}\,dt' + \int_{0}^{\infty} f_1(t)e^{-st}\,dt \\ &= F_2(-s) + F_1(s) \end{aligned} \tag{6.49}$$

where

$$F_1(s) = \mathscr{L}\{f_1(t)\}$$

$$F_2(s) = \int_{0}^{\infty} f_2(-t)e^{-st}\,dt = \mathscr{L}\{f_2(-t)\}$$

The function $f_2(-t)$ is the mirror image of $f_2(t)$ about the vertical axis at $t = 0$. We now have a method for obtaining $F(s)$ using tables for one-sided transforms. Merely break $f(t)$ into its positive and negative time portions. Take the mirror reflection of the negative time portion $f_2(-t)$ and find its one-sided Laplace transform $F_2(s)$. Now replace s by $-s$ for the negative time portion and add $F_2(-s)$ to the one-sided Laplace transform for the positive time portion.

Example 6.8 Find the Laplace transform of $f(t)$, where $f(t) = e^{-\alpha t}\xi(t) + e^{\beta t}\xi(-t)$, as shown in Figure 6.6. Clearly

$$f_1(t) = e^{-\alpha t}\xi(t)$$

$$f_2(t) = e^{\beta t}\xi(-t)$$

The mirror reflection of $f_2(t)$ about the vertical axis at $t = 0$ is

$$f_2(-t) = e^{-\beta t}\xi(t)$$

and so

$$F_2(s) = \mathscr{L}\{f_2(-t)\} = \frac{1}{s + \beta}, \qquad -\beta < \text{Re}(s)$$

Thus, $F_2(-s) = 1/(-s + \beta)$, Re $(s) < \beta$. The positive time portion $f_1(t)$ has a transform

$$F_1(s) = \frac{1}{s + \alpha}, \qquad -\alpha < \text{Re}(s)$$

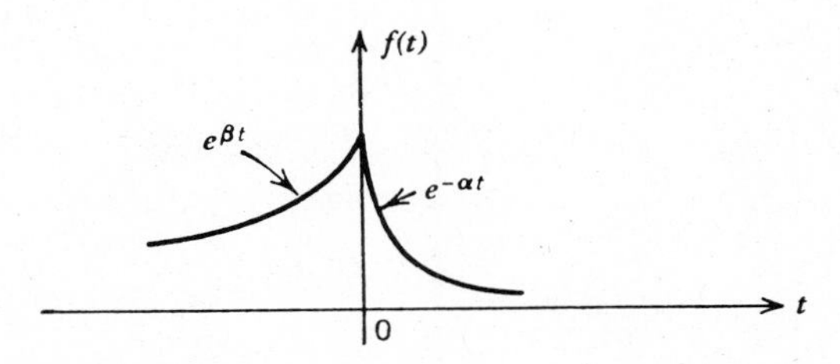

Figure 6.6

The two-sided transform of $F(s)$ is therefore

$$\mathscr{L}_b\{f(t)\} = F_1(s) + F_2(-s) = \frac{1}{s+\alpha} + \frac{1}{-s+\beta}$$

$$= \frac{(\alpha+\beta)}{(s+\alpha)(-s+\beta)}, \qquad -\alpha < \sigma < \beta$$

■

Example 6.9 Find the inverse Laplace transform of $F(s) = (2s+3)/(s+1)(s+2)$ if the convergence regions are:

1. $-2 < \text{Re}(s) < -1$
2. $\text{Re}(s) > -1$
3. $\text{Re}(s) < -2$.

We first break $F(s)$ into partial fractions.

$$F(s) = \frac{2s+3}{(s+1)(s+2)} = \frac{A}{s+1} + \frac{B}{s+2}$$

Evaluating A and B, we obtain

$$A = (s+1)F(s)|_{s=-1} = \left.\frac{2s+3}{s+2}\right|_{s=-1} = 1$$

$$B = (s+2)F(s)|_{s=-2} = \left.\frac{2s+3}{s+1}\right|_{s=-2} = 1$$

and so

$$F(s) = \frac{1}{s+1} + \frac{1}{s+2}$$

1. For the convergence region $-2 < \text{Re}(s) < -1$ we see that the pole at $s = -1$ lies to the right of the convergence region and the pole at $s = -2$ lies to the left of the convergence region. The first term, therefore, gives rise to a time function nonzero for negative t and the second term gives rise to a time function nonzero for positive t. In the notation of (6.49), we have that

$$F_1(s) = \frac{1}{s+2}, \qquad F_2(-s) = \frac{1}{s+1}$$

The positive time portion of $f(t)$ is therefore

$$f_1(t) = \mathscr{L}^{-1}\{F_1(s)\} = e^{-2t}\xi(t)$$

The mirror reflection of the negative time portion of $f(t)$ is the inverse transform of $F_2(s) = 1/(-s+1)$. Thus

$$f_2(-t) = \mathscr{L}^{-1}\{F_2(s)\} = \mathscr{L}^{-1}\left\{\frac{1}{-s+1}\right\} = -e^{t}\xi(t)$$

and so

$$f_2(t) = -e^{-t}\xi(-t)$$

Thus,

$$f(t) = -e^{-t}\xi(-t) + e^{-2t}\xi(t)$$

2. For the convergence region Re $(s) > -1$, we see that both poles in $F(s)$ lie to the left of the convergence region. Therefore $f(t)$ must be nonzero for positive t only: that is, $f(t)$ is causal. Thus, $F_2(-s) = 0$ and so

$$f(t) = (e^{-t} + e^{-2t})\xi(t)$$

3. For the convergence region Re $(s) < -2$, both poles in $F(s)$ lie to the right of the convergence region. Thus $f(t)$ is nonzero for negative t only. Therefore, in the notation of (6.49),

$$F_1(s) = 0, \qquad F_2(-s) = \frac{1}{s+1} + \frac{1}{s+2}$$

and so

$$f_2(-t) = \mathscr{L}^{-1}\{F_2(s)\} = \mathscr{L}^{-1}\left\{\frac{1}{-s+1} + \frac{1}{-s+2}\right\}$$

$$= -e^{t}\xi(t) - e^{2t}\xi(t)$$

This relation implies that

$$f(t) = f_2(t) = (-e^{-t} - e^{-2t})\xi(-t)$$

■

This example demonstrates how the "same" Laplace transform $F(s)$ can represent various time functions, depending on the region of convergence. The inversion process merely involves identifying poles to the left and right of the convergence region. By our previous discussions, poles to the left of the convergence region give rise to positive time functions and those to the right of the convergence region give rise to negative time functions. By converting that part of $F(s)$ with singularities to the right of the convergence region into a positive time function (replacing s by $-s$) we are able to use one-sided transform tables to evaluate two-sided transforms. In physical applications, the correct region of convergence can generally be obtained by physical reasoning. Time functions that approach $\pm\infty$ as $t \to \pm\infty$ do not exist physically. Thus, it is a fairly easy matter to consider all the possible convergence regions and rule out those which give rise to time functions that approach $\pm\infty$ as $t \to \pm\infty$. Only one "practical" time function will remain.

To summarize, the process of inverting two-sided Laplace transforms using one-sided transform pairs consists of these steps:

1. Expand $F(s)$ in a partial fraction expansion.
2. Identify those poles to the right of the convergence region. These terms give rise to the negative time portion of $f(t)$ and are denoted as $F_2(-s)$.
3. Replace $-s$ by s in $F_2(-s)$. Invert $F_2(s)$ to obtain the mirror image of the negative time portion of $f(t)$, denoted as $f_2(-t)$.
4. Finally, replace $-t$ by t to obtain $f_2(t)$. This term is added to the positive time portion of $f(t)$, $f_1(t)$, to obtain the complete time function.

6.6 APPLICATIONS OF LAPLACE TRANSFORMS—DIFFERENTIAL EQUATIONS

In Chapter 1, the basic models developed to describe linear systems were linear differential or difference equations with constant coefficients. If we have a linear time-invariant system with input $u(t)$ and output $y(t)$, the basic model that we have used to describe the system is

$$b_n \frac{d^n}{dt^n} y(t) + b_{n-1} \frac{d^{n-1}}{dt^{n-1}} y(t) + \cdots + b_0 y(t) = a_m \frac{d^m}{dt^m} u(t) + a_{m-1} \frac{d^{m-1}}{dt^{m-1}} u(t) + \cdots + a_0 u(t) \quad (6.50)$$

We can solve this differential equation by using the Laplace transform. Multiplying both sides of (6.50) by e^{-st} and integrating each term from $-\infty$ to ∞, we obtain

$$b_n \int_{-\infty}^{\infty} y^{(n)}(t)e^{-st}\,dt + \cdots + b_0 \int_{-\infty}^{\infty} y(t)e^{-st}\,dt$$
$$= a_m \int_{-\infty}^{\infty} u^{(m)}(t)e^{-st}\,dt + \cdots + a_0 \int_{-\infty}^{\infty} u(t)e^{-st}\,dt \quad (6.51)$$

Denote the Laplace transforms for $y(t)$ and $u(t)$ by $Y(s)$ and $U(s)$, respectively. Equation 6.51 can then be written, using the Laplace transform for the derivative, as

$$b_n Y(s)s^n + \cdots + b_0 Y(s) = a_m U(s)s^m + \cdots + a_0 U(s)$$

Solving for $Y(s)$, we have

$$Y(s) = \left(\frac{a_m s^m + a_{m-1}s^{m-1} + \cdots + a_0}{b_n s^n + b_{n-1}s^{n-1} + \cdots + b_0}\right)U(s) \quad (6.52)$$

Equation 6.52 is an algebraic equation in s that can readily be solved for $Y(s)$. $Y(s)$ can then be expanded in partial fractions and the inverse transform taken to find $y(t)$. Equation 6.52 is a conceptually useful equation also. The system is characterized by the ratio of polynomials in s: that is, $(a_m s^m + \cdots + a_0)/(b_n s^n + \cdots + b_0)$. We term this ratio of polynomials the system transfer function or system function $H(s)$:

$$H(s) = \frac{a_m s^m + a_{m-1}s^{m-1} + \cdots + a_0}{b_n s^n + b_{n-1}s^{n-1} + \cdots + b_0} \quad (6.53)$$

Thus the transform of the output of a linear time-invariant system is always given by multiplying $H(s)$ by the Laplace transform of the input $U(s)$: that is,

$$Y(s) = H(s) \cdot U(s)$$

By definition, the output of a linear system when the input is an impulse function is the impulse response of the system. Recall that the Laplace transform of an impulse is 1. Thus, for an impulse input, the output $y(t)$ is $h(t)$, the impulse response, and

$$Y(s) = H(s) \cdot 1 \quad (6.54)$$

Equation 6.54 merely states that $H(s)$ is the Laplace transform of the impulse response of the system, that is,

$$H(s) = \mathscr{L}_b\{h(t)\}$$

Notice that in going from (6.51) to (6.52), we assumed all initial conditions as zero. This assumption is always made in finding the system transfer function $H(s)$.

One of the advantages of the transform method is the automatic inclusion of initial conditions into the solution process. To demonstrate this property, consider the next example.

Example 6.10 Consider the simple RLC series network shown in Figure 6.7. There is an initial current i_L flowing in the inductor and initial voltage v_c across the capacitor. The switch is closed at $t = 0$. The differential equation for the system is

$$e(t) = Ri(t) + L\frac{di(t)}{dt} + \frac{1}{C}\int_{-\infty}^{t} i(t')\,dt' \tag{6.55}$$

Transforming this equation, we obtain

$$E(s) = RI(s) + L(sI(s) + i_L) + \frac{1}{C}\left(\frac{I(s)}{s} - \frac{v_c}{s}\right)$$

Notice that the signs on the initial conditions depend on the assumed direction for $i(t)$ relative to the flow of the current in L and the voltage across C. Rearranging terms yields

$$I(s)\left[R + Ls + \frac{1}{Cs}\right] = E(s) - Li_L + \frac{v_c}{Cs} \tag{6.56}$$

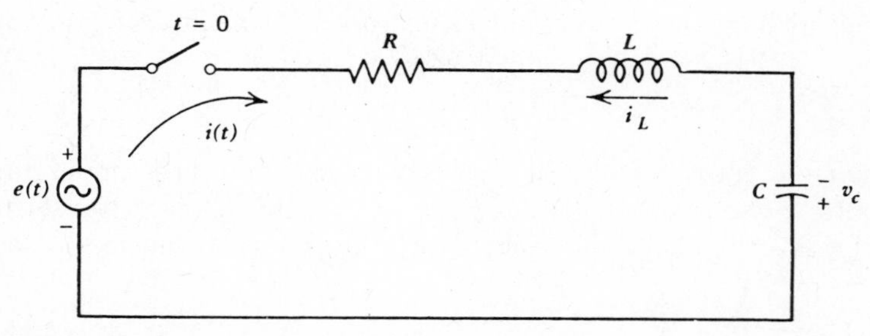

Figure 6.7

In (6.56), the bracketed terms are characteristic of the system, while the terms $E(s) - Li_L + (v_c/s)$ are characteristic of the input. The initial energy storage terms v_c and i_L act as forcing functions on the system [in conjunction with $E(s)$]. We find $i(t)$ by finding the inverse transform of $I(s)$ given by

$$I(s) = \frac{E(s) - Li_L + \dfrac{v_c}{Cs}}{R + Ls + \dfrac{1}{Cs}}$$

The Laplace transform is used almost exclusively to solve differential equations like (6.50) because the initial conditions are easily incorporated into the solution. However, the Fourier transform can be used in the same way. The main differences lie in the restricted class of functions that can be Fourier transformed and in the way initial conditions must be introduced. The initial conditions, when the Fourier transform is used, are lumped with the input forcing function in the following manner. Assume that we have a system described by the differential equation

$$a_3 \frac{d^3y(t)}{dt^3} + a_2 \frac{d^2y(t)}{dt^2} + a_1 \frac{dy(t)}{dt} + a_0 y(t) = u(t), \qquad t > 0 \tag{6.57}$$

In order to obtain a complete solution to (6.57), we need three initial conditions, such as $y(0)$, $y^{(1)}(0)$, $y^{(2)}(0)$. Assuming a nonzero solution for $t > 0$, we define the solution to be $y_1(t)$, where

$$y_1(t) = y(t)\xi(t)$$

thus we have that

$$\begin{aligned} \frac{dy_1(t)}{dt} &= y^{(1)}(t)\xi(t) + y(t)\,\delta(t) \\ &= y^{(1)}(t)\xi(t) + y(0)\,\delta(t) \end{aligned}$$

Similarly

$$\frac{d^2y_1(t)}{dt^2} = y^{(2)}(t)\xi(t) + y^{(1)}(0)\,\delta(t) + y(0)\,\delta^{(1)}(t)$$

$$\frac{d^3y_1(t)}{dt^3} = y^{(3)}(t)\xi(t) + y^{(2)}(0)\,\delta(t) + y^{(1)}(0)\,\delta^{(1)}(t) + y(0)\,\delta^{(2)}(t)$$

Substituting in (6.57), we write the original differential equation as

$$a_3 \frac{d^3 y_1(t)}{dt^3} + a_2 \frac{d^2 y_1(t)}{dt^2} + a_1 \frac{dy_1(t)}{dt} + a_0 y_1(t)$$

$$= u(t) + [a_1 y(0) + a_2 y'(0) + a_3 y^{(2)}(0)]\, \delta(t)$$

$$+ [a_2 y(0) + a_3 y^{(1)}(0)]\, \delta^{(1)}(t) + a_3 y(0)\, \delta^{(2)}(t) \quad (6.58)$$

This process has placed the initial conditions of the differential equation at time $t = 0$ on the right-hand side as part of the forcing function. We can now take the Fourier transform of (6.58) and proceed in much the same manner as with the Laplace transform process. ■

Example 6.11 A system is described by the differential equation

$$y^{(1)}(t) + ay(t) = e^{-\alpha t}\xi(t)$$

with the initial condition $y(0) = c$. We rewrite the differential equation to include the initial condition c on the right-hand side. Thus

$$y^{(1)}(t) + ay(t) = e^{-\alpha t}\xi(t) + c\, \delta(t)$$

Taking Fourier transforms on both sides yields

$$j\omega Y(\omega) + aY(\omega) = \frac{1}{\alpha + j\omega} + c$$

Solving for $Y(\omega)$, we obtain

$$Y(\omega) = \frac{c}{a + j\omega} + \frac{1}{(a + j\omega)(\alpha + j\omega)}$$

$$= \frac{c}{a + j\omega} + \frac{A}{a + j\omega} + \frac{B}{\alpha + j\omega}$$

where

$$A = \left. \frac{1}{\alpha + j\omega} \right|_{j\omega = -a} = \frac{1}{\alpha - a}$$

$$B = \left. \frac{1}{a + j\omega} \right|_{j\omega = -\alpha} = \frac{1}{a - \alpha}$$

and so

$$Y(j\omega) = \frac{c}{a + j\omega} + \frac{1}{\alpha - a}\left(\frac{1}{a + j\omega} - \frac{1}{\alpha + j\omega}\right)$$

Therefore

$$y(t) = \left\{ce^{-at} + \frac{1}{\alpha - a}(e^{-at} - e^{-\alpha t})\right\}\xi(t)$$

This problem can also be solved by using the Laplace transforms. Transforming the original differential equation, we have

$$sY(s) - c + aY(s) = \frac{1}{s + \alpha}$$

Solving for $Y(s)$, we obtain

$$\begin{aligned} Y(s) &= \frac{1}{(s + \alpha)(s + a)} + \frac{c}{s + a} \\ &= \frac{1}{\alpha - a}\left(\frac{1}{s + a} - \frac{1}{s + \alpha}\right) + \frac{c}{s + a} \end{aligned}$$

The inverse transform is therefore

$$y(t) = \left\{ce^{-at} + \frac{1}{\alpha - a}(e^{-at} - e^{-\alpha t})\right\}\xi(t)$$

which is the same as the result obtained using the Fourier transform. ■

6.7 STABILITY IN THE *s*-DOMAIN

One of the important questions that often must be answered in the analysis of a system concerns the stability of the system. In Chapter 3, we defined a stable system as one which had a bounded output for a bounded input. Obviously, this concept of stability is important if we are going to operate a system for any period of time. Stability, as we showed in Chapter 3, is a characteristic of the system and does not depend on the input to the system. In the state-variable formulation, we determined stability based on knowledge of the eigenvalues of the system matrix **A**. The eigenvalues of the matrix **A** allow us to determine the response of the natural modes of the system and thus determine stability. Stability can also be determined in the frequency or s domain.

The transfer function of a system $H(s)$ is the Laplace transform of the impulse response $h(t)$. We have seen that either $H(s)$ or $h(t)$ characterizes the system in the sense that we can find the output to any input given $H(s)$ or $h(t)$. Thus, because the stability of a system is characteristic of the system itself, independent of the input, we should be able to use $H(s)$ to determine stability.

Assume that $H(s)$ is the transfer function of a causal time-invariant system and is the ratio of two polynomials in s, as in (6.53). The transformed output is

$$Y(s) = \left(\frac{a_m s^m + a_{m-1} s^{m-1} + \cdots + a_0}{b_n s^n + b_{n-1} s^{n-1} + \cdots + b_0} \right) U(s) \tag{6.59}$$

Consider the time response of (6.59). If we perform a partial fraction expansion of (6.59), the denominator polynomial of $H(s)$, $b_n s^n + \cdots + b_0$, can factor into a variety of terms.

1. Simple poles of the form $c/(s + a)$. This form corresponds to a simple pole at $s = -a$. If a is positive, the pole is in the left half of the s plane, as shown in Figure 6.8. The corresponding time response is $ce^{-at}\,\xi(t)$, and as t increases, the time response dies away to zero. If a is negative, so that the pole is in the right half of the s-plane, then the time response increases without bound as t increases. Thus, a stable system must have real-valued poles of $H(s)$ in the left half of the s-plane. Repeated simple-poles also give rise to exponentially damped time responses for poles in the left half of the s-plane.

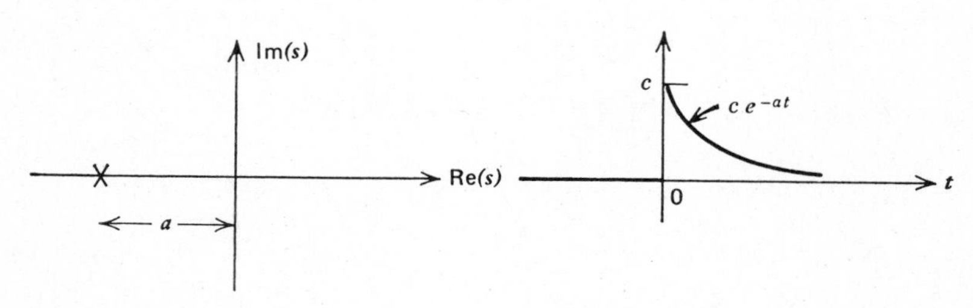

Figure 6.8

2. Complex conjugate poles of the form $c/[(s + \alpha)^2 + \omega^2]$. This term can be factored into two terms

$$\frac{c_1}{(s + \alpha + j\omega)} + \frac{c_1^*}{(s + \alpha - j\omega)}$$

This term thus represents a pair of complex conjugate poles. If α is positive, the poles are in the left half of the s-plane as shown in Figure 6.9. The corresponding time response is of the form $(ce^{-at}/\omega) \sin \omega t \, \xi(t)$, as shown in Figure 6.9. If α is negative, thereby placing the poles in the right half of the s plane, then the time function is $(ce^{|\alpha|t}/\omega) \sin \omega t \, \xi(t)$. Again we see that left-plane poles have time functions that die away as t increases and right plane poles correspond to time functions that increase without bound as t increases.

3. Complex conjugate poles of the form $c/(s^2 + \omega_0^2)$. Terms of this form represent complex conjugate poles on the $j\omega$-axis. The corresponding time function is $(c/\omega_0) \sin \omega_0 t$. In this case, there is no exponential damping, and so the response does not die away as t increases. It may appear at first glance that the time response does not increase as t increases. However, if the system is forced with a sinusoid of the same frequency ω_0, then a double complex conjugate pair of poles results, and $Y(s)$ has a term of the form $[1/(s^2 + \omega_0^2)]^2$. This term gives rise to a time response

$$\left(\frac{1}{2\omega_0^3}\right)(\sin \omega_0 t - \omega_0 t \cos \omega_0 t)$$

which increases without bound as t increases. Physically, we are exciting a natural resonance of the system with an input at precisely the resonant frequency. Because there is no loss ($\alpha = 0$) associated with this mode of the system, the output grows without bound. This is often called a marginally stable system. The same kind of considerations apply to a simple pole at the origin. This term gives rise to

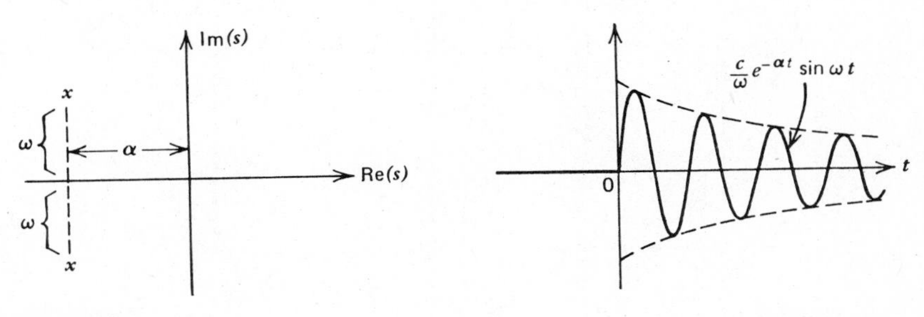

Figure 6.9

a time function that is a step. If this system is forced with a step, then a ramp output will result. Notice that if $H(s)$ has repeated poles on the $j\omega$-axis, terms occur of the form $[1/(s^2 + \omega^2)^2]$ that lead to time functions growing without bound.

To summarize, a causal, time-invariant lumped parameter system with transfer function $H(s)$ is stable if

1. All poles of $H(s)$ are in the left half of the s-plane.
2. The degree of the denominator polynomial of $H(s)$ is greater than or equal to the degree of the numerator polynomial.

The last consideration is required in order to rule out terms like s^{m-n}, a differentiator of degree $m - n$. If an input like $\sin \omega t$ were used to excite the system, then $y(t)$ would include a term $\omega^{m-n} \sin \omega t$. This term could be made as large as desired by increasing the input frequency ω.

Example 6.12 How is the stability of the system in Figure 6.10 affected by the amplifier gain g? This system is known as a feedback system. It has the very desirable property that the total response is not substantially affected by parameter changes within the feedback loop. The overall transfer function of the system is $H(s) = Y(s)/U(s)$ and can be calculated as follows. The output of the amplifier (in the transform domain) is $gY(s)$; thus, the input to the box containing $G(s)$ is

$$E(s) = U(s) - gY(s)$$

The output of the box containing $G(s)$ is, of course, $Y(s) = E(s)G(s)$. Therefore

$$\begin{aligned} Y(s) &= E(s)G(s) \\ &= [U(s) - gY(s)]G(s) \\ &= U(s)G(s) - gY(s)G(s) \end{aligned}$$

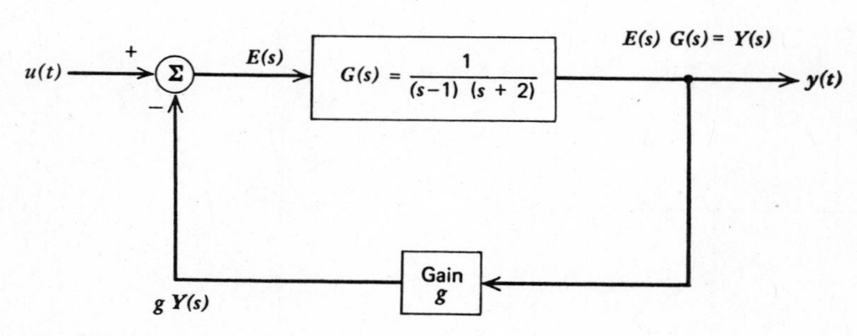

Figure 6.10

To find $H(s)$ we form the ratio $Y(s)/U(s)$. Thus

$$U(s)G(s) = Y(s) + gY(s)G(s) = Y(s)[1 + gG(s)]$$

and so

$$H(s) = \frac{Y(s)}{U(s)} = \frac{G(s)}{1 + gG(s)}$$

Stability of $H(s)$ depends only on the zeros of the denominator $1 + gG(s)$.

$$1 + gG(s) = 1 + g\frac{1}{(s-1)(s+2)} = \frac{(s-1)(s+2)+g}{(s-1)(s+2)}$$

The zeros of the denominator are thus determined by the roots of

$$(s+2)(s-1) + g = 0$$

or

$$s^2 + s - 2 + g = 0$$

The roots are thus

$$s_1, s_2 = -\frac{1}{2} \pm \sqrt{\frac{1 - 4(-2+g)}{4}} = -\frac{1}{2} \pm \sqrt{\frac{9}{4} - g}$$

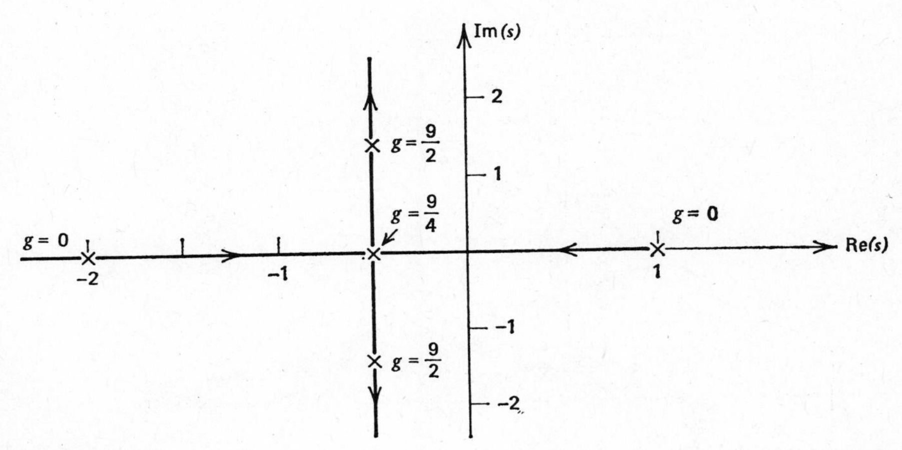

Figure 6.11

For $g < \frac{9}{4}$ the roots s_1 and s_2 are real. For $g > \frac{9}{4}$, the roots become complex conjugates. For $g = \frac{9}{4}$ there is a double root on the real axis of the s-plane at $\text{Re}(s) = -\frac{1}{2}$. If we were to plot a locus of the roots in the s-plane as g varies, we would obtain the graph of Figure 6.11. As we increase the gain of this feedback system, we actually stabilize an initially unstable system. The opposite phenomenon can also occur: that is, increasing the gain can make a stable feedback system unstable. ■

6.8 NONCAUSAL SYSTEMS AND INPUTS

In the analysis of most systems, the assumption is made that both the input $u(t)$ and the impulse response function $h(t)$ are causal functions. A noncausal impulse response implies that there exists some output before an impulse is applied. This interpretation thus leads one to term noncausal impulse responses as physically nonrealizable. Suppose that $h(t)$ is as shown in Figure 6.12. If we wish to find the output of this system at some time t resulting from an input $u(t)$, we can convolve $h(t)$ and $u(t)$ as shown graphically in Figure 6.12. Figure 6.12 shows why $h(t)$ is

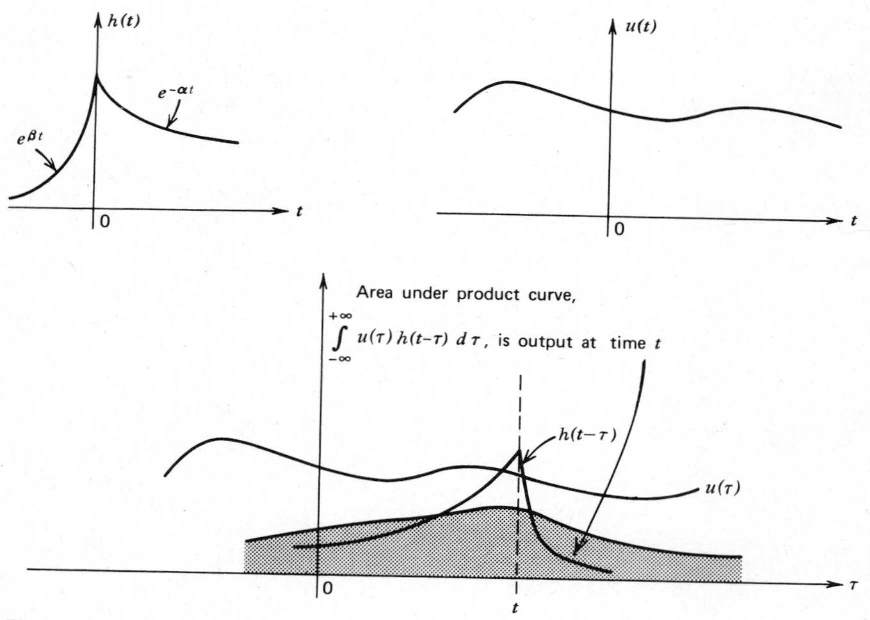

Figure 6.12

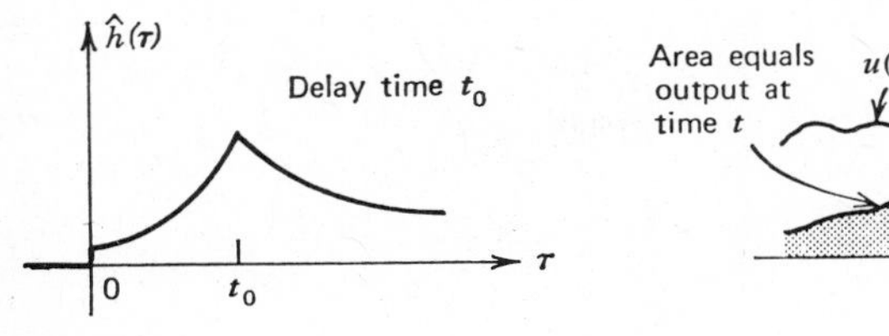

Figure 6.13

called unrealizable. To obtain the output at time t, we must weight future inputs of the input $u(t)$. If the processing is to occur in real time, this approach, of course, is impossible. However, there are many applications in which we can use noncausal impulse responses provided that

1. we can store the data $u(t)$ on a tape or in some other fashion or
2. we are willing to accept a time delay in obtaining the output $y(t)$.

There are many applications in which one processes stored data. For example, stored data is often used in seismic processing or in processing signals from space probes. In these cases, there is no reason not to use noncausal impulse responses to process the input data $u(t)$. In other cases, if we are willing to accept a time delay, of say t_0, after which we obtain the output, we can approximate noncausal impulse responses as closely as we wish. Thus, suppose we wish to approximate the impulse response of Figure 6.12. If we were to use a translated version of $h(t)$, as shown in Figure 6.13, truncated at $t = 0$, the output at time t would be obtained as shown in Figure 6.13. Notice that the actual output at time t is a good approximation to the noncausal output at time $(t - t_0)$. In other words, if we are willing to accept a delay of t_0, we can obtain a good approximation to the response of a noncausal system. This discussion thus serves to justify the use of noncausal impulse-response functions in practical applications. The two-sided Laplace transform is the appropriate transform tool to use for these noncausal systems. The Fourier transform can also be used to analyze such systems if we restrict our interest to functions that are absolutely integrable on $(-\infty, \infty)$, as the next example demonstrates.

Example 6.13 A system with impulse response $h(t)$ given by

$$h(t) = e^{-t}\xi(t) + e^{+2t}\xi(-t)$$

is excited by a signal $u(t)$ given by

$$u(t) = e^{-2t}\xi(t)$$

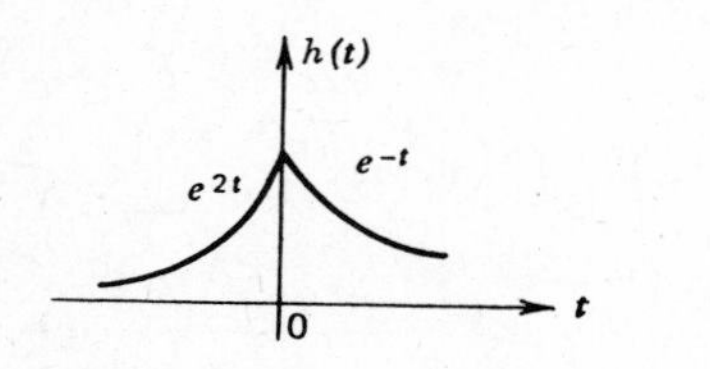

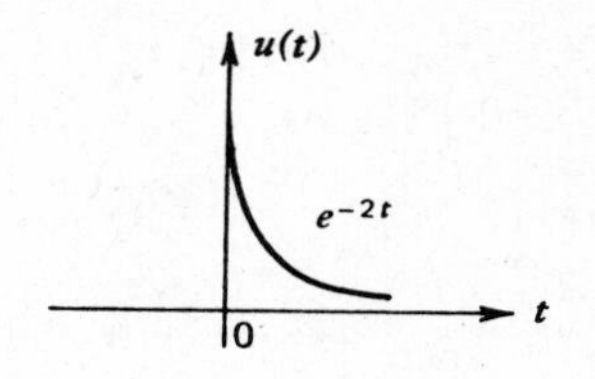

Figure 6.14

What is the output response? The input signal and impulse response are shown in Figure 6.14. The output $y(t)$ is given by

$$y(t) = u(t) * h(t)$$

which in the Fourier transform domain is

$$Y(j\omega) = U(j\omega)H(j\omega)$$

The system function $H(j\omega)$ is

$$\begin{aligned}
H(j\omega) = \mathscr{F}\{h(t)\} &= \int_{-\infty}^{\infty} [e^{2t}\xi(-t) + e^{-t}\xi(t)]e^{-j\omega t}\,dt \\
&= \int_{-\infty}^{0} e^{2t}e^{-j\omega t}\,dt + \int_{0}^{\infty} e^{-t}e^{-j\omega t}\,dt \\
&= \frac{1}{2 - j\omega} e^{(2-j\omega)t}\Big|_{-\infty}^{0} + \frac{-1}{1 + j\omega} e^{-(1+j\omega)t}\Big|_{0}^{\infty} \\
&= \frac{1}{2 - j\omega} + \frac{1}{1 + j\omega} = \frac{3}{(1 + j\omega)(2 - j\omega)}
\end{aligned}$$

Similarly, $U(j\omega)$ is

$$U(j\omega) = \frac{1}{2 + j\omega}$$

The transformed output is therefore

$$\begin{aligned}
Y(j\omega) = U(j\omega)H(j\omega) &= \frac{3}{(1 + j\omega)(2 - j\omega)(2 + j\omega)} \\
&= \frac{A}{1 + j\omega} + \frac{B}{2 - j\omega} + \frac{C}{2 + j\omega}
\end{aligned}$$

where

$$A = \left. \frac{3}{(2 - j\omega)(2 + j\omega)} \right|_{j\omega = -1} = 1$$

$$B = \left. \frac{3}{(1 + j\omega)(2 + j\omega)} \right|_{j\omega = 2} = \frac{1}{4}$$

$$C = \left. \frac{3}{(1 + j\omega)(2 - j\omega)} \right|_{j\omega = -2} = -\frac{3}{4}$$

and so

$$Y(j\omega) = \frac{1}{1 + j\omega} + \frac{\frac{1}{4}}{2 - j\omega} + \frac{-\frac{3}{4}}{2 + j\omega}$$

To find the inverse transform, we eliminate the ambiguity in which terms yield positive or negative time functions by requiring all time functions to be absolutely integrable. Because the term $(\frac{1}{4})/(2 - j\omega)$ has an inverse transform $\frac{1}{4}e^{2t}$, which approaches $+\infty$ as $t \to +\infty$ and 0 as $t \to -\infty$, we identify this term as a negative time function. The inverse transform is, by this process,

$$y(t) = \frac{1}{4} e^{2t}\xi(-t) + \left(e^{-t} - \frac{3}{4} e^{-2t}\right)\xi(t)$$

Suppose we now work this problem using the two-sided Laplace transform. The transformed output is

$$Y(s) = U(s)H(s)$$

where

$$U(s) = \mathscr{L}_b\{u(t)\} = \frac{1}{s + 2}, \qquad \sigma > -2$$

$$H(s) = \mathscr{L}_b\{h(t)\} = \frac{1}{s + 1} + \frac{1}{-s + 2}, \qquad -1 < \sigma < 2$$

$$= \frac{3}{(s + 1)(-s + 2)}$$

Thus

$$Y(s) = H(s)U(s) = \frac{3}{(s+1)(-s+2)(s+2)}$$

$$= \frac{A}{s+1} + \frac{B}{-s+2} + \frac{C}{s+2}$$

where

$$A = 1, \qquad B = \frac{1}{4}, \qquad C = -\frac{3}{4}$$

Thus

$$Y(s) = \frac{1}{s+1} + \frac{\frac{1}{4}}{-s+2} + \frac{-\frac{3}{4}}{s+2}, \qquad -1 < \sigma < 2$$

The poles at $s = -1$ and $s = -2$ lie to the left of the convergence region $-1 < \mathrm{Re}\,(s) < 2$ and so correspond to positive time functions. Similarly, the pole at $s = 2$ lies to the right of the convergence region and so corresponds to a negative time function. One could also reason physically, as we did in the case of the Fourier transform, to determine which terms in the partial fraction expansion of $Y(s)$ correspond to positive and negative time functions. The inverse transform is again

$$y(t) = \frac{1}{4} e^{2t}\xi(-t) + \left(e^{-t} - \frac{3}{4} e^{-2t}\right)\xi(t)$$ ■

6.9 TRANSIENT AND STEADY-STATE RESPONSE OF A LINEAR SYSTEM

The complete response of a system to any forcing function is composed of a transient portion that is characteristic of the system and a steady-state portion that depends on both the system and the forcing function. These ideas were first introduced in Chapter 3 in conjunction with the time-domain solution of differential equations. These concepts can also be illustrated in the frequency domain.

Suppose that we apply $u(t)$ to a system with impulse response $h(t)$. The output $y(t)$ in the transform domain is

$$Y(s) = U(s)H(s)$$

Let the transfer function of the system be

$$H(s) = \frac{N_1(s)}{D_1(s)} = \frac{N_1(s)}{(s - p_1)(s - p_2) \cdots (s - p_n)}$$

and the transformed input be

$$U(s) = \frac{N_2(s)}{D_2(s)} = \frac{N_2(s)}{(s - q_1)(s - q_2) \cdots (s - q_m)}$$

Then

$$Y(s) = \frac{N_1(s)N_2(s)}{(s - p_1) \cdots (s - p_n)(s - q_1) \cdots (s - q_m)}$$

We can expand $Y(s)$ in partial fractions as

$$Y(s) = \frac{c_1}{s - p_1} + \cdots + \frac{c_n}{s - p_n} + \frac{k_1}{s - q_1} + \cdots + \frac{k_m}{s - q_m}$$

and obtain the inverse transform

$$y(t) = \sum_{i=1}^{n} c_i e^{p_i t} + \sum_{i=1}^{m} k_i e^{q_i t} \tag{6.60}$$

The sum $\sum_{i=1}^{n} c_i e^{p_i t}$ in (6.60) is the transient response of $y(t)$. Notice that these terms result from the singularities of $H(s)$. [In general, the coefficients c_i depend on $H(s)$ and $U(s)$.] The transient response is a linear combination of terms oscillating at the natural frequencies of the system. In other words, the transient response results from the natural modes of the system. The sum $\sum_{i=1}^{m} k_i e^{q_i t}$ is called the steady-state portion of the response $y(t)$. These terms arise from the singularities in $U(s)$, and the coefficients k_i depend on both $H(s)$ and $U(s)$. In stable systems, the terms in the first sum die away with increasing t, while the steady-state portion may contain terms that are nonzero indefinitely. The impulse response $h(t)$ is obtained for an impulse input which in the s domain is an input of $U(s) = 1$.

Thus, $h(t)$ is a particular linear combination of the natural modes of the system. The transient and steady-state portions of $y(t)$ correspond to what we termed the homogeneous and nonhomogeneous solutions, respectively, for the differential equation describing the system.

Example 6.14 A voltage of $10 \cos 4t\xi(t)$ is applied to the network shown in Figure 6.15. What is the output voltage $v_0(t)$? The transfer function of this system is $H(s)$, given by

$$H(s) = \frac{V_0(s)}{V_i(s)} = \frac{1/Cs}{R + 1/Cs} = \frac{1}{RCs + 1} = \frac{1}{s + 1}$$

The Laplace transform of the input waveform is

$$V_i(s) = \mathscr{L}\{10 \cos 4t\xi(t)\} = \frac{10s}{s^2 + 16}$$

Thus the transform of the output voltage $v_0(t)$ is

$$V_0(s) = H(s)V_i(s) = \frac{10s}{(s^2 + 16)(s + 1)}$$

If we expand the right-hand side in partial fractions, we obtain

$$V_0(s) = \frac{As + B}{s^2 + 16} + \frac{C}{s + 1} \tag{6.61}$$

where

$$C = (s + 1)V_0(s)|_{s=-1} = \left.\frac{10s}{s^2 + 16}\right|_{s=-1} = -\frac{10}{17}$$

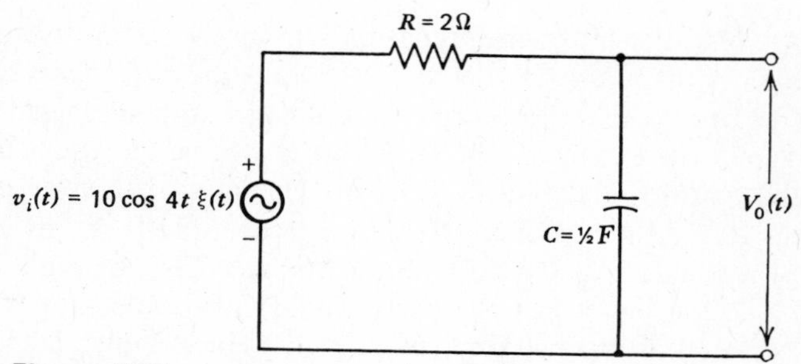

Figure 6.15

Thus, clearing (6.61) of fractions yields

$$10s = (As + B)(s + 1) - \frac{10}{17}(s^2 + 16)$$

$$= As^2 + Bs + As + B - \frac{10}{17}s^2 - \frac{160}{17}$$

Equating coefficients of like powers of s yields

$$A - \frac{10}{17} = 0, \qquad A = \frac{10}{17}$$

$$\frac{10}{17} + B = 10, \qquad B = \frac{160}{17}$$

and so

$$V_0(s) = \frac{\frac{10}{17}s + \frac{160}{17}}{s^2 + 16} - \frac{\frac{10}{17}}{s + 1} \tag{6.62}$$

The first term on the right-hand side of (6.62) is the steady-state term, and the second term is the transient component. Taking inverse transforms, we obtain

$$v_0(t) = \mathscr{L}^{-1}\{V_0(s)\} = \mathscr{L}^{-1}\left\{\frac{-\frac{10}{17}}{s + 1} + \frac{\frac{10}{17}s}{s^2 + 16} + \frac{\frac{160}{17}}{s^2 + 16}\right\}$$

$$= \underbrace{-\frac{10}{17}e^{-t}}_{\text{transient part}} + \underbrace{\frac{10}{17}\cos 4t + \frac{40}{17}\sin 4t}_{\text{steady-state part}}, \qquad t > 0$$

One can also obtain the steady-state component directly via the Fourier transform. Recall that a time-invariant linear system forced with a sinusoid yields as an output a sinusoid of the same frequency, with only the phase and amplitude of the input sinusoid being modified. The manner in which the input is modified is completely specified by the system transfer function $H(j\omega)$. The steady-state response to a sinusoid of angular frequency ω has magnitude $|H(j\omega)|$ and is shifted in phase by $\angle H(j\omega)$. In our case,

$$H(j\omega) = \frac{1}{1 + j\omega}$$

For angular frequency of $\omega = 4$,

$$H(j4) = \frac{1}{1 + j4} = \frac{1 - j4}{17} = \sqrt{\left(\frac{1}{17}\right)^2 + \left(\frac{4}{17}\right)^2} \angle \tan^{-1}\left(\frac{-4}{1}\right)$$

$$= \frac{1}{\sqrt{17}} \angle -76°$$

Hence, the steady-state response to an input of 10 cos 4t is

$$\frac{10}{\sqrt{17}} \cos(4t - 76°) = \frac{10}{\sqrt{17}} \cos 4t \cos 76° + \frac{10}{\sqrt{17}} \sin 4t \sin 76°$$

$$= \frac{10}{17} \cos 4t + \frac{40}{17} \sin 4t$$

which is the same result obtained by the Laplace transform process. ■

6.10 FREQUENCY RESPONSE OF LINEAR SYSTEMS

The last example suggests that there is an intimate connection between the system's transfer function $H(s)$ and the steady-state response of the system to a sinusoidal input, that is, the system's frequency response. If we examine the previous example, we see that, in fact, the frequency response for a system with transfer function $H(s)$ is simply $H(s)$ evaluated along the $j\omega$-axis.

This result can also be seen by the following argument. Assume that $H(s)$ is the transfer function of a stable causal system. The transformed output for an input $u(t) = \sin \omega t$ is

$$Y(s) = U(s)H(s) = H(s)\left(\frac{\omega}{s^2 + \omega^2}\right) = \frac{N(s)}{D(s)}\left(\frac{\omega}{s^2 + \omega^2}\right)$$

If we expand $Y(s)$ in partial fractions, we have

$$Y(s) = H(s)\omega\left(\frac{1}{(s + j\omega)(s - j\omega)}\right)$$

$$= \frac{N(s)}{D(s)} \omega \frac{1}{(s + j\omega)(s - j\omega)}$$

$$= \frac{A}{s + j\omega} + \frac{B}{s - j\omega} + \frac{\hat{N}(s)}{D(s)}$$

where

$$A = H(s)\omega \frac{1}{s - j\omega}\bigg|_{s=-j\omega} = \frac{H(-j\omega)\omega}{-2j\omega} = -\frac{H(-j\omega)}{2j}$$

$$B = H(s)\omega \frac{1}{s + j\omega}\bigg|_{s=j\omega} = \frac{H(j\omega)\omega}{2j\omega} = \frac{H(j\omega)}{2j}$$

and $D(s)$ is the denominator polynomial of $H(s)$. Therefore

$$Y(s) = -\frac{H(-j\omega)}{2j(s + j\omega)} + \frac{H(j\omega)}{2j(s - j\omega)} + \frac{\hat{N}(s)}{D(s)}$$

Because $H(s)$ has singularities only in the left-hand half of the plane, the term $\hat{N}(s)/D(s)$ represents transient terms in the time domain that die away as $t \to \infty$. Thus, the steady-state output is

$$\begin{aligned} y_{ss}(t) &= \mathscr{L}^{-1}\left\{-\frac{H(-j\omega)}{2j(s + j\omega)} + \frac{H(j\omega)}{2j(s - j\omega)}\right\} \\ &= -\frac{H(-j\omega)e^{-j\omega t}}{2j} + \frac{H(j\omega)e^{j\omega t}}{2j} \end{aligned} \tag{6.63}$$

Let $H(j\omega)$ be written in polar form as

$$H(j\omega) = |H(j\omega)|e^{j\theta(j\omega)}$$

Then

$$H(-j\omega) = |H(-j\omega)|e^{j\theta(-j\omega)} = |H(j\omega)|e^{-j\theta(j\omega)}$$

Substituting for $H(j\omega)$ and $H(-j\omega)$ in (6.60) then yields

$$\begin{aligned} y_{ss}(t) &= \frac{|H(j\omega)|e^{j\theta(j\omega)}e^{j\omega t} - |H(j\omega)|e^{-j\theta(j\omega)}e^{-j\omega t}}{2j} \\ &= \frac{|H(j\omega)|[e^{j(\theta(j\omega) + \omega t)} - e^{-j(\theta(j\omega) + \omega t)}]}{2j} \\ &= |H(j\omega)| sin\,(\omega t + \theta(j\omega)) \end{aligned} \tag{6.64}$$

Equation 6.64 states that if the input $u(t)$ is sin ωt, then the steady-state output $y(t)$ is $|H(j\omega)|$ sin $[\omega t + \theta(\omega)]$. The functions $|H(j\omega)|$ and $\theta(\omega)$ are the amplitude and angle functions obtained from the transfer function $H(s)$ by substituting $j\omega$ for s. This is an alternate way of establishing that the frequency response of a linear-time invariant system is the magnitude and phase of $H(j\omega)$. This relationship between the frequency response of a system and the transfer function of a system makes it easy to obtain the steady-state response for a sinusoidal input.

6.11 LAPLACE TRANSFORM ANALYSIS OF CAUSAL PERIODIC INPUTS TO LINEAR SYSTEMS

In applications, it is often useful to consider the Laplace transform of a causal periodic function $f(t)$. Assume that $f(t)$ is periodic with period T and that $f(t)$ is causal: that is, $f(t) = 0, t < 0$ and $f(t) = f(t + T), t > 0$. We can write the Laplace transform of $f(t)$ as

$$\begin{aligned}\mathscr{L}\{f(t)\} &= \int_0^\infty f(t)e^{-st}\,dt \\ &= \sum_{n=0}^\infty \int_{nT}^{(n+1)T} f(t)e^{-st}\,dt \\ &= \sum_{n=0}^\infty e^{-nTs} \int_0^T f(\tau)e^{-s\tau}\,d\tau, \qquad \text{for } \tau = t - nT\end{aligned}$$

The properties of a geometric series allow us to evaluate the sum $\sum_{n=0}^\infty e^{-nTs}$ as

$$\sum_{n=0}^\infty e^{-nTs} = \frac{1}{1 - e^{-sT}}$$

Thus

$$\mathscr{L}\{f(t)\} = \frac{\int_0^T f(t)e^{-st}\,dt}{1 - e^{-sT}} \tag{6.65}$$

Example 6.15 Find the Laplace transform of the periodic waveform shown in Figure 6.16. Substituting into (6.65) yields

$$\begin{aligned}F(s) &= \frac{1}{1 - e^{-sT}}\left[\int_0^{T/2} 1 \cdot e^{-st}\,dt + \int_{T/2}^T -1 \cdot e^{-st}\,dt\right] \\ &= \frac{1}{1 - e^{-sT}}\left[\frac{1 - e^{-sT/2}}{s} + \frac{e^{-sT} - e^{-sT/2}}{s}\right] \\ &= \frac{1 - e^{-sT/2}}{s(1 + e^{-sT/2})}\end{aligned}$$

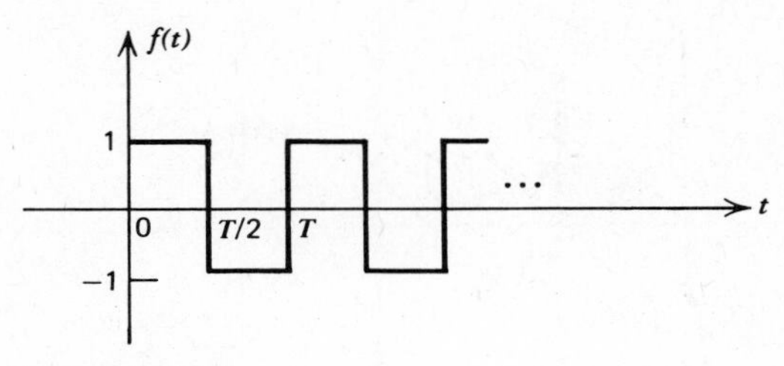

Figure 6.16

Another way of arriving at (6.65) is to recognize that we can write a periodic function $f(t)$ as

$$f(t) = f_1(t) + f_1(t - T)\xi(t - T) + f_1(t - 2T)\xi(t - 2T) + \cdots \tag{6.66}$$

where

$$\begin{aligned} f_1(t) &= f(t), \qquad t \in [0, T] \\ &= 0, \qquad \text{otherwise} \end{aligned}$$

Taking the Laplace transform of (6.66) gives

$$\begin{aligned} \mathscr{L}\{f(t)\} &= F_1(s) + F_1(s)e^{-sT} + F_1(s)e^{-2sT} + \cdots \\ &= F_1(s)[1 + e^{-sT} + e^{-2sT} + \cdots] \\ &= \frac{F_1(s)}{1 - e^{-sT}} \end{aligned} \tag{6.67}$$

where

$$F_1(s) = \mathscr{L}\{f_1(t)\} = \int_0^T f(t)e^{-st}\, dt$$

■

Example 6.16 Find the transform of the saw-tooth waveform shown in Figure 6.17. The period of $f(t)$ is T. Hence, using (6.67), we have

$$\begin{aligned} \mathscr{L}\{f(t)\} &= \frac{1}{1 - e^{-sT}} \int_0^T te^{-st}\, dt = \frac{1}{1 - e^{-sT}} \left[\frac{e^{-st}}{s^2}(-st - 1) \Big|_0^T \right] \\ &= \frac{1 - (1 + sT)e^{-sT}}{s^2(1 - e^{-sT})} \\ &= \frac{(1 + sT)(1 - e^{-sT}) - sT}{s^2(1 - e^{-sT})} = \frac{1 + sT}{s^2} - \frac{T}{s(1 - e^{-sT})} \end{aligned}$$

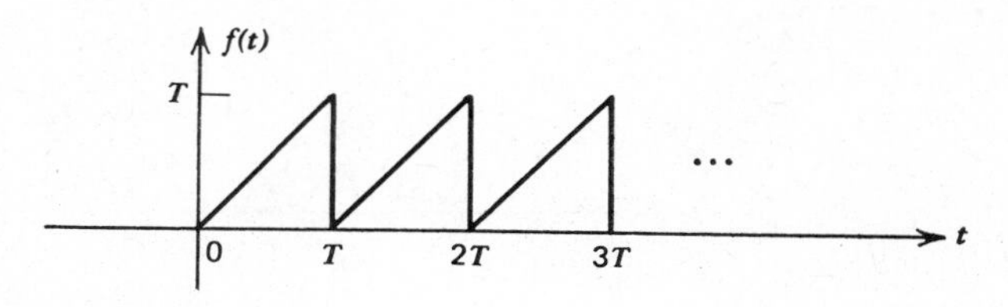

Figure 6.17

Example 6.17 Find the steady-state output current waveform of the *RLC* circuit shown for a periodic saw-tooth input voltage waveform of Figure 6.17 with $T = 1$. Assume an initial current of 1 ampere flowing in the inductor as shown in Figure 6.18. The differential equation describing this system is

$$e(t) = Ri(t) + L\frac{di(t)}{dt} + \frac{1}{C}\int_0^t i(t')\,dt'$$

Taking Laplace transforms of both sides yields

$$3I(s) + (sI(s) - 1) + \frac{2}{s}I(s) = \frac{1+s}{s^2} - \frac{1}{s(1 - e^{-s})}$$

Solving for $I(s)$,

$$I(s) = \frac{s^2 + s + 1}{s(s+1)(s+2)} - \frac{1}{(s+1)(s+2)(1 - e^{-s})}$$

The first term of this equation can be expanded in partial fractions to yield

$$\frac{1}{2} - e^{-t} + \frac{3}{2}e^{-2t}$$

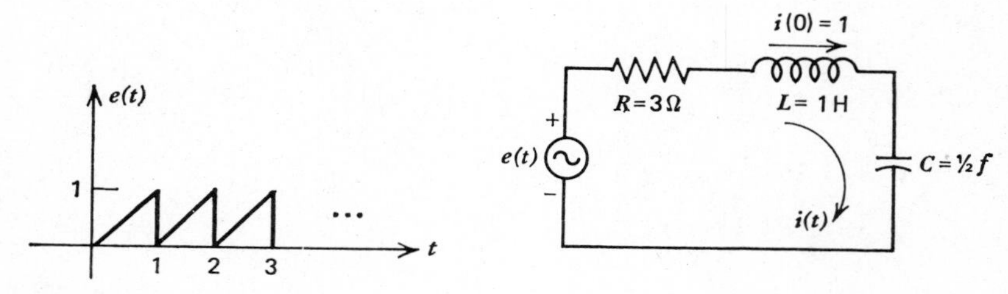

Figure 6.18

To find the inverse of the second term, we have

$$\frac{1}{(s+1)(s+2)(1-e^{-s})} = \left(\frac{1}{s+1} - \frac{1}{s+2}\right)\left(\frac{1}{1-e^{-s}}\right)$$
$$= \frac{1}{(s+1)(1-e^{-s})} - \frac{1}{(s+2)(1-e^{-s})} \tag{6.68}$$

Consider the expression

$$\frac{1}{(s+a)(1-e^{-sT})} = \frac{1}{s+a}(1 + e^{-sT} + e^{-2sT} + e^{-3sT} + \cdots)$$
$$= \frac{1}{s+a} + \frac{e^{-sT}}{s+a} + \frac{e^{-2sT}}{s+a} + \cdots$$

Consider the inverse of the $(n+1)$st term. The contribution of $1/(s+a)$ is e^{-at}. The exponential e^{-nsT} translates the time function e^{-at} to the right by nT and then truncates it to the left of $t = nT$. When this is done for each term, we obtain

$$f(t) = e^{-at} + e^{-a(t-T)}\xi(t-T) + e^{-a(t-2T)}\xi(t-2T) + \cdots$$

Over the interval $[nT, (n+1)T]$, we thus have for $f(t)$

$$f(t) = e^{-at}[1 + e^{aT} + e^{2aT} + \cdots + e^{naT}] \tag{6.69}$$

We can sum the bracketed term in (6.69), using the properties of the geometric series, to obtain

$$f(t) = e^{-at}\left[\frac{(e^{aT})^{(n+1)} - 1}{e^{aT} - 1}\right]$$
$$= \frac{e^{-a[t-(n+1)T]}}{e^{aT}-1} - \frac{e^{-at}}{e^{aT}-1}, \qquad t \in [nT, (n+1)T]$$

Returning to (6.68), we have that $T = 1$, so that over the interval $n < t < n+1$,

$$\mathscr{L}^{-1}\left\{\frac{1}{(s+1)(1-e^{-s})} - \frac{1}{(s+2)(1-e^{-s})}\right\}$$
$$= \left(\frac{e^{-(t-(n+1))}}{e-1} - \frac{e^{-t}}{e-1}\right) - \left(\frac{e^{-2(t-(n+1))}}{e^2-1} - \frac{e^{-2t}}{e^2-1}\right), \qquad n < t < n+1$$

The total solution is therefore

$$y(t) = \frac{1}{2} - e^{-t} + \frac{3}{2}e^{-2t} + \left(\frac{e^{-2t}}{e^2 - 1} - \frac{e^{-t}}{e - 1}\right) + \frac{e^{-[t-(n+1)]}}{e - 1} - \frac{e^{-2[t-(n+1)]}}{e^2 - 1}$$

$$= \underbrace{\left(\frac{3}{2}e^{-2t} - e^{-t} - \frac{e^{-t}}{e - 1} + \frac{e^{-2t}}{e^2 - 1}\right)}_{\text{transient}} + \underbrace{\left(\frac{1}{2} + \frac{e^{-[t-(n+1)]}}{e - 1} - \frac{e^{-2[t-(n+1)]}}{e^2 - 1}\right)}_{\text{steady-state}},$$

$$n < t < n + 1$$

Thus the steady-state current in the $(n + 1)$st interval is

$$y_{ss}(t) = \frac{1}{2} + \frac{e^{-[t-(n+1)]}}{e - 1} - \frac{e^{-2[t-(n+1)]}}{e^2 - 1}, \qquad n < t < n + 1$$

■

6.12 RELATIONSHIP OF THE *Z*-TRANSFORM TO THE FOURIER AND LAPLACE TRANSFORMS

We consider next the connection between the Fourier and Laplace transforms of a continuous function and the Z-transform of a sequence. Let $f(t)$ be a continuous function that is sampled at the time instants $\{\ldots, -T, 0, T, 2, T, \ldots\}$. The sampled version can be represented as

$$f_s(t) = f(t) \sum_{n=-\infty}^{\infty} \delta(t - nT)$$

$$= \sum_{n=-\infty}^{\infty} f(nT)\, \delta(t - nT) \tag{6.70}$$

where we have used the property that $f(t)\,\delta(t) = f(0)\,\delta(t)$ to equate $f(t)\,\delta(t - nT)$ to $f(nT)\,\delta(t - nT)$. The sampled version of $f(t)$ has the Laplace transform

$$F_s(s) = \mathscr{L}_b\{f_s(t)\} = \mathscr{L}_b\left\{\sum_{n=-\infty}^{\infty} f(nT)\,\delta(t - nT)\right\}$$

$$= \sum_{n=-\infty}^{\infty} f(nT)\mathscr{L}_b\{\delta(t - nT)\}$$

Now $\mathscr{L}_b\{\delta(t)\} = 1$, and by the time-shift theorem, $\mathscr{L}_b\{\delta(t - nT)\} = e^{-nsT}$. Thus we obtain

$$F_s(s) = \sum_{n=-\infty}^{\infty} f(nT)e^{-snT} \tag{6.71}$$

If we now make the substitution $z = e^{sT}$, then $F_s(s)$ becomes

$$F_s(s)|_{z=e^{sT}} = \sum_{n=-\infty}^{\infty} f(nT)z^{-n} = F(z) \tag{6.72}$$

where $F(z)$ is the Z-transform of the sequence of samples of $f(t)$, that is, $\{f(nT)\}$, $n = 0, \pm 1, \pm 2, \ldots$.

From this discussion, we see that the Z-transform may be viewed as the Laplace transform of the sampled time function $f(t)$ (with an appropriate change of variable), or quite independently as the generating function for a sequence $\{f_n\}$ that assumes the values $f(nT)$ for $n = \ldots, -2, -1, 0, 1, 2, \ldots$. We note that with $z = e^{+sT}$, the complex s-plane maps into the complex z-plane. Under this mapping, the imaginary axis, Re $(s) = 0$, maps onto the unit circle $|z| = 1$ in the z-plane. Also, the left-hand half-plane Re $(s) < 0$ corresponds to the interior of the unit circle $|z| = 1$ in the z-plane. This correspondence is shown schematically in Figure 6.19.

If we restrict s to the $j\omega$-axis in the s-plane, then $F_s(s)$ becomes $F_s(j\omega)$, the Fourier transform of the sampled function $f_s(t)$. From (6.71), we obtain

$$F_s(j\omega) = \sum_{n=-\infty}^{\infty} f(nT)e^{-j\omega nT} = F(z)|_{z=e^{+j\omega T}} \tag{6.73}$$

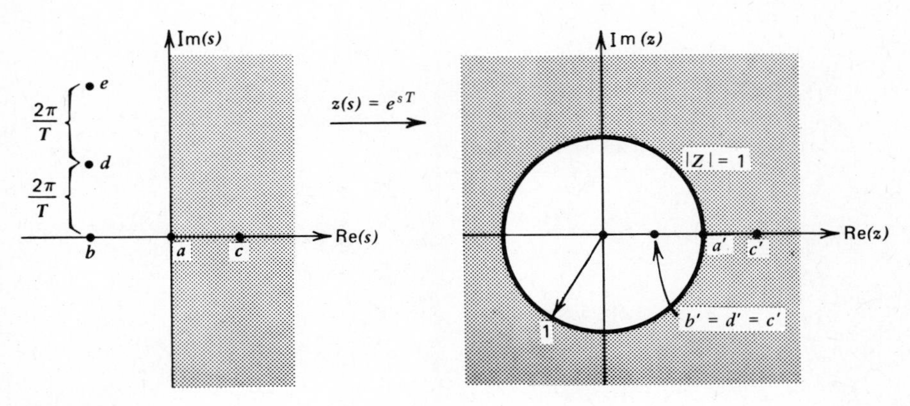

Figure 6.19

Notice that $F_s(j\omega) = F_s(j\omega + 2\pi/T)$: that is, $F_s(j\omega)$ is periodic with period $2\pi/T$. Now, if $s = j\omega$, in the z-plane we have that $z = e^{+j\omega T}$. Therefore $F_s(j\omega)$ can be obtained by evaluating $F(z)$ where z is restricted to the unit circle. One period of $F_s(j\omega)$ can be obtained by evaluating $F(z)$ once around the unit circle $|z| = 1$.

6.13 SUMMARY

The Laplace transform is a generalization of the Fourier transform in which the frequency variable s is a complex variable, $\sigma + j\omega$. Through this generalization, we increase the class of functions that can be transformed at the cost of dealing with a more abstract, less physical representation. As with other transform methods, we can often solve a problem modeled by a linear differential equation by a table look-up.

The Laplace transform analysis of a system is useful in examining the system's stability directly by finding the roots of the system's transfer function. This transform method is also a convenient way to formulate transient problems in which initial energy storage is nonzero at time $t = 0$, because the initial conditions at $t = 0$ are easily incorporated into the transform equations.

PROBLEMS

6.1. Find the inverse Laplace transforms of the following transforms. Assume that $f(t)$ is causal.

(a) $\dfrac{1}{(s+b)^4}$ *Answer*: $\dfrac{t^3 e^{-bt}}{6}$

(b) $\dfrac{s}{(s^2+a^2)^2}$ *Answer*: $\dfrac{t}{2a} \sin at$

(c) $\dfrac{e^{ST}}{(s+1)^3}$ *Answer*: $\dfrac{(t-T)^2 e^{-(t-T)}}{2} \xi(t-T)$

(d) $\dfrac{2}{(s^2+1)^2}$ *Answer*: $\sin t - t \cos t$

(e) $\dfrac{1}{s^4 - a^4}$ *Answer*: $\dfrac{1}{2a^3} (\sinh at - \sin at)$

(f) $\dfrac{1}{s(s^2 + a^2)^2}$ *Answer*: $\dfrac{1}{a^4}(1 - \cos at) - \dfrac{1}{2a^3} t \sin at$

(g) $\dfrac{s}{s^4 + 4a^4}$ *Answer*: $\dfrac{1}{2a^2} \sin at \sinh at$

6.2. Find the Laplace transform of the following time functions.

(a) $\xi(t - T)$ *Answer*: $\dfrac{e^{-sT}}{s}$

(b) $f(t) = \begin{cases} \sin t, & 0 < t < \pi \\ 0, & \text{otherwise} \end{cases}$ *Answer*: $\dfrac{1 + e^{-\pi s}}{s^2 + 1}$

(c) $\dfrac{1}{a^2}(1 - \cos at)\xi(t)$ *Answer*: $\dfrac{1}{s(s^2 + a^2)}$

(d) $\dfrac{1}{2a^3}(\sin at - at \cos at)\xi(t)$ *Answer*: $\dfrac{1}{(s^2 + a^2)^2}$

(e) $\left(\dfrac{1 - \cos bt}{t}\right)\xi(t)$ *Answer*: $\dfrac{1}{2}\left\{6 \cot^{-1}\left(\dfrac{s}{b}\right) + s \ln \dfrac{s^2}{s^2 + b^2}\right\}$

6.3. A linear system is described by the following differential equation. This system is forced with an input as shown in the graph. Find the output of the system.

$$\frac{d^2y(t)}{dt^2} + \frac{3dy(t)}{dt} + 2y(t) = u(t), \qquad y(0) = 0, \qquad y^{(1)}(0) = 1$$

Answer: $(e^{-t} - e^{-2t})\xi(t) + \dfrac{1}{2}\{1 - 2e^{-(t-1)} + e^{-2(t-1)}\}\xi(t - 1)$

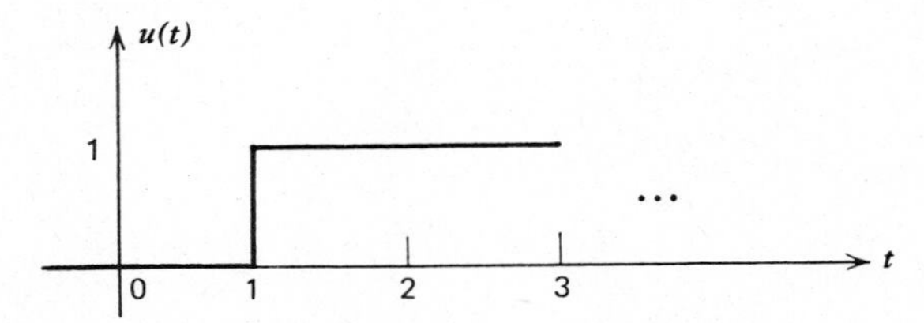

6.4. Consider the Laplace transform $F(s)$ given by

$$F(s) = \frac{1}{(s - 3)(s - 2)(s + 1)}$$

Find four possible time functions that have this Laplace transform. State the region of convergence for each.

6.5. Find the output voltage in the following circuit for all t in response to an input voltage e^{-t}.

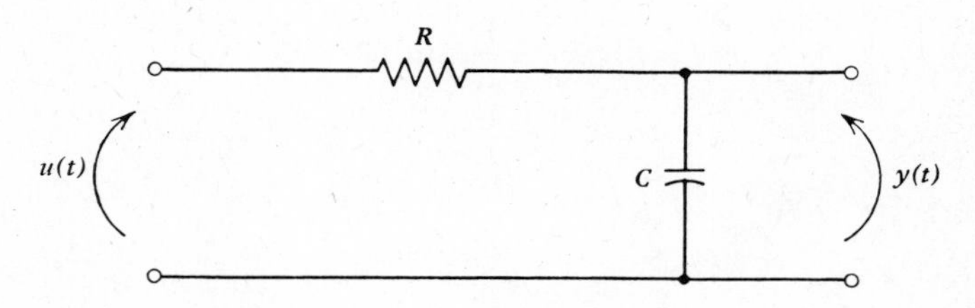

6.6. John Brightfellow has been asked to determine the contents of a black box (containing an electrical circuit). John decides to find the system's frequency response by forcing the system with a sinusoid. He applies a sinusoidal voltage and measures a direct current output. What is the circuit?

6.7. A voltage $e^{-t}\xi(t)$ is applied to a circuit. The response voltage is a ramp $t\xi(t)$. What is the circuit?

6.8. We wish to compare two systems, A and B, as shown below. The two systems are two methods of performing integration. In system A, the response $r(t)$ is integrated and then sampled to obtain a sequence $y_A(nT)$. In system B, the response $r(t)$ is first sampled and then these samples are summed (the discrete analog of integration) to obtain a sequence $y_B(nT)$. Assume that the input $u(t)$ is an impulse $\delta(t)$ and that $T = 1$ and $a = 0.2$. Obtain closed form expressions for $y_A(nT)$ and $y_B(T)$. Compare these two sequences for $n = 0, 1, 2, 3, 4$. Are the systems equivalent? Suppose you could use a single delay element in the discrete integrator. Design a better discrete integrator (not limited to the specific input of this problem).

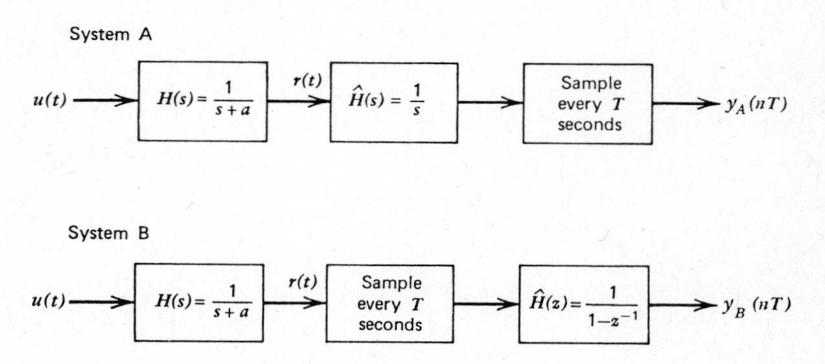

6.9. Find the impulse response of the following continuous-time system. Is the system stable?

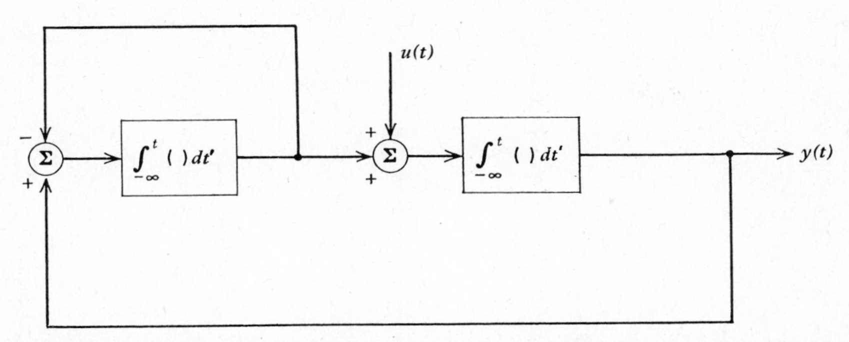

6.10. Is the feedback system shown below stable if the gain g is zero: that is, with no feedback? Plot the locus of poles in the s-plane for the overall system for both positive and negative values of g. For what range of g is the feedback system stable?

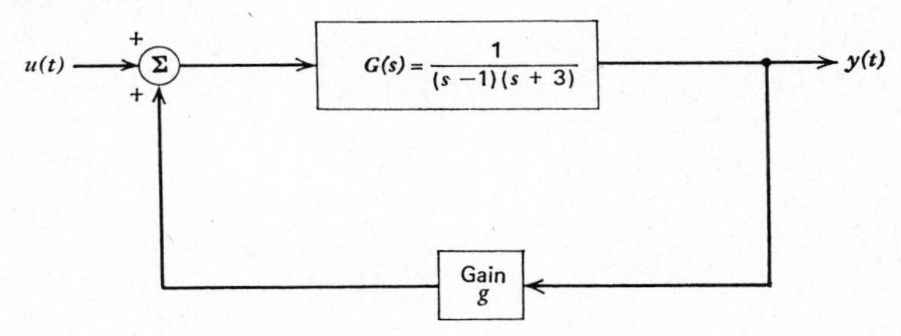

6.11. Assume that we have a transfer function of the form

$$H(s) = \frac{a_0 + a_1 s + \cdots + a_m s^m}{b_0 + b_1 s + \cdots + b_n s^n}$$

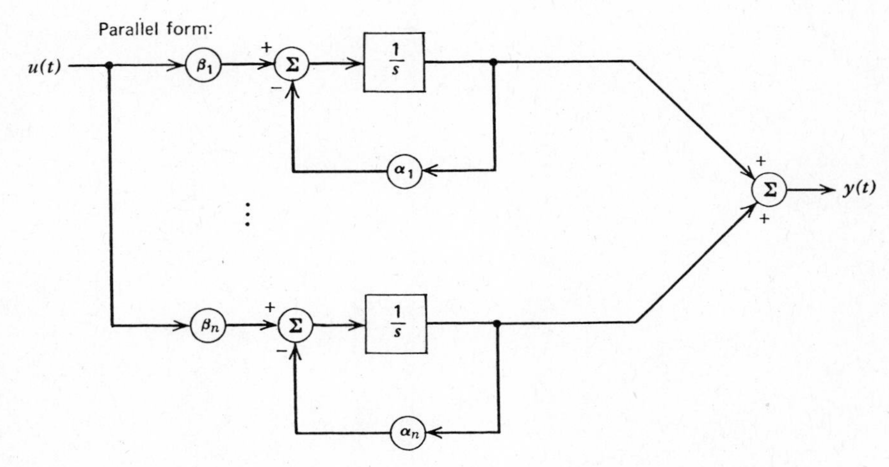

Cascade form:

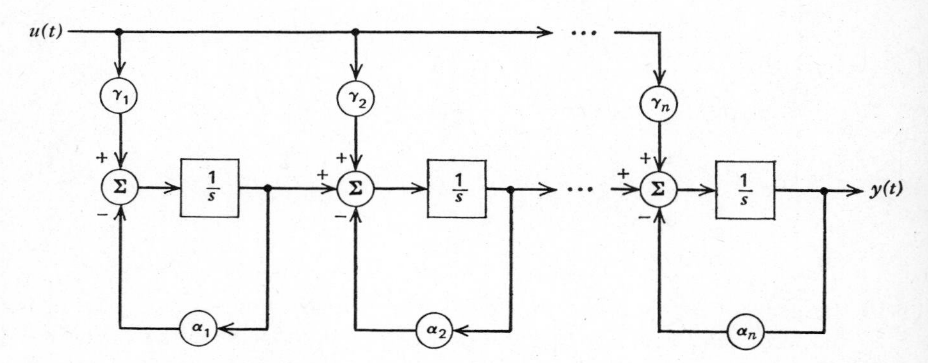

Assume that the poles of $H(s)$ are real and distinct. Show that $H(s)$ can be realized in the two forms shown above. Develop a method for computing the constants α_i, β_i, and γ_i.

6.12. What is the Laplace transform of the waveform shown below?

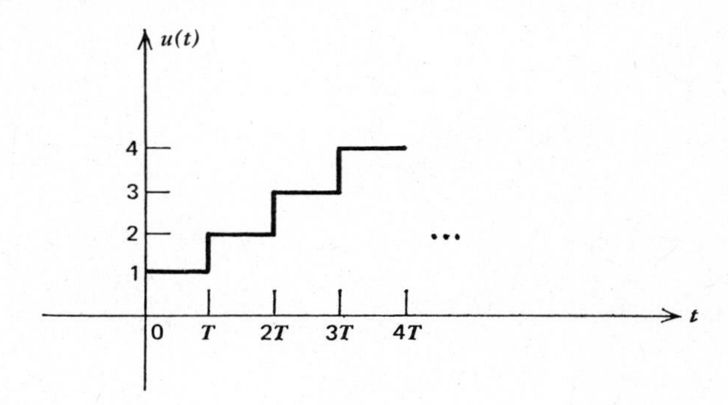

Answer: $U(s) = \dfrac{1}{s(1 - e^{-sT})}$

CHAPTER 7
AN INTRODUCTION TO THE DESIGN OF DIGITAL FILTERS

In the design of communication, control, and telemetry systems, we must often include filters to shape the spectrum of various signals. Because of the widespread use of digital computers and special purpose digital hardware, digital filtering has become an important technique in many applications that include speech, radar, sonar, seismic, and biomedical signal processing.

This chapter introduces some simple methods for designing digital filters. In Chapters 2 and 4 we analyzed discrete-time systems and investigated the frequency response function $H(e^{j\theta})$. We have the theory we need to calculate $H(e^{j\theta})$ from a block diagram, difference equation, impulse response, or transfer function description of a discrete-time system. Now we wish to extend this analysis to answer the following questions:

1. Given some desired filtering operation, how do we synthesize a discrete-time system whose frequency response achieves the desired goal?

2. If we process continuous-time signals by using the cascade of an analog-to-digital converter, a digital filter, and a digital-to-analog converter, how do we relate the digital filter transfer function $H(e^{j\theta})$ to the frequency response of the cascade combination $H(j\omega)$?

Digital filters are usually classified as either *finite duration impulse response* (FIR) or *infinite duration impulse response* (IIR) filters. FIR filters are characterized by a finite length impulse response or, equivalently, by a transfer function in the form of a polynomial

$$H(z) = a_0 + a_1 z^{-1} + \cdots + a_M z^{-M} \tag{7.1}$$

IIR filters, on the other hand, are characterized by transfer functions in the form of a rational function or ratio of polynomials

$$H(z) = \frac{\sum_{i=0}^{M} a_i z^{-i}}{1 + \sum_{i=1}^{N} b_i z^{-i}} \tag{7.2}$$

They possess an impulse response that, in general, is nonzero for all $n > 0$. FIR filters are always stable. In the IIR case stability is assured if the poles of $H(z)$ are within the unit circle in the z-plane. There are distinct design methods for each filter class. We shall begin our discussion with a design method for FIR filters.

7.1 DESIGN OF FIR DIGITAL FILTERS

In this development we extend concepts introduced in Chapters 2 and 5. This design procedure, called the window method, is only one of many methods used in the design of FIR filters. Suppose that we wished to implement a discrete-time filter whose transfer function, $H(e^{j\theta})$, is an approximation of a desired transfer function $H_d(e^{j\theta})$. $H_d(e^{j\theta})$ might, for example, represent an ideal lowpass filter. In the synthesis of FIR filters, our aim is to design the impulse-response sequence $\{h(n), n_1 \leq n \leq n_2\}$ such that $H(e^{j\theta})$ is close in some sense to $H_d(e^{j\theta})$. That is,

$$\begin{aligned} H(e^{j\theta}) &= \sum_{n=n_1}^{n_2} h(n) e^{-jn\theta} \\ &\cong H_d(e^{j\theta}) \end{aligned} \tag{7.3}$$

One often-used measure of closeness is the least mean square error approximation. We seek $H(e^{j\theta})$ so that the error ε is minimized where ε is defined as

$$\varepsilon = \frac{1}{2\pi} \int_{-\pi}^{\pi} |H_d(e^{j\theta}) - H(e^{j\theta})|^2 \, d\theta \tag{7.4}$$

Recalling that $H(e^{j\theta})$ is given by

$$H(e^{j\theta}) = \sum_{n=n_1}^{n_2} h(n)e^{-jn\theta} \tag{7.5}$$

we see that the minimization of ε is the classical problem in the theory of Fourier series. Equation 7.4 corresponds to (5.3) with basis functions $\{e^{-jn\theta}\}$. The coefficients $h(n)$ that minimize (7.4) are the Fourier coefficients

$$h_n = \frac{(H_d, e^{jn\theta})}{(e^{-jn\theta}, e^{jn\theta})}$$

$$= \frac{\int_{-\pi}^{\pi} H_d(e^{j\theta})e^{jn\theta}\, d\theta}{\int_{-\pi}^{\pi} e^{-jn\theta}e^{jn\theta}\, d\theta} = \frac{1}{2\pi}\int_{-\pi}^{\pi} H_d(e^{j\theta})e^{jn\theta}\, d\theta \tag{7.6}$$

Because we are assuming n takes on only N nonzero values, there is an error given by

$$\varepsilon = \frac{1}{2\pi}\int_{-\pi}^{\pi} |H_d(e^{j\theta}) - H(e^{j\theta})|^2\, d\theta \tag{7.7}$$

By using Parseval's relationship we can express (7.7) as

$$\varepsilon = \sum_{n=-\infty}^{\infty} |h_d(n) - h(n)|^2 \tag{7.8}$$

The coefficients $h_d(n)$ are, of course, the Fourier coefficients for $H_d(e^{j\theta})$. We can minimize ε in (7.8) by setting $h(n)$ equal to $h_d(n)$ for $n_1 \le n \le n_2$. The resulting error is thus

$$\varepsilon = \sum_{n \notin [n_1, n_2]} |h_d(n)|^2$$

$$= \sum_{n=-\infty}^{n_1-1} |h_d(n)|^2 + \sum_{n=n_2+1}^{\infty} |h_d(n)|^2 \tag{7.9}$$

In the common case where $H_d(e^{j\theta})$ is real-valued (for example, the lowpass characteristic to be treated below), the coefficients $h_d(n)$ will be symmetric about $n = 0$. We then generally choose $n_1 = -n_2$ so that the $h(n)$ coefficients will also be symmetric. The length of the filter then is $N = n_2 - n_1 + 1 = 2n_2 + 1$, and ε may be written as

$$\varepsilon = \sum_{n=-\infty}^{-(N-1)/2-1} |h_d(n)|^2 + \sum_{n=(N-1)/2+1}^{\infty} |h_d(n)|^2$$

$$= 2 \sum_{n=(N-1)/2+1}^{\infty} |h_d(n)|^2 \tag{7.10}$$

Example 7.1 Suppose we wish to design an FIR filter of length 11 that approximates a low pass filter with a cutoff frequency of $f_0 = 250$ Hz for a sampling frequency of $f_s = 1000$ Hz. In terms of normalized frequency θ, the cutoff frequency is $\theta_0 = 2\pi f_0/f_s = \pi/2$. Thus our desired frequency function is

$$H_d(e^{j\theta}) = \begin{cases} 1, & \theta < \dfrac{\pi}{2} \\ 0, & \theta > \dfrac{\pi}{2} \end{cases} \tag{7.11}$$

Expanding $H_d(e^{j\theta})$ in a Fourier series on the interval $[-\pi, \pi]$ yields Fourier coefficients $h_d(n)$ given by

$$h_d(n) = \frac{1}{2\pi} \int_{-\pi/2}^{\pi/2} 1 \cdot e^{jn\theta} \, d\theta$$

$$= \frac{1}{2} \operatorname{sinc}\left(\frac{n\pi}{2}\right), \qquad n = 0, \pm 1, \pm 2, \ldots$$

For an FIR filter of 11 coefficients we choose $h(n)$ equal to $h_d(n)$ for $n = 0, \pm 1, \ldots, \pm 5$. That is,

$$h(n) = \frac{1}{2} \operatorname{sinc}\left(\frac{n\pi}{2}\right), \qquad n = 0, \pm 1, \ldots, \pm 5$$

$$= \left\{\frac{1}{5\pi}, 0, -\frac{1}{3\pi}, 0, \frac{1}{\pi}, \underset{\uparrow}{\frac{1}{2}}, \frac{1}{\pi}, 0, -\frac{1}{3\pi}, 0, \frac{1}{5\pi}\right\} \tag{7.12}$$

The impulse response of (7.12) is noncausal. To obtain a causal response we can shift the impulse response so that $h(-5)$ becomes $h(0)$, $h(-4)$ becomes $h(1)$, etc.

This process delays the output by an amount of the shift. The response $H(e^{j\theta})$ becomes instead $H(e^{j\theta})e^{-5j\theta}$. We can show that $H(e^{j\theta})e^{-5j\theta}$ is causal by simply expanding term by term

$$\begin{aligned}\hat{H}(e^{j\theta}) &= e^{-5j\theta}H(e^{j\theta}) \\ &= e^{-5j\theta}\sum_{n=-5}^{5} h(n)e^{-jn\theta} \\ &= e^{-5j\theta}\{h(-5)e^{j5\theta} + h(-3)e^{j3\theta} + h(-1)e^{j\theta} + h(0) + \cdots\} \\ &= h(-5) + h(-3)e^{-2j\theta} + h(-1)e^{-4j\theta} + \cdots \end{aligned} \tag{7.13}$$

Equation 7.13 is the frequency response of a causal filter with $h(-5)$ corresponding to the $n = 0$ term. This impulse response is sketched in Figure 7.1a. One implementation of this filter is shown in Figure 7.1b. The resulting frequency-response function is shown in Figure 7.1c. The error of this approximation is

$$\begin{aligned}\varepsilon &= \sum_{n=-\infty}^{-6}\left|\frac{1}{2}\operatorname{sinc}\left(\frac{n\pi}{2}\right)\right|^2 + \sum_{n=6}^{\infty}\left|\frac{1}{2}\operatorname{sinc}\left(\frac{n\pi}{2}\right)\right|^2 \\ &= \frac{1}{2}\sum_{n=6}^{\infty}\left|\operatorname{sinc}\left(\frac{n\pi}{2}\right)\right|^2\end{aligned}$$

As seen in this example, the resulting transfer function $H(e^{j\theta})$, although the best fit to $H_d(e^{j\theta})$ in the sense of minimizing the mean squared error between the two, may exhibit relatively large under- and overshoots in its approximation. As an aid in understanding this effect, let us write $h(n)$ as the product of $h_d(n)$ and a window sequence $w(n)$. In the case of simple truncation as in Example 7.1 the window function is

$$w(n) = \begin{cases} 1, & 0 \le n \le 10 \\ 0, & \text{otherwise} \end{cases} \tag{7.14}$$

writing

$$h(n) = h_d(n)w(n) \tag{7.15}$$

we obtain

$$\begin{aligned}H(e^{j\theta}) &= \sum_{n=-\infty}^{\infty} h_d(n)w(n)e^{-jn\theta} \\ &= \sum_{n=-\infty}^{\infty}\left[\frac{1}{2\pi}\int_{-\pi}^{\pi} H_d(e^{j\alpha})e^{jn\alpha}\,d\alpha\right]w(n)e^{-jn\theta}\end{aligned} \tag{7.16}$$

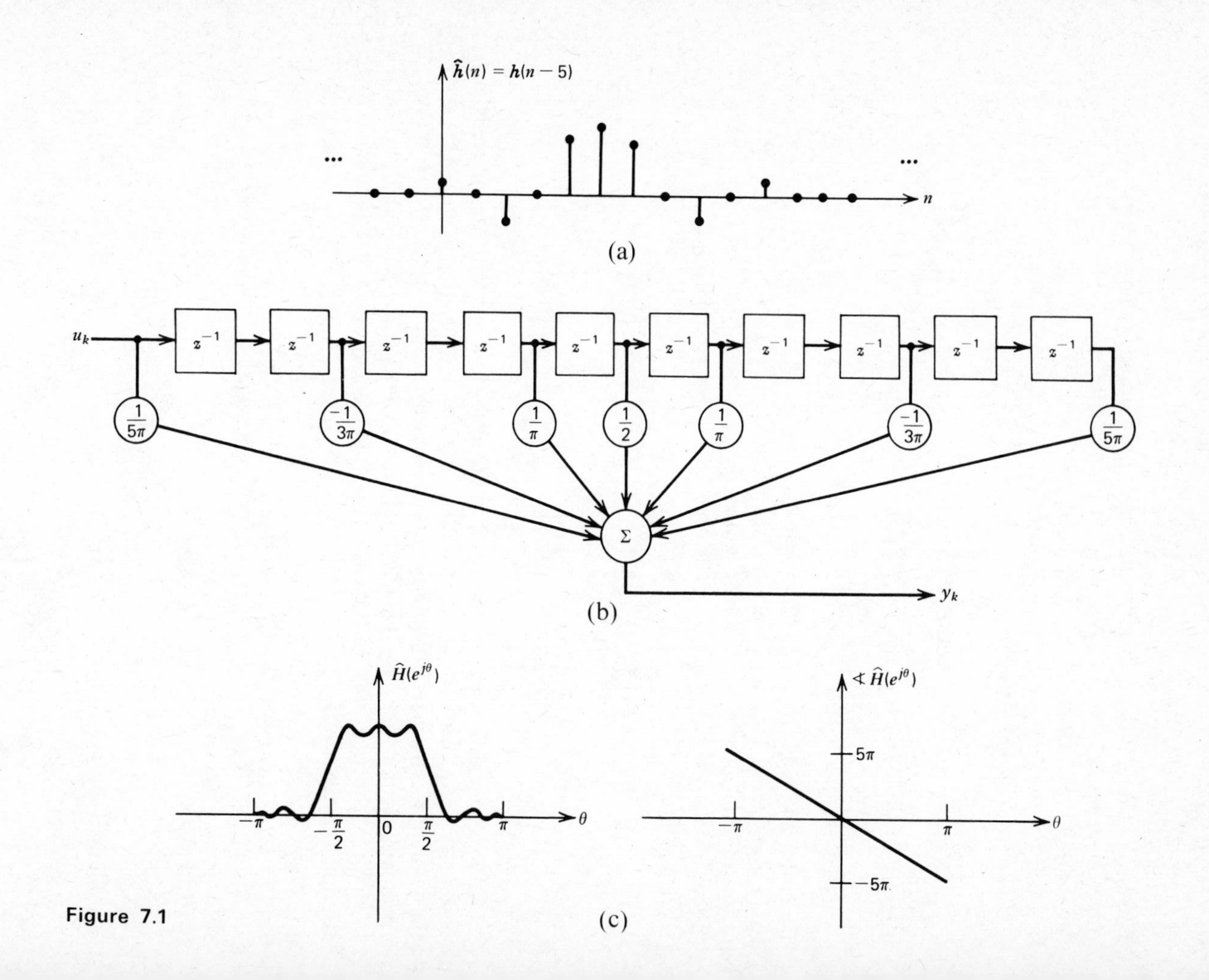
$\hat{h}(n) = h(n-5)$
n
(a)
u_k
z^{-1}
$\frac{1}{5\pi}$
$\frac{-1}{3\pi}$
$\frac{1}{\pi}$
$\frac{1}{2}$
$\frac{1}{\pi}$
$\frac{-1}{3\pi}$
$\frac{1}{5\pi}$
Σ
y_k
(b)
$\hat{H}(e^{j\theta})$
$-\pi$
$-\frac{\pi}{2}$
0
$\frac{\pi}{2}$
π
θ
$\measuredangle \hat{H}(e^{j\theta})$
5π
-5π
$-\pi$
π
θ
(c)

Figure 7.1

Interchanging the summation and integration operations, we obtain

$$H(e^{j\theta}) = \frac{1}{2\pi}\int_{-\pi}^{\pi} H_d(e^{j\alpha})\left[\sum_{n=-\infty}^{\infty} w(n)e^{-j(\theta-\alpha)n}\right] d\alpha$$

$$= \frac{1}{2\pi}\int_{-\pi}^{\pi} H_d(e^{j\alpha})W(e^{j(\theta-\alpha)})d\alpha \tag{7.17}$$

which expresses $H(e^{j\theta})$ as the *convolution* of the desired response $H(e^{j\theta})$ with the transform $W(e^{j\theta})$ of the window sequence $w(n)$. Note that, since $H_d(\cdot)$ and $W(\cdot)$ are periodic, this is a *circular* convolution (compare with the discussion in Section 5.6).

The result obtained in (7.17) is shown schematically in Figure 7.2, which illustrates the convolution of $H_d(e^{j\theta})$ with the function $W(e^{j\theta})$ given by

$$W(e^{j\theta}) = \sum_{n=0}^{10} 1 \cdot e^{-jn\theta}$$

$$= \frac{1 - e^{-j11\theta}}{1 - e^{-j\theta}}$$

$$= \frac{e^{-j(11/2)\theta}}{e^{-j(\theta/2)}} \cdot \frac{e^{j(11/2)\theta} - e^{-j(11/2)\theta}}{e^{j(\theta/2)} - e^{-j(\theta/2)}}$$

$$= e^{-j5\theta} \sin\left(\frac{11\theta}{2}\right) \Big/ \sin\left(\frac{\theta}{2}\right) \tag{7.18}$$

Here we see how the oscillations in $W(e^{j\theta})$ give rise to the ripples in the resulting transfer function $H(e^{j\theta})$. From these results we see immediately that $W(e^{j\theta})$ should approximate the impulse $2\pi\delta(\theta)$ in order that $H(e^{j\theta})$ be close to $H_d(e^{j\theta})$. This, unfortunately, implies that the window sequence $w(n)$ must be long; that is,

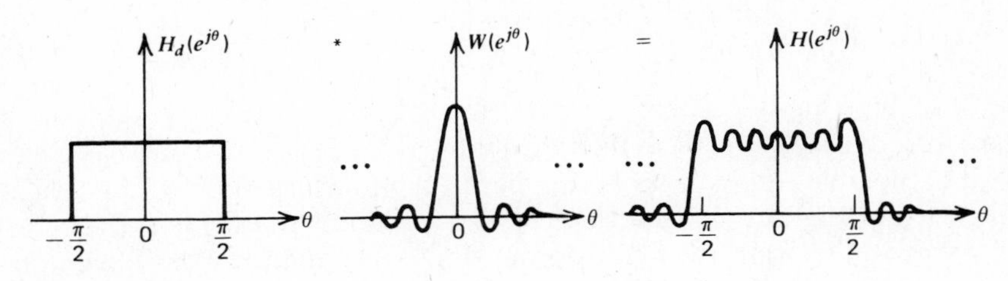

Figure 7.2

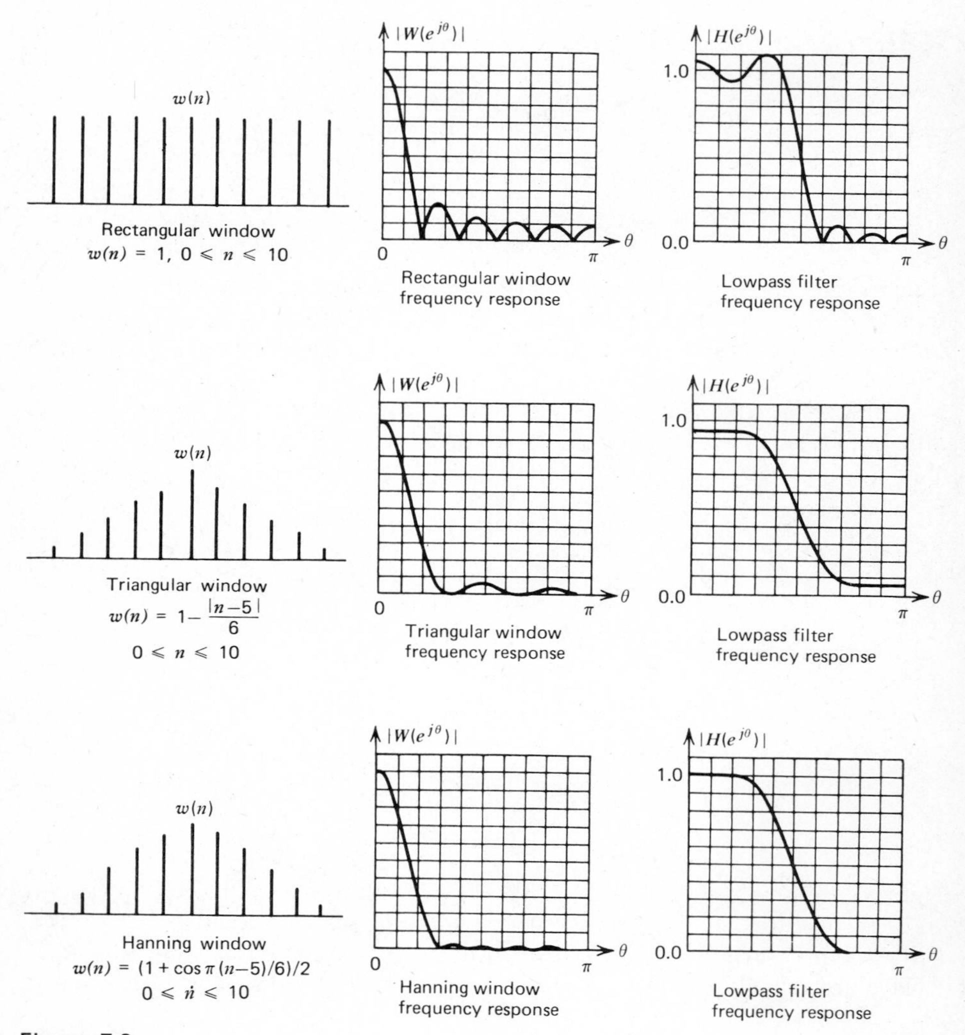

Figure 7.3

we must include many coefficients in our filter. For a fixed filter length, the best we can do is to shape $W(e^{j\theta})$ so that we obtain as narrow a main lobe as possible with an acceptable ripple in the side lobes of the function $W(e^{j\theta})$. This trade-off is illustrated in Figure 7.3 where representative choices of $w(n)$ (all of length 11) are shown together with their transforms $W(e^{j\theta})$ and the low-pass filter transfer function $H(e^{j\theta})$ that they yield. In each design, the parameters of Example 7.1

were used, with the filter impulse response taken as $h(n) = w(n) \cdot h_d(n)$. The choice of an appropriate window depends strongly on the particular application at hand. No single window function is optimal for all situations.

This design method yields filters that typically are implemented with the non-recursive structure shown in Figure 7.1*b*, which resembles a tapped delay line with the output formed as a weighted sum of delayed input samples. This method of design is quite general and can be applied to the approximation of any desired transfer function. However, many terms may be required to obtain an $H(e^{j\theta})$ that is a reasonable approximation to $H_d(e^{j\theta})$, especially in cases where $H_d(e^{j\theta})$ has discontinuities. In the next sections we consider the design of IIR filters that are implemented using recursive structures with feedback of past output values. In the realization of discontinuous frequency functions, IIR structures generally require fewer coefficients than FIR filters.

7.2 DESIGN OF IMPULSE INVARIANT IIR FILTERS

The central idea in the design of impulse invariant filters is that we should choose the impulse response $h(n)$ of our discrete-time filter to equal a sampled version of the impulse response $\hat{h}(t)$ of a desired continuous-time model filter. That is, we set

$$h(k) = T\hat{h}(kT) \tag{7.19}$$

where the amplitude scale factor T is included to make the filter gain independent of the sampling rate. We begin with a specification in terms of $\hat{H}(s)$, the Laplace transfer function corresponding to the continuous-time impulse response $\hat{h}(t)$. The transfer function $\hat{H}(s)$ might represent, for example, a Chebychev or maximally flat approximation to an ideal filter transfer function and would be obtained from standard network synthesis procedures. We require that $\hat{H}(s)$ be rational, that is, a ratio of polynomials in s, with the numerator of lower order than the denominator:

$$\hat{H}(s) = \frac{\alpha_0 + \alpha_1 s + \cdots + \alpha_M s^M}{\beta_0 + \beta_1 s + \cdots + \beta_N s^N}, \qquad N > M \tag{7.20}$$

We begin by expanding $\hat{H}(s)$ in its partial fraction form:

$$\hat{H}(s) = \frac{c_1}{s - p_1} + \frac{c_2}{s - p_2} + \cdots + \frac{c_N}{s - p_N} \tag{7.21}$$

where $p_1 \cdots p_N$ are the poles of $\hat{H}(s)$. We assume for now that no pole is repeated. From this relation we can write the impulse response $\hat{h}(t)$ as a sum of functions

$$\hat{h}(t) = \hat{h}_1(t) + \hat{h}_2(t) + \cdots + \hat{h}_N(t) \tag{7.22}$$

where $\hat{h}_i(t)$ is the inverse transform of the term $c_i/(s - p_i)$, namely,

$$\begin{aligned} \hat{h}_i(t) &= \mathscr{L}^{-1}\left\{\frac{c_i}{s - p_i}\right\} \\ &= c_i e^{p_i t} \xi(t) \end{aligned} \tag{7.23}$$

Now $h(n)$ is written as a corresponding sum of sequences

$$h(n) = h_1(n) + h_2(n) + \cdots + h_N(n) \tag{7.24}$$

where

$$\begin{aligned} h_i(n) &= T\hat{h}_i(nT) \\ &= \begin{cases} 0, & n < 0 \\ Tc_i e^{p_i nT}, & n \geq 0 \end{cases} \end{aligned} \tag{7.25}$$

Taking the Z-transform of this sum term by term, we have

$$H(z) = H_1(z) + H_2(z) + \cdots + H_N(z) \tag{7.26}$$

where

$$\begin{aligned} H_i(z) &= Z\{h_i(n)\} \\ &= \sum_{n=-\infty}^{\infty} h_i(n) z^{-n} \\ &= \sum_{n=0}^{\infty} Tc_i (e^{p_i T})^n z^{-n} \\ &= \frac{Tc_i}{1 - e^{p_i T} z^{-1}}, \qquad |z| > |e^{p_i T}| \end{aligned} \tag{7.27}$$

We now have the required digital filter in partial fraction form, where each term $c_i/(s - p_i)$ in the partial fraction expansion of $\hat{H}(s)$ has been replaced by the term $Tc_i/(1 - z^{-1}e^{p_i T})$ in $H(z)$. Thus each simple pole p_i of $\hat{H}(s)$ appears as a simple pole $e^{p_i T}$ of $H(z)$. Note that if Re $\{p_i\} < 0$, then $|e^{p_i T}| < 1$. Hence poles in the *left* half

s-plane map into poles *inside* the unit circle in the z-plane, as we require for a stable causal filter. The design for nonrepeated roots is summarized as

$$\hat{H}(s) = \frac{c_1}{s - p_1} + \frac{c_2}{s - p_2} + \cdots + \frac{c_N}{s - p_N} \tag{7.28a}$$

$$H(z) = \frac{c_1 T}{1 - e^{p_1 T} z^{-1}} + \frac{c_2 T}{1 - e^{p_2 T} z^{-1}} + \cdots + \frac{c_N T}{1 - e^{p_N T} z^{-1}} \tag{7.28b}$$

Example 7.2 Let us apply this technique to the design of a low-pass filter. First, we shall find the digital equivalent of the RC network of Figure 7.4. We select the cutoff frequency to be one fourth the sampling frequency, as in Example 7.1.

$$\omega_0 = \frac{1}{RC} = \frac{1}{4}\left(\frac{2\pi}{T}\right) = \frac{\pi}{2T}$$

$$\hat{H}(s) = \frac{1/RC}{s + (1/RC)}$$

Thus,

$$\hat{H}(s) = \frac{\pi/2T}{s + \pi/2T}$$

and hence the digitized version of $H(s)$ is

$$H(z) = \frac{T(\pi/2T)}{1 - e^{-\pi/2} z^{-1}}$$

$$\cong \frac{1.57}{1 - 0.21 z^{-1}}$$

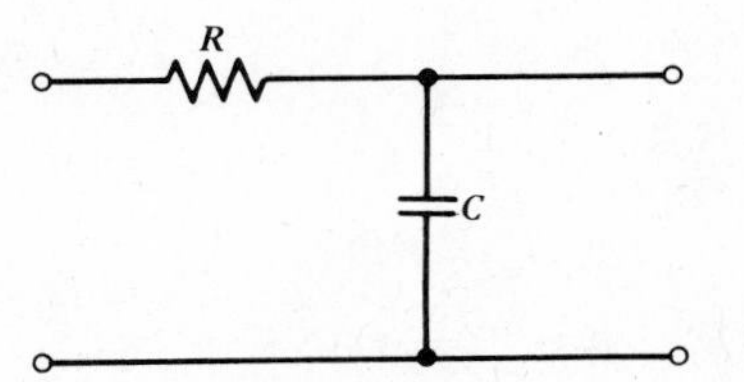

Figure 7.4

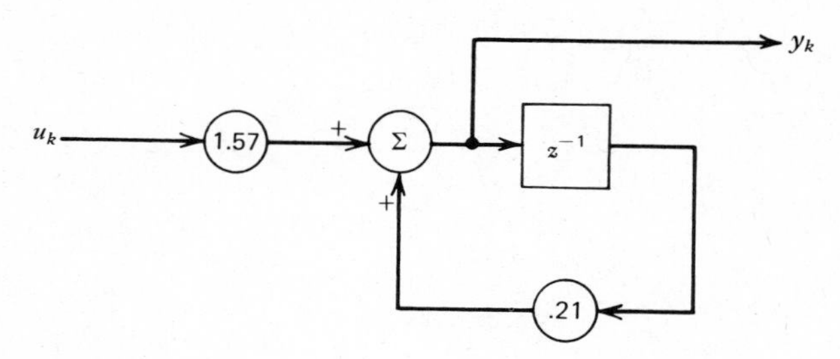

Figure 7.5

The difference equation is found from

$$(1 - 0.21z^{-1})Y(z) = 1.57U(z)$$

Taking inverse transforms, we have

$$y_k = 1.57u_k + 0.21y_{k-1}$$

which might be realized as in Figure 7.5. ■

Example 7.3 As a second example, let us construct the digital equivalent of a three-pole maximally flat (Butterworth) filter. We know that the continuous filter with unit dc gain has the transfer function

$$\hat{H}(s) = \frac{-p_1 p_2 p_3}{(s - p_1)(s - p_2)(s - p_3)}$$

where, from Figure 7.6

$$p_1 = -\omega_0$$

$$p_2 = -\omega_0\left(\frac{1 - j\sqrt{3}}{2}\right) = p_3^*$$

Expanding in partial fractions, we have

$$\hat{H}(s) = \frac{c_1}{s - p_1} + \frac{c_2}{s - p_2} + \frac{c_3}{s - p_3} \tag{7.29}$$

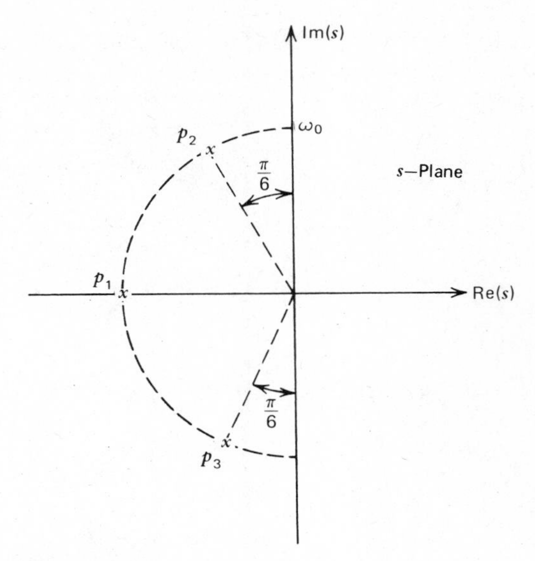

Figure 7.6

with

$$c_1 = \frac{-p_1 p_2 p_3}{(p_1 - p_2)(p_1 - p_3)} = \omega_0$$

$$c_2 = \frac{-p_1 p_2 p_3}{(p_2 - p_1)(p_2 - p_3)} = \frac{2\omega_0}{-3 + j\sqrt{3}}$$

$$c_3 = \frac{-p_1 p_2 p_3}{(p_3 - p_1)(p_3 - p_2)} = c_2^* = \frac{2\omega_0}{-3 - j\sqrt{3}}$$

Now the digital filter is found by applying the substitution

$$\frac{c}{s - p} \to \frac{cT}{1 - e^{pT} z^{-1}}$$

term by term to (7.26), yielding

$$H(z) = \frac{T\omega_0}{1 - e^{-\omega_0 T} z^{-1}} - \frac{2T\omega_0}{3 + j\sqrt{3}} \frac{1}{1 - z^{-1} e^{-(\omega_0 T/2)(1 - j\sqrt{3})}}$$
$$- \frac{2T\omega_0}{3 - j\sqrt{3}} \cdot \frac{1}{1 - z^{-1} e^{-(\omega_0 T/2)(1 + j\sqrt{3})}}$$

We wish, of course, to realize our digital filter using only real-valued coefficients. We can do so by combining the last two terms of $H(z)$ to give

$$H(z) = \frac{\omega_0 T}{1 - e^{-\omega_0 T} z^{-1}} - \omega_0 T \cdot \frac{1 - z^{-1} e^{-(\omega_0 T/2)}(\cos\alpha + (\sqrt{3}/3)\sin\alpha)}{1 - 2z^{-1} e^{-(\omega_0 T/2)}(\cos\alpha) + z^{-2} e^{-\omega_0 T}}$$

where

$$\alpha = \frac{\omega_0 T\sqrt{3}}{2}$$

If we take $\omega_0 = \pi/2T$, as in the previous two examples, we have

$$\alpha = \frac{\sqrt{3}\pi}{4} \text{ radians} = 78^\circ$$

and

$$H(z) \cong \frac{1.57}{1 - 0.21z^{-1}} - \frac{1.57 - 0.55z^{-1}}{1 - 0.19z^{-1} + 0.21z^{-2}}$$

As written above, $H(z)$ could be realized in the parallel form shown in Figure 7.7. If a cascade or direct form is preferred, the two terms could be combined, yielding

$$H(z) \cong \frac{0.58z^{-1} + 0.21z^{-2}}{(1 - 0.21z^{-1})(1 - 0.19z^{-1} + 0.21z^{-2})}$$

which might be realized in the cascade form of Figure 7.8.

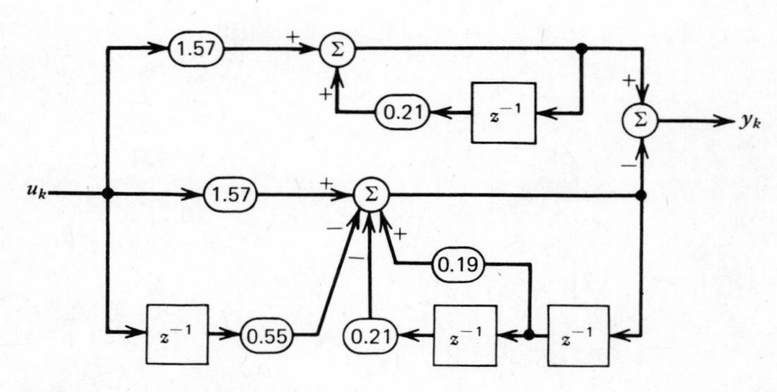

Figure 7.7

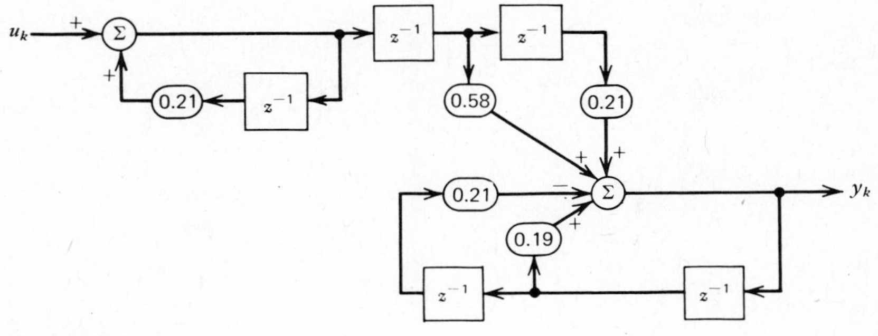

Figure 7.8 ■

As can be seen in this example, the impulse invariance implementation involves a certain amount of computation. Because the digital filter corresponding to the cascade $\hat{H}_1(s)\hat{H}_2(s)$ is not the cascade of the digital filters $H_1(z)$ and $H_2(z)$ corresponding to $\hat{H}_1(s)$ and $\hat{H}_2(s)$ individually, we must first obtain a partial fraction expansion of $\hat{H}(s)$. Second, we apply (7.28) to generate $H(z)$ term by term, and last, we must recombine the terms of $H(z)$ if we wish to realize our digital filter in cascade form.

Suppose now that $\hat{H}(s)$ has a repeated root, which gives rise to terms of the form $c/(s-p)^r$ in the partial fraction expansion. Taking the inverse Laplace transform, we find the corresponding impulse response component to be

$$\begin{aligned}\hat{h}_i(t) &= \mathscr{L}^{-1}\left\{\frac{c}{(s-p)^r}\right\}\\ &= \frac{c}{(r-1)!}\,t^{r-1}e^{pt}\xi(t)\end{aligned}$$

The sampled version is

$$h_i(n) = \begin{cases}\dfrac{cT}{(r-1)!}(nT)^{r-1}e^{pnT}, & n \geq 0\\ 0, & n < 0\end{cases}$$

with the Z-transform

$$\begin{aligned}H(z) &= \frac{c}{(r-1)!}\,T^r\sum_{n=0}^{\infty} n^{r-1}(e^{pT}z^{-1})^n\\ &= \frac{cT^r}{(r-1)!}\left(-z\frac{d}{dz}\right)^{r-1}\frac{1}{1-e^{pT}z^{-1}}\end{aligned}$$

Table 7.1

$\hat{H}_i(s)$	$H_i(z), \quad a = e^{pT}$
$\dfrac{c}{s-p}$	$\dfrac{Tc}{1-az^{-1}}$
$\dfrac{c}{(s-p)^2}$	$T^2 \dfrac{caz^{-1}}{(1-az^{-1})^2}$
$\dfrac{c}{(s-p)^3}$	$\dfrac{T^3c}{2}\left\{\dfrac{az^{-1}}{(1-az^{-1})^2} + \dfrac{2a^2z^{-2}}{(1-az^{-1})^3}\right\} = \dfrac{T^3caz^{-1}(1+az^{-1})}{2(1-az^{-1})^3}$
$\dfrac{c}{(s-p)^4}$	$\dfrac{T^4c}{6}\left\{\dfrac{az^{-1}}{(1-az^{-1})^2} + \dfrac{6a^2z^{-2}}{(1-az^{-1})^3} + \dfrac{6a^3z^{-3}}{(1-az^{-1})^4}\right\} = \dfrac{T^4caz^{-1}(1+4az^{-1}+a^2z^{-2})}{6(1-az^{-1})^4}$

The first few of these transforms are listed in Table 7.1 for $r = 1$ (the isolated pole we treated first), 2, 3, and 4. Again, to design our digital filter by the impulse invariance method, we first find the partial fraction expansion of $\hat{H}(s)$ and then write $H(z)$ as a term-by-term substitution. Each pole of $\hat{H}(s)$ at $s = p$ gives rise to a pole of $H(z)$ at $z = z = e^{pT}$ of the same multiplicity. In Table 7.1, the substitution $a = e^{pT}$ was made for notational simplicity. Both the partial fraction and combined forms of $H_i(z)$ are given.

Example 7.4 If we cascade two identical RC filters separated by an isolating amplifier as in Figure 7.9, we obtain the transfer function

$$\hat{H}(s) = \left[\frac{1/RC}{s + (1/RC)}\right]^2 = \frac{\omega_0^2}{(s + \omega_0)^2}$$

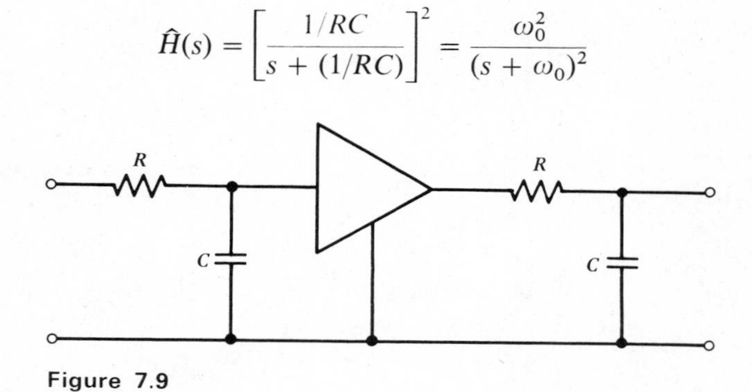

Figure 7.9

The equivalent digital filter transfer function $H(z)$ is found from Table 7.1 as

$$H(z) = \frac{T^2\omega_0^2 e^{-\omega_0 T} z^{-1}}{(1 - z^{-1}e^{-\omega_0 T})^2}$$

Again, with $\omega_0 = \frac{1}{4} \times (2\pi/T)$, we have

$$H(z) = \frac{T^2(\pi/2T)^2 e^{-\pi/2} z^{-1}}{(1 - e^{-\pi/2}z^{-1})^2} \cong \frac{0.51z^{-1}}{(1 - 0.21z^{-1})^2}$$

Note that $H(z)$ cannot be found simply by cascading the two identical digital filters of Example 2. ■

Let us now compare the desired frequency response and the actual frequency response for the three filters we have designed. Figure 7.10 shows the measured

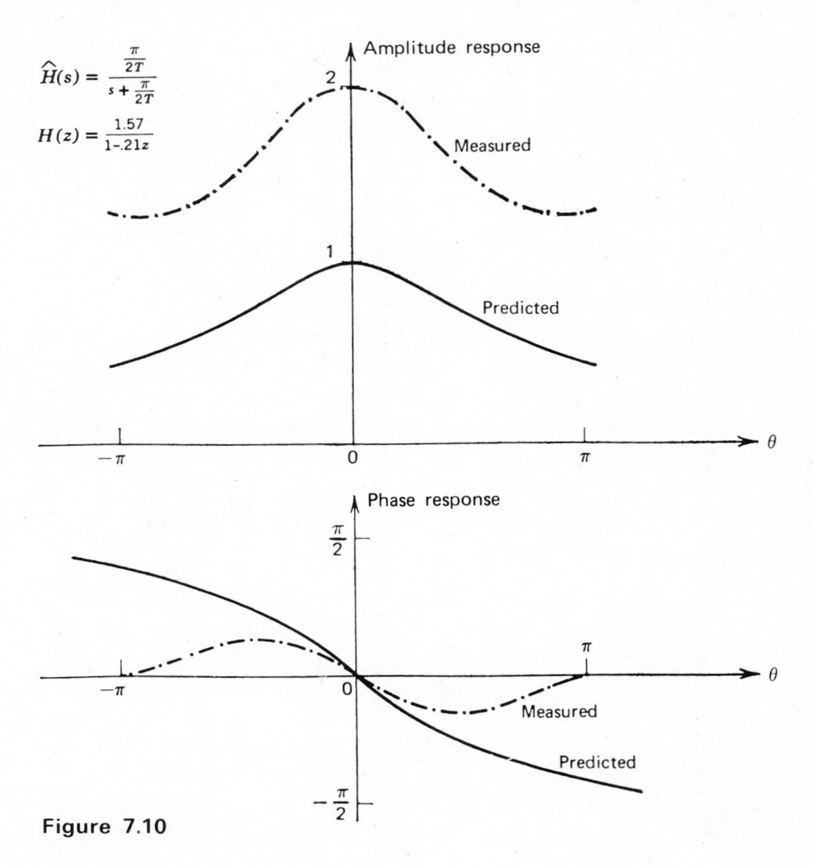

Figure 7.10

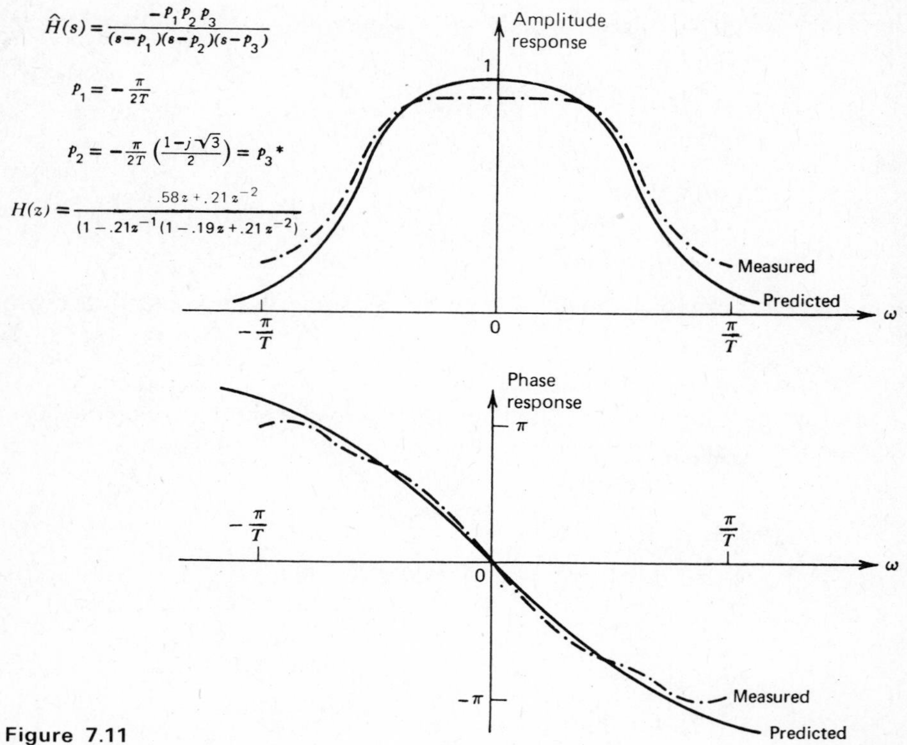

Figure 7.11

transfer function for the digital filter of Figure 7.5 from Example 7.2 compared with the desired $\hat{H}(j\omega)$. Note that the dc transmission† is $H(1) = 1.99$, rather than $\hat{H}(0) = 1$. Deviations in amplitude and phase can be seen at other frequencies. Figures 7.11 and 7.12 show the predicted and measured transfer functions for the filters of Examples 7.3 and 7.4. Again we see a discrepancy in the measured transfer functions.

One source of this discrepancy is the sampling implicit in (7.19). To relate $H(e^{j\theta})$ and $\hat{H}(j\omega)$, we write

$$H(e^{j\theta}) = H(z)|_{z=e^{j\theta}}$$

$$= \sum_{n=-\infty}^{\infty} h(n)e^{-jn\theta}$$

$$= T \sum_{n=-\infty}^{\infty} \hat{h}(nT)e^{-jn\theta} \tag{7.30}$$

† $\{y_k\}$ is found from the convolution of $\{u_k\}$ with $\{h_k\}$. With $u_k = 1$, all k, y_k is given by $y_k = \sum_m h_m u_{k-m} = \sum_m h_m = H(z)|_{z=1}$.

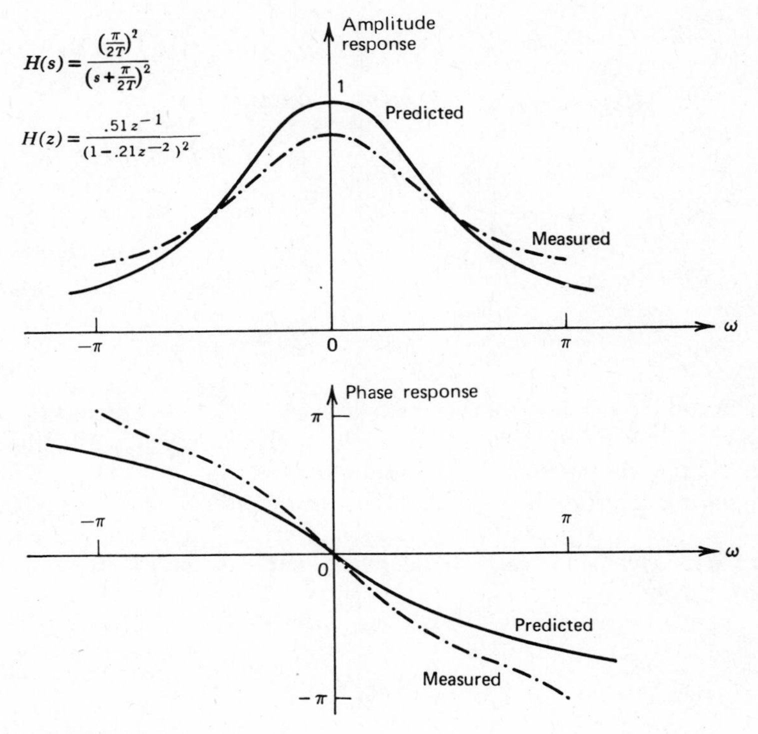

Figure 7.12

Setting $\theta = 2\pi f/f_{\text{samp}} = \omega T$, we obtain

$$H(e^{j\theta}) = T \sum_{n=-\infty}^{\infty} \hat{h}(nT)e^{-jn\omega T} \tag{7.31}$$

This last result can be interpreted as the Fourier transform of the sampled time function $\hat{h}_s(t) = \hat{h}(t) \sum_{k=-\infty}^{\infty} \delta(t - kT)$:

$$\mathscr{F}\left\{\hat{h}(t) \sum_{k=-\infty}^{\infty} \delta(t - kT)\right\} = \int_{-\infty}^{+\infty} \hat{h}(t) \sum_{k=-\infty}^{\infty} \delta(t + kT)e^{-j\omega t}\, dt$$

$$= \sum_{k=-\infty}^{\infty} \hat{h}(kT)e^{-j\omega kT} \tag{7.32}$$

Now using the relation

$$\mathscr{F}\left\{\sum_{k} \delta(t - kT)\right\} = \frac{2\pi}{T} \sum_{n} \delta\left(\omega - \frac{n2\pi}{T}\right),$$

we obtain the result

$$
\begin{aligned}
H(e^{j\theta})|_{\theta=\omega T} &= T\mathscr{F}\left\{\hat{h}(t)\sum_{k=-\infty}^{\infty}\delta(t-kT)\right\} \\
&= T\cdot\frac{1}{2\pi}\hat{H}(j\omega) * \frac{2\pi}{T}\sum_{n=-\infty}^{\infty}\delta\left(\omega-\frac{n2\pi}{T}\right) \\
&= \sum_{n=-\infty}^{\infty}\hat{H}\left(j\omega-j\frac{n2\pi}{T}\right)\Bigg|_{\theta=\omega T} \\
&= \sum_{n=-\infty}^{\infty}\hat{H}\left(j\frac{\theta}{T}-j\frac{n2\pi}{T}\right) \qquad (7.33)
\end{aligned}
$$

The conclusion we reach is that we should expect $H(e^{j\theta})$ to equal an *aliased* or *wound up* version of $\hat{H}(j\theta/T)$, rather than $\hat{H}(j\theta/T)$ itself. In fact, we find in Figures 7.11 and 7.12 that this is exactly what we do measure.

Comparing $\sum_n \hat{H}[j(\theta/T) - jn(2\pi/T)]$ with the measured results of Example 7.2, however, points to an additional source of error. These results are sketched in Figure 7.13. Here, the real and imaginary parts of the functions have been

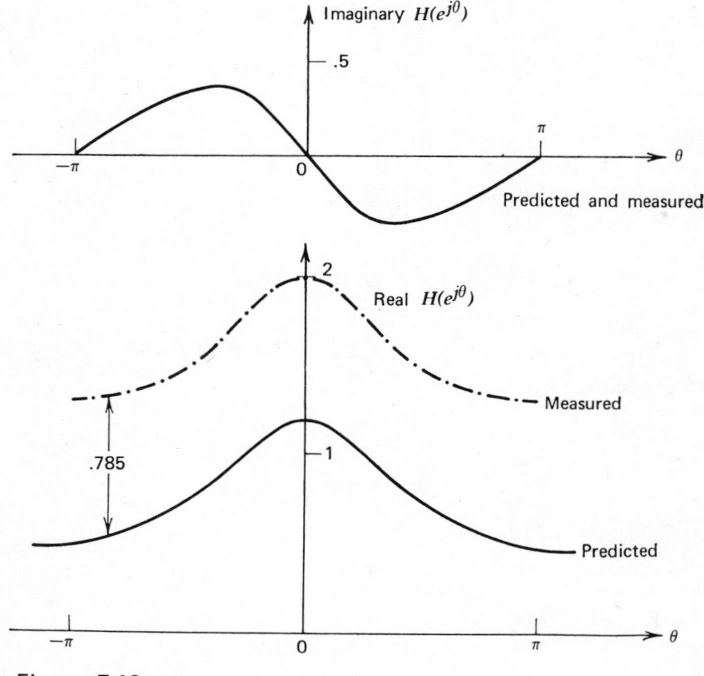

Figure 7.13

plotted, rather than the magnitude and phase. These plots show that the measured transfer function exceeds the predicted curve by the constant amount 0.785 (i.e., $\pi/4$) for all θ. To determine the source of this error, we return to (7.32) repeated below.

$$H(e^{j\theta})|_{\theta=\omega T} = \sum_{k=-\infty}^{\infty} h(k)e^{-jk\omega T}$$

$$= T \sum_{k=-\infty}^{\infty} \hat{h}(kT)e^{-jk\omega T}$$

$$= T\mathscr{F}\left\{\hat{h}(t) \sum_{k=-\infty}^{\infty} \delta(t-kT)\right\}$$

For the filter of Example 7.2,

$$\hat{h}(t) = \omega_0 e^{-\omega_0 t}\xi(t)$$

and $\hat{h}(t)$ is discontinuous at $t = 0$: hence, we should *not* have used the relation

$$\hat{h}(t)\delta(t) = \hat{h}(0)\delta(t)$$

in the $k = 0$ term above, because this relation holds only if h is continuous at 0.

To correct this error, we can again take, as in Chapter 3,

$$\delta(t) = \lim_{a\to 0} \frac{1}{a} p\left(\frac{t}{a}\right)$$

where

$$\int_{-\infty}^{\infty} p(t)\, dt = 1$$

Here, however, we restrict $p(t)$ to be an *even* function:

$$p(-t) = p(t)$$

For example, $p(t)$ might be a rectangular or Gaussian function centered at $t = 0$. Now if $f(t)$ has a jump discontinuity at $t = 0$ but is otherwise continuous in the neighborhood of 0, and the testing function $\theta(t)$ is continuous at 0, we can argue that

$$\lim_{a\to 0} \int_{-\infty}^{\infty} f(t)\frac{1}{a} p\left(\frac{t}{a}\right)\theta(t)\, dt = \lim_{a\to 0} \frac{f(0^+)+f(0^-)}{2} \int_{-\infty}^{\infty} \frac{1}{a} p\left(\frac{t}{a}\right)\theta(t)\, dt$$

and thus that we should take $f(t)\delta(t)$ to be

$$f(t)\delta(t) = \frac{f(0^+) + f(0^-)}{2}\,\delta(t) \tag{7.34}$$

where

$$f(0^+) = \lim_{\Delta \to 0} f(t + |\Delta|)$$

$$f(0^-) = \lim_{\Delta \to 0} f(t - |\Delta|)$$

are the limits of $f(t)$ at 0 from the right and left, respectively. [Note that this result is consistent with our previous result if $f(t)$ is continuous at 0.]

Now, applying (7.34), we have the corrected relation

$$\begin{aligned} H(e^{j\theta}) &= T \cdot \mathscr{F}\left\{ \sum_{k=-\infty}^{\infty} \hat{h}(kt)\delta(t - kT) \right\} \\ &= T \cdot \mathscr{F}\left\{ \hat{h}(t) \sum_{k=-\infty}^{\infty} \delta(t - kT) + \frac{\hat{h}(0)}{2}\,\delta(t) \right\} \\ &= T\,\frac{\hat{h}(0)}{2} + \sum_{n=-\infty}^{\infty} \hat{H}\left(\frac{\theta}{T} - \frac{n2\pi}{T}\right), \qquad \theta = \omega T \end{aligned} \tag{7.35}$$

where we have written

$$\hat{h}(t)\delta(t) = \frac{\hat{h}(0)}{2}\,\delta(t)$$

to account for the discontinuity of $\hat{h}(t)$ at $t = 0$. For the filter treated in Example 7.2, the constant term is

$$\frac{T}{2}\,\hat{h}(0) = \frac{T}{2}\,\omega_0 = \frac{T}{2}\,\frac{\pi}{2T} = \frac{\pi}{4} \cong 0.785$$

which is just the deviation we have measured. For the filters of Examples 7.3 and 7.4, $h(0) = 0$ and thus, this effect was not observed. In fact, $\hat{h}(0) = 0$ whenever the order of the denominator of $\hat{H}(s)$ is at least two greater than the order of its numerator (in Equation 7.17, $n \geq m + 2$). This fact follows from the initial value theorem

$$\hat{h}(0) = \lim_{s \to \infty} s\hat{H}(s)$$

If the denominator is of order one greater than the numerator order, the constant term $(T/2)\hat{h}(0)$ can be eliminated from the transfer function $H(e^{j\theta})$ by taking

$$h(n) = \begin{cases} \dfrac{T}{2}\hat{h}(0), & n = 0 \\ T\hat{h}(nT), & n \neq 0 \end{cases} \tag{7.36}$$

which gives for $\hat{H}(s) = c/(s - p)$ the digital equivalent

$$H(z) = \frac{Tc}{1 - az^{-1}} - \frac{Tc}{2}$$

$$= \frac{Tc}{2} \cdot \frac{1 + az^{-1}}{1 - az^{-1}}$$

with $a = e^{-pT}$ (compare with the first pair of Table 7.1).

Example 7.5 Let us repeat the design of Example 7.2, including the modification described above. From

$$\hat{H}(s) = \frac{\pi/2T}{s + \pi/2T}$$

with

$$\hat{h}(0) = \lim_{s \to \infty} s\hat{H}(s) = \frac{\pi}{2T}$$

we have

$$H(z) = \frac{T \cdot (\pi/2)T}{2} \cdot \frac{1 + e^{-\pi/2}z^{-1}}{1 - e^{-\pi/2}z^{-1}}$$

$$= \frac{\pi}{4} \cdot \frac{1 + e^{-\pi/2}z^{-1}}{1 - e^{-\pi/2}z^{-1}}$$

$$\cong 0.785 \cdot \frac{1 + 0.21z^{-1}}{1 - 0.21z^{-1}}$$

with a pole at $z = 0.21$ and a zero at $z = -0.21$. The corresponding difference equation is

$$y_k - 0.21y_{k-1} = 0.785u_k + 0.165u_{k-1} \qquad \blacksquare$$

Let us summarize the impulse invariance method. Beginning with the Laplace transfer function, $\hat{H}(s)$, of a desired continuous-time filter, we first find the partial fraction expansion of $\hat{H}(s)$ and then apply the equivalent pairs of Table 7.1 term by term to generate the digital filter transfer function $H(z)$.

$$\frac{c}{s-p} \rightarrow \frac{Tc}{1 - az^{-1}}, \qquad a = e^{pT} \tag{7.37}$$

If the desired impulse response, $\hat{h}(t)$, is nonzero at $t = 0$, the pair

$$\frac{c}{s-p} \rightarrow \frac{Tc}{2}\frac{1 + az^{-1}}{1 - az^{-1}}, \qquad a = e^{pT} \tag{7.38}$$

should be used. If $\hat{h}(0) = 0$, the two pairs will give the same results. The transfer function that will be realized is

$$H(e^{j\theta}) = \sum_{n=-\infty}^{\infty} \hat{H}\left(j\frac{\theta}{T} - j\frac{n2\pi}{T}\right) \tag{7.39}$$

In general, this expression will be a satisfactory approximation to $\hat{H}(j\theta/T)$ for filters whose denominators are of considerably higher order than their numerators or whose highest break frequency is a small fraction of the sampling frequency, $1/T$ Hz. In any case, the actual transfer function can be evaluated from (7.39) before any design efforts are carried out.

7.3 THE BILINEAR TRANSFORM METHOD

To avoid the problem of transfer function aliasing or "winding up" which is inherent in the impulse invariance design, we shall develop an alternate approach in this section. Rather than sampling the impulse response of a model continuous-time filter, we aim to map the transfer function $\hat{H}(s)$ directly to $H(z)$ by finding some function $z = \zeta(s)$ such that with $H(z) = H[\zeta(s)] \triangleq \hat{H}(s)$, we shall obtain a discrete-time filter with the desired properties. We shall find that this synthesis approach, while more abstract than the others, yields a procedure that is simple to apply and yields filters with superior performance for many applications.

Let us first list the properties that the mapping function $\xi(\cdot)$ should possess:

1. $z = \zeta(s)$ should be a 1-to-1 mapping such that the inverse $s = \zeta^{-1}(z)$ exists.
2. The s-plane $j\omega$-axis should map to the z-plane unit circle so that $H(e^{j\theta}) = \hat{H}(j\omega)$ for $e^{j\theta} = \zeta(j\omega)$.
3. The frequency points corresponding to d.c. should coincide, that is, $s = 0$ should map to $z = 1$: $1 = \zeta(0)$.
4. In order that stable causal models lead to stable causal discrete-time filters, the left half s-plane should correspond to the inside of the unit circle, so that poles will be correctly mapped.

A 1-to-1 function that maps circles and lines in one plane to circles and lines in another, as required by condition (2), is the bilinear transform:

$$z = \frac{a + bs}{c + ds} \tag{7.40a}$$

also written as†

$$cz + dsz - bs - a = 0 \tag{7.40b}$$

To satisfy condition (3) we must have $a = c$, and to satisfy (4) we must have c and d of opposite sign ($|z| < 1$ if $s < 0$ for s real-valued). To satisfy condition (2) we must have $|z| = 1$ for $s = j\omega$; that is,

$$\left| \frac{a + j\omega b}{c + j\omega d} \right| = 1$$

A simple set of constants that satisfies all of these requirements is $a = b = c = 1$ and $d = -1$, yielding

$$z = \zeta(s) = \frac{1 + s}{1 - s} \tag{7.41a}$$

with the inverse

$$s = \zeta^{-1}(z) = \frac{z - 1}{z + 1} \tag{7.41b}$$

† Note that this equation is linear in s if z is considered constant, and *vice versa*; hence the name *bilinear*.

To relate the continuous- and discrete-time frequency parameters, we write

$$s = j\omega = \frac{e^{j\theta} - 1}{e^{j\theta} + 1} = \frac{e^{j\theta/2}(e^{j\theta/2} - e^{-j\theta/2})}{e^{j\theta/2}(e^{j\theta/2} + e^{-j\theta/2})}$$

$$= j\,\frac{(e^{j\theta/2} - e^{-j\theta/2})/2j}{(e^{j\theta/2} + e^{-j\theta/2})/2}$$

$$= j \tan\left(\frac{\theta}{2}\right) \qquad (7.42)$$

which yields the result

$$\omega = \tan\left(\frac{\theta}{2}\right) \qquad (7.43a)$$

$$\theta = 2 \tan^{-1} \omega \qquad (7.43b)$$

The relationship between ω and θ represents a distortion or *warping* of the frequency axis, as shown in Figure 7.14. In mapping $\hat{H}(s)$ into $H(z)$, we prewarp the frequency axis using (7.43a). This procedure then compensates for the distortion that is later introduced in mapping the rational approximation $\hat{H}(s)$ into $H(z)$.

We shall now summarize the bilinear transform design method and present some examples of its use (Figure 7.15). To synthesize a discrete-time filter with this method, we first select our critical angles on the unit circle, then map these points to the s-plane using (7.43a), synthesize a continuous-time model using

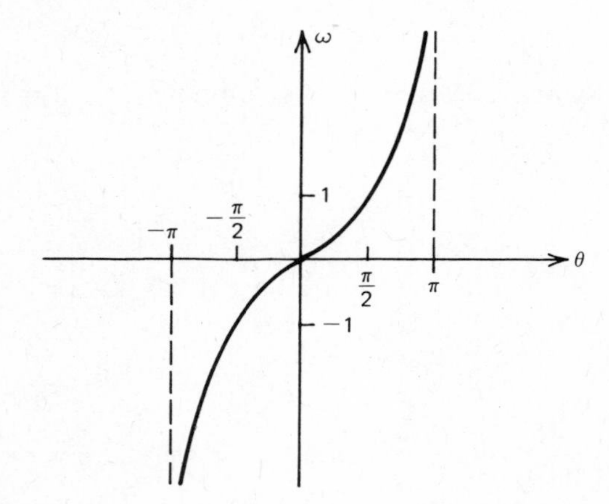

Figure 7.14

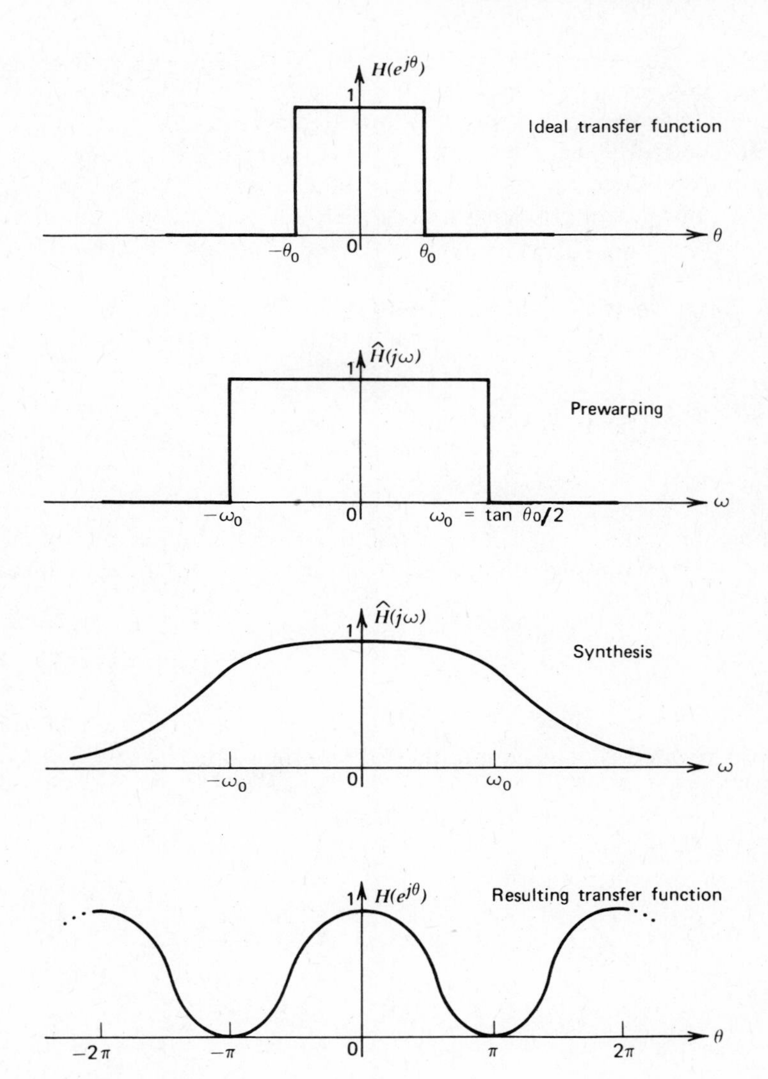

Figure 7.15

standard approximation techniques, and finally map the s-plane model to the z-plane filter using (7.42a). The resulting transfer function is given by

$$H(e^{j\theta}) = \hat{H}\left[j \tan\left(\frac{\theta}{2}\right)\right] \tag{7.44}$$

Note that absolutely no aliasing of the transfer function takes place. The only effect that is present in (7.44) is the warping of the frequency axis, which had been anticipated and precompensated.

Examples 7.6 through 7.9 that follow treat various designs for a lowpass filter with cutoff frequency equal to $\frac{1}{4}$ of the sampling frequency. The reader should compare these results with those of Examples 7.1 through 7.5, which treated the same design problem using the FIR and impulse invariance methods.

Example 7.6 We define the ideal transfer function, as before, as

$$H(e^{j\theta}) = \begin{cases} 1, & |\theta| < \dfrac{\pi}{2} \\ 0, & \dfrac{\pi}{2} < |\theta| < \pi \end{cases}$$

Prewarping the frequency axis, we find that the s-plane cutoff frequency should be $\omega_0 = \tan(\pi/4) = 1$. In this example we shall take a first order approximation for $\hat{H}(s)$, with

$$\hat{H}(s) = \frac{1}{s+1}$$

This approximation is mapped into the discrete-time filter

$$\begin{aligned} H(z) &= \hat{H}\left(\frac{z-1}{z+1}\right) \\ &= \frac{1}{(z-1)/(z+1)+1} \\ &= \frac{z+1}{2z} \\ &= \frac{1}{2} + \frac{z^{-1}}{2} \end{aligned}$$

In this case, we have obtained a 2-point FIR filter with the transfer function

$$H(e^{j\theta)} = \frac{1}{2} + \frac{1}{2} e^{-j\theta}$$

as shown in Figure 7.16.

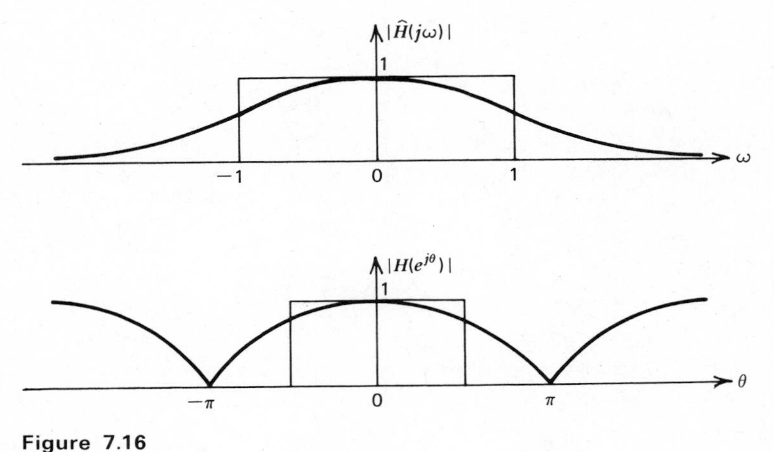

Figure 7.16

Example 7.7 If we take the same filter specifications as in the above example, but in this case realize $\hat{H}(s)$ as two identical cascaded sections as we did in Example 7.4, we have

$$\hat{H}(s) = \left(\frac{1}{s+1}\right)^2$$

yielding

$$\begin{aligned} H(z) &= \left[\frac{1}{(z-1)/(z+1)+1}\right]^2 \\ &= \left(\frac{z+1}{2z}\right)^2 \\ &= \left(\frac{1+z^{-1}}{2}\right)^2 \end{aligned}$$

with

$$H(e^{j\theta}) = \left(\frac{1+e^{-j\theta}}{2}\right)^2$$

The magnitude of this transfer function is sketched in Figure 7.17. Note that cascaded sections of $\hat{H}(s)$ are transformed into cascaded sections of $H(z)$. This property will often facilitate the design procedure.

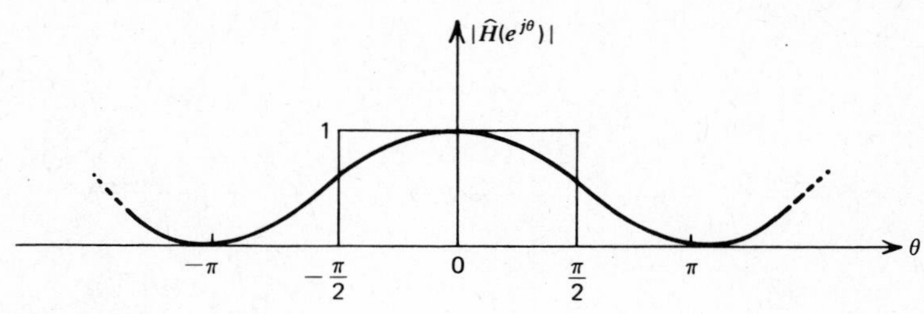

Figure 7.17

Example 7.8 Let us now use a three-pole Butterworth transfer function for $\hat{H}(s)$. As in Example 7.3 we have (again, with $\omega_0 = \tan \pi/4 = 1$)

$$\hat{H}(s) = \frac{-p_1 p_2 p_3}{(s - p_1)(s - p_2)(s - p_3)}$$

where

$$p_1 = -\omega_0 = -1$$

$$p_2 = -\omega_0 \cdot \frac{1 - j\sqrt{3}}{2} = -\frac{1}{2}(1 - j\sqrt{3})$$

$$p_3 = p_2^* = -\omega_0 \cdot \frac{1 + j\sqrt{3}}{2} = -\frac{1}{2}(1 + j\sqrt{3})$$

Substituting for p_1, p_2 and p_3, we obtain the model transfer function

$$\hat{H}(s) = \frac{1}{(s + 1)(s + (1 - j\sqrt{3})/2)(s + (1 + j\sqrt{3})/2)}$$

$$= \frac{1}{(s + 1)(s^2 + s + 1)}$$

The corresponding digital filter is

$$H(z) = \hat{H}\left(\frac{z - 1}{z + 1}\right) = \frac{1}{((z-1)/(z+1)+1)[((z-1)/(z+1))^2 + (z-1)/(z+1)+1]}$$

$$= \frac{(z + 1)^3}{2z(3z^2 + 1)}$$

$$= \frac{1 + z^{-1}}{2(3 + z^{-2})}$$

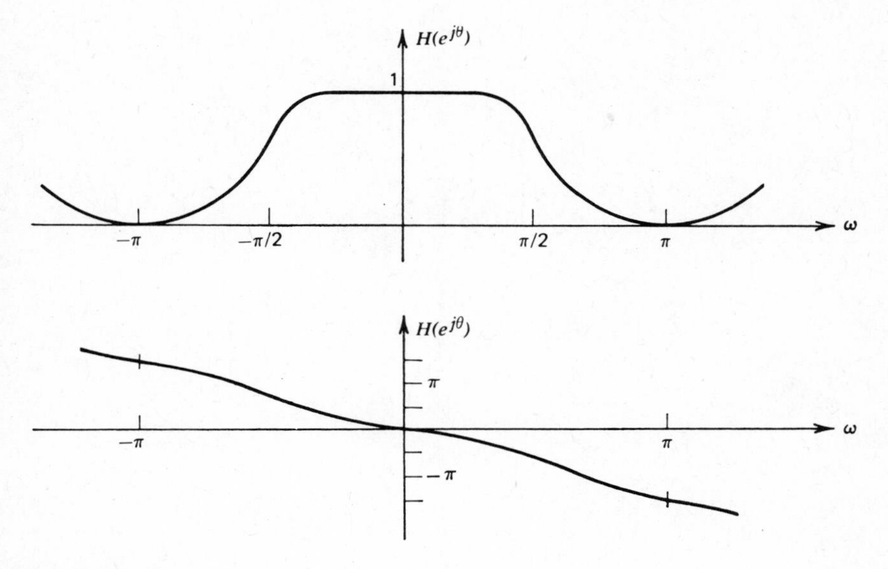

Figure 7.18

with the corresponding frequency response

$$H(e^{j\theta}) = \frac{1 + e^{-j\theta}}{2(3 + e^{-2j\theta})}$$

as sketched in Figure 7.18. ■

As can be seen from these examples, the bilinear transform method is easily implemented, especially in cases where the desired transfer function $H_1(s)$ takes the values 0 or 1 along the $j\omega$-axis (e.g., for low-pass, bandpass, etc., filters). In this case, we merely map the cutoff frequencies from one plane to another. The designed transfer function $\hat{H}$ can be made as close to the ideal characteristic as desired. Workers in the field of physical network synthesis have developed an extensive body of theory to solve this very problem. This theory is particularly applicable because the realized filter response $H(e^{j\theta})$ is, except for a warping of the frequency axis, identical to the model $H(j\omega)$.

Example 7.9 It should be apparent to the reader that the bilinear transform filter design procedure is well suited to computer implementation. A program written for digital filter design was used to generate a six-pole Chebychev low-pass filter with half-power frequency 1 (as in the previous examples) and 0.1

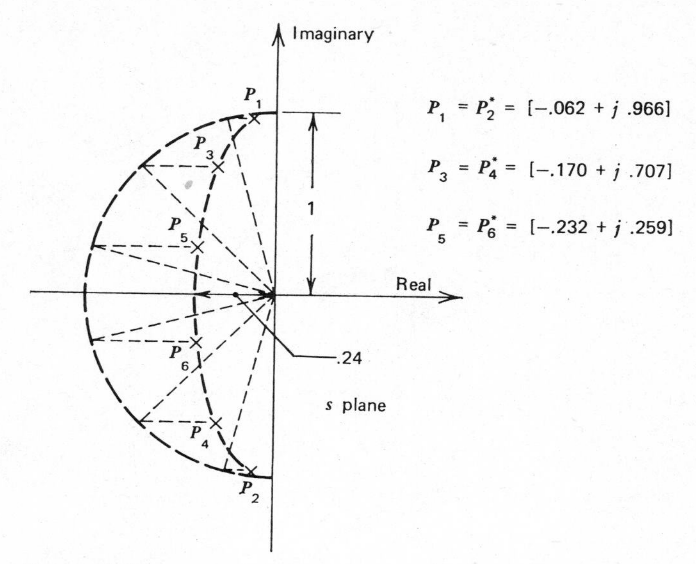

Figure 7.19

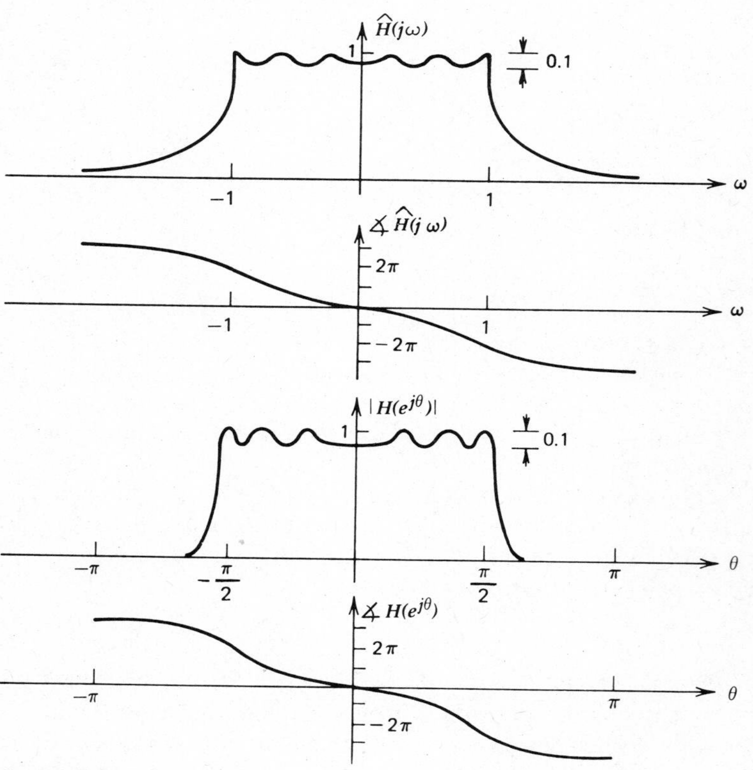

Figure 7.20

ripple in the passband. The appropriate s-plane poles are shown in Figure 7.19. The corresponding digital filter has the transfer function

$$H(z) = \frac{0.18 + 0.37z^{-1} + 0.18z^{-2}}{1.0 - 0.06z^{-1} + 0.88z^{-2}} \times \frac{0.20 + 0.40z^{-1} + 0.20z^{-2}}{1.0 - 0.50z^{-1} + 0.64z^{-2}} \times \frac{0.24 + 0.48z^{-1} + 0.24z^{-2}}{1.0 - 1.11z^{-1} + 0.41z^{-2}}$$

The frequency responses $\hat{H}(j\omega)$ and $H(e^{j\theta})$ for this filter are shown in Figure 7.20. Note the horizontal distortion that occurs in mapping $\hat{H}(j\omega)$ into $H(e^{j\theta})$. ■

7.4 CONTINUOUS-TIME FILTERING WITH DISCRETE-TIME SYSTEMS

To conclude this chapter, let us examine the mixed continuous- and discrete-time system shown in Figure 7.21. Here the continuous-time input $u(t)$ is sampled every T seconds by an analog-to-digital converter (ADC), the resulting sequence $u(nT)$ is passed through a discrete-time system with transfer function $H(z)$, and the output sequence $w(n)$ is passed through a digital-to-analog converter (DAC) to generate the continuous-time output $y(t)$. We wish to obtain an expression for the output spectrum $Y(j\omega)$ in terms of the input spectrum $U(j\omega)$, and to relate the equivalent continuous-time transfer function

$$H_{\text{eq}}(j\omega) = \frac{Y(j\omega)}{U(j\omega)} \tag{7.45}$$

to $H(e^{j\theta})$ from the discrete-time filter. We shall find that this investigation will help us to integrate the material from earlier portions of the text and will lend insight into the synthesis of discrete-time filters.

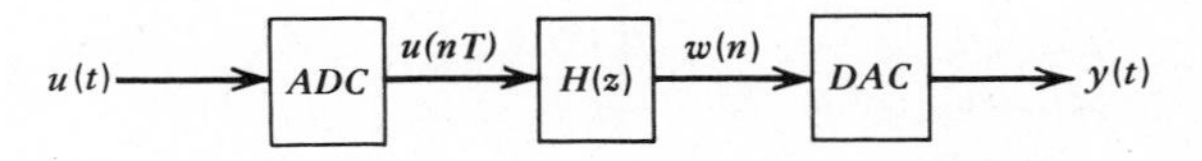

Figure 7.21

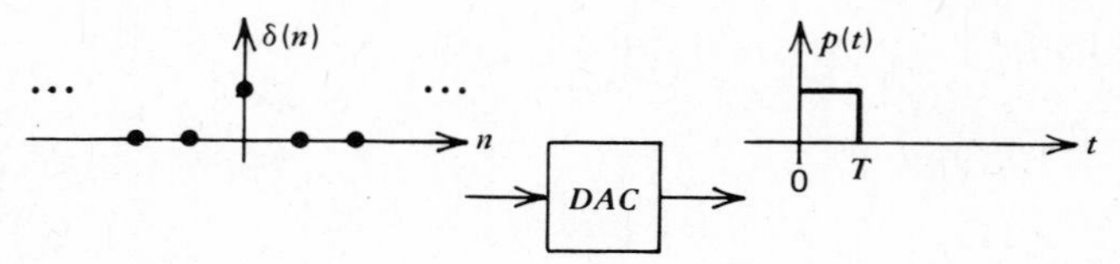

Figure 7.22

We begin by expressing $w(n)$ as a convolution of $u(nT)$ with $h(n)$, the discrete-time system impulse response:

$$\begin{aligned} w(n) &= u(nT) * h(n) \\ &= \sum_m u(mT)h(n-m) \end{aligned} \tag{7.46}$$

We now represent $y(t)$ as a superposition of weighted pulses:

$$y(t) = \sum_n w(n)p(t-nT) \tag{7.47}$$

where $p(t)$ is the continuous-time response of the DAC to a discrete-time impulse at its input. For example, if the DAC output is a voltage proportional to the digital input that is held constant for the time duration T, then $p(t)$ would be a rectangular pulse as shown in Figure 7.22. This particular DAC is termed a zero-order hold (ZOH). Other choices would be the linear point connector (LPC) discussed in the problems or more sophisticated interpolation devices. Combining (7.46) and (7.47), we have

$$y(t) = \sum_{n=-\infty}^{\infty} \sum_{m=-\infty}^{\infty} u(mT)h(n-m)p(t-nT)$$

with the transform

$$\begin{aligned} Y(j\omega) &= \int_{-\infty}^{\infty} \sum_{n,m=-\infty}^{\infty}\sum u(mT)h(n-m)p(t-nT)e^{-j\omega t}\,dt \\ &= \sum_{m=-\infty}^{\infty} \Bigg\{ u(mT)e^{-j\omega mT} \\ &\qquad \sum_{n=-\infty}^{\infty} \left[h(n-m)e^{-j\omega(n-m)T} \int_{-\infty}^{\infty} p(t-nT)e^{-j\omega(t-nT)}\,dt \right] \Bigg\} \\ &= \sum_{m=-\infty}^{\infty} \left\{ u(mT)e^{-j\omega mT} \sum_{n'=-\infty}^{\infty} \left[h(n')e^{-j\omega n'T} \int_{-\infty}^{\infty} p(t')e^{-j\omega t'}\,dt' \right] \right\} \\ &= \sum_{m=-\infty}^{\infty} u(mT)e^{-j\omega mT} H(e^{j\omega t})P(j\omega) \end{aligned} \tag{7.48}$$

In developing (7.48), we interchanged the order of summations and the integration. This interchange is allowable if the sums and integrals converge absolutely; that is, if $H(z)$ converges on the unit circle and $P(s)$ converges on the $j\omega$-axis. Now we recognize the sum involving $\{u(mT)\}$ as the Fourier transform of a sampled version of $u(t)$:

$$\begin{aligned}\sum u(mT)e^{-j\omega mT} &= \mathscr{F}\left\{\sum_m u(mT)\delta(t - mT)\right\} \\ &= \mathscr{F}\left\{u(t)\sum_m \delta(t - mT)\right\} \\ &= \mathscr{F}\{u_s(t)\} \\ &= \frac{1}{T}\sum_{k=-\infty}^{\infty} U\left(j\omega - j\frac{2\pi k}{T}\right) \end{aligned} \tag{7.49}$$

This yields the final result

$$Y(j\omega) = \frac{1}{T}H(e^{j\omega T})P(j\omega)\sum_{k=-\infty}^{\infty} U\left(j\omega - j\frac{2\pi k}{T}\right) \tag{7.50}$$

which expresses $Y(j\omega)$ as a filtered version of the aliased input spectrum. In the special case where $x(t)$ is bandlimited, so that no frequency overlapping occurs in the summation, and where we have good reconstruction: that is,

$$P(j\omega) \cong \begin{cases} T, & |\omega| < \dfrac{\pi}{T} \\ 0, & |\omega| > \dfrac{\pi}{T} \end{cases} \tag{7.51}$$

then only the $k = 0$ term of the summation is passed. Now (7.50) becomes

$$Y(j\omega) = H(e^{j\omega T})U(j\omega) \tag{7.52}$$

from which the equivalent transfer function is found to be

$$\begin{aligned} H_{eq}(j\omega) = \frac{Y(j\omega)}{U(j\omega)} &= H(e^{j\omega T}) \\ &= H(e^{j\theta})|_{\theta=\omega T} \end{aligned} \tag{7.53}$$

This result yields yet another interpretation of the discrete-time filter transfer function $H(z)$. The frequency response of the ADC-$H(z)$-DAC system in Figure 7.31, assuming a bandlimited input and perfect reconstruction, is found by evaluating $H(z)$ on the unit circle at an angle of $\theta = \omega T$. Alternatively, we can express θ as 2π times the ratio of the frequency of interest to the sampling frequency $f_{\text{samp}} = 1/T: \theta = 2\pi f/f_{\text{samp}}$.

Example 7.10 Suppose that we wished to limit the frequency content of a voice signal to the range $300 \leq f \leq 3000$ Hz. Assume an input bandwidth of 4 kHz and a sampling frequency of 8000 samples/second. Then our discrete-time filter should be designed to pass signals in the range of $\theta_1 = 2\pi \cdot 300/8000 = 0.236$ radians (13.5°) to $\theta_2 = 2\pi \cdot 3000/8000 = 2.36$ radians (135°). In a bilinear transform design of this filter, the model filter cutoff points would be $\omega_1 = \tan \theta_1/2 = 0.035$ and $\omega_2 = \tan \theta_2/2 = 2.414$. ■

7.5 CONCLUSION

In this chapter, we have investigated several methods for synthesizing discrete-time filters. In the synthesis of FIR filters we again encountered the concepts of Fourier analysis, using for our filter weights a window sequence times the coefficients in the Fourier series expansion of the desired transfer function. We gained insight into the choice of this window by expressing the resulting transfer function as a convolution of the desired transfer function with the window transform.

In designing IIR filters, we first examined the impulse invariance approach. Here we chose the discrete-time filter impulse response to equal a sampled version of the impulse response for a model continuous-time filter. The design process involves a partial fraction expansion and 1-1 substitution of a Z-transform term for each Laplace transform term. The resulting transfer function is a wound-up or aliased version of the model transfer function. This approach works well if the aliasing is not severe, and is often used in the discrete-time simulation of continuous-time systems.

The last design method involved the algebraic transformation of a model continuous-time filter to its z-plane equivalent. The resulting transfer function in this case has the same general character as the model transfer function, with a distortion of the frequency axis but no alteration of the amplitude characteristic. This approach is appropriate for the design of low-pass, bandpass, and similar filters whose desired transmission is either 1 (in the passbands) or 0 (in the stopbands).

Finally, we investigated the use of these discrete-time systems in the processing of continuous-time data. We obtained an expression for the system output spectrum as the product of the discrete-time system transfer function, the transform of the DAC impulse response, and the wound-up spectrum of the input. We saw that with a bandlimited input and ideal reconstruction device, the overall continuous-time transfer function was $H(e^{j\theta})$ with $\theta = 2\pi f/f_{\text{samp}}$.

PROBLEMS

7.1. (a) Find an expression for the coefficients $h(n)$ of an FIR approximation to the ideal bandpass characteristic

$$H_d(e^{j\theta}) = \begin{cases} 1, & \theta_1 \leq |\theta| < \theta_2 \\ 0, & \text{otherwise} \end{cases}$$

(b) Show that this result may be expressed as

$$h(n) = 2 \cos(\theta_0 n) \cdot h_1(n)$$

where $\theta_0 = (\theta_1 + \theta_2)/2$ and $h_1(n)$ is the impulse response of an FIR lowpass filter with bandwidth $(\theta_2 - \theta_1)/2$.

(c) Generalize this result: if

$$h_2(n) = A \cos(n\theta_0) h_1(n)$$

express $H_2(e^{j\theta})$ in terms of $H_1(e^{j\theta})$.

7.2. Devise at least three approaches to the design of an FIR filter that approximates

$$H_d(e^{j\theta}) = \begin{cases} 2, & 0 < |\theta| < \dfrac{\pi}{8} \\ 1, & \dfrac{\pi}{8} < |\theta| < \dfrac{\pi}{4} \\ 0, & \dfrac{\pi}{4} < |\theta| < \pi \end{cases}$$

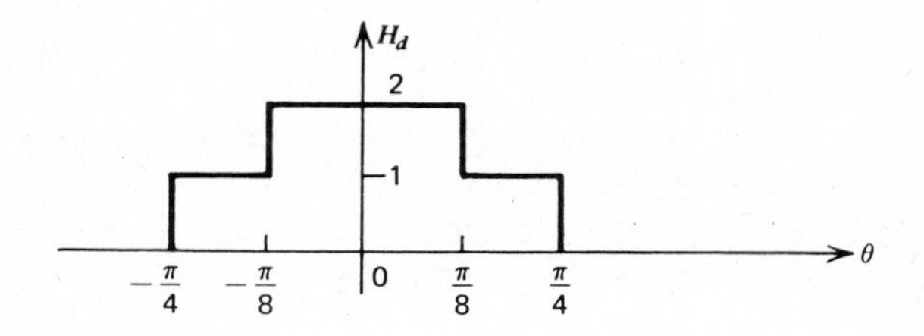

Show that the resulting expressions for $h(n)$ are identical.

7.3. Design an FIR highpass filter to approximate the characteristic

$$H_d(e^{j\theta}) = \begin{cases} 0, & |\theta| < \theta_0 \\ 1, & \theta_0 < |\theta| \le \pi \end{cases}$$

How is the resulting impulse response $h(n)$ related to the impulse response of an FIR lowpass filter with the same bandwidth?

7.4. Design an FIR approximation to an ideal differentiator with transfer function

$$H_d(e^{j\theta}) = j\theta, \qquad |\theta| < \pi$$

For a filter length N and a window of your choice, plot the resulting frequency response (a simple BASIC or FORTRAN program might be used).

7.5. Repeat Problem 7.4 for an FIR approximation to the Hilbert transform filter

$$H_d(e^{j\theta}) = -j \operatorname{sgn}(\theta) = \begin{cases} j, & -\pi < \theta < 0 \\ -j, & 0 < \theta < \pi \end{cases}$$

Again, plot the resulting frequency response.

7.6. Generate a procedure for designing a *step invariant* discrete-time filter with transfer function $H(z)$ from a continuous-time model with transfer function $\hat{H}(s)$. The appropriate design constraint is

$$g(n) = \hat{g}(nT)$$

where $g(n)$ is the discrete-time filter response to a step input sequence and $\hat{g}(t)$ is the continuous-time filter response to a continuous-time step input $\xi(t)$.

7.7. Evaluate the frequency response $W(e^{j\theta})$ corresponding to the triangular and Hanning windows shown in Figure 7.3. How do these responses compare with that of the rectangular window with respect to the sidelobe ripple and the main lobe width? Generalize your expressions for arbitrary length N.

7.8. Carry out the design of a 4-pole Butterworth approximation to an ideal low-pass filter having a cutoff frequency of $(\pi/2T)$ rad/sec: that is, one-fourth the sampling frequency. Use both the impulse invariance and bilinear transform methods, and compare the resulting transfer functions. Which filter has the sharper cutoff?

7.9. Would you anticipate any difficulty in performing the impulse invariance design of a highpass digital filter? In performing the bilinear transform design? Elaborate.

7.10. If $H(z)$ is the transfer function of a lowpass filter, what type of filter does $H(-z)$ represent? What type does $H(z^2)$ represent?

7.11. Plot the amplitude and phase responses of a filter with transfer function

$$H(z) = \frac{z - a}{z - 1/a}, \qquad 1 < a$$

What type of filter is this? Can you think of a use for it?

7.12. Sketch the pole-zero plots of the transfer function $H(z)$ for causal stable filters having the amplitude responses shown below.

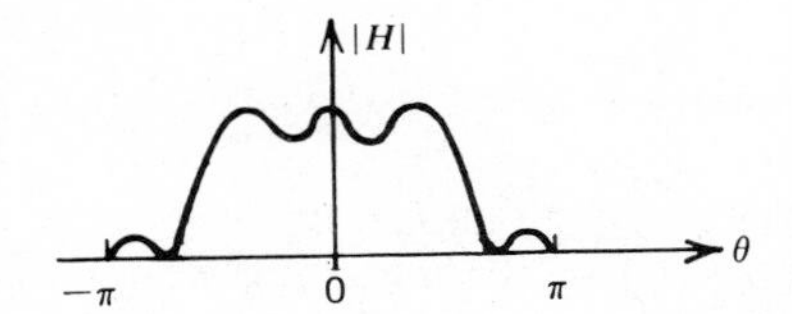

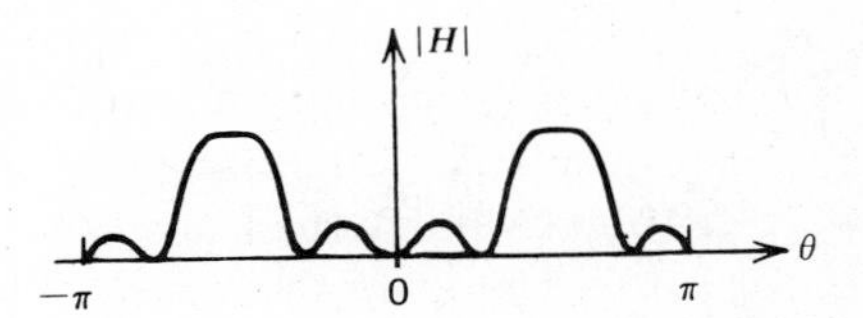

7.13. A Zero Order Hold (ZOH) is perhaps the most common form of reconstruction device, and is usually implemented with a simple digital-to-analog converter (DAC). This device puts out a constant voltage, proportional to the last digital input, for one sampling period as shown below.

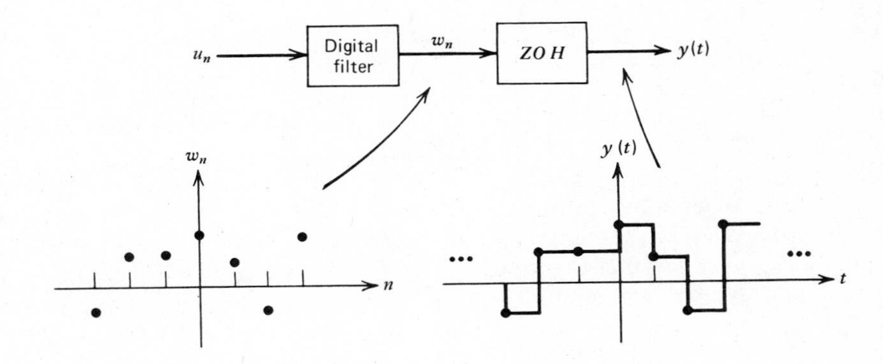

(a) What is the impulse response $p(t)$ of the ZOH (cf. Figure 7.22)?
(b) What is the corresponding frequency response $P(j\omega)$?
(c) Do you think that the ZOH makes a good reconstruction device? Why or why not?

7.14. The linear point connector (LPC) is another reconstruction device. The LPC output is formed by connecting samples of the digital filter output sequence with straight-line segments.

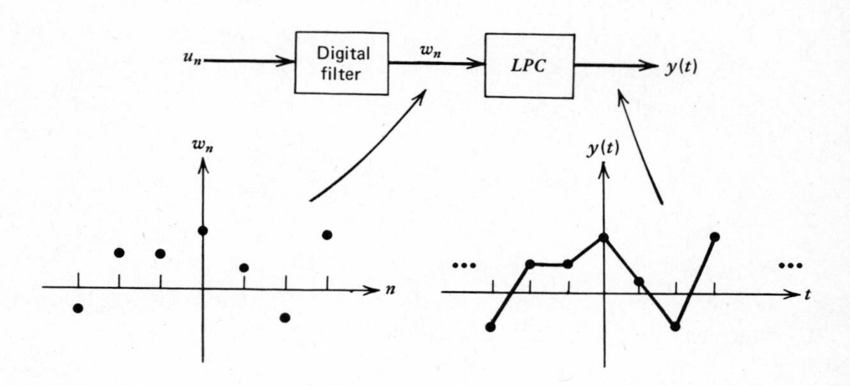

(a) Evaluate $p(t)$ for the LPC. Is this a realizable device? If not, how could you modify it to make it realizable?
(b) Evaluate the transfer function $P(j\omega)$ for the LPC.
(c) Compare the transfer functions for the LPC and ZOH. Which would make the better reconstruction device? Why?

7.15. Compare the operation of the integrator circuits below. The first is a continuous-time integrator; the second is a digital filter implementation that approximates an integral by the sum $w_k = \sum_{m=-\infty}^{k} u_m$.

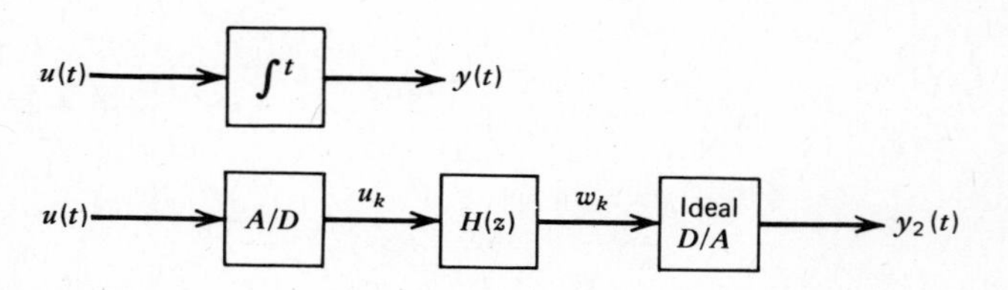

(a) What is the transfer function $H(z)$ of the digital filter?
(b) What is the continuous-time transfer function for each system? You may consider the input $u(t)$ to be bandlimited to π/T rad/sec.
(c) Under what conditions is the digital integrator a good approximation to the continuous-time integrator?

7.16. A seismic transponder signal is to be transmitted from a landed spacecraft back to Earth. Although the signal has frequency components up to 10 Hz, the scientists responsible for analyzing the data are interested in spectral components only up to 5 Hz. Therefore, to reduce the required data rate, it is planned to low-pass filter the signal, eliminating energy above 5 Hz; sample the filtered signal at 10 Hz; and send this sampled data back to Earth at a rate of 10 samples per second.

One way to implement this plan is to sample the original signal at 20 samples per second (twice the highest frequency), operate on the samples with a digital filter having a cutoff frequency of $f_{\text{samp}}/4$, and then subsample the digital filter output, transmitting only every other output sample. The continuous time signal is then reconstructed at the receiver by use of an ideal low-pass filter.

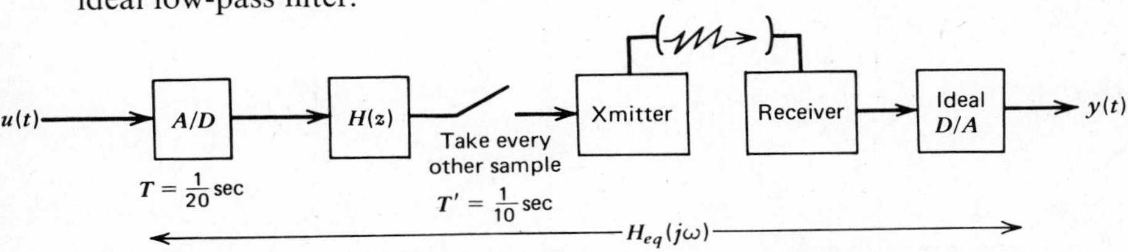

Assuming that an ideal DAC is available, will this method work? If $H(z)$ is the digital filter transfer function, what would be the equivalent continuous-time transfer function, $H_{\text{eq}}(\omega)$, from input to output? What would be the output from an input of $u(t) = \cos \omega t$, where $\omega/2\pi$ is not necessarily less than 20 Hz?

7.17. Although a bandlimited signal can in theory be reconstructed exactly from samples taken at slightly greater than twice the highest frequency, in practice it is difficult to realize a good approximation to the required ideal lowpass filter. At the same time, we would often like to take as few samples as possible as in the preceding problem, where we wish to minimize a required transmission rate.

To help resolve this problem, it has been proposed that a digital interpolation filter be used to effectively double the sampling frequency by inserting samples between those taken from the original signal. If $s(t)$ is the bandlimited signal and $\{s(kT)\}$ is the sequence of samples taken at a rate of $1/T = 2 \times f_{\max}$, then a new sequence $\{u_k\} = \cdots s(-T), 0, s(0), 0, s(T), 0, s(2T), \ldots$ would be formed and used as the input to a digital filter with transfer function $H(z)$. $H(z)$ is to be chosen such that the output sequence $\{w_k\}$ has the form

$$\{w_k\} = \cdots s(-T), s\left(-\frac{T}{2}\right), s(0), s\left(\frac{T}{2}\right), s(T), s\left(\frac{3T}{2}\right), \ldots$$

If this scheme is successful, we shall have inserted a guard band of $2\pi/T$ rad/sec with no additional signal energy into the spectrum of

$$w(t) = T \sum w_k \delta\left(t - \frac{kT}{2}\right)$$

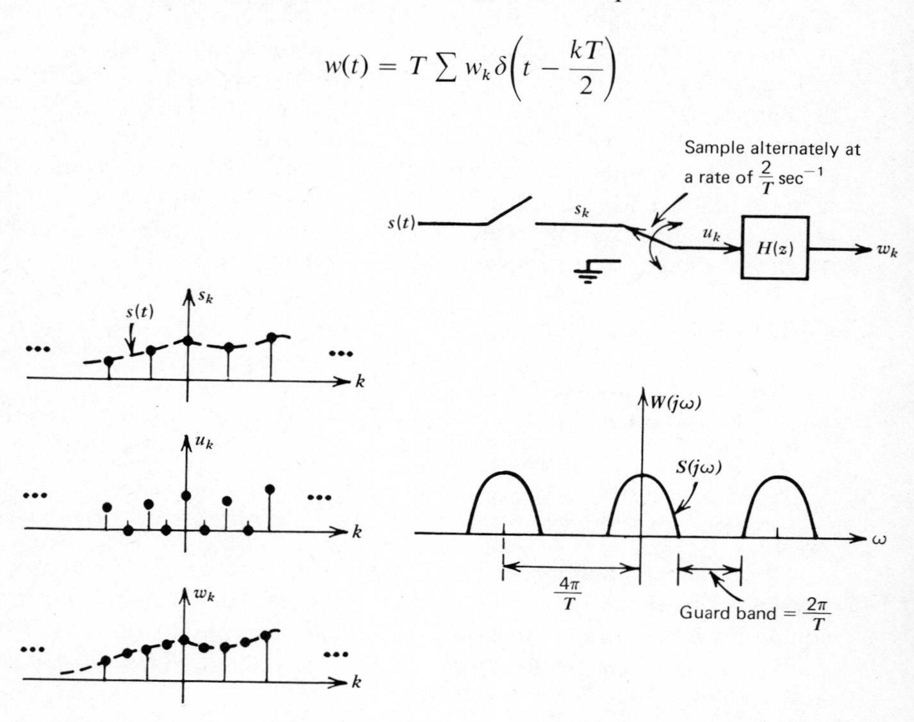

This would make the reconstruction filter much easier to build. Could such a digital filter be built? If so, how would you do it? If not, why not?

SUGGESTED READINGS

1. Bracewell, R. N., *The Fourier Integral and Its Applications*, McGraw-Hill, New York, 1978.

A comprehensive treatment of the Fourier transform at a more advanced level than presented here.

2. Director, S. W. and R. A. Rohrer, *Introduction to System Theory*, McGraw-Hill, New York, 1972.

This book contains a more comprehensive and extensive treatment of state variables for both continuous and discrete-time systems.

3. Freeman, H., *Discrete-Time Systems*, Wiley, New York, 1965.

An excellent discussion of discrete-time systems at the senior-graduate level.

4. Hildebrand, F. B., *Finite Difference Equations and Simulations*, Prentice-Hall, Englewood Cliffs, N.J., 1968.

Chapter 1 contains a complete discussion of linear difference equations.

5. Jury, E. I., *Theory and Applications of the Z-Transform Method*, Wiley, New York, 1964.

A complete and detailed treatment of Z-transforms and their applications.

6. Kuo, F. F. and J. F. Kaiser, *System Design by Digital Computer*, Wiley, New York, 1966.

Chapter 7 treats the topics of filter synthesis and presents additional examples. The selection of window functions for the design of nonrecursive filters is discussed in detail.

7. Lathi, B. P., *Signals, Systems, and Communications*, Wiley, New York, 1965.

This book contains a good discussion of convolution and the Fourier transform for continuous-time systems.

8. Liu, C. L. and J. W. S. Liu, *Linear System Analysis*, McGraw-Hill, New York, 1975.

A discussion of linear systems comparable to that presented here.

9. McGillem, C. D. and G. R. Cooper, *Continuous and Discrete Signal and System Analysis*, Holt, Rinehart and Winston, New York, 1974.

A discussion of linear systems comparable to that presented here.

10. Oppenheim, A. V. and R. Schafer, *Digital Signal Processing*, Prentice-Hall, Englewood Cliffs, N. J., 1975.

A complete discussion of digital filters written for first-year graduate students.

11. Papoulis, A., *The Fourier Integral and Its Applications*, McGraw-Hill, New York, 1962.

A senior-graduate-level text on the theory and applications of the Fourier transform.

12. Schwarz, R. J. and B. Friedland, *Linear Systems*, McGraw-Hill, New York, 1965.

A general treatment of continuous and discrete-time linear systems at a somewhat higher level than presented here.

13. Wylie, C. R., *Advanced Engineering Mathematics*, McGraw-Hill, New York, 1966.

A good general reference on the solution of differential equations, with some discussion of difference equations.

14. Zadeh, L. A. and C. A. Desoer, *Linear System Theory*, McGraw-Hill, New York, 1963.

A graduate-level text that contains a complete and rigorous discussion of linear systems.

APPENDIX A
EVALUATION OF GEOMETRIC SERIES

In this appendix, we derive closed-form expressions for various geometric series such as

$$\sum_{n=n_1}^{n_2} a^n = a^{n_1} + a^{n_1+1} + \cdots + a^{n_2-1} + a^{n_2}, \qquad n_2 \geq n_1$$

These series appear recurrently in our study of discrete-time systems.

We begin with $n_1 = 0$ and obtain

$$\sum_{n=0}^{n_2} a^n = 1 + a + a^2 + \cdots + a^{n_2}$$
$$= \begin{cases} \dfrac{1 - a^{n_2+1}}{1-a}, & a \neq 1 \\ n_2 + 1, & a = 1 \end{cases}$$

The closed form expression for $a \neq 1$ can be verified by multiplying both sides by $1 - a$ to obtain an identity. For the case $a = 1$, we note that there are $n_2 + 1$ terms in the series, with each term equal to 1.

Now, using these expressions, we obtain the relations

$$
\begin{aligned}
\sum_{n=n_1}^{n_2} a^n &= \sum_{0}^{n_2} a^n - \sum_{0}^{n_1-1} a^n \\
&= \begin{cases} \dfrac{1-a^{n_2+1}}{1-a} - \dfrac{1-a^{n_1}}{1-a}, & a \neq 1 \\ n_2 + 1 - n_1, & a = 1 \end{cases} \\
&= \begin{cases} \dfrac{a^{n_1} - a^{n_2+1}}{1-a}, & a \neq 1 \\ n_2 - n_1 + 1, & a = 1 \end{cases}
\end{aligned}
$$

where we take $0 \leq n_1 \leq n_2$.

To deal with an infinite sum, we let $n_2 \to \infty$ above, Now, if $|a| < 1$, we have

$$\lim_{n \to \infty} a^n = 0$$

and

$$
\begin{aligned}
\sum_{n=n_1}^{\infty} a^n &= \lim_{n_2 \to \infty} \sum_{n_1}^{n_2} a^n \\
&= \lim_{n_2 \to \infty} \left[\frac{a^{n_1}}{1-a} - \frac{a^{n_2+1}}{1-a} \right] \\
&= \frac{a^{n_1}}{1-a} \qquad \begin{matrix} |a| < 1 \\ n_1 \geq 0 \end{matrix}
\end{aligned}
$$

The corresponding sum for $a \geq 1$ does not converge. Letting $n_1 = 0$ and $n_1 = 1$, we obtain the special cases

$$\sum_{n=0}^{\infty} a^n = \frac{1}{1-a}, \qquad |a| < 1$$

$$\sum_{n=1}^{\infty} a^n = \frac{a}{1-a}, \qquad |a| < 1$$

Note that n_1 and n_2 are positive numbers in these expressions. To generalize the results for any n_1 and n_2, we first let $n_1 < 0 \le n_2$. Now we have

$$\sum_{n=n_1}^{n_2} a^n = \sum_{n=n_1}^{-1} a^n + \sum_{n=0}^{n_2} a^n$$

With $m = -n$ in the first sum on the right, we obtain the following sums over positive indices, which can be evaluated using the previous results:

$$\begin{aligned}
\sum_{n=n_1}^{n_2} &= \sum_{m=1}^{-n_1} \left(\frac{1}{a}\right)^m + \sum_{n=0}^{n_2} a^n \\
&= \frac{\left(\dfrac{1}{a}\right) - \left(\dfrac{1}{a}\right)^{-n_1+1}}{1 - \dfrac{1}{a}} + \frac{1 - a^{n_2+1}}{1 - a} \\
&= \frac{1 - a^{n_1}}{a - 1} + \frac{1 - a^{n_2+1}}{1 - a} \\
&= \frac{a^{n_1} - a^{n_2+1}}{1 - a}, \qquad a \ne 1
\end{aligned}$$

as before.

With $n_1 \le n_2 \le 0$, we again substitute $m = -n$ to obtain

$$\begin{aligned}
\sum_{n=n_1}^{n_2} a^n &= \sum_{m=-n_2}^{-n_1} \left(\frac{1}{a}\right)^m, \qquad 0 \ge -n_2 \ge -n_1 \\
&= \frac{(1/a)^{-n_2} - (1/a)^{-n_1+1}}{1 - \dfrac{1}{a}} \\
&= \frac{a^{n_2} - (1/a)a^{n_1}}{1 - \dfrac{1}{a}} \\
&= \frac{a^{n_1} - a^{n_2+1}}{1 - a}, \qquad a \ne 1
\end{aligned}$$

Thus, for $a \neq 1$ and $n_1 - n_2$, with no other restrictions on n_1 and n_2, we have

$$\sum_{n=n_1}^{n_2} a^n = \frac{a^{n_1} - a^{n_2+1}}{1-a}, \qquad a \neq 1$$

For $a = 1$, we observe that there are $n_2 - n_1 + 1$ terms in the sum for any $n_1 \leq n_2$, and obtain the value $n_2 - n_1 + 1$ for the sum. Alternatively, we can apply L'Hospital's rule to obtain

$$\begin{aligned}
\lim_{a \to 1} \sum_{n=n_1}^{n_2} a^n &= \lim_{a \to 1} \frac{a^{n_1} - a^{n_2+1}}{1-a} \\
&= \frac{d/da(a^{n_1} - a^{n_2+1})|_{a=1}}{d/da(1-a)|_{a=1}} \\
&= \frac{n_1 - (n_2+1)}{-1} \\
&= n_2 - n_1 + 1
\end{aligned}$$

The results of this appendix are summarized in Table A.1

Table A.1

1.	$\sum_{n=n_1}^{n_2} a^n = \begin{cases} \frac{a^{n_1} - a^{n_2+1}}{1-a}, & a \neq 1 \\ n_2 - n_1 + 1, & a = 1 \end{cases}$
2.	$\sum_{n=0}^{\infty} a^n = \frac{1}{1-a}, \quad \lvert a \rvert < 1$
3.	$\sum_{n=1}^{\infty} a^n = \frac{a}{1-a}, \quad \lvert a \rvert < 1$
4.	$\sum_{n=n_1}^{\infty} a^n = \frac{a^{n_1}}{1-a}, \quad \lvert a \rvert < 1$

Note: n_1 and n_2 may be either positive or negative in the expressions above, with $n_1 \leq n_2$.

APPENDIX B
THE SINC FUNCTION

The sinc function is defined as

$$\operatorname{sinc}(x) = \frac{\sin(x)}{x}$$

The sinc function is an even function, sinc $(-x)$ = sinc (x). The zeros are evenly spaced occurring at $\pm n\pi$, $n = 1, 2, 3, \ldots$. The following table lists values of sinc (x) for x between 0 and 10 in steps of 0.25.

x	sinc (x)	x	sinc (x)	x	sinc (x)	x	sinc (x)
0.00	1.000000						
0.25	0.989615	2.75	0.138785	5.25	−0.163606	7.75	0.128335
0.50	0.958851	3.00	0.047040	5.50	−0.128280	8.00	0.123669
0.75	0.908851	3.25	−0.033290	5.75	−0.088396	8.25	0.111830
1.00	0.841470	3.50	−0.100223	6.00	−0.046569	8.50	0.093939
1.25	0.759187	3.75	−0.152416	6.25	−0.005308	8.75	0.071397
1.50	0.664996	4.00	−0.189200	6.50	0.033095	9.00	0.045790
1.75	0.562277	4.25	−0.210585	6.75	0.066673	9.25	0.018798
2.00	0.454648	4.50	−0.217228	7.00	0.093855	9.50	−0.007910
2.25	0.345810	4.75	−0.210377	7.25	0.113528	9.75	−0.032771
2.50	0.239388	5.0	−0.191784	7.50	0.125066	10.00	−0.054402

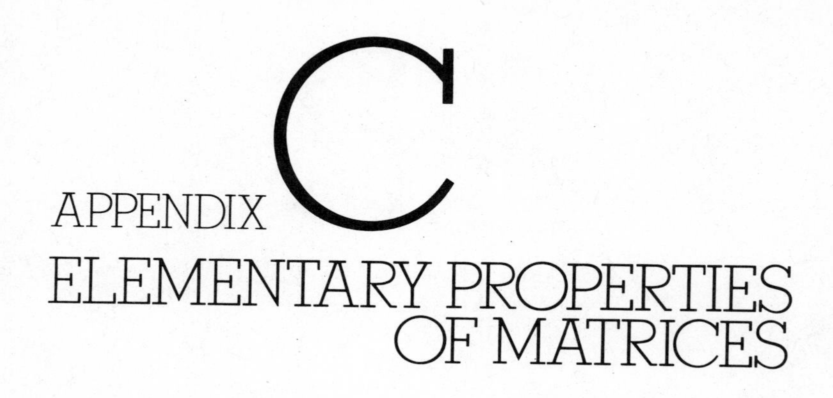

DEFINITIONS

1. An $m \times n$ *matrix* is a rectangular array of elements having m rows and n columns. We denote matrices by boldface capital letters.

$$\mathbf{A} = \begin{bmatrix} a_{11} & a_{12} & \cdots & a_{1n} \\ a_{21} & a_{22} & \cdots & a_{2n} \\ a_{m1} & a_{m2} & \cdots & a_{mn} \end{bmatrix} = [a_{ij}]$$

2. An $(n \times 1)$ matrix is a *column vector* or *vector*. A $(1 \times n)$ matrix is a *row vector* or *vector*. We shall use the convention that vectors are column vectors.
3. A *diagonal matrix* is an $(n \times n)$ matrix such that all elements not on the main diagonal (from upper left to lower right) are zero.
4. The *identity matrix* **I** is a diagonal matrix with all its diagonal elements equal to one.
5. The *transpose*, $\mathbf{A}^T$, of a matrix **A** is formed by interchanging the rows and columns.

6. The *determinant* of an $(n \times n)$ matrix $\mathbf{A}$ is a scalar obtained by calculating the parallelepiped hypervolume specified by considering each row as an n vector in n-dimensional space. We denote the determinant of $\mathbf{A}$ as $|\mathbf{A}|$.
7. If $|\mathbf{A}| = 0$, then $\mathbf{A}$ is *singular*. If $|\mathbf{A}| \neq 0$, then $\mathbf{A}$ is *nonsingular*.
8. The *cofactor* of the element a_{ij} of the square matrix $\mathbf{A}$ is $(-1)^{i+j}$ times the determinant of the matrix formed by omitting the ith row and jth column of the original matrix $\mathbf{A}$. For example, if

$$\mathbf{A} = \begin{bmatrix} a_{11} & a_{12} & a_{13} \\ a_{21} & a_{22} & a_{23} \\ a_{31} & a_{32} & a_{33} \end{bmatrix}$$

then the cofactor

$$a_{12} = (-1)^3 \left| \begin{bmatrix} a_{21} & a_{23} \\ a_{31} & a_{33} \end{bmatrix} \right| = a_{23}a_{31} - a_{21}a_{33}$$

9. The *inverse matrix*, $\mathbf{A}^{-1}$, of the square matrix $\mathbf{A}$ is defined by the equation

$$\mathbf{A}\mathbf{A}^{-1} = \mathbf{A}^{-1}\mathbf{A} = \mathbf{I}$$

The inverse matrix can be calculated using

$$\mathbf{A}^{-1} \doteq \left[\frac{\text{cofactor } a_{ij}}{|\mathbf{A}|} \right]^T$$

The inverse is found by replacing each element of $\mathbf{A}$ with its cofactor, transposing, and dividing by $|\mathbf{A}|$.

Another method of finding inverses is to attach the $n \times n$ identity to the $n \times n$ matrix $\mathbf{A}$ to form a matrix $\mathbf{A} \mid \mathbf{I}$. Now use elementary row operations to change $\mathbf{A}$ into $\mathbf{I}$. The inverse matrix $\mathbf{A}^{-1}$ will now have replaced the original identity matrix, that is, $\mathbf{A} \mid \mathbf{I}$ is now $\mathbf{I} \mid \mathbf{A}^{-1}$. The elementary row operations are:

(i) interchange of two rows,
(ii) multiplication of any row by a nonzero scalar,
(iii) replacement of the ith row by the sum of the ith row and p times the jth row $(i \neq j)$.

10. A square matrix $\mathbf{A}$ is *nilpotent* of degree p if $\mathbf{A}^P = 0$ and $\mathbf{A}^{P-1} \neq 0$.
11. The *rank* of a matrix $\mathbf{A}$ is the number of dimensions spanned by the rows (or columns) considered as n (or m) vectors. The rank of an $n \times n$ nonsingular matrix is n. The rank of an $n \times n$ singular matrix is less than n.

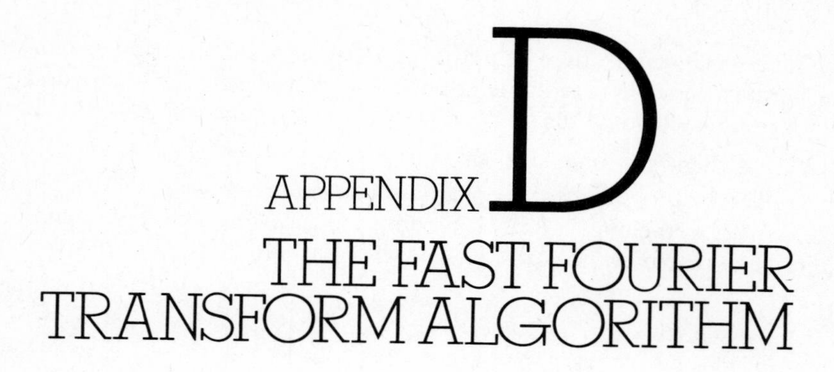

APPENDIX D THE FAST FOURIER TRANSFORM ALGORITHM

The FORTRAN listing here† is for calculating the discrete Fourier transform using the Fast Fourier Transform algorithm. In the subroutine FFT (A, M, N), A is a complex array of dimension N that initially contains the input sequence $u(n)$ and on return contains the transform $U(k)$. The dimension of the FFT is $N = 2^M$.

```
      SUBROUTINE FFT(A,M,N)
      COMPLEX A(N),U,W,T
      N = 2**M
      NV2 = N/2
      NM1 = N - 1
      J = 1
      DO 30 I = 1,NM1
      IF  (I.GE.J) GO TO 10
      T = A(J)
      A(J) = A(I)
      A(I) = T
10    K = NV2
20    IF (K.GE.J) GO TO 30
```

† Based on J. W. Cooley et al., "The Fast Fourier Transform and its Applications," *IEEE Trans. Education* **12**, pp. 27–34, March 1969.

```
      J = J - K
      K = K/2
      GO TO 20
30    J = J + K
      PI = 3.141592653589793
      DO 50 L = 1,M
      LE = 2**L
      LE1 = LE/2
      U = CMPLX (1.,0.)
      W = CMPLX (COS(PI/FLOAT(LE1)), -SIN(PI/FLOAT(LE1))
      DO 50 J = 1,LE1
      DO 40 I = J,N,LE
      IP = I + LE1
      T = A(IP)*U
      A(IP) = A(I) - T
40    A(I) = A(I) + T
50    U = U*W
      RETURN
      END
```

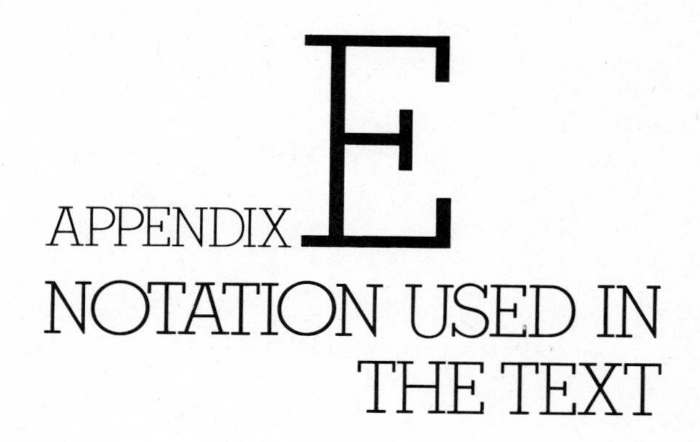

APPENDIX E NOTATION USED IN THE TEXT

FUNCTIONS

1. Unit step function: $\xi(t) = \begin{cases} 1, & t > 0 \\ 0, & t < 0 \end{cases}$

2. Signum function: sgn $(t) = \begin{cases} 1, & t > 0 \\ -1, & t < 0 \end{cases}$

3. Dirac delta function: $\int_{-\infty}^{\infty} f(t)\delta(t - t_0)\, dt = f(t_0), \qquad f(t)$ continuous at t_0

4. Sinc function: sinc $(x) = \sin x/x$

5. Sine integral: si $(u) = \int_0^u \text{sinc}(x)\, dx$

6. Binomial coefficient: $\binom{n}{m} = \dfrac{n!}{(n-m)!m!}$

GLOSSARY

CONVENTIONS

1. The symbol $|\quad|$ means the magnitude of the complex quantity of the time function contained within.
2. The symbol arg ( ) means the phase angle of the complex quantity contained within.
3. The symbol $*$ denotes convolution, that is,

$$u * h = \int_{-\infty}^{+\infty} u(\tau)h(t - \tau)\, d\tau$$

or $$u * h = \sum_{k=-\infty}^{\infty} u_k h_{n-k}$$

4. The symbol (,) denotes an inner product, that is,

$$(f, g) = \int_{-\infty}^{+\infty} f(t)g^*(t)\,dt$$

or

$$(f, g) = \sum_k f_k g_k^*$$

5. The symbol $*$ as a superscript denotes complex conjugate, that is, if $g = a + jb$, then $g^* = a - jb$.
6. The symbol $\leftrightarrow$ is used to relate a function or sequence and its transform, that is, $x_k \leftrightarrow X(z)$ or $f(t) \leftrightarrow F(j\omega)$.
7. The symbol Re ( ) means the real part of the complex quantity of the function contained within.
8. The symbol Im ( ) means the imaginary part of the complex quantity of the function contained within.
9. A superscript h denotes that the function (sequence) is the homogeneous solution of a differential (difference) equation.
10. A superscript p denotes that the function (sequence) is the particular solution of a differential (difference) equation.
11. A number superscript in parentheses is used to denote higher order derivatives of the function, for example,

$$y^{(3)}(t) = \frac{d^3 y(t)}{dt^3}$$

12. Boldface letters denote a vector or a matrix.
13. The notation D^n is used to denote an nth order derivative, that is,

$$D^n[y(t)] = \frac{d^n y(t)}{dt^n}$$

14. The notation $S^{\pm n}$ is used to denote an nth order shift, that is,

$$S^{\pm n}[y_k] = y_{k \pm n}.$$

INDEX